HELIOPHYSIC:

Evolving Solar Activity and the Climates of Space and Earth

Heliophysics is a fast-developing scientific discipline that integrates studies of the Sun's variability, the surrounding heliosphere, and the environment and climate of planets. Over the past few centuries, our understanding of how the Sun drives space weather and climate on the Earth and other planets has advanced at an ever increasing rate. The Sun is a magnetically variable star and, for planets with intrinsic magnetic fields, planets with atmospheres, or planets like Earth with both, there are profound consequences.

This volume, the third and final book in a series of three heliophysics texts, focuses on long-term variability from the Sun's decade-long sunspot cycle and considers the evolution of the planetary system from a climatological perspective over its life span of some ten billion years. Topics covered include the dynamo action of stars and planets from their formation to their demise; the evolution of the solar spectral irradiance and the response of Earth's troposphere, ionosphere, and magnetosphere; the internal and external sources of cosmic rays and their modulation by the heliospheric magnetic field; and planetary habitability subject to internal and external drivers of the climate system. In addition to its utility as a textbook, it also constitutes a foundational reference for researchers in the fields of heliophysics, astrophysics, plasma physics, space physics, solar physics, aeronomy, space weather, planetary science, and climate science. Additional online resources, including lecture presentations and other teaching materials, can be accessed at www.cambridge.org/9780521112949.

The three volumes in the Heliophysics series are:

I *Heliophysics*: *Plasma Physics of the Local Cosmos*
II *Heliophysics*: *Space Storms and Radiation – Causes and Effects*
III *Heliophysics*: *Evolving Solar Activity and the Climates of Space and Earth*

CAROLUS J. SCHRIJVER is an astrophysicist studying the causes and effects of magnetic activity of the Sun and of stars like the Sun, and the coupling of the Sun's magnetic field into the surrounding heliosphere. He obtained his doctorate in physics and astronomy at the University of Utrecht in the Netherlands in 1986, and has since worked for the University of Colorado, the US National Solar

Observatory, the European Space Agency, and the Royal Academy of Sciences of the Netherlands. Dr Schrijver is currently principal physicist at Lockheed Martin's Advanced Technology Center, where his work focuses primarily on the magnetic field in the solar atmosphere. He is an editor or editorial board member of several journals including *Solar Physics*, *Astronomical Notices*, and *Living Reviews in Solar Physics*, and has co-edited three other books.

GEORGE L. SISCOE received his Ph.D. in physics from the Massachusetts Institute of Technology (MIT) in 1964. He has since held positions at the California Institute of Technology, MIT, and the University of California, Los Angeles – where he was Professor and Chair of the Department of Atmospheric Sciences. He is currently a Research Professor in the Astronomy Department at Boston University. Professor Siscoe has been a member and chair of numerous international committees and panels and is on the editorial board of the *Journal of Atmospheric and Solar Terrestrial Physics*. He is a Fellow of the American Geophysical Union and the second Van Allen Lecturer of the AGU, 1991. He has authored or co-authored over 300 publications that cover most areas of heliophysics.

HELIOPHYSICS

Evolving Solar Activity and the Climates of Space and Earth

Edited by

CAROLUS J. SCHRIJVER
Lockheed Martin Advanced Technology Center

GEORGE L. SISCOE
Boston University

CAMBRIDGE UNIVERSITY PRESS
Cambridge, New York, Melbourne, Madrid, Cape Town,
Singapore, São Paulo, Delhi, Tokyo, Mexico City

Cambridge University Press
The Edinburgh Building, Cambridge CB2 8RU, UK

Published in the United States of America by Cambridge University Press, New York

www.cambridge.org
Information on this title: www.cambridge.org/9780521130202

First published 2010
First paperback edition 2011

A catalogue record for this publication is available from the British Library

Library of Congress cataloguing in publication data
Heliophysics : evolving solar activity and the climates of space and
Earth / edited by Carolus J. Schrijver, George L. Siscoe.
p. cm.
ISBN 978-0-521-11294-9 (Hardback)
1. Solar activity. 2. Weather – Effect of solar activity on. 3. Heliosphere (Astrophysics)
4. Solar-terrestrial physics. I. Schrijver, Carolus J. II. Siscoe, George L. III. Title.
QB524.H456 2010
523.7–dc22
2010022868

ISBN 978-0-521-11294-9 Hardback
ISBN 978-0-521-13020-2 Paperback

Additional resources for this publication at www.cambridge.org/9780521130202

Contents

The plates are to be found between pages 242 and 243.

Preface

Over the past few centuries, our awareness of the couplings between the Sun's variability and the Earth's environment, and perhaps even its climate, has been advancing at an ever increasing rate. The Sun is a magnetically variable star and, for planets with intrinsic magnetic fields, planets with atmospheres, or planets like Earth with both, there are profound consequences and impacts. Today, the successful increase in knowledge of the workings of the Sun's magnetic activity, the recognition of the many physical processes that couple the realm of the Sun to our galaxy, and the insights into the interaction of the solar wind and radiation with the Earth's magnetic field, atmosphere, and climate system have tended to differentiate and insularize the solar, heliospheric, and geospace sub-disciplines of the physics of the local cosmos. In 2001, the NASA Living With a Star (LWS) program was initiated to reverse that trend.

The recognition that there are many connections within the Sun–Earth systems approach has led to the development of an integrated strategic mission plan and a comprehensive research program encompassing all branches of solar, heliospheric, and space physics, and aeronomy. In doing so, we have developed an interdisciplinary community to address this program. This has raised awareness and appreciation of the research priorities and challenges among the LWS scientists and has led to observational and modeling capabilities that span traditional discipline boundaries. The successful initial integration of the LWS sub-disciplines, under the newly coined term "heliophysics", needed to be expanded into the early education of scientists. This series of books is intended to do just that: aiming at the advanced undergraduate and starting graduate-level students, we attempt to teach heliophysics as a single intellectual discipline. Heliophysics is important both as a discipline that will deepen our understanding of how the Sun drives space weather and climate at Earth and other planets, and also as a discipline that studies universal astrophysical processes with unrivaled resolution and insight possibilities. The goal of this series is to provide seed materials for the development of new researchers and new scientific discovery.

Richard Fisher, Director of NASA's Heliophysics Division
Madhulika Guhathakurta, NASA/LWS program scientist

Editors' notes

This volume is the third of a three-part series of texts (and an on-line problem set) in which experts discuss many of the topics within the vast field of heliophysics. The texts reference the other volumes by number:

I Plasma Physics of the Local Cosmos
II Space Storms and Radiation: Causes and Effects
III Evolving Solar Activity and the Climates of Space and Earth

The project is guided by the philosophy that the many science areas that together make up heliophysics are founded on common principles and universal processes, which offer complementary perspectives on the physics of our local cosmos. In these three volumes, experts point out and discuss commonalities and complementary perspectives between traditionally separate disciplines within heliophysics.

Many of the chapters in the volumes of this series have a pronounced focus on one or several of the traditional sub-disciplines within heliophysics, but we have tried to give each chapter a trans-disciplinary character that bridges gaps between these sub-disciplines. Some chapters compare stellar and planetary environments, others compare the Sun to its sister stars or compare planets with one another, while others look at general abstractions such as magnetic field topology or magnetohydrodynamic principles that are applicable to several areas.

The vastness of the heliophysics discipline precludes completeness. We hope that our selection of topics helps to inform and educate students and researchers alike, thus stimulating mutual understanding and appreciation of the physics of the universe around us.

The chapters in this volume were authored by the teachers of the heliophysics summer school following the outlines provided by the editors. In the process of integrating these contributions into this volume, the editors have modified or added segments of the text, included cross references, pointed out related segments of text, introduced several figures and moved some others from one chapter to another, and attempted to create a uniform use of terms and symbols (while allowing some differences to exist to remain compatible with the discipline's literature usage). They bear the responsibility for any errors that have been introduced in that editing process.

Additional resources

The texts were developed during summer schools for heliophysics held over three successive years at the facilities of the University Corporation for Atmospheric Research in Boulder, Colorado, and funded by the NASA Living With a Star program. Additional information, including text updates, lecture materials, (color) figures and movies, and teaching materials developed for the school can be found at www.vsp.ucar.edu/Heliophysics. Definitions of many solar-terrestrial terms can be found via the index of each volume; a comprehensive list can be found at www.swpc.noaa.gov/info/glossary.htm.

Heliophysics

helio-, prefix, on the Sun and environs; from the Greek *helios.*
physics, n., the science of matter and energy and their interactions.

Heliophysics is the

- *comprehensive new term for the science of the Sun–solar-system connection.*
- *exploration, discovery, and understanding of our space environment.*
- *system science that unites all of the linked phenomena in the region of the cosmos influenced by a star like our Sun.*

Heliophysics concentrates on the Sun and its effects on Earth, the other planets of the solar system, and the changing conditions in space. Heliophysics studies the magnetosphere, ionosphere, thermosphere, mesosphere, and upper atmosphere of the Earth and other planets. Heliophysics combines the science of the Sun, corona, heliosphere, and geospace. Heliophysics encompasses cosmic rays and particle acceleration, space weather and radiation, dust and magnetic reconnection, solar activity and stellar cycles, aeronomy and space plasmas, magnetic fields and global change, and the interactions of the solar system with our galaxy.

From NASA's *Heliophysics. The New Science of the Sun–Solar-System Connection: Recommended Roadmap for Science and Technology 2005–2035.*

1

Interconnectedness in heliophysics

Carolus J. Schrijver and George L. Siscoe

1.1 Introduction

The volumes on heliophysics, of which this is the third, emphasize universal processes for which some basic physical phenomenon manifests itself in a variety of circumstances throughout the local cosmos and beyond. The topics range from the variability of the star next to which we live to the distant interstellar medium, via planetary environments including, in particular, geospace in which a magnetic field and atmosphere shield us from most of the dangerous consequences of solar variability – taking us from solar flares, coronal mass ejections, and their associated energetic particles, via the dynamic interplanetary medium, to magnetospheric, ionospheric, and tropospheric consequences. This volume in particular emphasizes interconnectedness, which manifests itself in three different guises that appear, often implicitly, throughout the text.

First, there is the interconnectedness by the universal processes themselves: magnetohydrodynamics, radiative transfer, networks of chemical reactions, magnetic-field dynamics and topology, particle acceleration, shocks, turbulence, etc., pervade all three volumes. Second, we see the interconnectedness in the very evolution of the solar system, from the formation of the central star and its orbiting planets, to the impact of the star on planetary habitability, and the eventual demise of the solar system as we know it (Fig. 1.1). Third, there are many connections between a variety of research disciplines, each of which is advanced by what is learned from other disciplines, thus providing mutual support in their quest for deeper understanding.

This introductory chapter lights up the stage upon which the stories in this volume unfold by introducing some of the key components of each of these three faces of interconnectedness. The extensive network of linkages from one scientific discipline to another emphasizes how important it is to have a systems point of view, both for the individual researchers (to which we hope these books on

Heliophysics: Evolving Solar Activity and the Climates of Space and Earth, eds. Carolus J. Schrijver and George L. Siscoe. Published by Cambridge University Press. © Cambridge University Press 2010.

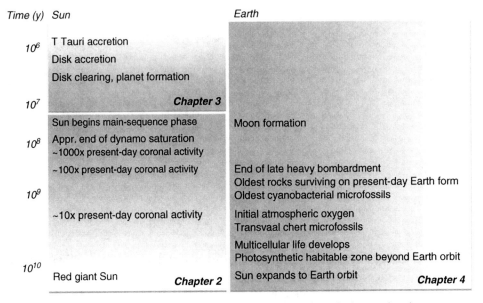

Fig. 1.1. Key events in the history of the Sun and Earth, pointing to the chapters in this volume in which these are discussed.

heliophysics will contribute) and for the various research disciplines and their supporting government agencies that must work together, recognizing that the advance of any one discipline occurs within the context of the combined activities in all of science and technology; a case for this was made eloquently by Stokes (1997), who pointed out that it is the combination of a broad spectrum of activities in pure science, applied engineering, and applied science that best stimulates societal advances.

We explore a few of the emerging themes in the sections below.

1.2 Field–plasma–neutral interaction

The interconnectedness of magnetic fields and ionized matter is pervasive throughout heliophysics. This field–plasma coupling is observable in many different domains, ranging from the Earth's ionosphere to the Sun's surface, and the effects reach even further from the deep interior of Sun and planets to the outermost reaches of the heliosphere. Substantial problems in this area continue to hamper progress in heliophysics, regardless of the ubiquity of field–plasma interactions, despite the possibility of both remote-sensing and *in-situ* observations (although by no means as ubiquitous as the processes themselves), and notwithstanding the rapid advances in computational (radiative) magnetohydrodynamics.

Some of the key problems in this research area involve the acceleration of ions and electrons to energies that vastly exceed the characteristic kinetic energies of the parent plasma population, and the role of a variety of waves in particle diffusion and acceleration (see Vol. II). Large solar flares and their associated coronal mass ejections, for example, appear to distribute energy over thermal, non-thermal, and bulk kinetic energies roughly equally in order of magnitude (see Vol. II, Chapter 5); clearly, particle acceleration is an integral part of the processes associated with destabilization and reconnection of magnetic field throughout heliophysics.

The coupling with neutral components within the plasma introduces complications not only in terms of electrical conductivity and drag forces (e.g. Vol. I, Chapter 12 and Vol. II, Chapter 12), but also can shift the ionization balance by collisions between the component species. The latter, of course, also leads to a network of chemical reactions that are well studied within the upper atmosphere of Earth and other planets (Chapters 13 and 16), in which non-local coupling by radiative transfer plays an important role (as it does in the solar near-surface layers).

Advances are being made by combining geospace and heliospheric studies (that offer at least some *in-situ* observing), solar remote-sensing observations, laboratory plasma physics, and many-particle or hybrid particle–MHD simulations with powerful computer facilities. Some of these connections are made in the *Heliophysics* series (particularly in Vol. II), but most of this area of interactions between scientific disciplines is beyond what can be covered in these few textbooks, having its own rich literature with branches connecting many of the physical sciences.

1.3 Transport of angular momentum and energy

Transport of angular momentum through the coupling of distant concentrations of mass occurs either through gravitational tides, by magnetic stresses, or by flows. Gravitational coupling has obviously played an important part in the spin-orbit synchronization of the Earth's single moon. This coupling continues to be important as a stabilizer for the direction of the Earth's spin axis (Chapter 4), even as it causes the precession of that axis with associated climatic effects (Chapters 11 and 12).

Tidal forces also act significantly on Jupiter's moon Europa and Saturn's moon Enceladus, in which they appear to result in liquid water in their interiors (Chapter 4), which makes these moons interesting objects to study from an exo-biological perspective. Tidal spin-orbit coupling also leads to the formation of short-period, highly active binary stars (like the so-called RS CVn type binary systems; see Chapter 2). Gravitational tidal coupling is generally well understood, and thus the transfer of rotational energy can be estimated with reasonable accuracy.

Angular momentum transport via the magnetic field is important in the coupling of protostars and young T Tauri stars to their surrounding disks and magnetized stellar winds (Chapter 3 in this volume, and Vol. I, Chapter 9). After the early formation phases of a planetary system, the loss of stellar angular momentum continues through a stellar wind, leading to magnetic braking of the stellar rotation and the concomitant gradual decrease in stellar activity with age (Chapter 2). In tidally interacting binaries with one or more magnetically active components, the loss of spin angular momentum by a stellar wind drains the orbital angular momentum reservoir, eventually leading to the merger of the component stars, leaving an old but rapidly spinning single star (like FK Comae). The theory of the loss of angular momentum by a stellar wind is very much incomplete, but observations of stellar clusters and binary stars whose ages can be estimated from evolutionary phases of the component stars, measurements of the solar wind for several decades throughout the heliosphere, and data on the stellar equivalents of the heliospheric bow shock have established a fair level of empirical knowledge (Chapter 2).

Angular momentum transport by flows inside astrophysical bodies is the cause of their near-rigid rotation with latitude and depth of the solar interior (e.g. Chapter 5). But the models of the full convective envelopes of stars and giant planets need to advance significantly before we can use their results in, for example, magnetohydrodynamic dynamo models in which the non-rigid rotation and other large-scale circulations appear to be crucial (Chapters 5 and 7).

Transport of kinetic and electromagnetic energy is of such obvious importance throughout heliophysics that it needs no mention here. Understanding it well enough that we can formulate accurate self-consistent models of, for example, photosphere-to-corona, stratosphere-to-magnetosphere, or Sun-to-planet energy transfer is, of course, another matter that is only partially discussed in these volumes.

1.4 Dynamo action

Dynamo action or, more specifically, the generation and maintenance of a (generally varying) magnetic field against diffusive decay (see Vol. I, Chapter 3 and Chapters 5, 6, and 7 in this volume), is a widespread phenomenon throughout the universe, from entire galaxies to the scale of planets, spanning a range of scales of $\mathcal{O}(10^{14})$. The stellar dynamo is critical to the very formation of stars and their planetary systems (Chapter 3) and to the shielding of planetary atmospheres against, for instance, ablation by stellar winds (Chapter 4) and the effects of energetic particles (galactic cosmic rays and solar energetic particles), and interplanetary manifestations of coronal mass ejections (see Chapter 9 and various chapters in Vols. I and II).

Astrophysical and planetary dynamo theory and modeling (Chapters 5, 6, and 7) clearly have much in common, both in their basic processes and – at least in the most active stars – perhaps even in the domains of their characteristic magnetic Reynolds numbers, as argued in Chapter 7. The latter suggests a universal scaling property (Section 7.6.5) that measures "saturated" dynamo action (Section 2.3.2) in terms of the available energy from either nuclear fusion in stellar cores or nuclear fission and residual thermal energy in planetary interiors. This similarity in astrophysical theaters of dynamo action enables the direct application of numerical magnetohydrodynamic models at currently achievable magnetic Reynolds numbers to planets and at least some types of stars alike; such a direct comparison between models and observations may lead to an explanation of the generally dipolar field of the coolest dwarf stars (Section 2.7) and it raises the possibility of irregular field reversals in such stars similar to what is found for the geodynamo and its models (Section 7.2.2.3).

Despite remarkable advances in the numerical modeling of astrophysical dynamos, a comprehensive dynamo theory with predictive capabilities has yet to be developed: describing the Sun's decadal and multi-decadal patterns in its magnetic activity, or the characterization of stellar activity levels based on the stars' fundamental properties, remain beyond current capabilities. Here, both astrophysical and geophysical empirical data provide critical input. Studies of populations of Sun-like stars have revealed the scaling of stellar magnetic activity with rotation rate, multi-decade observations of stars like the Sun are beginning to reveal a richness of long-term variability well beyond the historical records of solar variability, and analysis of the stratification of isotope abundances and chemical composition of ice deposits spanning multiple centuries are expanding our view of solar–heliospheric activity by hundreds of years up to possibly hundreds of thousands of years beyond historical records (Chapters 2, 7, 11, and 12). Moreover, understanding the formation history of stars, as described in Chapter 3, relies on improved understanding of stellar dynamos (and, likely, dynamos within the innermost, warm parts of accretion disks).

Improved understanding of all these areas – star formation, stellar activity, star–planet connections, magnetospheric shielding, ..., and the underlying dynamo action – are advancing jointly through their interconnectedness. The need for a correct interpretation of the empirical data that are in principle available reaches even further into apparently distant scientific disciplines: e.g. the proper reading of ^{10}Be (half life of 1.5 Myr) concentrations found in arctic and antarctic ice sheets (Chapter 11), or ^{14}C (half life 5.7 kyr) to ^{12}C isotope ratios used in tree-ring studies, requires that one also understand how these isotopes are formed (which involves cosmic-ray physics, discussed in Chapter 9, and nuclear physics) and how they are transported and deposited (bringing in atmospheric transport processes, paleoclimatology, and the geological and atmospheric processes involved in precipitation of the ejecta from volcanic eruptions – discussed in Chapters 4, 15, 11, and 12 – as well as the meandering transport of galactic cosmic rays against the turbulent out-flowing solar wind – discussed in Chapters 8 and 9): just imagine, as you read this volume, the chain of processes from the acceleration of, say, a proton to become a member of the high-energy galactic cosmic-ray population to the storage of the ^{14}C isotope in living tissue, including yours, and how the use of the ^{14}C to ^{12}C isotope ratio in archeological dating studies is entangled with understanding of the variability of solar activity over the past millennia.

1.5 Extreme events and habitability

We encounter power laws throughout heliophysics, such as in the energy distribution of solar and stellar flares (Vol. II, Chapter 5, and Chapter 2 in this volume), solar coronal mass ejections (Vol. II, Chapter 6), energetic particles and galactic cosmic rays (Vol. II, Chapter 9, and Chapter 9 in this volume), and in the flux distribution of solar active regions (Chapter 2). In many of these cases, there is insufficient empirical statistical knowledge to establish whether there are cutoffs to these distributions for the largest or most-energetic events. Yet, although very large events are rare, they may not be rare enough to ignore their potential effects on evolutionary time scales of planets or even on the more ephemeral manifestation of life on planets. For example, the energy distribution of flares on active stars (Section 2.3.3) continues up to at least three orders of magnitude larger than the largest solar flares documented historically. One observation suggests that events another three orders of magnitude larger are possible on what appears to be a run-of-the-mill G1.5 V star, i.e. a star very much like our own Sun, at least in spectral type.

Extrapolation of the observed power-law frequency distribution of energies, and estimation of the equivalent radiation dose at ground level by energetic particles associated with such events, suggests that if such extreme events occur on a star of solar-like activity, they could have impacts on life on Earth on geological time scales. We do not currently know what limits the upper end of the spectrum of flare energies, how the energy spectrum of solar energetic particles changes towards extreme events, or how to reliably convert the energy in energetic particles to damage potential in biological tissue. But that flares can be larger than observed by our present-day instrumentation seems very likely, if only because of the estimated magnitude of the large white-light flare reported on in 1859 by Carrington and Hodgson (see Vol. II, Chapters 2 and 5). Such flares, and their associated coronal mass ejections (suggested by sporadic reports made centuries ago of what could be aurorae at unusually low latitudes), would have substantial impact on our infrastructure in space (see Vol. II, Chapter 2).

It appears both possible and instructive to find records of such large events in, for example, isotope ratios and chemical abundances in ice cores, or perhaps in sediment layers with unusually stable stratifications (see e.g. Usoskin, 2008, and McCracken *et al.*, 2001, for a discussion of past solar energetic particle events based on lunar samples and ice cores, which suggests a break in the power-law spectrum that would make extremely large events very rare, if not absent altogether). The combination of radiological studies of geological and snow-precipitation records, continued stellar observations of an ensemble of Sun-like stars, and advancing the modeling of active-region generation, magnetic destabilization, particle acceleration mechanisms, etc., likely will provide better constraints on such large events of importance to the habitability of planets and, in particular, to the protection of humanity's space assets by improved forecasting of extreme events.

Other "extreme events" described in this volume include the destabilization of the Kuiper belt leading to heavy bombardment and orbital modifications for the planets (Chapters 3, 4, and 11); rapid global climate changes subject to changes in atmospheric greenhouse conditions (Chapters 4 and 12); impacts of large comets (such as the large impact off the Yucatan peninsula some 65 Myr in Earth's past, or those observed on Jupiter in 1994 and 2009); and, of course, large volcanic eruptions (Chapter 12). We do not discuss potential events external to the heliosphere here, but refer to Vol. II, Section 2.6, for a brief discussion of some types of such events.

1.6 Our remarkable, remarkably sensitive environment

In 1960, Frank Drake devised an expression to estimate the number N_C of advanced civilizations within our galaxy, here shown relative to the number of stars N_* in the galaxy:

$$n_C \equiv \frac{N_C}{N_*} = \left(\frac{r_* \tau_C f_i}{N_*} \right) f_p N_p f_\ell, \qquad (1.1)$$

where r_* is the characteristic rate of star formation within the galaxy (estimated to be $\mathcal{O}(10 \, \mathrm{yr}^{-1})$), f_p is the fraction of stars with planets of which N_p per star might be able to support life, f_ℓ is the fraction of those planets that actually will develop life during the life time of the star, of which a fraction f_i develop intelligent life that persists for a time τ_C. Here, we omit a factor that quantifies how likely such civilizations would in principle be able to communicate with other such civilizations.

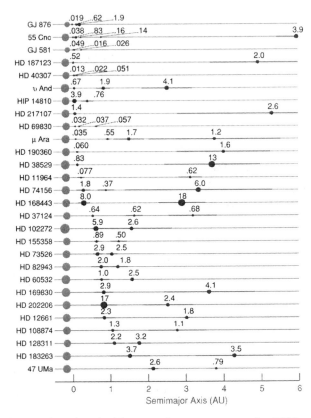

Fig. 1.2. Of some three hundred known planetary systems by 2009, at least 30% are known to contain more than one planet. This diagram shows the semimajor axes (relative to the radius of the Earth's orbit around the Sun) and minimum masses (in units of Jupiter masses) for the planets in multi-planet systems. The diameters of the symbols for the central stars scale with the cube root of stellar mass. (From Wright *et al.*, 2009. Reproduced by permission of the AAS.)

The recent and continuing rapid advances in instrumentation have led to the discovery of hundreds of exoplanets to date (Fig. 1.2),[†] and we can anticipate many more discoveries in the near future by, for example, NASA's Kepler mission that was launched in March of 2009. These studies suggest that $f_p = \mathcal{O}(1)$.

Chapter 4 argues that because microbial life started so soon after the Late Heavy Bombardment of the Earth (Fig. 1.1), the fraction f_ℓ of planets that are in principle capable of supporting terrestrial-like life and that develop such life is also likely to be, very roughly – to astronomical standards – of order unity.

The characteristic number of planets per star that are capable in principle of supporting Earth-like life, N_p, at some time in its history is hard to estimate with our

---

[†] See www.exoplanets.eu for a comprehensive list.

present-day knowledge. But because the width of the habitable zone (Chapter 4) is fairly large – at least in order of magnitude – compared to the size of a planetary system, we may not be too far off if we set $N_p = \mathcal{O}(1)$.

In view of these arguments, many of which are discussed within this volume, we may argue that the product $f_p N_p f_\ell$ is likely not to be very much smaller than unity. Making even a very rough estimate of n_C is, however, subject to the very large uncertainty on the value of $\tau_C f_i$. This estimate for the characteristic lifetime of intelligent civilizations that form on a fraction of all life-bearing planets is the most uncertain number in Eq. (1.1). We have no basis in either statistical empirical knowledge or in validated social–biological–ecological studies to have much confidence in any number we assign to $\tau_C f_i$. Thinking about this number is of interest from a heliophysical perspective, though, because upper limits to τ_C are likely dictated by events of extraterrestrial origin: impacts by large bodies, major stellar flares and mass ejections, destabilization of planetary orbits, atmospheric ablation, planetary drift relative to the habitable zone, and more exotic, extra-heliospheric phenomena such as supernovae; these and other processes that contribute to the long-term habitability of the Earth led to the title of a book that reviews these and other processes: *Rare Earth* by Ward and Brownlee (2000).

If we are pessimistic about the longevity of civilizations, and set $\tau_C \sim 10^3$ yr, or if alternatively we take a characteristic time scale between disastrous impacts of $\tau_C \sim 10^8$ yr, estimates for n_C would have upper limits spanning values of 1 in $3 \times 10^7/f_i$ to 1 in $300/f_i$. This vast range of values for n_C has obviously profound consequences for our likelihood of finding intelligent life in Earth's vicinity, compounded by the problem of establishing a realistic value for the fraction f_i of planets on which intelligent life develops in the first place.

1.7 System complexity

Before we end this introductory chapter, we point out one more thing: heliophysics is a discipline filled with sensitive non-linear interdependencies and feedback pathways. One example of that is found in the modeling of climate responses to the $\sim 0.1\%$ variability in total solar irradiance (Chapter 12), in which it may be that the effect of the variable radiative energy deposited within the stratosphere (a very small fraction of the total; Chapters 10 and 14) is amplified to have consequences throughout the troposphere and even the global ocean circulation. Another minute effect is found in the modeling of the differential rotation profile within the solar convective envelope, where pole–equator temperature differences of ~ 10 K are discussed for a region with a characteristic temperature of one million kelvin (Chapter 5).

These effects sometimes occur where large opposing effects very nearly, but not quite, cancel out, as in a planetary climate system where energy input and output very nearly balance through a variety of channels that involve a range of time scales (from the near-immediate balance in the troposphere to the slow response of the deep oceans). This presents validation problems for such models requiring high fidelity of the codes, intercomparison of multiple codes or multiple realizations of the system in the real world, either by intercomparing multiple planets or many stars, or by comparison with long records of one system preferably under a variety of conditions of internal and external drivers (see Chapters 2 and 12). These problems go hand in hand with opportunities for discovery, because the near-cancellation of effects generally depends on time and place. High-resolution climate modeling, for example, suggests that, whereas solar driving may play a relatively minor role in global climate change in the past half century, the 11-year fingerprint of this driving may be more (or less) pronounced in any given local record than in globally averaged records (see Chapter 16).

Heliophysics is the science of all the processes within the local cosmos, from distant past to distant future, that govern the evolution of the Earth, the Sun, and the planet-bearing heliosphere in its entirety. Whereas heliophysics is not, per se, aiming to evaluate the Drake expression, it is interesting to contemplate how much of the physics discussed in this volume and in the preceding two needs to be combined into setting a heliophysical upper limit to $\tau_C f_i$ or in validation of highly complex non-linear numerical models.

We hope you enjoy the voyage through space and time offered in this volume, and invite you to think how to optimize the interaction of the many scientific disciplines that are explicitly mentioned and implicitly required in the making of this story.

2

Long-term evolution of magnetic activity of Sun-like stars

Carolus J. Schrijver

This chapter describes the magnetic activity of the Sun throughout most of its history. We begin with a summary of the activity of the Sun as described in historical records, i.e. going back at most a few centuries. The picture of the Sun in time can be completed by complementing solar observations with stellar data. This chapter focuses on the period from a solar age of roughly 0.1 Gyr onward. The formation of the Sun and its solar system and the very earliest phases of their joint evolution are discussed in Chapters 3 and 4.

The Sun's magnetic activity[†] and the associated changes in its extended atmosphere evolve on time scales that range from minutes up to billions of years. The coupling of these scales requires that we describe the patterns of the solar magnetic field from the smallest currently observable structures to the global dipole field; for the longest time scales, we need to include a discussion of stellar evolution. This chapter therefore summarizes the observationally determined properties of the solar magnetic field before discussing the variations in solar spectral radiance and the coupling of the solar coronal field to the heliosphere. These descriptions are based on the historical records of this activity that extend over only a few decades for high-energy particle and electromagnetic radiation, over about a century for its magnetic field, and four centuries for sunspot records. The discussion of solar activity on time scales beyond a few centuries necessarily relies on biological and geophysical records (discussed in Chapters 11 and 12). We need to look at stars other than the Sun to learn about solar activity for the distant past (after the formative phase discussed in Chapter 3) as well as the future on time scales of hundreds of millions of years to billions of years.

[†] The term "activity" is used both to describe the general existence of magnetically heated outer atmospheres in the Sun and sun-like (or "cool") stars, and to describe specific processes that include flux emergence, flaring, and eruptions. Thus "stellar activity" can refer to the existence of magnetic field in their photospheres, to a characteristic emission level in chromospheric to coronal radiative losses, or to signatures of starspots or atmospheric flares observed anywhere in their spectra.

Heliophysics: Evolving Solar Activity and the Climates of Space and Earth, eds. Carolus J. Schrijver and George L. Siscoe. Published by Cambridge University Press. © Cambridge University Press 2010.

The basic solar terminology used here was introduced in Vol. I, Chapter 8 (see e.g. Table 8.1 for definitions of photosphere, chromosphere, and corona, and characteristic physical properties of these domains for the solar atmosphere). The free on-line journal *Living Reviews in Solar Physics*[†] provides reviews of a variety of heliophysical topics, including long-term variability of the Sun and stars, solar and stellar winds, and the Sun–climate link. For a comprehensive discussion of solar and stellar magnetic activity, see Schrijver and Zwaan (2000) and Foukal (2004). Güdel (2007) provides an extensive description of the long-term evolution of solar activity and its consequences. Ribas *et al.* (2005) focus on the solar radiance from 1 Å to 1700 Å over the lifetime of the Sun from about 0.1 Gyr to 7 Gyr. To contrast the activity of a Sun-like star with that of a much cooler red main-sequence (or dwarf) star, see Guinan and Engle (2009). A comparative description of magnetic fields in cool stars (with a convective envelope) and warm stars (with a radiative zone immediately below their photosphere) is given by Donati and Landstreet (2009).

2.1 A brief history of the Sun: past, present, future

The stellar magnetic field played an important role in the entire history of the formation of the solar system and the evolution of life on Earth. From the evidence we have, based on both solar and stellar observations, we can piece together what most likely happened (see Fig. 1.1 for a summary representation of some of the evolutionary stages). Let me first present a summary scenario before reviewing the observational and theoretical evidence supporting it in this chapter and in Chapter 3.

In the initial phases of the contraction of the gaseous cloud out of which the Sun and the planets formed, the magnetic field was instrumental in transport of angular momentum out of the core to the outer regions of the cloud. Without the expulsion of the bulk of that initial angular momentum, no star could have formed because centrifugal forces would have dominated over the pull of the gravitational field (see Chapter 3). The details of this strong decrease in angular momentum in the initial star-forming processes, and in particular what role planetary systems play in this (for example, Jupiter's orbital angular momentum is approximately 10^5 times the rotational angular momentum of the Sun) are areas of active study.

Once a protostar and a surrounding extended gaseous disk had formed, solar magnetic activity truly began to manifest its significance. In the first 10 million years or so, the Sun's strong magnetic field coupled to the inner domains of the pre-planetary disk. The star continued to accumulate mass from the disk for some

[†] LRSP: http://solarphysics.livingreviews.org/

time, even though the disk did not extend to the stellar surface: the magnetic field of the rapidly rotating star sweeps through the inner disk region, allowing matter to accrete primarily along the field. Accretion thus occurs in evolving columns connecting the disk to patches on the stellar surface. Where the material could not reach the surface, the gas pressure aided by the magnetic field channeled material into jets of gas, shooting away from the poles of the star, perpendicular to the disk.

Once planets formed around the young central star, they were subjected to intense X-ray and ultraviolet radiation hundreds to thousands of times more intense than at present. Moreover, they were embedded in a strong stellar wind and frequent magnetic storms. The activity level varied irregularly on time scales of years to decades, not yet cyclically as in many mature stars like the Sun. Magnetically active regions frequently emerged onto the surface, and with them dark starspots. Spots in young stars cover as much as half a hemisphere as opposed to the 0.5% maximum coverage in the present Sun (which is only rarely achieved, see Fig. 2.2). As a result, the stellar radiance varied with time by as much as about a factor of two. The rotation period of a star was only a few days in this early phase of its evolution, so planets orbiting the active young Sun were subjected to comparably large variations in irradiance on time scales of a few days in addition to the diurnal variations on their surface induced by their own rotation.

As the Sun reached an age of a billion years, its magnetic activity had subsided strongly. The spot coverage decreased, and relatively more smaller clusters of magnetic flux surrounded the smaller or less common spots. These small clusters are somewhat brighter than the surrounding stellar surface (which is why they were named faculae). At this age, dark spots and bright faculae compensated each other almost exactly, so that the total radiance averaged over a solar rotation or more varied little even though the magnetic activity and the related highly variable energetic radiation continued. The irregular variations in long-term activity made way for quasi-regular cyclic variations on a time scale of several years to about a decade.

The evolution of the solar internal structure in the first billions of years is expected to be associated with a gradual increase of several tens of percent in the total irradiance for the orbiting planets. Whether mass loss through a strong solar wind countered some of this anticipated change in solar radiance is still being studied.

As life began to develop on Earth, the effects of the bright faculae began to outweigh the effects of the dark sunspots on total solar irradiance, so that overall the Sun actually brightened somewhat when there were more spots (as it continues to do at present), although short-lived dips in irradiance occur when spots cross the central portions of the solar disk. Multi-year averages of the irradiance of the orbiting planets in that phase vary by a few tenths of a percent or less. But sometimes the magnetic activity subsided for many decades, and the radiance possibly changed somewhat more strongly. One such episode, referred to as the Maunder minimum from 1645 to 1715 CE, coincides with the Earth's seventeenth-century Little Ice Age: crops failed in Northern Europe, and people were skating on London's Thames in June. Earlier, during a solar Grand Maximum in activity in the late Middle Ages, temperatures on Earth reached a peak (but see Chapter 12). At present, anthropogenic effects dominate climate change (see Chapter 12).

In the next few billion years of the Sun's evolution, its magnetic activity will continue to decline gradually. In the meantime, the total solar irradiance for Earth will increase by a factor of about two over the next $\sim$5 Gyr, after which the Sun will rapidly – on an astronomical scale – swell into a red giant star. Whether or not it will expand to just within or just beyond the Earth's orbit really is a moot point: any atmosphere will have evaporated off the Earth long before then (see Chapter 4).

2.2 Present-day solar activity

2.2.1 Photosphere

2.2.1.1 Bipole emergence properties

The patterns in the solar surface magnetic field range in scale over a factor of at least 10 000, from the global dipole to the smallest scale that is currently observable ($\sim$150 km, which – to put telescope power in perspective – is equivalent to $\sim$1 cm at 250 km). The time scales for the evolution of these patterns differ by a factor of over 10^6 if we compare the decade-long swing in the global dipole field to the minutes-long changes in the granular convection. The largest scales are dominated by changes associated with the solar cycle, while the smallest scales appear to be largely independent of the cycle. The intermediate scales form a continuum that smoothly connects the extremes in scale, with smoothly changing properties along the size spectrum of the magnetic features.

The Sun's magnetic field emerges from within the interior in the form of flux bundles, whose arched, frayed tops breach the solar surface (the photosphere) to form photospheric cross sections that are referred to as bipolar regions (see the example in Fig. 2.1), with arches of magnetic field that reach into the solar outer atmosphere above it that are discussed later in this chapter.

The largest of these bipolar regions are referred to as "active regions"; these should, by an inconsistently applied definition, contain one or more dark pores or sunspots, and typically have an absolute flux of $\sim$10^{20} Mx up to about 3×10^{22} Mx. Smaller regions, with lower fluxes, are referred to as "ephemeral regions", but that distinction is not based on a strict definition of either size or flux, and is rather fuzzily used in the literature. The statistical properties of regions with less than $\sim$10^{18} Mx remain poorly known as these regions are short-lived, ubiquitous, and

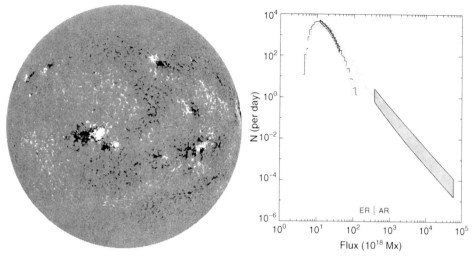

Fig. 2.1. (Left) First solar magnetic map (magnetogram) of the current millennium, taken by SOHO's MDI on 2001/01/01 00:03 UT. The magnetogram (with white/black for negative/positive line of sight polarity) shows a variety of active regions, embedded in patches of largely unipolar enhanced supergranular network, mixed-polarity quiet Sun regions, and low-flux polar caps (weak at this near-maximum phase of the cycle, and weakened further in the line-of-sight flux map because of center to limb effects). (Right) Distribution function of emerging magnetic bipolar regions on the Sun, showing the emergence frequency per day per flux interval of 10^{18} Mx, estimated for the entire solar surface. The shaded region on the right envelops the range of the active-region spectrum for solar cycle 22 (for half-year intervals around sunspot minimum and maximum). The histograms on the left are for the ephemeral regions; the shaded band shows where observations are least affected by spatial (lower cutoff) and temporal (upper cutoff) biases. The spectrum for regions below $\sim 10^{19}$ Mx has yet to be determined; the cutoff here is caused by the limited resolution of the SOHO/MDI magnetograph. (Figure from Hagenaar *et al.*, 2003.)

intricately mixed, with sizes down to the instrumental detection limits even for current state-of-the-art observatories; their cycle variability (if any) has yet to be measured.

The frequency distribution of the absolute magnetic fluxes Φ of active regions (Fig. 2.1) can be approximated by a power law such that $N(\Phi)d\Phi = a(t)\Phi^{-1.9\pm0.1}d\Phi$ (Harvey and Zwaan, 1993). Because the best-fit slope is likely larger than -2, an integral of the fluxes over the frequency spectrum would be dominated by the smallest scales, but as the uncertainty range allows the value to be below -2, which would lead to the opposite conclusion, we would need to know if any cutoffs exist on the large or small side of the spectrum. For the smallest regions, this has not yet been adequately studied (and may in practice be limited by the available observational resolution), while for the largest regions the statistics of observed regions does not (yet) suffice to decide on the existence of an upper limit. But with the power-law index close to -2, clearly both the few large and the many small regions contribute significantly to the overall budget of magnetic flux that emerges onto the solar surface.

The function $a(t)$ describes the variation through the sunspot cycle; $a(t)$ is essentially independent of flux for $\Phi \gtrsim 3 \times 10^{20}$ Mx. Its time dependence is characterized by a fairly rapid rise phase lasting $\sim$3–5 yr and a more gradual decay phase of $\sim$6–9 yr. The total active region flux and total sunspot area are statistically linearly related, so that the rapid rise and the more gradual decay of the activity cycle are also readily seen in sunspot coverage diagrams such as the bottom panel in Fig. 2.2.

The global ephemeral-region spectrum appears to smoothly extend the active-region spectrum towards the smallest scales for which statistical studies have been performed to date. But in extending this spectrum, there is a significant change in the dependence on the sunspot cycle: the cycle variation for the emergence frequency of ephemeral regions with fluxes around 10^{19} Mx is weak – no more than $\sim$20% – and apparently in anti-phase with the cycle for the active regions (Hagenaar *et al.*, 2003).

At the onset of a sunspot cycle, active regions preferentially appear at midlatitudes of about 30° (see Fig. 2.2), with infrequent emergences at higher latitudes,

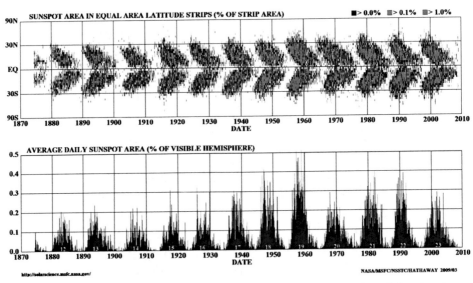

Fig. 2.2. Butterfly diagram showing sunspot latitudes (top) and total fractional area coverage (bottom) as a function time for the last six magnetic cycles. (Figure courtesy of D. Hathaway.)

although never in recorded history in the polar caps above 60°. Successively later generations of regions emerge progressively closer to the equator (Spörer's rule). Towards the end of a characteristic sunspot cycle, when its regions appear typically close to the equator, the next sunspot cycle may already have started at mid-latitudes. Successive cycles generally overlap for up to about two years. In contrast, some sunspot cycles are rather clearly separated from their successor at times of extended cycle minima, such as around 1911 and 2008. A latitude–time diagram of sunspots, active regions, or magnetic flux density reveals a relatively symmetrical pattern about the equator, starting at mid-latitudes and progressing towards the equator, resembling a succession of pairs of wings, from which such diagrams obtained their characteristic name: the butterfly diagram.

Large active regions (>10 square degrees) emerge with a characteristic spread about the mean emergence latitude in the butterfly diagram of about 10°. For smaller regions, the spread increases smoothly with decreasing size. The small ephemeral regions show at most a weak latitude dependence, emerging all over the solar disk from equator to poles.

Large active regions emerge with their polarities separated in the direction of rotation, and these polarities are therefore generally referred to as "leading" and "following". The leading polarity of the vast majority of active regions in the northern hemisphere is of one polarity, while that in the southern hemisphere is of the opposite polarity (Fig. 2.1; Hale's rule). These patterns reverse from sunspot cycle to sunspot cycle, so that the mean period of the magnetic cycle is 22.2 yr (important in understanding the modulation of galactic cosmic rays, discussed in Chapter 9).

The line connecting the centers of gravity of the two polarities in an active region tends to be inclined towards the equator in the direction of rotation by about 4° (Joy's law). The mean tilt angle shows a weak latitude dependence, but is independent of the size or flux of the regions. The scatter about the mean tilt angle, however, increases towards smaller regions, with ephemeral regions of $<10^{19}$ Mx being essentially randomly oriented (e.g. Hagenaar *et al.*, 2003).

Overall, the properties of solar bipolar regions change smoothly from the ubiquitous, randomly oriented, cycle-independent ephemeral regions to the active regions with a clear preference for a migrating emergence latitude, a pronounced tilt angle, and the clear pattern of the sunspot cycle. The differences between ephemeral and active regions are sometimes interpreted as indicative of different sources: a locally operating turbulent dynamo versus the global cycle dynamo. The two populations smoothly transition into each other, however, and stellar observations suggest that the least active Sun-like stars have magnetic activity levels significantly below the quietest solar supergranular network (although possibly the internetwork field may persist in such stars; see Section 2.6), so that even the apparently unvarying ephemeral-region population likely depends on stellar rotation (see Section 2.3.2).

Successive cycles have markedly different strengths (with jumps that not infrequently exceed a factor of two; see bottom panels of Fig. 2.2 and Fig. 2.5) and durations (roughly from 9 to 14 yr).

2.2.1.2 Active region structure and evolution

Large magnetic bipoles, called active regions, emerge into the photosphere with little warning. The largest of them may be detected helioseismically a few hours prior to emergence, but the signal is (at least currently) measurable only when the rising flux bundle is already close to the surface: in reality, the emergence is faster than the time it takes to compile and process the helioseismic data. It has been argued that, as the emergence of flux is not preceded by a diverging plasma flow, the flux bundles rise because of their own buoyancy, not because they are embedded in a large convective upwelling.

The emergence of flux bundles from within the convective envelope is a topic of research in which rapidly advancing computational resources stimulate progress (e.g. Cheung *et al.*, 2008, and references therein). Qualitatively, the common assumption is that a flux bundle that wraps around the Sun's radiative interior at or near the bottom of the convective envelope somehow destabilizes and forms a bulge that then continues to rise through the 200 Mm thick convective envelope because of the buoyancy (e.g. Archontis *et al.*, 2004; Murray *et al.*, 2006, and references therein) that is a consequence of the fact that magnetic field adds to pressure (see Vol. I, Section 6.4), so that the interior of a tube is less dense than the exterior when in pressure balance. The very large contrast in gas pressure of some 4.5 orders of magnitude between the surface and the bottom of the convective envelope should cause isolated flux bundles to expand by a comparable factor during their rise, which would imply fibril diameters $\mathcal{O}(1 \text{ km})$, which is some 10^5 times smaller than the local pressure scale height at great depth that should be the characteristic scale of convection. Part of this anticipated expansion can be countered by draining of material within the flux tubes back into the deep interior along the tube into its deepest segments. A dominant effect that limits expansion at least near the surface is argued to be the winding of the field around itself in the shape of flux ropes. Numerical models of the dynamo do not have the resolution to address the formation of flux ropes on the relevant scales, and flux-emergence models starting with pre-configured tubes (e.g. Fan *et al.*, 1998, and references therein) require windings that would, at least on initial emergence, be quite strong compared to the preponderantly east–west orientation of emerging flux. At present, the formation and rise of the flux bundles that form active regions after emergence remain major unsolved problems in solar (and stellar) physics.

Another problem concerning flux emergence that impacts dynamo modeling is related to their preferential orientation: in order for the flux bundles to emerge with the predominant east–west orientation, there should be relatively little impact of the turbulent convection throughout the rise phase. This is estimated to require field strength at the base of the convective envelope of 10^{4-5} G (e.g. Fan, 2009), which exceeds the equipartition field strength of less than 10^4 G that might be expected from a simple interaction of convective flows and embedded magnetic field (see Fig. 2.1 in Schrijver and Zwaan, 2000). It remains to be understood how the solar dynamo manages to create such strong fields in flux tubes or, if it does not, how the weaker flux bundles manage to rise to the surface with little deflection in latitude (subject to the Coriolis force) or scatter in orientation (subject to convective motions).

Upon emergence, the active-region flux is initially distributed over a multitude of small bipoles that suggest that the flux bundle has been frayed by the interaction with the convective flows, causing a mixed-polarity pattern as the rope is seen in cross section at the photosphere (Cheung *et al.*, 2008). Shortly after emergence, much of the field strengthens and becomes nearly vertical to the surface. This is a very important property that enables surface flux dispersal modeling, as discussed in Section 2.2.1.4.

This strengthening of the field is caused by the radiative cooling of the plasma that it contains, causing plasma to slide back to deeper layers, strengthening the field in this process of evacuation or "convective collapse" (e.g. Cheung *et al.*, 2008). The maximum strength of the field is set by the balance of the near-photospheric gas pressure with the magnetic pressure within an idealized completely evacuated flux bundle; for the Sun, this is 2–3 kG. The Sun's surface magnetic field has a range of values, reaching from this maximum value in large flux tubes that make spots, pores, and faculae, down to below the detection limit; in stars with higher or lower photospheric pressures (depending on surface gravity and temperature), the maximum values will be higher or lower accordingly. This behavior makes it essential to differentiate between intrinsic field strengths B of concentrated flux bundles and the average magnetic flux density φ that is an average over all such flux bundles contained within a resolution element; this is sometimes oversimplified into a single characteristic average field strength B_c within a resolution element and a corresponding matching area filling factor f such that $\varphi \equiv f B_c$.

The undulation of the emerging field strands through the photosphere causes mass-loaded dips to alternate with largely evacuated bulges into the outer atmosphere. It remains to be understood how this field eventually manages to rise into the chromosphere and corona, but reconnection between the trans-photospheric field segments at or near the photosphere may be one way in which the field loses its plasma anchors (Lites, 2008).

The emergence phase of active regions lasts no longer than 5 days even for the largest, most complex regions (Harvey and Zwaan, 1993). This is followed by a phase of maturity, in which active regions evolve slowly, but remain coherent magnetic entities for several days for small active regions up to one or two months for the largest regions. The average flux density in mature active regions is about $|\varphi| \equiv f B_c \approx 100\,\text{Mx/cm}^2$ during this phase, regardless of region size (Schrijver and Harvey, 1994). The existence of a characteristic mean magnetic flux density for mature active regions in the context of diffusive flux dispersal has yet to be understood; it may be that circulating flows (perhaps set in motion by the enhanced radiative losses by the faculae of active regions) play a role in temporary flux confinement, but how such flows evolve as regions age has yet to be established (e.g. Hurlburt and DeRosa, 2008; Hindman *et al.*, 2009).

It is worth noting that existing active regions are preferred sites for subsequent flux emergence: sequences of active-region emergences form active-region nests that can persist for up to six months. Altogether, almost one in every two active regions emerges within the confines of a preexisting active region (Harvey and Zwaan, 1993).

2.2.1.3 Flux concentrations and radiance

The mixed polarities in the emerging regions eventually sort themselves out into the adjacent leading and following patches (see Fig. 2.1). In that process, small, dark pores often form by the clustering of even smaller concentrations of magnetic flux (the relatively bright faculae), and large sunspots form by the coalescence of pores or small spots, some with complete and some with partial penumbrae around them. The cross section and viewing angle determine if a flux concentration is dark or bright, i.e. if it acts as a heat leak or as a heat block for radiance (see Section 2.4). A large concentration (with a radius R exceeding a few photospheric pressure scale heights, i.e. larger than $\sim 0.5\,\text{Mm}$ for the Sun) shows up dark, because its field with $B \gtrsim 1\,\text{kG}$ is strong enough to suppress convection, so that heating from below and radiative cooling into space settle at a lower equilibrium temperature in the interior of the tubes. Concentrations comparable to, or smaller than the pressure scale height ($R \lesssim 0.1\,\text{Mm}$) are bright (called – appropriately – faculae), particularly when viewed at an angle away from disk center, because their evacuated interiors enable us to view radiation leaking in through their sides from the hotter layers below the surrounding photosphere (e.g. Zwaan and Cram, 1989). Intermediate to the faculae and the spots or pores lie the magnetic knots; in these structures the darkened interior and the brightened flux-tube walls cancel so that no particular structure is seen in broad-band visible light. The mix of spots, pores, knots, and faculae determines if a star is brighter or dimmer with changing activity (see Section 2.4).

2.2.1.4 *Flux dispersal*

After a phase of coherence that lasts for days to weeks for small to large active regions, respectively, the regions begin to diffuse into their surroundings. In the solar case, the effective diffusion constant is dominated by far by the random walk of magnetic flux caused by the evolving convection; the diffusion associated with the plasma's resistivity is far smaller owing largely to the large dimensions of the field structures (with a conductivity several times that of copper on Earth). In the top layers of the solar convection zone, the magnetic diffusivity is of order $\eta = 7 \times 10^{-3}$ km^2/s (see Table 5.1 for values throughout the solar convective envelope), so that the diffusive time scales would range from ~ 0.2 yr for a tube with a radius of 100 km, to ~ 300 yr for a typical sunspot. The convection, in contrast, results in an effective dispersal coefficient of $\eta_e \sim 300$ km^2/s, i.e. about 50 000 times larger.

The large-scale dispersal of magnetic field over the solar surface can be described as a combination of three characteristics of the plasma flows: (1) the latitude-dependent differential rotation (leading to a pole–equator lap time of about 180 days; see Fig. 2.3); (2) a slow ($\lesssim 15$ m/s) but persistent meridional poleward advection; and (3) the diffusive supergranular random walk (with $\eta_e \sim 300$ km^2/s). Models show this to work successfully for individual active regions on time scales of months (e.g. Sheeley *et al.*, 1983), and for the global Sun for time scales up to about a decade (e.g. Schrijver *et al.*, 2002). This flux dispersal is a key ingredient of the so-called Babcock–Leighton dynamo mechanism discussed in Section 6.2.2.

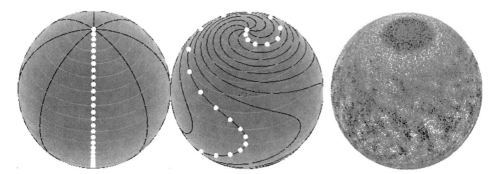

Fig. 2.3. (Left and center) Visualization of the effects of differential rotation and equator-to-pole meridional flow for Sun-like conditions: lines of equal longitude (with markers) are distorted into a spiral pattern. The center panel shows the distorted lines after 3 months. (Right) Simulated magnetogram for a star like the Sun, but with an active-region emergence rate 30 times larger, based on the model by Schrijver and Title (2001). The simulated star is shown from a latitude of 40° to better show the polar-cap field structure. (Figures from Schrijver, 2002.)

The flux dispersal process can be described quantitatively with remarkable success as a diffusion of a signed scalar field. This is a consequence of the fact that the magnetic field in the photosphere is mostly vertical to the surface (owing to the buoyancy of the evacuated field concentrations): the magnetic induction equation (Eq. (3.18) in Vol. I) is readily rewritten as a surface diffusion equation if the field is assumed to be strictly vertical. Note that the diffusion description is nearly, but not quite, linear: the diffusive dispersal coefficient is smaller for larger flux concentrations than for smaller ones (Schrijver, 2001), presumably because the larger concentrations resist distortion more effectively, so that the effective dispersal is lower when activity is high and vice versa. The dependence of the dispersal coefficient on the flux in the concentrations has consequences that are expected to be particularly pronounced for very active stars (see Section 2.6).

The ensemble properties of the active-region source combined with the three components of the plasma flows that transport the flux after emergence suffice to understand most of the behavior of the large-scale solar field, including the evolution of the solar global dipole and quadrupole moments for the surface field (see Fig. 2.4), which together dominate the evolution of the quiescent heliospheric field (Section 2.2.4). The few-degree equatorward tilt of active-region dipole axes causes the trailing polarity to have a slightly larger probability of reaching the pole on the corresponding hemisphere, while the leading polarity has a larger probability of crossing the equator. The antisymmetry in polarity patterns of emerging flux on the two hemispheres (Hale's polarity rule) thus leads to a buildup of the trailing polarities in the corresponding polar regions over a sunspot cycle. As these patterns reverse in the subsequent sunspot cycle, the polar-cap fields first cancel against the incoming flux of opposite polarity, and eventually a new polar cap accumulates of the opposite polarity (see Schrijver and Title, 2001, and references therein); a snapshot of this process as modeled for a young, more active Sun is shown in Fig. 2.3(right). The polar caps thus function as capacitors for the trailing polarity of the active regions, varying nearly in anti-phase with the solar cycle: weak polar caps near cycle maxima, and strong polar caps near cycle minima.

The solar polar-cap flux at cycle minimum is comparable to the flux contained in a single large active region. Consequently, the solar photospheric magnetic field has a pronounced high-order structure to it in which strong low-order components (important for the coupling to the heliosphere) may either form as a result of the moderate flux in the opposite-polarity polar caps separated by a solar diameter at cycle minimum, or by the ensemble of weakly tilted active regions at cycle maximum. The gradual alternating weakening and strengthening of the polar-cap fields does therefore not dominate the heliospheric field by itself. In fact, the combined effect of the polar caps and the active regions is that the heliospheric field shows relatively little variation over the years (e.g. Schrijver and DeRosa, 2003). The overall

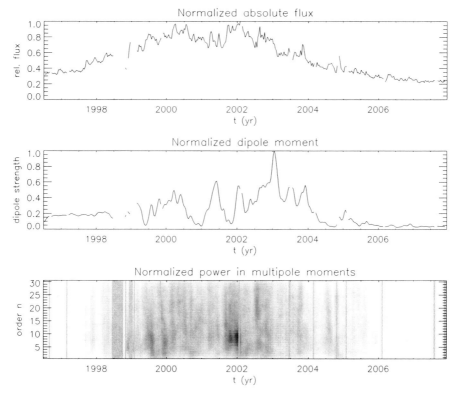

Fig. 2.4. The multipole components of the solar magnetic field. The top panel shows the normalized unsigned magnetic flux on the Sun, and the middle panel shows the strength of the dipole component at the solar surface, each sampled once every 10 days. The curves in these two panels are shown only for those times at which magnetogram data were available. The bottom panel shows the relative power in the multipole components for order n, normalized to the dipole strength ($n \equiv 1$), stronger for darker shading; light grey areas show intervals with missing data. Compare this diagram with Fig. 7.1.

change of the large-scale field looks like a rather erratically tilting and spinning magnet (as can be inferred from the few snapshots in Fig. 8.1 in Vol. I), in which the dipole moment, for example, fluctuates markedly during the cycle maximum and much less during minimum. The dipole component is comparable in power to that in dozens of higher spherical-harmonic orders (see Fig. 2.4).

The capacitor function of the polar caps may not be quite perfect. Simulations of solar activity and the resulting polar and heliospheric field revealed that the variations in strength from one cycle to the next should have resulted in periods in which no polar-field reversals should have occurred. This is neither consistent with the indirect observables of solar-activity records in, for example, ice cores (see Chapter 12), nor with direct observations over the past two cycles (Schrijver *et al.*,

2002). The explanations of this lack of hysteresis in the high-latitude field include the possibility that the meridional flow is modulated in step with the dynamo strength (Wang *et al.*, 2002) – with cause and consequence not yet sorted out – or that the scalar diffusion approximation fails subtly. The latter appears unavoidable: the numerical cancellation between opposite polarities in the random-walk flux dispersal model may be perfect, but on the real Sun this is associated with field inclinations and eventual subduction or expulsion of horizontal field, which violates the assumptions of strict verticality needed to convert the induction equation into a diffusion equation. This may be particularly important for the ubiquitous weak field component, which may be involved in 3D flux transport. Baumann *et al.* (2006) argue that the 3D transport effect has an associated diffusion coefficient of order 50–$100\,km^2/s$, which would lead to an effective decay of a large-scale field on a time scale of 5–$10\,yr$, thus preventing secular drifts in the polar field. Understanding these effects is important, for example, to the modeling of magnetic field on the surfaces of stars (see Section 2.6), and to our understanding of the variations in the heliospheric field and the associated changes in the galactic cosmic rays discussed in Chapter 9.

2.2.2 *Forecasting the solar sunspot cycle*

Shortly after the discovery of sunspots during the European Renaissance, the Sun remained relatively clear of sunspots for some seven decades, until the beginning of the eighteenth century (discussed further in Section 2.8). Since then, a quasi-regular sunspot cycle has dominated the decadal variations in solar activity (Fig. 2.5).

The modulation of the solar sunspot cycle (and its associated variations in cosmic-ray fluxes – see Chapter 9 – and spectral radiance – see Chapter 10) continue to puzzle dynamo theorists. The basic nature of the oscillatory phenomenon is captured fairly well by dynamo concepts (Chapters 5 and 6), but the variability from cycle to cycle (and particularly the phenomenon of modified activity during the Maunder minimum) remains a mystery.

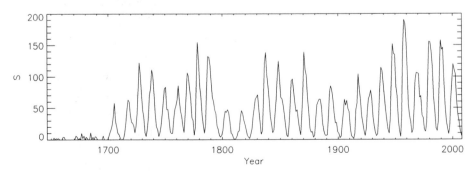

Fig. 2.5. Yearly averaged sunspot numbers. (Data from http://sidc.oma.be/.)

In the absence of an *ab-initio*, self-contained, self-consistent theory for the cycle modulation, a variety of (quasi-)empirical methods have been tested to forecast the strength of sunspot cycles, with methods including some modeling, analyses of solar "climate" trends, spectral analyses, and even neural networks (Pesnell, 2008). The forecasts for Cycle 24 (starting in 2008–9) cover the entire range of historically observed values, but validating any of those forecasts that turn out statistically consistent with the observations will take multiple cycles. This is further discussed in Section 6.3.5.

2.2.3 The upper solar atmosphere

2.2.3.1 Magnetic structure and dynamics

The magnetic configuration at the photospheric level is composed of a sea of mixed-polarity, low-flux patches with quasi-random inclination relative to the vertical combined with an ensemble of compact strong-field regions with near-vertical orientation that is contained in the downflow lanes of the supergranular convection with a characteristic length scale of 20 to 30 Mm (see Section 5.2 for a discussion of solar convective scales and Section 5.2.2 for supergranulation in particular). The field of the weak, mixed-polarity internetwork component forms a high-order multipole in which the plasma pressure almost everywhere exceeds the magnetic pressure (high-β; see Eq. (3.11) in Vol. I). The strong-field concentrations expand rapidly with height into a canopy above which the plasma is predominantly low-β. The canopy reaches over the solar surface at a height of about 2000 to 4000 km, above which the field becomes much less structured in both strength and direction.

The domains above active regions are low-β up to at least a few times the scale size of the region (Gary, 2001). Above much of the quiet Sun – defined pragmatically as the areas surrounding the active regions – the plasma β will generally be below unity, but in many places β is of order unity (Schrijver and Van Ballegooijen, 2005). With increasing height, the high-order multipole field weakens rapidly, until the lowest-order multipoles dominate. The plasma density also decreases with increasing height with a typical pressure scale height of order 50–150 Mm near the surface, and higher in the higher-temperature atmosphere above it. The combined effect is that the plasma β is low within the active-region corona, climbs to around unity at 100–300 Mm, and eventually reaches a low-β state again which it retains up to a height of order 10 solar radii. Beyond that distance, the outflowing solar wind maintains an open field out to the distant reaches of the heliosphere (Section 2.2.4).

The evolving magnetic field in the solar corona readily reconnects. On small scales, the granule-scale random walk of the photospheric footpoints of the field leads to a mild twist of neighboring coronal field lines, but true braiding is absent,

suggesting that reconnection untangles the field on the time scale of the granulation, i.e. approximately 10 minutes. On the largest scale, beyond the scales of
the active regions and the quiet-Sun filaments, the field can be approximated by
a potential field subject to the dynamic pressure effects of the solar wind only at
great heights. On intermediate scales, the situation is less clear. Newly emerging
active regions reconnect with other near or distant active regions on a time scale
as short as a fraction of a day (Longcope *et al.*, 2005), relaxing towards a potential
state. The injected and induced electrical currents do not entirely disappear, however: large-scale current systems in quiet-Sun filament configurations persist for
months, in a balance of decay by dissipation and buildup by shearing flows (e.g.
Mackay and van Ballegooijen, 2006).

2.2.3.2 *Heating and radiation*

The magnetic field in the solar atmosphere is associated with the transport and
dissipation of non-thermal energy; about one part in 10^4 of the Sun's luminosity is radiated from the quiet chromosphere (e.g. Avrett, 1981), and an order of
magnitude less than that from the corona (e.g. Vaiana *et al.*, 1976). For the most
active stars, in contrast, a total of about 1% of the luminosity can be converted into
outer-atmospheric heating (e.g. Vilhu and Walter, 1987)

Within some 2000 to 6000 km of the surface layers with temperatures of about
5800 K, we find the highly dynamic chromosphere. At temperatures of about
10 000 K to 20 000 K, neutral hydrogen is still very abundant, while other atomic
species are typically singly or doubly ionized. Thermal emission in strong spectral
lines from this domain cannot readily escape, and is frequently either re-absorbed
or scattered.

Above the chromosphere, the temperature rises rapidly to 1–4 MK, with much
higher temperatures during flares (discussed in Vol. II, Chapter 5). This domain, the
corona, is optically thin except at radio frequencies (dependent on field strength).
Between the corona and chromosphere (see Fig. 10.2 for example images) lies a
particularly poorly understood "transition region", in which thermal conduction
leads to a very steep temperature gradient. Without effects of plasma dynamics,
ambipolar diffusion, or suprathermal and energetic particles, this domain would be
only some tens of kilometers thick along a field line. But these processes cause
the effective observed thickness of the transition region to be multiple times larger
than expected based on classical electron heat conduction alone.

When measured for relatively large areas – i.e. when averaging over an
ensemble of similar atmospheric components – the radiative losses from the
outer atmosphere increase with the magnetic flux density at the base. A variety of heating mechanisms has been proposed for the chromosphere, the corona,

or – for many scenarios – both (e.g. Narain and Ulmschneider, 1996). Non-thermal energy is likely deposited into the corona in the form of electrical currents that are the result of the motion of the field's photospheric footpoints that are moved about by convective flows. The cascade of such currents to smaller scales, and the details of the eventual dissipation continue to be debated, as is the relative importance of wave dissipation. For the chromosphere, the situation is even less clear: waves of both predominantly magnetic and predominantly acoustic nature have been proposed to play a dominant role, but numerical simulations suggest that electrical currents and reconnection phenomena contribute if not dominate.

With the high degree of structure in the magnetic field within the chromosphere, different mechanisms may dominate in different environments. For example, above weak-field regions over the interior of supergranular cells, pure acoustic waves continue to be a leading contender for the cause of the radiative losses with a good agreement between observations and hydrodynamic models; the (possibly dominant) contribution associated with a small-scale, locally acting turbulent dynamo has yet to be estimated. This emission is referred to as the "basal emission" (Schrijver, 1992, see Section 2.6), because it is the lowest chromospheric emission level observed for stars as a whole and for patches of the Sun seen at moderate spatiotemporal resolution; the basal emission depends on stellar surface temperature and is weakly dependent on surface gravity.

The chromospheric and coronal emissions above the basal level scale as power laws with each other and with the average magnetic flux density of the underlying field: $F_i \propto |\varphi|^{b_i}$. The power-law index b_i between radiative and magnetic flux densities (Fig. 2.6) appears to be an essentially monotonic function of the formation temperature of the radiation observed, increasing from about 0.5 for chromospheric emission from $\sim$15 000 K plasma to just over unity for X-ray emission from $\sim$3 MK plasma; these power laws hold over a contrast in X-ray surface flux densities from $100\times$ below the quiet Sun to $100\times$ above the active Sun, spanning a total of nearly five orders of magnitude (much of which will be covered by the Sun over its lifetime, see Section 2.4).

The chromospheric and coronal heating of the Sun and of stars like the Sun are a function only of the magnetic flux density (note that despite the non-linearity of the relationships, spatially resolved solar observations and disk-averaged stellar observations line up to within the instrumental uncertainties, see Schrijver and Zwaan, 2000, and Schrijver and Title, 2005). In other words, once the magnetic field is in the stellar atmosphere, the dissipation of that energy and the distribution of the energy over the outer-atmospheric domains are independent of stellar properties: stars with masses from about $0.09\,M_\odot$ (equivalent to about

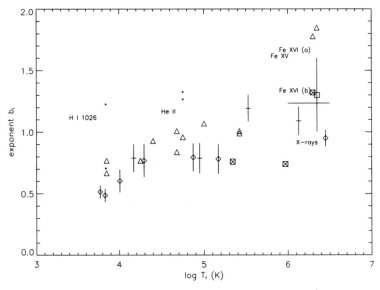

Fig. 2.6. Exponents b_i in the power-law relationships $F_i \propto |\varphi|^{b_i}$ between solar and stellar radiative flux densities and the photospheric magnetic flux density, as a function of the temperature of formation T_f of the radiative diagnostic of stellar magnetic activity. The exponents are derived from data either observed with some angular resolution or disk-averaged for the Sun as a star. Note that the optically thick H I and He II lines deviate from the pattern formed by the other ions in this diagram. (Figure adapted after Schrijver and Zwaan, 2000, with added data from Fludra and Ireland, 2008.)

90 Jupiter masses) to a few solar masses, with radii of $<0.5\,R_\odot$ to $>50\,R_\odot$, and with coronal X-ray flux densities ranging over a factor of 10^5 all adhere to the same scaling relationship within the measurement uncertainties and the intrinsic stellar variability.

This is not to say that the atmospheres are similar from star to star: more active stars and evolved stars have coronae with a persistently hot component above 10 MK, very active stars have components with coronal plasma densities of order $10^{11-12}\,\mathrm{cm}^{-3}$ in addition to a component similar to the solar range of $10^{8-9}\,\mathrm{cm}^{-3}$ (remember that the corresponding volume emissivities scale with the square of the density, so these structures have brightness contrasts of factors of $\mathcal{O}(10^6)$ per unit volume with quiescent solar structures). These structural differences are part of the overall process leading to the so-called flux–flux relationships, of which Fig. 2.7(left) is an example.

The near-linear relationship between the coronal X-ray fluxes and the magnetic fluxes has an interesting side effect: one can readily compare elements of the solar corona (such as so-called "coronal bright points", many of which are associated with ephemeral regions, or the coronae of entire active regions) with

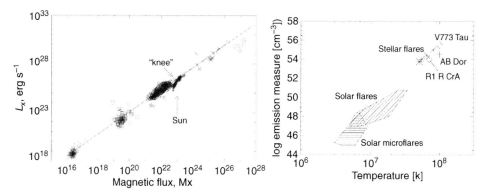

Fig. 2.7. (Left) X-ray luminosity L_X vs. total unsigned magnetic flux for solar and stellar objects. Dots: quiet Sun. Squares: X-ray bright points (associated with small pairs of opposite polarity in quiet Sun, either over recently emerged ephemeral regions or over chance encounters of concentrations of flux of opposite polarity). Diamonds: solar active regions. Pluses: solar disk averages. Crosses: G, K, and M-type dwarf stars. Circles: T Tauri stars. Solid line: power-law approximation $L_X \propto \Phi^{1.15}$. (From Pevtsov *et al.*, 2003. Reproduced by permission of the AAS.) (Right) Emission measure, EM, versus electron temperature, T, of solar flares and four stellar flares (asterisks), a protostellar flare (diamond), a T Tauri stellar flare (diamond), and a stellar flare on AB Dor (K0 IV zero-age main-sequence single star; plus sign). (Figure from Shibata and Yokoyama, 2002; see also Fig. 5.13 in Vol. II.)

disk-integrated quantities. And that enables us to line up the Sun with its stellar counterparts. The result is a near-linear relationship in which both local heating flux densities and surface areas are involved, spanning X-ray luminosities together over almost 12 orders of magnitude (Fig. 2.7(left)). Such diagrams are more appropriately made for flux densities, i.e. normalized to a unit surface area, which still range over five orders of magnitude in X-ray flux densities (Schrijver and Zwaan, 2000).

2.2.4 Coupling of Sun and heliosphere and the solar wind

The weak field in the high solar corona is stretched out into an effectively open field into the heliosphere by the plasma pressure that drives the solar wind (e.g. Vol. I, Chapter 9). The mass loss associated with the solar wind is estimated to be about 3×10^{-14} solar masses per year (which, by the way, amounts to half the rate of mass loss that is associated with the luminosity of the Sun resulting from the equivalence of mass and energy).

The open-field regions map back into the solar corona to regions called "coronal holes" because of their very faint X-ray emission. Apparently, the coronal energy deposition that is converted into heat in closed field lines works differently for open field. Here, most of the available energy (comparable to that of the closed-field corona over quiet Sun) goes into the acceleration of the solar wind.

The overall configuration of the high coronal field, and even the plasma distribution within it (as measured from the electron-scattered photospheric light above the solar limb), can be modeled successfully with the MHD approximation (e.g. Riley *et al.*, 2001). An even simpler approximation is often used because it is far more quickly and easily computed; this is the so-called potential-field source-surface (PFSS) approximation. The PFSS model assumes the field to become radial at some distance from the Sun as a result of the wind's pressure. A suitable value for the source-surface radius is $R_{SS} \approx 2.5$ (measured from Sun center). If the coronal field is modeled assuming that it is potential between the observed radial component at the surface and the assumed radial field at the source surface, the locations of coronal holes and quiescent solar-wind conditions can be modeled with remarkable success (e.g. Wang and Sheeley, 1993; see Riley *et al.*, 2006, for an assessment of relative merits of MHD and PFSS modeling).[†]

The success of the PFSS approximation is in part a consequence of the fact that most of the active regions emerging onto the solar surface carry relatively little net current (meaning not enough to significantly distort the coronal field), while for those that do, this current decays on a time scale of one to at most a few days (e.g. Schrijver, 2005, 2007). The result is that the Sun's magnetic field is partitioned into a closed volume (that generally resides underneath a single band that undulates in latitude as it wraps around the Sun) and a small number of open-field regions that carry the solar wind (and the heliospheric field and the Sun's angular momentum, as described below). An example of this field structure is shown in Fig. 8.1 of Vol. I. Note that the rotation of this slowly evolving PFSS field geometry above the differentially rotating surface readily explains the rotation patterns of coronal holes and the 27.0-day (Bartels) periodicity in recurrent space weather phenomena as the Parker-spiral current sheet (see Vol. I, Chapter 9) sweeps by the Earth, riding on top of the coronal field (see e.g. Schrijver, 2005, for a comprehensive discussion; see Wang and Sheeley, 1993, for a detailed explanation of how a coronal hole can appear to be unaffected by differential rotation for multiple rotations if its magnetic environment is dominated by a single strong active region to whose rotation rate it is synchronized, or can appear to rotate differentially if multiple magnetically active regions at different latitudes significantly contribute to the shape the open-field region).

[†] Note that the ratio of Earth's radius to the radius of the core–mantle boundary, $\sim$1.9 (see Chapter 7), is comparable to the radius of the solar "source surface" to the solar radius, usually set to 2.5. It is important to keep this in mind when comparing, for example, maps of the high-coronal field at the source surface, which contain only relatively low-order fields, to maps of the geomagnetic surface field.

2.3 Stellar activity

2.3.1 A brief review of stellar structure and evolution

Stellar evolution is a vast area of research, and much has been written on that topic; we refer to Rose (1998), as an entry point into that literature. Here, we introduce only some principles, terminology, and properties needed within the present context.

In the strict definition, a star is a self-gravitating body in which gravity is countered by gas pressure that is maintained by nuclear fusion balancing the loss of thermal energy through the stellar surface. Before a star forms, a contracting cloud forms opaque but still nebulous Herbig–Haro objects associated with collapsing clouds, and then pre-main-sequence T Tauri stars (the subject of Chapter 3). Once a balance between contraction and internal pressure has been found, stars are on the "main sequence", where they spend by far the largest fraction of their lifetime. The term main sequence refers to the well-defined clustering of stars in any one of a variety of Hertzsprung–Russell diagrams, in which the stellar luminosity or a logarithmic equivalent (the "magnitude") is plotted against the surface temperature or some filter ratio that measures the relative brightness in differently colored filters (often the $B - V$ value is used, referring to the *B*lue and *V*isible magnitudes, respectively). Examples of such brightness–color diagrams (often referred to as H–R diagrams) are shown in Figs. 2.8 and 2.9(left). Stars are generally characterized by their color, or an equivalent descriptor of their spectral properties called "spectral type" (see the top of Fig. 2.8).

When stars run out of hydrogen fuel in their cores, they evolve off the main sequence in the H–R diagram (see Fig. 2.9(left)) to become giant or supergiant stars. Their eventual fate depends on their mass: low-mass stars fade into ever-cooling white dwarfs, heavier stars eject some of their outer layers, while very heavy stars become supernovae and leave neutron stars or black holes behind. Objects that are too light to sustain hydrogen fusion during any stage of their evolution (although they may have phases with deuterium fusion) are called "brown dwarfs", which have masses of $\lesssim 75\ M_{\mathrm{Jup}}$. These cool very slowly, taking billions of years to lose their thermal energy. Even cooler objects merge into the realm of (heavy) jovian planets.

Before stars reach the main sequence, they migrate through the H–R diagram from the top right (as red giants), initially moving down (to become red subgiants), then curving towards the main sequence (increasing their temperature to become orange, yellow, white, or even blue stars) with a much weaker change in their luminosity than during their initial contraction phase (see Chapter 3). All stars cooler than a surface temperature of about 10 000 K have a "convective envelope", or mantle, immediately below their surfaces, and the coolest stars, be they young or

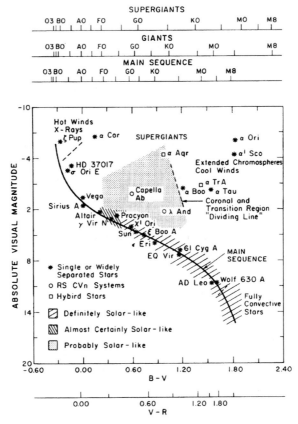

Fig. 2.8. A Hertzsprung–Russell diagram showing stars with substantial magnetic activity in shaded or hatched domains, which are distinguished in groups of solar-likeness as indicated in the legend (by Linsky, 1985). The main sequence where stars spend most of their lifetime fusing hydrogen into helium in their cores is indicated by a solid curve; well above that lies the domain of the supergiant stars, with the giant star domain in between. Also indicated is the region where massive winds occur and hot coronal plasma is apparently absent. Some frequently studied stars (both magnetically active and non-active) are identified by name. The axes above the main panel show the spectral types for supergiant, giant, and main-sequence stars for the corresponding spectral color index $B - V$ or corresponding $V - R$ index.

old, are fully convective. All of these stars make up the ensemble of "cool stars". Beneath the convective mantles, if any, lie the "radiative interiors" in which energy is transported diffusively by photons; fusion occurs within this interior in the deep "core" of main-sequence stars (see Fig. 2.10 for a graphic comparison – not to scale – of internal structure along the main sequence).

The evolutionary time scales are a sensitive function of mass. A star with a mass of, say, three solar masses evolves towards the main sequence in a few

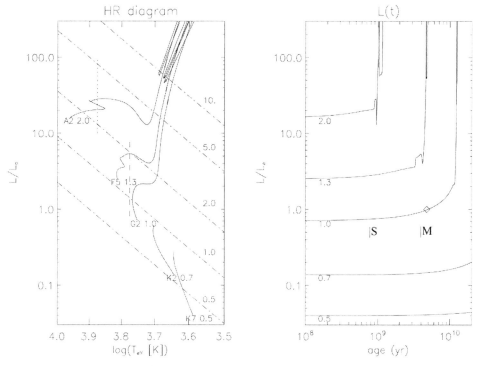

Fig. 2.9. Evolutionary diagrams for luminosity, surface temperature, and age from the mature, main-sequence phase onward. The diagram on the left relates the stellar luminosity (in present-day solar units) with the surface effective temperature (K) in a Hertzsprung–Russell diagram (data from Pietrinferni *et al.*, 2004; for an initial helium of $Y = 0.2734$ and "metal" abundance (i.e. everything heavier than helium) of $Z = 0.0198$). Evolutionary tracks start on the "zero-age main sequence" (ZAMS), and are labeled with the spectral type on the main sequence and the stellar mass (in solar units). The dashed line segment indicates where dynamo action reaches its full strength; for shallower convective envelopes (warmer stars), the activity level weakens until it has dropped by a factor of 100 at the dotted line segment relative to a Sun-like star at the same angular velocity (from Schrijver, 1993). The slanted dashed-dotted lines indicate stellar radii, with labels in solar units. The diagram on the *right* shows the evolution of the stellar luminosity with stellar age (years since ZAMS). The diamond shows the present-day Sun (see Fig. 4.5 for details on the Sun's red-giant phase). The approximate ages for which the oldest fossils of single-cell microbial life (S) and multi-cellular plants and animals (M) have been found on Earth are indicated (cf., Chapter 4).

million years. On the main sequence, where they stay for only $\sim0.4\,\mathrm{Gyr}$, these stars have no magnetic activity, and they only resume magnetic activity after they evolve off the main sequence when they develop convective envelopes again for another 100 million years or so, until they rapidly evolve into what eventually explodes as a supernova. A star of solar mass remains magnetically active to some degree throughout the $\sim10\,\mathrm{Gyr}$ that it spends on the main sequence, and during the

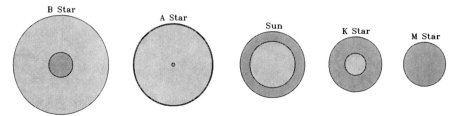

Fig. 2.10. Schematic representation of the radiative (light grey) and convective (dark grey) internal structure of main-sequence stars. The thickness of the outer convection zone for the A-star is here greatly exaggerated; drawn to scale it would be thinner than the black circle delineating the stellar surface on this drawing. Relative stellar sizes are also not to scale: a B0 V star has a radius of $\sim 7.5\,R_\odot$, and an M0 V star has a radius of $\sim 0.6\,R_\odot$, i.e. 12 times smaller.

subsequent ~ 0.8 Gyr giant phase (e.g. Maeder and Meynet, 1988; its maximum radius may reach ~ 0.99 AU, and the maximum luminosity is likely to be around $5200\,L_\odot$; e.g. Sackmann *et al.*, 1993) until it evolves into an ever-cooling white dwarf. A $0.2\,M_\odot$ M9 brown dwarf takes ~ 1 Gyr (Sills *et al.*, 2000) merely to contract to the main sequence, changing little in effective temperature as it descends in the H–R diagram.

During this evolution, the stellar luminosity and its associated color (or effective) temperature continually change. Examples of evolutionary changes are given in Fig. 2.9 for stars from $0.5\,M_\odot$ to $2.0\,M_\odot$.

The young Sun should have been some 25% fainter at the start of the Archean era (at $\sim 3.8 \times 10^9$ yr, when life is presumed to have originated) than the current mature Sun according to stellar-evolution models (e.g. Gough, 1981). This should have resulted in a much cooler Earth, covered in ice. Yet, geological records show that there was liquid water on Earth even in the first billion years of its existence. How this could happen continues to be studied (see a discussion and references by Gaidos *et al.*, 2000). The greenhouse effect as a result of the high concentration of carbon dioxide may have compensated the lower energy input from the young Sun. Alternatively, the Sun may have been significantly more massive, and therefore brighter, early in its life; if there was substantial mass loss in a strong wind in the first billion years, this paradox would be resolved. Sackmann and Boothroyd (2003) study the problem of the "faint, young Sun", by comparing evolutionary scenarios with present helioseismic measurements of the internal structure of the Sun. They conclude that a more massive young Sun (brighter, and with somewhat tighter planetary orbits) is compatible with the present internal structure for a Sun with a mass up to about 1.07 present solar masses. Their best-fit model, starting with a mass of 1.07 solar masses, has an initial irradiance at Earth that is 5% higher than at present (compared to 50% lower for the Standard Solar Model), would subsequently decrease to about 10% lower, and then increase again to the present value.

2.3.2 Solar and stellar activity

The defining properties of stellar magnetic activity are the existence of variable coronal (X-ray) and chromospheric (UV–optical) emission. These characteristics are observed for a wide variety of stars. Magnetically active stars include all the stars with sustained hydrogen fusion in their cores, i.e. the so-called main-sequence stars, that have convective envelopes immediately below their surfaces. In single stars or in wide binaries, the activity level measured by emission from the chromospheres or coronae of these stars, or by the coverage by starspots, increases monotonically with increasing angular velocity to rotation periods as short as a few days. Rather than using the rotation period per se, however, studies of the rotation–activity relationships frequently use the Rossby number:

$$R_o = \frac{v}{2\Omega \sin(\theta)\ell} \sim \frac{P_{\text{rot}}}{\tau_{\text{conv}}}, \qquad (2.1)$$

which is defined such that it measures the relative importance of the inertial to Coriolis forces ($\mathbf{v} \cdot \nabla \mathbf{v}$ and $\mathbf{\Omega} \times \mathbf{v}$, respectively) acting on parcel of plasma of scale ℓ moving with velocity $\mathbf{v}$ in a rotating system with angular velocity Ω. The central expression is a definition that includes the latitude, which is often neglected when estimating the global effect of rotation. When, moreover, the convective turnover time scale $\tau_{\text{conv}} = \pi \ell/v$ for characteristic length scales and velocities of the deepest (largest and slowest) convective motions in a stellar convective envelope is introduced, the commonly used final expression results. When using the Rossby number, the activity is seen to increase with rotation up to a value of $R_o \sim 0.1$ (see Fig. 2.11(upper), which uses the symbol N_R for the Rossby number).

For even more rapidly rotating stars, the activity reaches a saturation level, and for stars with rotation periods of only a fraction of a day, supersaturation sets in, with activity decreasing with increasing angular velocity. It appears that, when proceeding towards shorter rotation periods, the coronal activity saturates first, followed by chromospheric activity, and finally by starspot coverage. This has led to the suggestion that different processes set in at successively shorter rotation periods: centrifugal stripping of the high corona (Jardine and Unruh, 1999; Ryan *et al.*, 2005), saturation of the level to which non-thermal heating can be extracted from the near-surface convection or deposited into the chromosphere (e.g. Vilhu, 1987), and finally saturation of the dynamo process itself (e.g. O'dell *et al.*, 1995; see the discussion of a scaling relationship that potentially connects planetary dynamos and saturated stellar dynamos in Section 7.6.5) possibly by the coupling of the

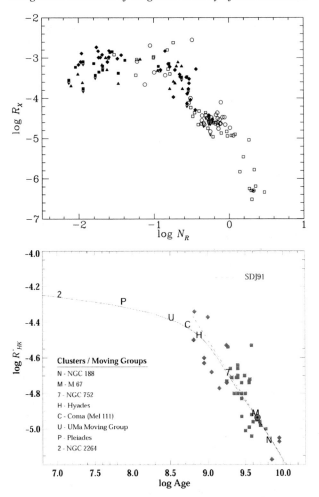

Fig. 2.11. (Upper) Relationship between coronal X-ray emission relative to the stellar luminosity, $R_X = L_X/L_{bol}$ – shown vertically – and Rossby number $N_R \equiv R_o$ (Eq. 2.1) – shown horizontally – for F5 through M5 main-sequence stars in four clusters and a selection of field stars. Symbols: filled triangles for IC 2391 (~40 Myr), filled squares for α Per (~70 Myr), filled diamonds for Pleiades (~115 Myr), open circles for Hyades (~600 Myr), and open squares for field stars. The Sun (⊙) is located near the low end of the activity measure for the sample of stars in this figure. (Figure from Patten and Simon, 1996; cluster ages in the figure as in that work, in the caption as updated in the compilation by Denissenkov *et al.*, 2009.) (Lower) Activity–age relationship for main-sequence stars for the chromospheric Ca II H, K emission for field stars (diamonds and squares) and average properties for associations of young stars. (Figure from Baliunas *et al.*, 1998.)

magnetic field and the plasma flows (see Section 6.2.1.2 and Fig. 6.3) or because the Coriolis force changes the large-scale circulation patterns that are involved in efficient dynamo action.

Stars warmer than the Sun have shallower convective envelopes. Their magnetic activity is markedly suppressed compared to cooler stars with the same rotation period. This has been argued to be either because of the shallowness of their envelope or because of the short average turnover time of convection resulting in little influence of the Coriolis force that otherwise would introduce a preferential direction into the system (e.g. Noyes *et al.*, 1984a; Schrijver, 1993). By spectral type F2, significant magnetic activity is observed, which rapidly increases in efficiency towards G0 as the convective envelope becomes deeper and the deep convective motions approach or exceed the rotation period.

Magnetic fields are observed along the main sequence as far as we have been able to identify and apply Zeeman sensitive spectral lines, i.e. down to at least M9.5 (e.g. Liebert *et al.*, 1999; Reid *et al.*, 1999; Fleming *et al.*, 2003; Berger *et al.*, 2008). At that point we have already reached the brown dwarfs, i.e. astrophysical objects that are too small to have sustained hydrogen fusion in their cores.

For stars above the main sequence, activity is seen both in stars that have recently formed and are still contracting to the main sequence (pre-main-sequence stars, which include fully convective T Tauri stars) and stars that have exhausted their core hydrogen supply and are moving away from the main sequence, once again en route to a fully convective giant phase, now sustained by nuclear fusion of helium and heavier elements in either their core or in shells surrounding a burned out core.

During their main-sequence phases, cool stars exhibit a variety of activity patterns. A clear activity cycle, as exhibited by the Sun ever since the Maunder minimum, is relatively rare, even for solar analogs: only roughly 60% of solar-like stars show a clear activity cycle (e.g. Baliunas *et al.*, 1998), and the reasons for this and for those that set the cycle duration are still being researched (see Fig. 2.12 and e.g. Saar and Brandenburg, 1999).

A few Sun-like stars in the solar neighborhood are so-called flat-activity stars, showing no clear cycle at all, yet they rotate with a period similar to that of the Sun. Such stars have been argued to be in a state similar to the solar Maunder minimum in the early 1600s; see Section 2.8 for further discussion.

2.3.3 Flares and eruptions

Frequently, emerging flux bundles on the Sun are observed to carry strong, concentrated, net electrical currents, or exhibit rotational motions in the photosphere or even in the corona. These motions suggest that the entire system supports a bulk electrical current. Such currents lead to non-potentiality of the overlying magnetic

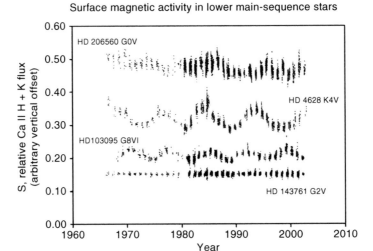

Fig. 2.12. Examples of chromospheric activity cycles (as observed in the H and K resonance lines of singly ionized Ca, or Ca II H+K). Surface magnetic activity records of four stars on or near the lower main sequence from a survey begun by O. C. Wilson in 1966 at Mount Wilson Observatory. Surface magnetism is measured as the ratio of the flux in the emission cores of singly ionized calcium lines in the violet (the Fraunhofer H and K lines at 393.3 and 396.8 nm) and photospheric flux in nearby regions of the spectrum, necessarily integrated over the unresolved stellar disks. The strength of the H and K fluxes increases as the coverage by and intensity of magnetic surface features increases; on the Sun the H and K fluxes vary nearly in phase with the sunspot cycle. The four records show the counterpart of the Sun approximately 2 billion years ago (upper curve, HD 206860; $P = 4.7$ d), and then three Sun-like stars, which show records similar to the present-day Sun, HD 4628 ($P = 38$ d), HD 103095 ($P = 31$ d, or $P = 60$ d (Frick *et al.*, 2004)) and HD 143761 ($P = 21$ d). Both HD 4628 and HD 103095 display decadal periodicities similar to the sunspot cycle. The star HD 143761 may be in a state like the Sun's Maunder minimum. The star HD 103095 is an extremely old (approximately 10 billion years) metal-deficient subdwarf, and is shown as an example of the persistence of decadal magnetic activity cycles in a star of extreme age compared to the Sun. The spectral types are listed next to each record's star name. Arbitrary vertical shifts in the average value of the H and K relative fluxes have been applied in order to show the records without overlap; the offsets are 0.0 (HD 143761), 0.02 (HD 103095), 0.09 (HD 4628) and 0.15 (HD 206860). (Courtesy S. L. Baliunas.)

field, and are often involved in flares and eruptions (see Vol. II, Chapters 5 and 6). Solar flares define power laws in spectra of frequency, N, versus peak brightness or overall energy, L. The spectrum of $N(L)$ can be approximated by a power law $N(L)\mathrm{d}L \propto L^{\gamma}\mathrm{d}L$ with $\gamma \approx -2$; the value of γ reported in the literature depends on the instrument and wavelengths used, and on the sample used (active region flares, EUV quiet-Sun brightenings, etc.), and ranges from about -2.4 to -1.5. The flare energies studied range from $\sim 10^{24}$ ergs to $\sim 10^{32}$ ergs.

The relatively small solar flares drown into a quasi-steady background emission if the Sun is observed as a star. It is not surprising, therefore, that stellar flare spectra are limited to large flares that stand out above the surface-integrated X-ray fluxes. Audard *et al.* (2000), for example, analyze observations of F through M type main-sequence stars to find ubiquitous power laws with power-law indices near $\alpha_f = -2$ (with a possible mild steepening from cool to warm stars). Flare X-ray luminosities in their sample range up to 10^{35} ergs, i.e. up to ~ 1000 times brighter than the largest solar flares, with no evidence for a cutoff energy. They find that flare frequencies for energies exceeding 10^{32} ergs scale proportionally to the time-averaged X-ray emission, saturating as the X-ray activity saturates (see Section 2.3.2), and contribute some 10% of the total X-ray luminosity. Their best-fit results, adopting $\alpha_f = -2$, and using a characteristic solar X-ray luminosity of 3×10^{27} erg cm^{-2} s^{-1} (e.g. Judge *et al.*, 2003), support a scaling for the frequency of large flares with energy E_f exceeding a threshold value of E_{32}^* (in units of 10^{32} ergs, characteristic of a large solar flare) of

$$N^* \left(E_f > E_{32}^* \right) \approx 0.26 \left(\frac{L_X}{L_\odot} \right)^{0.95 \pm 0.1} \left(\frac{1}{E_{32}^*} \right) \; / \text{day}. \quad (2.2)$$

Schaefer *et al.* (2000) discuss the occurrence of super-flares on Sun-like stars: the largest flare they observe has an energy of 2×10^{38} ergs, which is over a million times more energetic than the largest solar flares (and on the solar scale might register as an X80 event). That flare appears to have occurred on (i.e. corresponded to a position in the sky at) a single, moderately rapidly rotating main-sequence star S Fornacis, with spectral type G1 V, i.e. very similar to the present-day Sun. We do not know, at present, if there is a maximum for flare energies on the Sun, but are fortunate that the really energetic ones are very infrequent: with Eq. (2.2), we find for stars like the present-day Sun that flares with energies exceeding 10^{35} ergs are likely to occur once per decade, and – assuming there is no upper cutoff to flare energies – those with energies exceeding 10^{38} ergs may occur only once every 10^4 yr. When the Sun was only 0.1 Gyr old, and similar to EK Dra (see Table in Fig. 2.14), flares with energies exceeding 10^{35} ergs would likely have occurred once per week, and those with energies exceeding 10^{38} ergs may have occurred about once per decade.[†]

[†] The impact of flare-related (X)(E)UV emission on the Earth's outermost atmosphere is discussed in Vol. II, Chapter 12, and in Chapters 10, 13, and 16. Smith *et al.* (2004), for example, study coupling of different wavelengths within the Earth's atmosphere: they describe how the irradiation of the atmosphere, for Archean and present-day conditions, by ionizing radiation leads to a cascade of energy through secondary ionization processes which leads to a UV flux at ground level much higher than expected based on the optical thickness at a given wavelength.

It appears that quiescent activity and flaring activity on stars scale with each other, as also seen in the rise and fall of quiescent and impulsive heating through the solar cycle. One result of this is that more-active stars have a stronger high-temperature component, so that the effective X-ray "color temperature" or spectral hardness increases with activity. It also appears that larger flares are associated with higher characteristic temperatures, going from solar microflares to large flares on very active cool stars; one example of this dependence is shown in Fig. 2.7(right).

2.4 Spots, faculae, network field, and spectral radiance

Sunspots cause a readily observable decrease in the solar irradiance at Earth as they rotate across the solar disk. This effect can be used to measure starspot coverage on stars. Other methods include the use of molecular spectral lines (formed preferentially in the cooler starspot atmospheres relative to the surrounding warmer photospheres), or the analysis of deformed spectral lines as starspots move from the blue-shifted to the red-shifted part of the line during their disk passage. The latter method can even give information on the spot latitudes: equatorial spots have a larger Doppler swing than high-latitude spots, with the net change depending also on the tilt of the stellar rotation axis in the sky.

Starspots (see e.g. Schrijver, 2002, for a comparison of sunspot and starspot properties and for references) in active stars have been measured to cover up to ~70% of the surface (in the short-period contact binary VW Cephei), compared to a characteristic cycle maximum coverage of 0.1–0.2% for the Sun. Starspot coverage typically decreases with decreasing stellar atmospheric activity as measured by, for example, chromospheric and coronal emissions.

Sunspots are surrounded by smaller magnetic concentrations, called faculae, that are relatively bright compared to the surrounding photosphere (Section 2.2.1.3). This population counteracts the effects of the dark spots, but as the faculae are more uniformly distributed around the Sun (and, we expect, around stars) they do not cause much of a rotational modulation, but rather a slow increase and decrease of overall solar irradiance in step with the sunspot cycle. This is also seen on other stars of solar-like activity (Fig. 2.13; see Hall *et al.*, 2009, for a discussion of this trend for a selection of solar analogs). For very active stars, however, the increased relative occurrence of starspots causes the stellar radiance to vary weakly in anti-phase with overall magnetic activity, and stars of intermediate activity show little radiance variability on time scales of years to decades at all.

The variations in solar spectral irradiance over time have been estimated, e.g. by Ribas *et al.* (2005) who estimate the X-ray, EUV, FUV, and UV emissions of a young Sun-like star that has just arrived at the main sequence (zero-age main sequence, or ZAMS); Guinan *et al.* (2003); Telleschi *et al.* (2005); and by Güdel *et al.* (1997), who look at the different behavior of the high-temperature (12–30 MK)

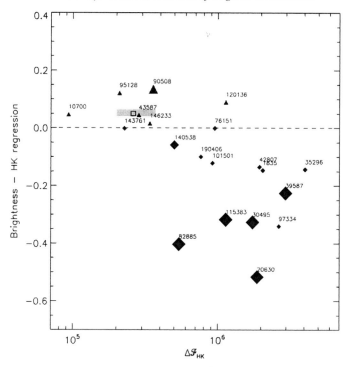

Fig. 2.13. Slope of the regression of photometric brightness variation versus chromospheric (Ca II H+K) emission variation, plotted as a function of average chromospheric activity level as measured by the Ca II H+K excess flux density: stars above the dashed line increase in brightness with activity (the Sun's range in activity is indicated by the shaded box; note that the vertical position of the Sun in this diagram is based on bolometric variations, whereas for the other stars it is based on the b and y Strömgren passbands), while stars below the dashed line decrease in brightness with increasing activity. The stars HD 10700 (τ Ceti, G8 V) and HD 143761 (ρ CrB, G2 V) are considered solar analogs in a state possibly like that of the Sun's Maunder minimum. (Figure from Hall *et al.*, 2009; see also Radick *et al.*, 1998. Reproduced by permission of the AAS.)

component compared to the Sun-like 1–5 MK component. Figure 2.14 summarizes the measured spectral irradiance in several passbands versus estimates of stellar age.

2.5 Activity, rotation, and loss of angular momentum

The primary stellar property that determines the level of magnetic activity is the rate of rotation. The rate at which a star spins is influenced by the evolutionary changes in (1) the moment of inertia, (2) the angular momentum loss through a stellar wind, and (3) the angular momentum exchange in tidally interacting binaries.

Sun-like stars of different ages

Name	Spectral type	P_{rot} (days)	Age (Gyr)
EK Dra	G1.5 V	2.7	0.1*
π^1 UMa	G1.5 V	4.9	0.3
χ^1 Ori	G1 V	5.2	0.3
κ^1 Cet	G5 V	9.2	0.7
β Com	G0 V	12.0	1.6
Sun	G2 V	25.4	4.6
β Hyi	G2 IV	~ 28	6.7

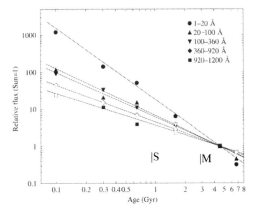

* Järvinen *et al.* (2007) report an age of 0.03–0.05 Gyr.

Fig. 2.14. Spectral radiance versus age of solar-type stars (identified in the table on the left, with spectral type, rotation period, and estimated age), in solar units. Measurements are shown by filled symbols; missing data (open symbols) are derived from power-law fits (solid lines) for passbands from 1 to 1200 Å. The approximate ages for which the oldest fossils of single-cell microbial life (S) and multi-cellular plants and animals (M) have been found on Earth are indicated (see Chapter 4). (Figure adapted after Ribas *et al.*, 2005.)

(1) The evolutionary changes in the global moment of inertia are readily computed from stellar evolutionary models (see the example in Fig. 2.15). These changes amount to several orders of magnitude during the first tens of millions of years of a star (when magnetic coupling with surrounding accretion disks is also important, see Chapter 3) and the final fraction of a Gyr. During the main-sequence phase, they are generally negligibly small compared to the loss of angular momentum through the outflowing wind.

(2) The outflowing stellar wind is coupled to the stellar magnetic field, which introduces a relatively long arm over which the stellar wind can extract angular momentum, so that it eventually carries far more than its own specific angular momentum. The torque on the star is applied by the magnetic field into the stellar interior, and the rapid convective motions cause the angular momentum to be extracted from the entire convective envelope. How much radial and differential rotation this sets up within the convective envelope remains under active study (see e.g. Denissenkov *et al.*, 2010, who argue that this depends on stellar mass as well as initial rotation rate), but the argument is generally made that the convective envelope spins down as a whole. The coupling to the radiative interior underneath it occurs somehow by coupling to a primordial field, wave exchange, or slow flows (see Section 5.5.5). In rapidly evolving stars with shallow convective envelopes this may lead to a (temporary) strong differential rotation between envelope and

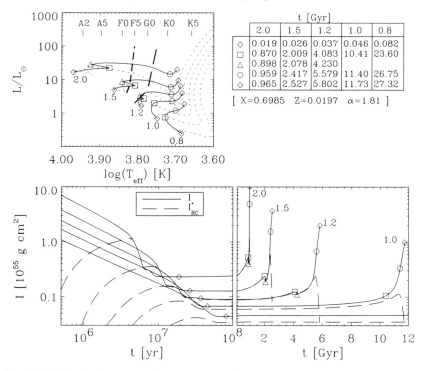

Fig. 2.15. Evolutionary tracks (top panel) for late-type stars of various masses, from the pre-main sequence to main sequence (dotted curves in the top panel), and from there to the base of the giant branch (solid curves in the top panel). The diamonds indicate the zero-age main sequence (at the lower-left end of the solid curves) and the end of model computations. Stellar masses are given in units of the solar mass. The dashed-dotted curve marks the onset of envelope convection. Ages at selected points along the tracks are listed in the table in the top right of this figure for stellar masses indicated in the top row. The evolutionary variations of moments of inertia of the entire star (solid curves) and of the radiative interior (dashed curves) are shown in the lower panels. (Figure from Charbonneau *et al.*, 1997.)

interior. For the Sun, however, helioseismic measurements have shown that interior and envelope rotate at very nearly the same rate, with the interior matching the angular velocity of the differentially rotating envelope at a latitude of about 30° (see Fig. 5.1b).

The angular momentum loss leads to a spin down relative to the evolution in which only the total moment of inertia I is evolved as the star ages. During the main-sequence phase, I changes little (Fig. 2.15), so that most of the change of P_{rot} with age past the first Gyr (during which dynamo saturation is important) is the result of the loss of angular momentum. During this evolutionary phase, the rotation rate depends on age t as

$$\Omega_{rot} \propto t^{1/2} \tag{2.3}$$

(e.g. Skumanich, 1972). For the present-day Sun, the time scale of angular momentum loss is ~ 1 Gyr.

For a small sample of stars, the bow shock formed by the interaction of the stellar wind with the surrounding interstellar medium results in a measurable signal, from which mass loss rates can be estimated. This suggests that the energy flux density that powers the stellar winds scales essentially linearly with the magnetic flux density at the surface, as do the coronal radiative losses (Section 2.2.3.2).

(3) Finally, let us look at binary stars for some interesting aspects of stellar activity. Even though the Sun is a single star, there are interesting lessons to be learned from close binaries. The gravitational tides in binaries with periods of order a week or less (depending on stellar masses and radii) are so strong that the orbital and rotational periods of these stars are synchronized on time scales much less than the main-sequence lifetime. Because any cool-star components of such binaries lose angular momentum through their wind, they will tend to spin down, but the tidal coupling replenishes the lost rotational angular momentum from the reservoir of orbital angular momentum. This causes the orbital separation to shrink, the locked orbital and rotational periods to decrease, and – counterintuitively – the activity to increase with age until eventually the stars merge into a single, rapidly rotating but old star (forming the class of FK Comae stars; see e.g. Bopp and Stencel, 1981). It is the population of tidally interacting binaries and the existence of such old rapid rotators that unambiguously showed us that activity is related causally to rotation, and only indirectly to stellar age. Interestingly, the tidally interacting binaries are even more active than their single counterparts at a given rotation rate (Rutten, 1987); the cause of this "overactivity" remains to be understood.

2.6 Dynamos: polar spots, small-scale field, and flux dispersal

There are a few properties of stellar dynamos that we have yet to touch on to complete the information needed to assemble the description of the solar activity over its 10 Gyr lifetime given in Section 2.1.

The dispersal of magnetic field on stars other than the Sun is subject to the same transport mechanisms as on the Sun. The surface differential rotation is constrained by observations of stars, using either Doppler imaging techniques or time-dependent rotational modulations; it appears that the differential rotation is remarkably insensitive to stellar rotation, as for example the pole–equator lap times that have been inferred are in most cases similar to the ~ 180 days observed for the Sun. Unfortunately, that is the only quantity for which we have observational guidance: the meridional flow (even its direction) is unconstrained, and as there is as yet no successful quantitative model for the supergranular scale of convection, we do not know how the flux dispersal coefficient scales with convection.

Another major uncertainty is the source pattern of the stellar active regions: we know neither the flux spectrum nor the latitudinal distribution. Interestingly, applying solar properties to stellar simulations yields results that are compatible with the stellar observables. This reveals, for example, that increasing the frequency of active-region emergence leads to strong polar caps, which can be so strong that magnetic field may coagulate into polar spots (e.g. Schrijver and Title, 2001; examples of such simulations are shown in Fig. 2.3(right) and 2.16(left)). But simulations of flux emergence suggest that in rapidly rotating stars the Coriolis force likely deflects emerging flux to higher latitudes, so that bipolar regions emerge close to the stellar poles (see the example in Fig. 2.16, and the work by Holzwarth *et al.*, 2007, and references therein).

The latter studies have been stimulated by the observations that active stars commonly have starspots at high latitudes if not, in fact, covering their poles (Fig. 2.16 (right); also Fig. 3.9). Very few sunspots have ever been recorded poleward of 45° in latitude (Fig. 2.2). The numerical experiments described above show that the mere existence of polar starspots requires either a high-latitude emergence of stellar active regions (perhaps as a consequence of the Coriolis forces associated with the rapid stellar rotation), or a much faster poleward meridional advection to carry spots there before they decay (Holzwarth *et al.*, 2007), or may form as polarcap fields become so strong by the large amount of flux emerging on active but otherwise solar-like stars that they spontaneously form pores and spots (Schrijver

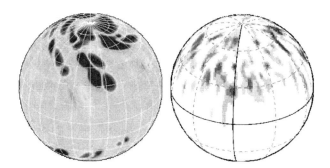

Fig. 2.16. (Left) Radial magnetic field simulated with a surface flux transport model assuming solar-like transport parameters, but for a flux emergence rate that is 30× solar, and a larger latitudinal range for flux emergence, combined with a meridional flow peaking at 100 m/s (∼5× solar). (Right) Observed radial magnetic field distribution for the rapidly rotating star AB Doradus ($P_{\mathrm{rot}} = 0.51$ d). (From Holzwarth *et al.*, 2007. Reproduced with permission © Wiley-VCH Verlag GmbH & Co. KGaA.) See Color Plate 1.

and Title, 2001); probably all three of these mechanisms play some role, but their relative effects remain to be established.

The oldest, slowest-rotating stars are the least magnetically active. Their X-ray emission is extremely weak or absent and their chromospheric emission reaches a minimal, "basal" level below which no stars are observed. The basal emission level from a solar-type star is encountered on the Sun only over the interior of the supergranular network (Schrijver, 1995). That is to say, the solar chromosphere in its most inactive state (such as during an extended cycle minimum) still outshines a basal-level star. Consequently, even much of the solar ephemeral region population vanishes as rotation slows further. Thus, a turbulent dynamo may be at work creating perhaps the chaotic and weak internetwork field, but the ephemeral region population is associated with rotation-driven dynamo action, despite its apparent lack of cycle modulation.

2.7 Fully convective stars, brown dwarfs, and beyond

Somewhere along the spectrum of stars, brown dwarfs, and exoplanets, a transition from an outflow-driven asterosphere to a field-shielded magnetosphere is expected to occur. Where that happens remains to be determined, but one can imagine what happens in that transition. Stars warmer than about M9 have signatures of a solar-like corona and asterosphere in which the atmospheric activity is powered by convective motions at and just below the surface that move the magnetic field. Coolward of M9, several things change: chromospheric and coronal emissions weaken markedly (e.g. Reiners and Basri, 2009; Basri, 2009), the surface magnetic field becomes more dipolar and the time scale for the evolution of the field patterns increases strongly (Donati *et al.*, 2008b), while the relatively cool plasma in the near-surface layers is largely neutral and has been argued to decouple from the magnetic field (Mohanty *et al.*, 2002). Consequently, at some point on the very cool side of the H–R diagram, we should find objects with a jovian-like environment in which a rotation-dominated magnetosphere, extracting power from the star's rotation, is enveloped in a magnetopause formed by the incoming wind of the interstellar medium (Schrijver, 2009).

2.8 The Maunder minimum state of solar and heliospheric activity

The period of remarkably low sunspot activity between 1645 and 1715 is generally referred to as "the Maunder minimum". Throughout this period, sunspots were rarely seen (despite near-complete daily observational coverage, see e.g. Hoyt and Schatten, 1996; Vaquero, 2007), with a characteristic frequency some two orders of magnitude below that typical of the past century. Associated with that low sunspot number was reduced auroral activity, particularly at latitudes below about 55° N (e.g. Siscoe, 1980; Letfus, 2000; McCracken *et al.*, 2004), suggesting a reduction in frequency of strong CMEs or current-sheet crossings, or both.

Although sunspots were scarce, they were not absent. One interesting aspect of the Maunder minimum sunspot data is the marked asymmetry in the butterfly diagram. During the last four decades of the Maunder minimum, sunspot records compiled by Ribes and Nesme-Ribes (1993) reveal that all but a small percentage of the reported spots occurred on the southern hemisphere, with a near-symmetric distribution resuming in the sunspot cycle from 1715 onward.

The multi-decadal change in the ^{10}B and ^{14}C isotope properties suggests that the overall transport of the galactic cosmic rays was modified throughout the Maunder minimum compatible with the interpretation that the heliospheric field was much weaker during the Maunder minimum (Chapters 9 and 11). Moreover, a magnetic cycle of some sort appears to have persisted throughout the Maunder minimum (e.g. McCracken *et al.*, 2004, and references therein; Miyahara *et al.*, 2004), as found from the study of ^{10}Be isotope concentrations in ice cores and ^{14}C to ^{12}C isotope ratios in tree rings (Chapter 11) that are indicative of a modulation of the galactic cosmic rays by the heliospheric magnetic field and solar wind (Chapter 9). The presence of both the 11-year and – more pronounced – 22-year cycles in the records (e.g. Miyahara *et al.*, 2004) can be interpreted in light of the transport theory of galactic cosmic rays, which is sensitive to the polarity pattern of the heliospheric magnetic field (Chapter 9).

The sunspot cycle starting around 1715 reached a group sunspot number only about one-fifth of that over the previous decades, but the increased auroral activity after 1716 – following one earlier mid-latitude auroral sighting in 1707 – immediately attracted attention (Siscoe, 1980). The 1715 eclipse was the first to describe the solar corona as structured after a series of observations noting only a "dim symmetric glow surrounding the eclipsed Sun" (see Foukal and Eddy, 2007, who also note that the 1706 and 1715 eclipse observations showed the existence of an extended chromosphere, manifested in the red flash, but by then the Sun was already climbing out of the deepest phase of the Maunder minimum).

One may, in principle, learn more about a Maunder minimum state by comparing solar properties with those of the ensemble of cool stars. In trying to do so, one immediately runs into the problem that we do not know enough about the solar and heliospheric conditions during the Maunder minimum to know what stars to include in the comparison, because a "Maunder minimum state" can be – and has been – defined in different ways for stars. One possible definition is that of Sun-like main-sequence stars that change from cycling to non-cycling or vice versa in the observational record; no such stars have yet been unambiguously identified (see e.g. Hall and Lockwood, 2004, who analyze the chromospheric Ca II H+K activity of 57 Sun-like stars, and find that approximately 15% of the sample shows little systematic year-to-year trend). A second possibility is to look for extremely inactive Sun-like stars; of these, there are very few, particularly if one is very careful to exclude stars from the samples that have evolved slightly off the main sequence (e.g. Wright, 2004).

A third possible definition of Maunder minimum stars yields the "flat activity" stars that show no significant inter-annual variation in observations extending over a decade or more. Such stars have activity levels comparable to that during solar minimum, as noted by e.g. Hall and Lockwood (2004) and Judge and Saar (2007). The latter study points at two particular candidate stars: HD 10700 (τ Ceti, G8V) and HD 143761 (ρ CrB, G2V; see Fig. 2.13).

The general picture forming from the above of the Sun during its Maunder minimum, is one of a star continuing a magnetic cycle, but with a strongly reduced amplitude, with an average activity level comparable to that of the Sun at sunspot minimum (with most activity associated with ephemeral regions), with a weaker heliospheric field and possibly weaker solar wind flux. Many details remain to be considered and understood, though, including the different temporal patterns in the ^{10}Be isotope concentrations and the sunspot number variations (perhaps influenced by climatic changes associated with volcanism, see Chapter 12), the strongly asymmetric spot distribution and possible change in cycle frequency (e.g. Beer *et al.*, 1998), and the apparently comparable galactic cosmic-ray fluxes during other, less-severe, extended minima (e.g. McCracken *et al.*, 2004).

3

Formation and early evolution of stars and protoplanetary disks

Lee W. Hartmann

3.1 How do stars form?

A star like the Sun begins its life as a dense concentration of molecular gas, called a cloud "core". The mechanisms by which molecular clouds of many solar masses break up into stellar mass pieces are a matter of debate; probably turbulence generated in the process of forming the cloud produces the denser fragments which accrete to form stars (e.g. Ballesteros-Paredes *et al.*, 2007; Bonnell *et al.*, 2007). Here, I will sidestep these issues and concentrate on the collapse of cold, dusty cores into stars. The essential point is that, given the large sizes of protostellar clouds, they almost certainly contain enough angular momentum to form disks of substantial size and mass; thus, a major part of the story of star formation involves moving matter from a disk into a small, spherical protostar.

To make a star of a given mass M from a gas with temperature T, gravity must overcome the pressure support; this means that the radius R of the protostellar cloud must exceed

$$R \lesssim \frac{GM}{c_s^2} = \frac{GM \mu m_H}{kT}, \tag{3.1}$$

where c_s is the sound speed and m_H is the mass of the hydrogen atom. Taking a mean molecular weight $\mu = 2.3$, appropriate for molecular hydrogen plus helium, and a typical cold molecular cloud temperature of $T = 10$ K, Eq. (3.1) implies that a solar mass star must collapse from a cloud of radius $R \sim 2 \times 10^4$ astronomical units (AU). We see pre-stellar dense concentrations of this size (Fig. 3.1) with properties such that they are likely to be on the verge of gravitational collapse. As these cloud cores have sizes $\sim 10^6$ times larger than the final radius of any resulting star, it is clear that virtually all of the angular momentum of the initial cloud must be transferred somewhere else; in general, it must be to a circumstellar disk. In this way, the formation of stars necessarily leaves behind material that can in principle form planets.

Heliophysics: Evolving Solar Activity and the Climates of Space and Earth, eds. Carolus J. Schrijver and George L. Siscoe. Published by Cambridge University Press. © Cambridge University Press 2010.

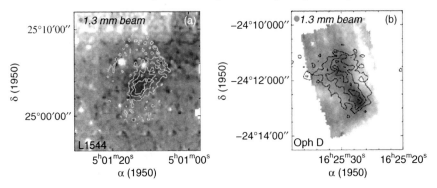

Fig. 3.1. Starless molecular cloud cores (protostellar clouds), as detected in mid-infrared extinction (greyscale) and 1.3 mm dust continuum emission (contours). (From Bacmann *et al.*, 2000.)

Alternatives to the disk solution for the angular momentum problem have fallen by the wayside. For example, the collapsing, magnetized cloud can emit Alfvén waves which can transfer angular momentum to the external medium (von Hoerner *et al.*, 1959), but this is unlikely to be very effective, as the cloud does not collapse unless it is "magnetically supercritical", which in effect means that the Alfvén speed is less than the infall velocity, making it difficult to propagate the signal out before collapse. Moreover, magnetic flux cannot be carried into the inner regions efficiently as the cloud collapses owing to the low ionization state of the material (Umebayashi and Nakano, 1990). Fragmentation into multiple star systems, including binary systems, can put most of the angular momentum into orbital motion, but a moment's thought shows that unless fragmentation occurs hierarchically on a multitude of scales, it is simply not possible to contract through the many orders of magnitude in radial scale without some other means of angular momentum loss. Finally, magnetically coupled disk winds have been invoked to shed angular momentum (e.g. Konigl, 1989). Models of this type also run into significant difficulties, including the low ionization state of much of the disk which makes it difficult for the magnetic field to affect much of the gas; but some variant of this process may well be crucial in establishing the initial angular momentum of the young Sun, as discussed further below in Section 3.5.

The accretion disk is basically an engine in which angular momentum is transferred outward to ever-decreasing amounts of material while the majority of the mass moves inward to the center (Lynden-Bell and Pringle, 1974). In the case of at least moderately ionized disks, it seems increasingly certain that magnetic turbulence provides the necessary angular momentum transport for accretion (Balbus and Hawley, 1998). The low ionization of protostellar disks is likely to render this mechanism ineffective over significant radial regions; gravitational torques

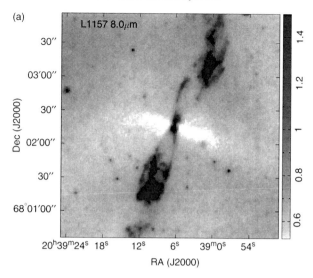

Fig. 3.2(a). An 8 μm image of an accreting low-mass protostar. The lighter, filamentary region running east–west (horizontally in the image) represents dust extinguishing the background radiation; this indicates that the densest, most massive region of the material falling in to make disk and star is far from spherically symmetric. The dark regions running north–south (top to bottom) are due to protostellar continuum emission reflected from dust and molecular emission lines excited by a high-velocity, bipolar outflow thought to be driven from the innermost regions of the proto-stellar accretion disk. (Courtesy John Tobin.)

can come to the rescue, moving most of the cloud mass into the central regions in any event (Section 3.2). However, gravitational torques alone will leave a sizeable amount of mass in the disk, of order 10–30% of the central star mass. As this is much larger than estimated by many techniques, and substantially more than assumed in many models of planet formation, it may be necessary for additional angular momentum transport to occur via magnetic turbulence. On longer time scales, the remaining disk gas is probably removed by some mechanism of ejection due to stellar X-rays and extreme-ultraviolet (EUV) heating (Section 3.8).

This picture of star formation has considerable observational support. Cold clouds of the mass and size indicated in Eq. (3.1) are seen in star-forming regions (e.g. Fig. 3.1), some with already growing protostars (Fig. 3.2(a)(a)). We also observe extended circumstellar disks around many young stars (Fig. 3.3). The masses of these disks are at least $\sim 1\%$ that of the central star; with radii of hundreds of AU, they clearly must contain most of the system angular momentum.

The implication of this picture is that most of the mass of a star must pass through its disk; that is, stars are most directly formed by disk *accretion*. As shown schematically in the right side of Fig. 3.2, disk accretion may not be steady if it cannot keep up with the infall to the disk; instead, early stellar evolution may be punctuated

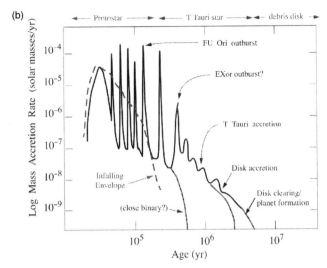

Fig. 3.2(b). Schematic diagram of a likely accretional history of a typical low-mass star. The dashed curve indicates the expected rate of infall of matter from the protostellar envelope (e.g. dense region indicated in the left-hand panel). The solid curve suggests a possible variation of accretion through the protostellar disk onto the central star, which may be steady at the earliest times but is subject to strong variations in accretion (so-called FU Ori outbursts). In this picture, material piles up in the disk due to the infall rate being higher than the disk can smoothly pass on to the central star; this leads to episodic bursts of accretion which drain the excess disk mass. Finally, after infall ceases, slower, more steady accretion occurs during the T Tauri phase, which may cease because either a binary companion or planets accrete the remaining mass. This results in "clearing" the disk, i.e. removing most of the small dust and apparently most of the gas. Finally, secondary production of small amounts of dust can occur during the debris disk stage, when solid bodies collide and shatter. (From Zhu *et al.*, 2009a.)

by outbursts of very rapid accretion followed by extended periods of slow mass addition. There is observational evidence for these accretional outbursts in the FU Orionis objects (Hartmann and Kenyon, 1996); their properties suggest that disks are likely to be quite massive, at least in early stages (Section 3.6).

3.2 Disks and angular momentum transport

The mechanisms that must transport disk material inward to form the star basically determine the mass distribution within the disk, which among other properties affects planet formation. Significant progress has been made in understanding accretion in astrophysical disks over the last two decades, but protostellar–protoplanetary disks pose special problems of their own, basically because it is uncertain whether and to what extent magnetic fields can couple to the highly neutral disk gas.

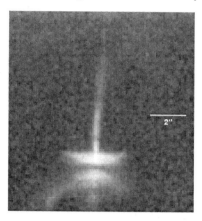

Fig. 3.3. Optical image of the accreting young star HH 30, showing the upper surfaces of its dusty disk in scattered light (the dark lane is due to dust extinction of the central star by the disk), along with an optical, high-velocity, bipolar jet. For scale, 2 arcsec = 280 AU. (From Burrows *et al.*, 1996.)

Initially, it was thought that some type of convective motion might be responsible for producing outward angular momentum transport in disks, but this does not seem to work. To see why, consider simple hydrodynamic interchange of material as shown in the left-hand panel of Fig. 3.4. The inner blob has less angular momentum, so moving it outward while moving the outer blob inward results in inward angular momentum transport. In general, it appears that convection can do little, if not move material the wrong way (Ryu and Goodman, 1992; Stone and Balbus, 1996). On the other hand, magnetic fields can perform the necessary outward transfer of angular momentum (Balbus and Hawley, 1998, and references therein). As shown in the middle panel of Fig. 3.4, if the magnetic field lines are thought of schematically as springs tying adjacent disk annuli together, then as differential rotation continually separates the regions "tied" to the field (e.g. evolution from dashed line to solid curve), the "springs" or field lines become stretched, and the resultant forces will work in the direction of spinning up the outer annulus while spinning down the inner annulus.

The magnetic fields shown in the top-down view of the middle panel of Fig. 3.4 cannot be stretched indefinitely; at some point there will be reconnection and diffusion as the flow becomes turbulent. A schematic picture of what happens is shown in the left side of Fig. 3.5, where the view is now of a meridional plane through the disk. An initially vertical field is perturbed radially (left drawing); these radial perturbations grow due to the shear in the disk (as in the projection of Fig. 3.4, middle drawing); and eventually the field lines become so stretched that they pinch off and develop into full turbulence, as shown in the numerical results in the right-most panel of Fig. 3.5.

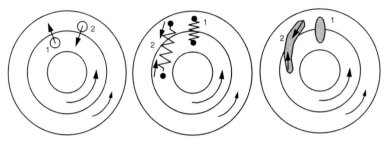

Fig. 3.4. Schematic treatment of angular momentum transfer in a shearing disk with angular velocity decreasing outwards. An initially radial field is perturbed radially (left panel); these radial perturbations grow due to the shear in the disk (middle panel). In the case of gravitational instability (right panel), an excess of material gets sheared out by the differential rotation; the gravitational attraction on the sheared excess (spiral arm pattern) exerts a restoring force in the same sense as the magnetic case, again transferring angular momentum outward. (From Hartmann, 2009a.)

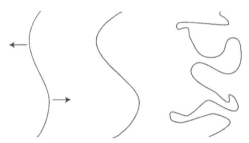

Fig. 3.5. Schematic development of an initially vertical magnetic field in the disk perturbed radially, as viewed in a meridional plane. The indicated perturbations result in the magnetic field becoming stretched radially as indicated in the middle part of the figure, as the field also becomes sheared by the disk angular velocity gradient (see the middle panel of Fig. 3.4). Eventually the stretching and shearing of the field results in diffusion and reconnection, resulting in turbulence as indicated in the right-most part of the figure.

Although there is currently some controversy over the efficiency of this "magnetorotational instability", or MRI, it seems very likely that it provides a sufficiently effective means of promoting accretion in astrophysical disks – provided, of course, that the magnetic field can couple effectively to the gas; there must be a sufficient population of ions and electrons to collide rapidly enough with neutral gas to make the MRI work. Protostellar disks are problematic in this regard: with much or most of their mass heavily shielded from ionizing radiation, and possessing temperatures far too low to effectively ionize even low-ionization potential metal atoms such as Na and K, it seems highly unlikely that the MRI can account for (at least low-mass) star formation on its own.

If other processes cannot transfer mass from the disk into the central star fast enough to keep up with the infall from the protostellar cloud, it appears that gravitational instability (GI) will provide the necessary angular momentum transport. As shown in the right-most part of Fig. 3.4, self-gravity causes local contraction and compression, which then becomes sheared due to the differential rotation. The gravitational attraction of one "end" of the spiral arm pulls on the other; this has the effect of accelerating the outer material at the expense of decelerating the inner material – i.e. transferring angular momentum outward. If all else fails, it appears that this mechanism will prevent most of the mass from remaining in the disk, but instead will allow accretion toward the central object.

For some time the possibility of gravitationally unstable disks in the T Tauri stage (roughly, low-mass stars with ages $\gtrsim 1$ My to ~ 10 My) was dismissed due to low disk masses estimated from mm-wave emission. This is problematic; as mm-wave fluxes are too large to be explained with typical interstellar medium dust, some growth in dust sizes from $\sim\mu$m to $\gtrsim 1$ mm is required, and why stop there? Calculations of dust opacities for plausible compositions indicate that the usual values assumed in estimating disk masses are near the upper range of what is possible, with much more parameter space available with much lower dust opacities (i.e. size distributions with most of the mass in bodies > 1 cm in radius; D'Alessio *et al.*, 2001). This means it is very likely that we have been underestimating disk masses by a factor that is hard to estimate (Hartmann *et al.*, 2006).

3.3 Disk winds

The other major potential mechanism of disk angular momentum transport is that of winds. It is now thought that most of the angular momentum of disks results in expansion of the outer disk rather than simply being lost in a wind; however, because low-mass stars become slowly rotating early in their existence (Section 3.5), it is quite possible that winds from the innermost disk regions play a central role in regulating the rotation of protostars.

Young stars with disks often eject powerful, collimated, bipolar winds or jets (see the *Protostars and Planets V* review volume, Reipurth *et al.*, 2007, for several review papers on this topic). These outflows are clearly the result of disk accretion. We can say this confidently because (a) young stars without disks do not show this phenomenon, and (b) mass ejection rates, as best we can determine, clearly scale with the accretion rate (Fig. 3.6). Indeed, in the case of the most powerful low-mass outflows – those of the FU Ori objects – accretion is the only energy source large enough to account for the necessary driving (Hartmann and Kenyon, 1996).

The high degree of collimation seen in many jets (e.g. Fig. 3.3) favors magnetic fields, as well-developed theory shows that rotating fields can provide the

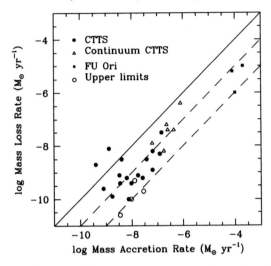

Fig. 3.6. Mass accretion rates in young stars vs. mass loss rates. Errors are probably factors of three or more in each coordinate. The solid line is $\dot{M}(wind) = \dot{M}(accretion)$; the dashed lines are wind mass loss rates of 10% and 1% of the mass accretion rate. Overall, the observations indicate that mass ejection is about 10% of the mass accretion rates, consistent with energetic requirements of driving the mass loss by accretion energy (see text). (From Calvet, 1998.)

necessary collimation (e.g. Shu *et al.*, 1994). Moreover, the observed outflows or jets are relatively cold; that is, the sound speeds of the gas are well below escape velocity, making thermal acceleration unimportant; and thus magnetic acceleration is not only attractive but probably necessary. What is not clear is whether *outer disk* regions exhibit outflows, at least at a sufficiently significant level to affect disk evolution.

The basic version of the cold, magnetically driven wind takes advantage of the rapid disk rotation to fling material outward (and later collimate it: I will not consider this aspect further; discussions of the basic mechanism can be found in Shu *et al.*, 1994). Near the disk it is assumed that the magnetic pressure is much larger than the gas pressure. In this limit, the magnetic fields are stiff at the launching region, i.e. corotation of the inner wind is assured. In this case the energy (Bernoulli) constant of the motion becomes

$$E = \frac{v_p^2}{2} + c_s^2 \ln \rho - \frac{R^2 \Omega_0^2}{2} - \frac{GM_*}{(R^2 + z^2)^{1/2}} = \frac{v_p^2}{2} + c_s^2 \ln \rho - \Phi_e, \quad (3.2)$$

where v_p is the poloidal velocity, Ω_0 is the (Keplerian) angular velocity of the disk in which the magnetic field is rooted, c_s is the (assumed isothermal) sound speed, and Φ_e is an effective potential term including the effects of rotation and magnetic fields. The behavior of the flow depends upon the form of Φ_e, which in turn depends upon the geometry of the flow.

In the case of a perfectly vertical field, perpendicular to the disk, any material that flows outward must be propelled initially by gas pressure; the Keplerian rotation is of course insufficient by itself to drive outflow. The atmospheric structure is nearly hydrostatic until one reaches a radial distance such that

$$c_s^2 \sim \frac{GM_*}{(R^2 + z^2)^{1/2}} , \tag{3.3}$$

in analogy with a Parker thermal wind (see Vol. I, Chapter 9). When the gas is cold, the flow "starts" only at large radii; the flow interior to this must pass through many scale heights of density, resulting in negligible outflow.

In contrast, a field line tipped away from the rotation axis can effectively drive a cold flow, taking advantage of the $R^2\Omega_0^2/2$ term in Eq. (3.2). Neglecting thermal pressure,

$$E = \frac{1}{2}v_p^2 + \Phi_e , \tag{3.4}$$

where the "effective" potential is

$$\Phi_e = -\frac{GM_*}{R_0}\left[\frac{1}{2}\frac{R^2}{R_0^2} + \frac{R_0}{(R^2 + z^2)^{1/2}}\right]. \tag{3.5}$$

Consider now a small displacement along the field line, with a coordinate given by s, and

$$ds^2 = dR^2 + dz^2 . \tag{3.6}$$

At the base of the flow, the disk material is rotating at the local Keplerian velocity. This is an equilibrium state, because $d\Phi_e/ds = 0$ at $z = 0$. However, if $d^2\Phi_e/ds^2 < 0$, this equilibrium is *unstable*; any small perturbation along the field line will result in an increased (outward) poloidal velocity from Eq. (3.4). If θ is the angle between the field line and the disk plane, the critical stability criterion

$$\frac{\partial^2\Phi_e}{\partial s^2} = 0 \quad (R = R_0, z = 0) \tag{3.7}$$

requires $\tan^2\theta_c = 3$, or $\theta_c = 60°$ (Blandford and Payne, 1982). Disk magnetic field lines that are tipped away from the rotation axis by an angle greater than 30° result in an unstable equilibrium, and rapid outflow will commence at the disk.

Using the basic theory of magnetocentrifugal acceleration, spatially resolved kinematics – expansion, rotation – of jets can be used to infer the origin of the outflow, below currently resolvable scales. Observations of jets using the Hubble Space Telescope have suggested that the source region for the observed optical jets is ∼0.2 to 2 AU (Bacciotti *et al.*, 2002; Anderson *et al.*, 2003; Coffey *et al.*, 2007).

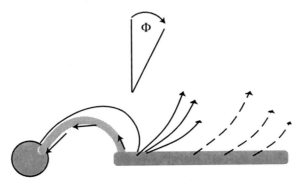

Fig. 3.7. Schematic structure for a connected system of accretion disk, stellar wind, and stellar magnetosphere. Magnetic fields that penetrate the disk inside the co-rotation radius (where the angular velocity of the rotating disk matches the angular velocity of the star) allow material to accrete (grey curves); fields penetrating the disk outside of corotation help provide a spindown torque (solid dark curve). In the X-wind model (Shu *et al.*, 1994), the wind arises from the disk just at corotation (arrows), while disk wind models involve mass loss from a wider range of disk radii (dashed arrows). Magnetic field lines pitched at angles $\Phi_e > 30°$ allow for rapid, cold mass loss (see text).

These estimates must be regarded as uncertain, as it is very difficult to detect the jet rotation; the analysis must assume no asymmetries in the flow, which may be questionable, given the probable presence of complex internal shocks needed to heat the radiating jet gas.

While outflows clearly emerge from the inner disk, there is little evidence for significant mass loss from outer disks, which could take away significant amounts of angular momentum (e.g. Pudritz and Norman, 1983; Konigl, 1989). In addition, there are difficulties with assuming that the disk wind dominates angular momentum transport even in the inner disk. Removing all the angular momentum by the wind involves removing all the accretion energy in the wind as well, leaving no remaining energy to radiate; but this is problematic, because some rapidly accreting pre-main-sequence disks are self-luminous (e.g. Zhu *et al.*, 2007). It seems more plausible that other mechanisms – the gravitational and magnetorotational instabilities – dominate the angular momentum transport of disks, with the winds being a byproduct of accretion. However, the slow rotation of low-mass protostars may require a powerful wind from the innermost regions to remove the final amount of angular momentum (Section 3.5).

3.4 What are young stars like?

Solar-type stars begin their lives with only modestly larger radii than main-sequence values. This is a consequence of (a) the need to have a significant gas opacity to trap thermal energy, and thus produce enough pressure to halt collapse, and (b) the fact that most of the energy of accretion is radiated outward rather than being trapped (e.g. Stahler *et al.*, 1980a, 1980b). Item (b) is ensured in general by the very high opacity of the protostar compared with the infalling material, and in particular by the angular momentum of the protostellar core, which makes much (most) of the material land first on the disk rather than onto the central star.

In the absence of energy input, the star contracts on the Kelvin–Helmholtz time scale

$$\tau_{kh} = \frac{3}{7} \frac{GM_*^2}{R_* L_*}, \tag{3.8}$$

where R_* is the protostellar radius and L_* its luminosity. This is basically the ratio of the internal energy divided by the rate at which energy is being lost, with the numerical coefficient set in this case by the assumption that the star is completely convective. More detailed calculations (e.g. Stahler *et al.*, 1980a, 1980b) indicate that, during protostellar accretion, the protostellar luminosity and radius have roughly those values that would yield a Kelvin–Helmholtz contraction time of the same order as the time scale for infall. In the case of the protostellar cloud described above, this time scale is $\sim R/c_s$, or a few times 10^5 yr.

For low-mass protostars, fusion of deuterium can play an important role in stopping protostellar contraction at early times. Deuterium fusion occurs at a significant rate when the central temperature reaches $\sim 10^6$ K; this results when $R_*/M_* \sim 5 R_\odot/M_\odot$ for a completely convective star. However, as D has a very low abundance, its fusion represents a significant energy source for only a modest time at low masses and very short times for higher-mass, higher-luminosity objects. The result is that stars of masses $\lesssim 0.5 M_\odot$ may be detected initially near the D main sequence in the Hertzsprung–Russell (H–R) diagram (see Section 2.3.1 and Figs. 3.8 and 2.8), but the youngest higher-mass objects will be found below this "birthline" (Stahler, 1988; Hartmann *et al.*, 1997). After D is exhausted, the solar-type star will then undergo Kelvin–Helmholtz contraction until it reaches the main sequence, as shown in Fig. 3.8.

Stellar ages for very young stars are estimated from Kelvin–Helmholtz contraction time scales. The accuracy of these estimates depends mainly on uncertainties in two quantities: the stellar mass and the "starting" radius for KH contraction (left-hand panel of Fig. 3.8). Masses are mostly estimated from theoretical evolutionary tracks, though progress is being made in calibrating these from binary orbits and disk rotation; currently there are significant uncertainties for the lowest-mass stars. For higher masses, calibrations are better but the starting radius or birthline position is uncertain, as it depends upon the precise thermal content of accreted matter rather than on the occurrence of D fusion (see Fig. 3.8). For solar mass stars, the upshot is that ages are uncertain by a factor of two or more for Kelvin–Helmholtz estimates ~ 1 Myr, and perhaps 30% at 10 Myr (Hillenbrand *et al.*, 2009).

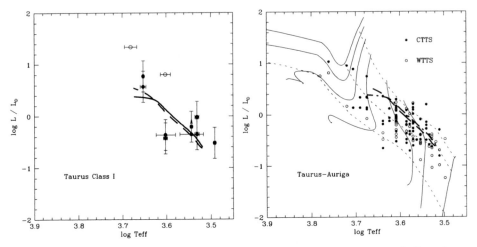

Fig. 3.8. Hertzsprung–Russell diagram positions of Taurus protostars (left) and young (T Tauri) stars (right). These plots of two observed quantities – the stellar luminosity L (in solar units) and the effective temperature T_{eff} – can be used directly to infer the stellar radius R via the equation $L = 4\pi R^2 \sigma T_{\mathrm{eff}}^4$, and indirectly the stellar mass via evolutionary tracks. (Left) Solid and dashed curves correspond to theoretical estimates of initial protostellar radii ("birthlines") as a function of effective temperature (which corresponds roughly to mass). The open circles denote objects in which most of the luminosity derives from accretion, not stellar photospheric radiation. The agreement between theory and observation is reasonably satisfactory given the uncertainties, showing that low-mass protostars do indeed begin their existence with radii only a few times larger than that of the Sun (see text). (Right) Standard stellar evolutionary tracks compared with observed H–R diagram positions of T Tauri stars in the Taurus–Auriga star-forming region. The dashed lines show approximate isochrones for 1×10^6 yr, and 1×10^7 yr, assuming contraction from very large radii, along with the birthlines of the left-hand panel. Ages of young solar-type stars are thus determined by the amount they have descended in the H–R diagram from the birthline, due to gravitational contraction (see text). (From Hartmann, 2009a.)

Stellar magnetic fields and activity are important for understanding the angular momentum "problems" treated in the following section. Great progress has been made in this area over the last decade; see reviews by Strassmeier (2009) for an overview of starspots on active stars, Johns-Krull (2009) and Hussain *et al.* (2009) for a summary on what has been found for surface magnetic fields of T Tauri stars, and in particular Donati *et al.* (2008a) for a brief review of the efforts to project magnetic field configurations from surface polarization measurements.

In brief, large areas of the photospheres of very young stars are covered with strong magnetic fields, with $B \sim 2$ kG and covering (or filling) factors of tens of percent. Polar dark spots seem to be typical, though there are significant spots at other latitudes, and the spot areas/fields are not axisymmetric – explaining why there is often substantial rotational modulation of the optical/near-IR stellar photospheric emission.

The properties of T Tauri magnetic fields, as inferred from Doppler imaging and polarimetry, are quite reminiscent of the behavior of older main-sequence (or post-main-sequence; RS CVn) stars with strong magnetic activity by virtue of their rapid rotation. It appears that, at some basic level of activity, there is a change of mode in which strong magnetic fields tend to be concentrated near (but not exactly at) rotational poles (see Section 2.6).

The variability of the rotationally-modulated starspot-produced light curves – on time scales of days, weeks, months, years (Herbst *et al.*, 2007) – indicates that the fields are not fossil in origin but are produced by some sort of stellar dynamo. Indeed, given that these stars are mostly or fully convective, it is theoretically expected that buoyant, turbulent flux loss mandates continual regeneration. There also appears to be some sort of "saturation" of magnetic activity (Section 2.3.2; also e.g. Preibisch *et al.*, 2005); that is, for a given stellar structure, magnetic heating as indicated by X-ray emission for instance reaches a plateau at some rotation rate; faster rotation then does not produce even more coronal emission.

Efforts have been made to use surface magnetic fields determined from Doppler polarimetry over many photospheric lines to project magnetic field strengths and topologies beyond the stellar surface. These efforts are fraught with difficulties – not least that the inversion of the data to model is not unique – but are suggestive nonetheless. Figure 3.9 shows the results for an estimate of the magnetic field structure of the accreting T Tauri star BP Tau (Gregory *et al.*, 2008). Moving from left to right in Fig. 3.9, one observes the polar spot dominated region, with inner field lines from the pole closing at the equator; on larger scales, there is something like a large-scale dipolar (though non-aligned) field, extending out to several stellar

Fig. 3.9. Inferred magnetic field structure of the classical T Tauri star BP Tau. Surface shading shows photospheric magnetic field strength; the three figures from left to right show estimated near-field closed, far-field closed, and open magnetic field lines. Red and blue tones indicate oppositely directed radial magnetic field strengths. (From Gregory *et al.*, 2008.) See Color Plate 2.

radii; and finally open field lines inferred from polar regions, reminiscent of polar coronal holes on the Sun.

The large-scale (dipolar) magnetic field strengths of these stars are important in understanding the interface between the accretion disk and the stellar photosphere. Johns-Krull (2009; see also references therein) makes the point that, while Zeeman *broadening* clearly demonstrates the existence of 2 kG photospheric fields over substantial areas of the star, the low measurements or upper limits of *polarization* suggest that there must be substantial flux reversals to cancel out the net polarization; this would seem to indicate that the fields are of higher order than dipole, and thus that the large-scale (dipolar) component may be relatively weak. On the other hand, the analyses performed by Gregory *et al.*, Donati *et al.*, etc. suggest some reversals on the photosphere (Fig. 3.9) but non-negligible large-scale fields nonetheless.

An important consequence of the large magnetic fields of pre-main-sequence stars is that the stellar magnetospheric pressure and torques truncate the accretion disks well above the stellar photospheres. Magnetospheres are certainly present, given the strong fields found empirically. Moreover, it is clear from observations that T Tauri stars accrete through their magnetospheres. The strong and highly velocity-broadened Hα emission line profiles of accreting T Tauri stars are convincingly explained by some type of quasi-radial infall (e.g. Hartmann *et al.*, 1994; Muzerolle *et al.*, 2001); this implies that the rapid rotation and slow radial drift of accreting material in the disk must be disrupted, most plausibly by the stellar magnetosphere (Fig. 3.10). The magnitude of the observed velocity linewidths can

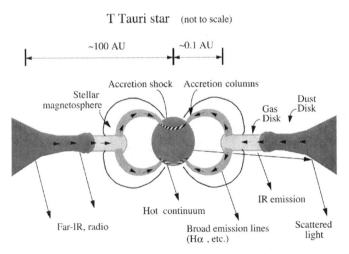

Fig. 3.10. Schematic representation of magnetospheric/disk interaction in low-mass, pre-main-sequence (T Tauri) stars, with diagnostics of specific regions labeled. (From Hartmann, 2009a.)

be explained only if the stellar magnetic field is strong enough to truncate the disk at least a few stellar radii above the photosphere, allowing the essentially freely infalling gas to develop a large gravitationally produced velocity.

In addition to broad emission lines, accreting T Tauri stars exhibit significant amounts of excess continuum emission at wavelengths running from the far-ultraviolet through the optical region. This ultraviolet–optical continuum emission is most plausibly explained as radiation produced in the accretion shock at the base of the magnetosphere, where the material in near-free fall comes to rest at the stellar photosphere (Calvet and Gullbring, 1998). As described in the previous paragraph, it appears that the disk must be magnetospherically truncated at a few stellar radii above the photosphere; this implies that most of the energy generated by accretion will be radiated in this accretion shock. Estimates of mass accretion rates for T Tauri stars are thus generally based on setting the ultraviolet–optical emission excess luminosity $L_{acc} \sim GM_* \dot{M} / R_*$.

3.5 What sets the initial angular momenta of stars?

One of the most striking problems of angular momentum transport is that very slowly rotating low-mass stars are produced by accretion from rapidly rotating disks. In general, T Tauri stars of masses $\lesssim 1 \, M_\odot$ rotate at rates from a few tens of percent to less than ten percent of their breakup values (Herbst *et al.*, 2007, and references therein). The problem of producing slowly rotating stars somewhat older is made much more difficult by the apparent requirement of spinning down the star at the same time it is accreting high-angular-momentum material (see below). Of course, if magnetic stellar winds were intrinsically powerful enough to spin down stars rapidly, there would be no problem; but spin down does not seem to be extremely rapid in non-accreting stars, at least not on time scales needed to explain the slow rotation in stars of ages ~ 1 Myr (Hartmann and Stauffer, 1989; Bouvier *et al.*, 1997).

One possible option is that the magnetospheric coupling between the star and its disk transfers the angular momentum outwards at the necessary rate. However, there are difficulties with applying this model. In the first place *accretion*, which is observed in essentially all T Tauri stars with detectable inner disks, basically requires magnetic field lines tied to disk material inside of corotation; this spins down the gas so that it can accrete, spinning up the star. Spindown of the accreting star requires magnetic fields connected to the disk outside of corotation; thus, to explain T Tauri stars one would like one set of stellar magnetic field lines to be connected inside of corotation, and another outside of corotation, and somehow balance the angular momentum addition due to the accretion with coupling to the outer disk. The Ghosh and Lamb model assumes that field lines penetrate the disk both inside and outside of corotation, so that spindown can occur even with accretion (see Koenigl, 1991, for the application of this model). Numerical simulations indicate that a quasi-steady state may be possible with a large enough turbulent diffusivity (e.g. Long *et al.*, 2005, 2007, and references therein), but whether such diffusivities are realistic is unknown.

Shu *et al.* (1994) criticized the Ghosh and Lamb model because the steady-state situation seems implausible. They avoided the problem of winding up of the magnetic field by assuming instead that the magnetic field penetrates the disk precisely at corotation. However, real T Tauri magnetic fields are not axisymmetric, by significant amounts (see above), and so the field cannot plausibly connect with the inner disk *precisely* at corotation at all longitudes even at a given instant of time. An even bigger problem is that inner disk radii originally thought to be consistent with corotation were estimated from infrared excesses; but these excesses show where the dust is being evaporated, not necessarily where the inner edge of the *gas* disk occurs (Muzerolle *et al.*, 2003; Fig. 3.10). Some interferometric results suggest that the gas disk does extend inward of the dust (Eisner *et al.*, 2007, 2009). In addition, near-infrared CO emission line profiles indicate more rapid rotation than expected if the disk is truncated at corotation (Carr, 2007). These estimates of inner gas disk radii are significantly inside of corotation, raising the question as to whether there is a strong enough large-scale magnetic field to effectively couple to the outer disk for spindown.

T Tauri magnetospheres are probably best thought of as a series of individual magnetic loops, not all of which are filled with accreting gas; this makes it easier to explain the very small covering factors of the hot (shocked) continuum regions on the stellar photosphere of order $\lesssim 1\%$ (Calvet and Gullbring, 1998). As at least some of the loops (if not most) must connect to the disk interior to corotation, it is almost certainly the case that magnetic field lines must tend to become twisted. Such twists rapidly lead to a "ballooning out" of closed field lines, with eventual opening up of field lines and possible ejection of mass, with reconnection following (van Ballegooijen, 1994; Lynden-Bell and Boily, 1994; Fig. 3.11).

Matt and Pudritz (2004) argued that this process of field-line blow-out strongly reduces the magnetic angular momentum transfer from the star to the disk. They then argued (Matt and Pudritz, 2005, 2008a, 2008b) that a stellar wind coupled to the stellar magnetic field must be responsible for T Tauri spindown. However, the required mass loss rates are so large as to require that a significant fraction of accretion energy be used to drive the stellar wind by some unspecified mechanism. Cranmer (2008) made a quantitative attempt to develop a mechanism by which accretion energy can be transferred to a stellar wind. In Cranmer's model, time-dependent accretion of clumps within closed magnetospheric loops excites magnetic waves, which propagate to open field lines and then add momentum

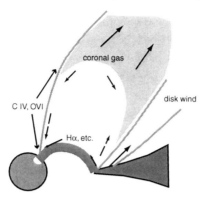

Fig. 3.11. Twister model of magnetic coupling and ejection in a star–disk system. (From Hartmann, 2009a.)

and energy to the stellar wind. The calculations are complex and difficult, so the efficiency of this mechanism is unclear.

The calculations of Matt and Pudritz (2004) did not take into account angular momentum loss accompanying the driving of mass loss as the field lines balloon outwards, as repeatedly shown in numerical simulations (e.g. Uchida and Shibata, 1984; Goodson and Winglee, 1999; Fig. 3.11). As shown in the calculations of Goodson and Winglee (1999), Matt *et al.* (2002), and others, a centrifugal flow is driven from the field lines connecting with the disk as they bend outward, while a hotter flow can result from material near the top of the loop. The bending outward of the field lines connected to the disk provides a favorable geometry for centrifugally driving outflow, even for regions interior to corotation.

I suggested a variant of this idea that might enhance the efficiency of angular momentum and mass loss (Hartmann, 2009b). The idea is that the loops which twist and bulge outward contain magnetically heated gas, as is inferred from the emission lines of infalling material (Fig. 3.10). This amounts to direct conversion of a fraction of the accretion energy, due probably to the twisting of the field lines. If the density is somewhat lower than in the typical accreting columns, the gas can easily be heated to coronal temperatures, so it can become a coronal wind. The net result is that once the field lines open out, gas heated by accretion energy is ejected, gas that originated in the disk and so does not need to be lifted off the star but starts out at a large radius, and that is connected *directly* by magnetic field lines to the star, ensuring spindown.

The disadvantage of this coronal twister model of angular momentum loss is that it is difficult to estimate its efficiency, both in terms of how much material is involved in the mass ejection and the duty cycle of such ejection – the frequency of field-line opening. However, as the radiative losses of accreting gas appear to constitute at least a small percentage of the accretion energy (Hartmann, 2009a), it seems reasonable to allot a significant amount of energy to heating large coronal loops, which can then bulge out and eject material.

The coronal twister model predicts that material will stop draining out of the loop once coronal temperatures are achieved. A characteristic temperature, whose isothermal sound speed corresponds to the Keplerian velocity at $3\,R_* = 6\,R_\odot$ from a $0.8\,M_\odot$ star, is roughly 1.8×10^6 K. Güdel and Telleschi (2007) and Günther and Schmitt (2008) note that there appears to be an excess of O VII X-ray emission relative to both O VIII and O VI, implying an excess of emission measure at temperatures ~ 1–2×10^6 K in accreting stars. The emission measures of O VII-emitting gas seem more consistent with mass loss rates of $10^{-10}\,M_\odot\,\mathrm{yr}^{-1}$ or lower, rather than the $10^{-9}\,M_\odot\,\mathrm{yr}^{-1}$ suggested by Matt and Pudritz (2008a, b). However, as this material is already at 2–3 R_*, it is perhaps not unreasonable to assume an Alfvén radius of order 10 R_*, in which case a mass loss rate of $10^{-10}\,M_\odot\,\mathrm{yr}^{-1}$ could compensate for angular momentum addition of accretion at $10^{-8}\,M_\odot\,\mathrm{yr}^{-1}$.

It is worth emphasizing that most of the angular momentum loss must occur during the protostellar phase, because that is when most of the mass is accreted and protostars are not rotating much more rapidly than T Tauri stars (Covey *et al.*, 2005). If most of the mass is rapidly accreted during outbursts (Fig. 3.2), the magnetosphere might end up crushed against the stellar surface. Whether something like the twister model would work in this case, or whether some different mechanism must be employed, is unclear.

3.6 Protoplanetary disks and gravity

The mechanisms of angular momentum transport determine the mass distribution within the protoplanetary disk. It is important to understand whether gravitational instabilities dominate this transport, in which case accretion onto the central star is likely to decay away with time, leaving a relatively massive disk behind; or whether another mechanism not tied to gravity can reduce disk mass distributions leading to the epoch of planet formation.

The one non-gravitational mechanism of angular momentum transport that we currently understand (at some level) is the magnetorotational instability (Section 3.2). It is possible that the upper layers of the otherwise cold disk can be non-thermally ionized by stellar X-rays and cosmic rays, to the extent that a significant amount of mass and angular momentum transport can occur (e.g. Gammie, 1996). If large amounts of the disk can be activated magnetically in this way, then the disk can behave essentially as a standard viscous disk, with most of the mass at large radii (e.g. Hartmann *et al.*, 1998). However, X-ray and cosmic-ray ionization are insufficient if small dust grains, which can absorb ions and electrons very efficiently, are not heavily depleted (Sano *et al.*, 2000). Furlan *et al.* (2006) found that fits to the observed Spitzer IRS (Infra-Red Spectrograph) spectra required levels of depletion of 10^{-2} to 10^{-3} from interstellar medium values of small dust; however, Sano *et al.* (2000) estimated that depletions of order 10^{-4} are needed for the MRI to operate robustly in upper disk layers.

As discussed earlier, it is plausible if not likely that protostellar disks are initially gravitationally unstable, given the need to accrete most of the mass of the central star through the disk and likely limited MRI transport in cold disks. If the MRI is inefficient, the disk could settle into a state of marginal gravitational instability, with the Toomre parameter

$$Q = \frac{c_{\mathrm{s}}\Omega}{\pi G \Sigma} \sim 1.4 \qquad (3.9)$$

(e.g. Boley *et al.*, 2006), where Ω is the Keplerian (presumed to be the epicyclic) angular frequency and Σ is the disk surface mass density. The Q parameter basically results from satisfying two conditions: one, that gravity can overcome resisting gas pressure forces; and two, that gravity is stronger than the effects of angular momentum in opposing collapse. Larger values of Q mean that the disk is gravitationally stable, while smaller values of Q indicate strong instability. In many instances disks tend to self-regulate; strong instabilities tend to produce heating via shocks that raise c_{s} and thus increase Q, until the sound speed rises sufficiently that the instabilities heating the gas begin to decay.

Even if the MRI is reasonably well activated by non-thermal ionization, it may easily be insufficient over the $1-10$ AU region to transport all the mass viscously; this could result in the general picture suggested by Gammie (1996), in which a "magnetically dead" zone of the disk is sandwiched radially by MRI-active regions at small and large radii.

To develop this further, consider estimates of the mass distribution of the solar nebula. Desch (2007) tried to update the so-called "minimum mass solar nebula" (MMSN) using the so-called "Nice model", which posits substantial outward migration of the outer giant planets from their original positions. The surface density distribution derived by Desch, shown in Fig. 3.12, is substantially above the original MMSN derived from the current positions of the giant planets, which in turn was already above the maximum $\Sigma \sim 100\,\mathrm{g\,cm^{-2}}$ estimated for non-thermal ionization by cosmic rays in the most optimistic scenario. While either version of the MMSN must be considered uncertain, the possibility that the solar nebula had a dead zone must clearly be considered. It is also interesting that the Desch estimate does not fall very far below the gravitational instability result (Eq. 3.9).

The consequence of a disk structure with a "dead zone", as described in the previous paragraph, may be highly time-variable accretion during the protostellar

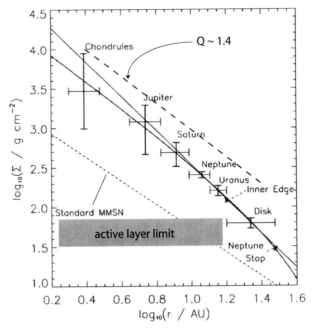

Fig. 3.12. Two estimates of the minimum mass solar nebula. The lower, light-dashed curve indicates the usual estimate, derived from the current position of the giant planets and accounting for the missing light elements; the solid curves show a higher estimate based on the initial positions of the giant planets assumed in a model that has substantial outward migration of the giant planets. Limits on the expected MRI-active surface density due to non-thermal ionization and on the surface density expected for a marginally gravitationally unstable disk (the dashed line showing the condition for the critical value of the "Toomre Q" parameter; see Eq. (3.9)) are also shown. (Modified from Desch, 2007. Reproduced by permission of the AAS.)

phase. As shown by Zhu *et al.* (2009a), GI can be relatively efficient in transferring mass inward at large disk radii but tends to become inefficient at small radii; conversely, the MRI becomes increasingly important at small radii, especially at high mass accretion rates. If matter moving inward under GI dissipates enough energy locally in the inner disk, it can "turn on" the MRI thermally, resulting in an onrush of mass onto the central star. This picture has been invoked to explain the FU Orionis outbursts (Armitage *et al.*, 2001; Zhu *et al.*, 2009b). During FU Ori outbursts, of order $10^{-2} M_{\odot}$ gets dumped onto the central low-mass star over time scales $\sim 10^2$ yr. It is difficult to explain the FU Ori outbursts without having a large amount of disk mass at a few AU, well above that of the standard MMSN.

The possibility of gravitational instability leads to the possibility of forming giant planets directly through gravitational fragmentation (Boss, 2003). This suggestion runs into difficulty, however, because a low Q is not enough; the disk must be able to cool on something like an orbital period to continue fragmenting (Gammie, 2001; Johnson and Gammie, 2003; Rice *et al.*, 2003); otherwise perturbations shear out and transport angular momentum instead. This poses a problem for protostellar disks because they are so cold, and thus do not cool rapidly (e.g. Rafikov, 2005; Boley *et al.*, 2006). The cooling time scale for an optically thick disk is roughly

$$t_c \sim \frac{\Sigma c_s^2}{\gamma - 1} \frac{4\tau_R}{3\sigma T_c^4},$$ (3.10)

where $\tau_R = \Sigma k_R / 2$ is the vertical Rosseland mean optical depth; this is basically the energy content divided by the blackbody radiation loss. Numerically, for temperatures below 170 K, one finds

$$t_c \Omega \sim 4 \times 10^4 (M/M_\odot)^{3/2} R_{10}^{-9/4}$$ (3.11)

(where R_{10} is a characteristic scale in units of 10 AU), which poses an obvious difficulty for fragmentation in that the cooling time far exceeds the Keplerian period. (Things change on distance scales $\sim$100 AU or larger, because the disk typically becomes optically thin, and thus cools much more rapidly than indicated by the above equation.)

Even if fragmentation could occur after infall ceases, one would still expect it to be more important early on, when the disk is more massive. It is not obvious how initial gravitational instability would explain the observed clearing of disks over millions of years.

3.7 Dust disk evolution

In the core accretion model for the formation of giant planets (Lissauer and Stevenson, 2007, and references therein), and in all models of terrestrial planet formation, dust grains grow from sub-micrometer sizes to thousands of kilometers. A starting point for thinking about how planets grow from disks is then considering observations of the evolution of disk dust, detected through its emission.

Figure 3.13 shows the estimated fractions of young stars in various groups with large dust disk excesses as a function of age. Results from differing wavelength regions are incorporated here; the open circles trace only emission from the inner disk, while the solid circles can trace dust emission at significant levels out to radii of order 10–20 AU. The overall result is that optically thick dust disks (with the opacity probably dominated by particles of μm size or a bit less) disappear on time scales of a few Myr. While less is known about the presence of gas in the inner 10–20 AU, clearing of small dust particles seems generally accompanied by removal or disappearance of gas as well (Lissauer and Stevenson, 2007). It is important to

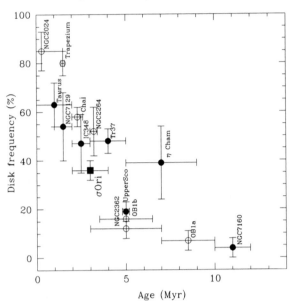

Fig. 3.13. Fraction of stars with near-infrared disk emission as a function of the age of the stellar group. Open circles represent the disk frequency for stars in the T Tauri mass range, derived using observations out to 2–3 μm; solid symbols represent the disk frequency as measured to 8 μm or beyond. (Modified from Hernández *et al.*, 2007; also Hartmann, 2009a.)

emphasize that there is no single time scale for disk clearing. Some (inner) disks disappear immediately, perhaps because of disk disruption by a binary companion; others take a few Myr; a small percentage last for 10 Myr.

The disk can "disappear" in one of three ways; mass can be accreted, ejected, or condensed into large bodies. It is difficult to accrete all the mass of the disk, as some must be left behind to take up the angular momentum; the outer disk is likely to expand over time and evolve on continually slower time scales (Hartmann *et al.*, 1998). Evaporation of the disk may be important, though it is thought to take place over longer time scales than this (Section 3.8). Perhaps the strongest evidence for coagulation into larger bodies is the detection, either through spectral energy distribution fitting and/or imaging, of systems with substantive outer disks but inner disk holes or gaps (Fig. 3.14). This is consistent with the idea that settling, grain growth, coagulation, and formation of large solid bodies occurs fastest in the inner disk, where the surface densities are largest.

Dust grains in the disk generally are thought to evolve to larger sizes, with a decreasing population of small grains with increasing age. During this overall growth, dust is expected to settle vertically and drift radially. This evolution of dust in size and position in the disk can reduce and ultimately eliminate infrared excess

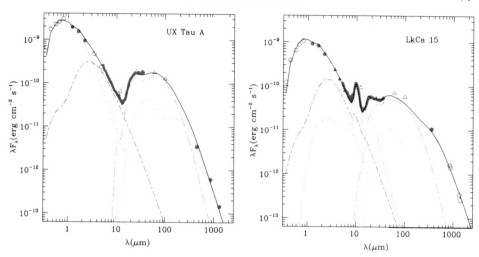

Fig. 3.14. Spectral energy distributions of two T Tauri stars that appear to have mostly optically thick dust disk emission, but with inner gaps, as indicated by the dip(s) in emission at wavelengths of $\sim 10\,\mu$m. The estimated size for UX Tau A of the gap is ~ 50 AU; LkCa 15 has a small amount of optically thin dust within a gap of outer radius of about 46 AU, consistent with mm-wave interferometry. The solid curves are total model fluxes to compare with observations; moving sequentially to longer wavelengths, the dotted curves show the stellar photosphere, long dot-long dash curves inner disk dust emission, dot-dash curves emission from the disk "wall" facing the star, and light dot-dashed curves outer disk emission. (Modified from Espaillat *et al.*, 2007. Reproduced by permission of the AAS.)

emission, consistent with the observed disappearance of dusty disk emission over millions of years (Fig. 3.13). In principle, dust growth can be extremely rapid. Dullemond and Dominik (2005) and Tanaka *et al.* (2005) considered the long-term evolution of dust particles in disks. In both treatments, the disk interior to about 10 AU becomes optically thin on time scales of 0.1 to 1 Myr, as dust particles settle and coalesce into larger bodies. The evolution of the mm fluxes is slower because of longer time scales of accumulation in the outer disk, with substantial reductions in mm-wave emission on time scales of 10 Myr. One might expect that turbulence would lengthen settling and growth time scales, but Dullemond and Dominik find faster growth due to turbulent mixing. The rapid clearing of the inner dust disk in these two sets of models is inconsistent with observations, leading Dullemond and Dominik to argue that ongoing dust destruction is required to repopulate the small particle distribution.

As particles grow in size, many effects converge to make evolution uncertain. For example, the difference in velocities between objects within an order of magnitude of meter size can result in their complete shattering or disruption. Turbulent eddies or whirlpools might help collect these objects at low velocities so that they can accrete (e.g. Dominik *et al.*, 2007), or alternatively disperse them more widely.

Can core accretion proceed fast enough to explain the observed disk clearing on time scales as short as 1–2 Myr? One problem is the formation of km-sized planetesimals from cm-sized objects. Such bodies are thought to be held together lightly – too large for effective sticking and too small for gravity to become important – and, as bodies of differing sizes have differing velocities due to gas drag, collisions between these objects might shatter them rather than build them up. Another problem is that the so-called Type I inward migration due to torques between the disk and the body is very rapid, making it important to grow quickly at ~ 1 Earth mass to avoid falling into the central star on a time scale $< 10^6$ yr (Papaloizou *et al.*, 2007, and references therein). These estimates have usually been made in the "minimum" MMSN (Fig. 3.12); the time scale for inward migration is inversely proportional to the surface density, so gravitationally unstable disks may pose even bigger problems in this regard.

3.8 Disk evaporation

As the planets are overabundant in heavy elements relative to the Sun, it is clear that most of the original gas in the solar nebula has been lost. Of course, some of it accreted into the Sun, but it is unlikely that all of this material was removed in this way. For some time it was thought that a powerful solar-type wind was responsible for gas removal from the nebula. However, we now realize that the strong mass loss we see is not a solar wind but a disk wind; more importantly, the wind material is ejected perpendicular to, not into, the disk (Fig. 3.3).

The high-energy radiation emitted by T Tauri stars provides a mechanism by which the gas of the disk can be evaporated rather than accreted. In this case, rather than generating stellar mass loss from the star via a coronal wind, one can generate disk mass loss from a much lower temperature wind because the material is ejected from much farther out in the gravitational potential field, where the escape velocity is very much smaller than at the stellar surface. Using the usual Parker wind formula (e.g. Eq. 3.3) and assuming photoionization and thus heating to a typical temperature of $\sim 10^4$ K, the sonic point (see Vol. I, Chapter 9) occurs for

$$R_s \sim \frac{GM_*}{2c_s^2} \sim 3.6\frac{M_*}{M_\odot}\ \text{AU}, \tag{3.12}$$

where the mean molecular weight is 0.67, appropriate for a gas of cosmic abundance with ionized hydrogen and neutral helium. Thus, ionizing photons have the potential for removing disk gas at radii of a few to ten AU from the central star.

To see the essential physics of the problem with a minimum of geometrical complication, assume that a volume of $4\pi R^3$ must be ionized, where R is a characteristic radius of escape. This estimate is justified because the gas must maintain its ionization over the disk to a distance comparable to its escape radius to flow out of the gravitational potential well. The balance between photoionization and recombination leads to

$$\Phi_i = 4\pi R^3 n_e n_p \alpha_B \, , \tag{3.13}$$

where Φ_i is the flux of ionizing photons from the central source, n_e and n_p are the electron and proton densities, respectively, and α_B is the Case B recombination rate for hydrogen. Assuming complete ionization of hydrogen, the mass loss rate is

$$\dot{M} \sim 10^{-9} \Phi_{i,41}^{1/2} R_{10}^{1/2} \, M_\odot \, \mathrm{yr}^{-1}, \tag{3.14}$$

where $\Phi_{i,41}$ is the Lyman continuum flux in units of $10^{41} \mathrm{s}^{-1}$ and R_{10} is a characteristic scale of the flow in units of 10 AU. This estimate illustrates the potential of photoevaporation to remove disk gas over evolutionarily interesting time scales. Much more sophisticated treatments of the outflow have been considered by Hollenbach *et al.* (1994), Clarke *et al.* (2001), Font *et al.* (2004), and Alexander *et al.* (2006a, b), but this illustrates the basic result. More detailed calculations of this combined evolution were presented in Alexander *et al.* (2006b), who find typical inner disk clearing on a time scale of a few Myr and full disk dispersion on time scales of order 10 Myr, assuming $\Phi_{i,41} = 10$. Similar results in a more ambitious calculation have been presented by Owen *et al.* (2010), who argue that X-ray heating is more important than extreme ultraviolet (EUV) emission in depleting gas disks.

Unfortunately, the true ionizing fluxes of young stars are not really known because interstellar absorption prevents direct detection. Models of the accretion shock predict little emission or no EUV (Lyman continuum) emission (Calvet and Gullbring, 1998) and so the EUV is probably dominated by emission due to stellar magnetic activity.

Ribas *et al.* (2005) used data from the Far Ultraviolet Spectroscopic Explorer and the International Ultraviolet Explorer, among other missions, to estimate short-wavelength emission fluxes from nearby solar-type stars. For the youngest star in their sample, EK Dra (age $\sim$ 100 Myr), they used the flux evolution in other wavelengths to scale to the Sun in the 360–920 Å band to arrive at an equivalent photon flux of about 4×10^{39} photons s^{-1}. If this flux can be taken to be typical of low-mass T Tauri stars, the result suggests evaporation of disks due to stellar magnetic activity occurs on time scales of order 10 Myr or more. Whether photoevaporation plays a major role in the strong disk evolution from 1 to 10 Myr remains unclear.

Disks close to a hot luminous star can be photoevaporated rapidly due not only to EUV (Lyman continuum) radiation but also by far-ultraviolet (FUV) ($\sim$1000 Å) radiation, which can heat the gas to temperatures $\sim$1000 K as electrons are driven off grains. The FUV radiation thus can drive a wind off the outer disk, and may be more important in many systems if most of the disk mass resides at large distances. We have clear examples of this disk photoevaporation by a massive cluster star in the Orion nebula (O'dell, 1998). Although the solar nebula appears to have been "polluted" by ejection from a supernova (Wadhwa *et al.*, 2007, and references therein), it is not clear that it was close enough to the massive star such that FUV radiation was important in evaporating solar system gas.

3.9 Exoplanets

The ultimate stage of disk evolution, in addition to accretion and photoevaporation, involves the growth of planetary bodies. We now know of over 300 exoplanets (see Fig. 1.2), with the number continually increasing. Of course, the first major surprise was the discovery of Jupiter-mass bodies at very small orbital radii (e.g. Fig. 3.15). This emphasized the almost certain necessity of inward *migration*, as it appears unlikely that disks can be sufficiently massive at 0.02–0.1 AU to form such objects (unless disks are gravitationally unstable all the way to the central star). The other major surprise was how eccentric most of the exoplanet orbits are (Fig. 3.15). These two features are probably related, especially if planet–planet scattering is responsible for much of the inward migration. Before discussing migration further, it is useful to consider how the planets would form in the first place.

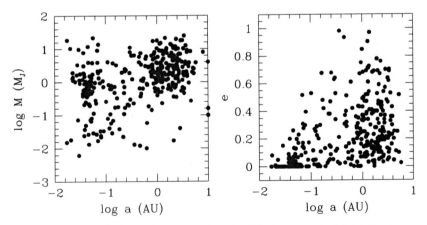

Fig. 3.15. Mass (left; in Jupiter masses) and orbital eccentricity (right) for known exoplanets vs. semimajor axis. (From http://exoplanet.eu.)

The two major scenarios of planet formation are those of core accretion and gaseous gravitational instability. The problems with gravitational fragmentation were discussed in Section 3.6. One strong piece of evidence for core accretion versus gaseous gravitational fragmentation is the strong dependence of exoplanet frequency on the metal abundances of the host star, which is to be expected if solid material provides the initial seeds for core formation. One must have accretion of solids to make terrestrial planets in any event.

In the core-accretion model for giant planet formation, solid bodies accumulate via collisions until the resulting core is sufficiently large that its gravity can pull in surrounding gas (see Lissauer and Stevenson, 2007, for a review). There is some concern that core accretion might proceed too slowly to explain the observed disk clearing on time scales as short as 1–2 Myr in significant numbers of stars. There are two potential bottlenecks in the process. One is the formation of km-sized planetesimals from cm-sized objects. Such bodies are thought to be held together lightly – too large for effective sticking and too small for gravity to become important – and, as bodies of differing sizes have differing velocities due to gas drag (as discussed in the preceding section), collisions between these objects might shatter them rather than build them up. In addition, the inward rapid migration of such bodies (see above) may require fast agglomeration, especially for Earth-sized objects (Ward, 1997), though studies suggest that this inference of rapid migration may not always be correct (e.g. Nelson, 2005; Paardekooper and Papaloizou, 2009). Various schemes of dust concentration might help avoid shattering by reducing relative motions and increasing densities, perhaps through vortices or eddies (e.g. Klahr and Bodenheimer, 2006) or in other turbulent structures (Cuzzi *et al.*, 2001; Lyra *et al.*, 2008).

Once km-sized planetesimals are made, collisions among them can lead to the building of terrestrial planets and giant planet cores. The remaining bottleneck is that of accumulating gas. The energy released by accretion of planetesimals and gravitational contraction of the envelope must be radiated by the outer envelope. If the opacity of the envelope is large, it must extend to large radii; in turn, this can limit the gas available for accretion, which must lie close enough that the tidal forces of the central star do not overcome the protoplanet's gravity. Hubickyj *et al.* (2005) showed that, with sufficiently massive cores, giant planets can form within 1 Myr for an opacity $\sim2\%$ of interstellar values. These authors attribute the reduction in opacity (dominated by dust) to rainout of solid materials in the planetary envelope; as grain growth almost certainly precedes core formation, reduced dust opacity is an extremely plausible assumption.

Our knowledge of extrasolar terrestrial planets is essentially non-existent at the time of this writing, but much more will be known in the next few years due to the

Kepler space mission, which will be able to detect the transits of terrestrial planets across the disks of many low-mass stars.

There is general agreement that terrestrial planets generally (fully) form later than the giant planets; gas drag is important in early stages but the final growth may well occur after gas removal from the disk. As summarized by Nagasawa *et al.* (2007), once growth to km-sized planetesimals has occurred, gravitational effects become important. At first the planetesimals grow by gravitational focusing; as they grow, eventually they excite or stir up other bodies, making their relative velocities larger and limiting accretion. The result is thought to be a set of "oligarchic" protoplanets with relatively similar masses (at least locally). After the oligarchs have swept up most of the available material, interactions between them dominate the subsequent evolution, with large impacts a major feature. This indicates that the final state of terrestrial planet systems is difficult to predict, as it is the result of chaotic growth.

Even after the terrestrial planets are essentially fully formed, significant system evolution can occur, simply because multi-body gravitating systems are generally not stable. A particularly interesting possibility is long-term evolution and migration due to interactions of an outer system of gas/ice giants with the planetesimals left over in the outer disk, objects formed in regions with such low densities that growth to large bodies was not possible. As discussed by Levison *et al.* (2007), there probably has been outward migration of at least Neptune in our outer solar system, based on the analysis of resonant structure in our own planetesimal system – the Kuiper belt. One possible mechanism for explaining this migration is giant planet–planetesimal interactions. Such gravitational perturbations can result

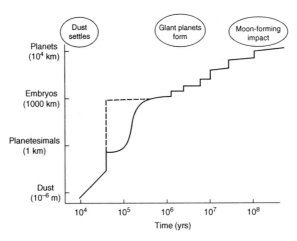

Fig. 3.16. Schematic scale–age diagram for major stages of planet formation. (From Lunine *et al.*, 2009.)

in the system becoming dynamically unstable, resulting in ejection and scattering of many planetesimals into high-eccentricity orbits; this has been suggested, in the so-called "Nice" model, as an explanation for the late heavy bombardment seen in the impact history of the Moon (see Chapter 4). An overall sequence of events for the solar system, and perhaps others, is shown schematically in Fig. 3.16.

3.10 Concluding remarks

The formation of a star like the Sun naturally results in a remnant rotating disk in which planets can form; this disk is likely to be fairly massive initially, with significant loss of mass due to accretion onto the star and ultimate photoevaporation of its gas. The resulting protostar is surprisingly slowly rotating, given that it must be formed by disk accretion. However, by present-day solar standards, these stars are rapid rotators, resulting in very strong magnetic activity; spots that modify the apparent stellar luminosity by as much as 10% or more, X-ray coronal emission at levels three orders of magnitude more powerful than the present Sun's, frequent enormous flares, etc. The strength of the (non-accretion-driven) stellar wind in this phase is unknown, but it would be surprising if it were not also orders of magnitude – perhaps three orders of magnitude – stronger than the present solar wind.

Magnetic phenomena are clearly important in producing the disk wind or jet, and are almost certainly implicated in controlling the formation of the star through disk accretion and in determining the initial angular momentum of the central star. It is likely that we have not yet taken advantage of the insights that have been derived from studies of solar magnetic activity to help solve some of the outstanding issues in the early evolution of low-mass stars and their protoplanetary disks. We should soon find just how common terrestrial planets are, with some insight into their orbital properties; the topic of solar–terrestrial interactions will then expand into new areas, probably many of which we do not foresee.

4

Planetary habitability on astronomical time scales

Donald E. Brownlee

4.1 Introduction

This chapter discusses general concepts of planetary habitability as well as major events in Earth's history that relate to habitability over its full past and future evolution. The Sun plays a determining factor for habitability in the solar system as other stars also are critical to habitability in other planetary systems. Stars provide well-known benefits to life, but they also cause life-ending processes such as the loss of oceans and the loss of planetary atmospheres.

4.2 Environmental limits for life as we know it

Because life has not been detected anywhere but on Earth, the nature of extraterrestrial life remains completely unknown. In light of this famous shortcoming we can still use environmental requirements for terrestrial organisms to estimate where organisms similar to life-as-we-know-it might plausibly exist elsewhere. While this can be criticized as being overly provincial, the Earth-biased approach provides a practical means to access the potential habitability of other worlds. For life based on complex interactions of compounds analogous to the biomolecules of life on Earth, it is relatively straightforward to set general constraints on environments that might support life similar to life-as-we-know-it.

Many of the environmental constraints are influenced by the central star in a planetary system, as the Sun does in our solar system. The Sun provides warmth and energy for photosynthesis, but it also influences many of Earth's fundamental atmospheric, oceanic, and biological processes. The evolution of the Sun over geologic time influences habitability and it also ultimately terminates the ability of Earth to support life.

On Earth, the limiting environmental requirements for life are quite different for microbes than for multicellular organisms or metazoans. Microbial organisms

Heliophysics: Evolving Solar Activity and the Climates of Space and Earth, eds. Carolus J. Schrijver and George L. Siscoe. Published by Cambridge University Press. © Cambridge University Press 2010.

are small but they have been highly successful in developing means of living in a diverse range of environments. For a variety of reasons, the ensemble of microbial organisms has proven capable of adapting to a broader range of environmental conditions than can be tolerated by multicellular organisms; consequently, there are surely many more planets in the universe that provide environments capable of supporting microbial life than there are bodies that could support larger organisms analogous to Earth's animals or plants. In consideration of the habitability of other bodies, it is prudent to consider two broad categories of life: those similar to Earth's microbes and those similar to animals or plants.

Microbes have dominated Earth history and have been here for over 3.5 billion years. Multicellular organisms are a more recent development and show up clearly in the fossil record only within the past 0.6 billion years. Although larger and in many ways more sophisticated, their environmental requirements are constrained relative to the needs of microbial organisms. While a specific microbial organism typically is adapted to live only in a limited environmental range, microorganisms in general have adapted to survive extreme ranges of environmental conditions. Microbial life is robust and different organisms can thrive in a wide range of thermal, chemical, and other physical conditions. For example, hyperthermophiles whose optimum growth occurs above 80 °C have been found to grow at 121 °C while psychrophiles, organisms that have optimal growth below 20 °C, have been found to grow at temperatures as low as −15 °C. Microorganisms are also known that can grow over extreme ranges of salinity, pressure, acidity (pH), water activity, metal content, and radiation levels. While these "extremeophiles" can live in what we consider to be severe environments, there are no specific super-organisms that can live or even survive in broadly ranging environments, say at both high and low temperatures. Any particular extremeophile species usually only grows over a very limited range of conditions. Often the adaptation to extreme environments is done by sophisticated communities of various types of interdependent microorganisms.

Compared to microbial life, multicellular organisms only survive in moderate conditions. For example, the extreme temperature limits for the growth of animals are bounded by temperature ranges beginning near freezing and extending to approximately 50 °C. Plants and animals took billions of years before they developed on Earth, in part because they require stable, constrained and sometimes very special environments. For example, the presence of atmospheric oxygen is a necessity for animals even though free oxygen is not chemically stable in an abiotic atmosphere containing water and nitrogen. It is also a toxic gas for organisms not specifically adapted to live with it and in general its presence degrades organic compounds by oxidation. Surface environments suitable for the survival of plants and animals do not exist anywhere in the solar system except on Earth. Even on Earth it is likely that these advanced organisms will only exist for ∼10% of the planet's total life, as argued below.

The environmental requirements for any type of life potentially involve many factors but a basic starting point is consideration of four important issues: (1) adequate temperature, (2) the presence of water, (3) chemical conditions that allow the survival of complex organic compounds, and (4) a source of energy to drive metabolic pathways. An important issue is where these conditions must be met. It is often assumed that they have to occur at or near a planetary surface, but if the energy requirements can be met, habitable conditions could occur in subsurface regions. Earth is the only body in the solar system that presently meets the four requirements at its surface, but many other bodies could be habitable for subsurface life: all planetary-sized bodies have warm interiors and most have liquid water at some depth. Mars, the larger moons, many asteroids, and even Pluto have, or had, warm, wet interior regions that might match the needs of life that does not rely on sunlight for energy.

4.3 The habitable zone and the effects of stellar heat

One of the most important requirements for life as we know it is water. The ability to retain surface water is the general basis of the concept of the habitable zone (HZ). As most commonly used, the habitable zone is an estimate of the range of distances from a star where an Earth-like planet can maintain surface water for extended periods of time. While a number of factors, including greenhouse gases, tilt of spin axis, planet composition, surface gravity, and cloud properties can be important for habitability, the primary factor considered for the habitable zone is the most fundamental, just the distance from the star (see Chapter 11). For the present-day Sun, the habitable zone is generally considered to be the range from just inside Earth's orbit to a region near or just beyond Mars' orbit. The inner boundary is where surface water is lost to space by either a runaway greenhouse effect or the "moist greenhouse" effect (see Section 4.10). In a full runaway, the surface temperature can exceed the critical point of water ($374\,°C$), i.e. the temperature where liquid water and steam have the same density and are not distinguishable from each other. Due to the extreme greenhouse warming caused by an ocean mass of water vapor, the surface temperatures on an Earth-like planet can reach the melting points of rocks. In comparison, the moist greenhouse is gentle and occurs when the partial pressure of water vapor at high altitudes becomes sufficiently elevated so that a substantial flux of water can be transported into the stratosphere and beyond. At high altitudes, water molecules are decomposed by UV photolysis and the liberated hydrogen ultimately escapes to space.

The outer edge of the habitable zone occurs when surface water freezes. A commonly quoted limit is 1.37 AU based on the onset of formation of carbon dioxide (CO_2) ice clouds (Kasting *et al.*, 1993). A more extended limit of 1.67 AU is based on the maximum greenhouse warming that could occur in a cloud-free CO_2–H_2O

atmosphere. The highest estimate and perhaps an upper limit is 2.4 AU based on a combination of cloud altitudes and particle sizes that could optimize radiative warming by CO_2 clouds (Mischna *et al.*, 2000).

The concept of the habitable zone is becoming increasingly important as Earth-like planets are beginning to be discovered around other stars. The first terrestrial-mass planet that appears likely to be in the habitable zone of another star is the planet Gliese 581d in the planetary system around the 0.3 solar mass star Gliese (or GJ) 581 (see Fig. 1.2). While the designation of the habitable zone for this star is somewhat complicated by the fact that this cooler star has a much different spectral energy emission than the Sun and that the planet is considerably more massive than the Earth, both the atmospheric models of Selsis *et al.* (2007) and the geophysical models of von Bloh *et al.* (2007) indicate that this planet may be able to sustain liquid water on its surface and hence is in the habitable zone.

While the habitable zone is normally considered to be defined by the presence of surface water, there are other ways to define a habitable zone. The surface water definition is reasonable for animal habitability but is likely to be overly restrictive for microbial life. If organisms can live in moist rock, entirely beneath a planetary surface and cut off from sunlight, then a much larger range of distances can be tolerated. For example, in the solar system, there are many bodies beyond Mars that could potentially harbor such life. Even with cryogenic surface temperatures, the interiors of planets and large moons are warmed with radiogenic heat from the decay of uranium, potassium, and thorium, and in some cases by heat left over from the bodies' formation or by heat associated with tidal effects from the gravitational interaction with other bodies. At some depth they have regions that are warm and wet.

Examples of organisms of the type that might be able to live in such conditions include archea and bacteria that were found in deep drill cores in the Colombia river basalt that appear to be deriving metabolic energy by chemical reactions not related to surface processes (Stevens and McKinley, 1995). Such deep life may derive energy from hydrogen produced by oxidation of Fe^{2+} or other sources (Chapelle *et al.*, 2002). For such organisms, the main requirements may be just liquid water and basalt or some other source of oxidizable iron that can produce hydrogen. Other examples of deep life include deep sea vents where hot "mineral-laden" water enters the base of the ocean from regions heated by either volcanic activity or the chemical energy liberated by chemical reactions such as the hydration of the anhydrous silicate olivine from the mantle. It has been suggested that life may have originated at these seafloor vents (Baross and Hoffman, 1985; Nisbet and Sleep, 2001).

Because the limits for subsurface life are so poorly known and because interior heat is generally not related to the intensity of sunlight or starlight, it is not appropriate to consider a habitable zone for deep subsurface life. Heat from a central star is not a major factor for subsurface microbial life unless the temperature is too high. Some of the most intriguing places where extraterrestrial life might be possible in the solar system are small, cold bodies that lie far outside the conventional habitable zone. These include bodies such as the jovian satellites Europa and Ganymede, Saturn's moon Titan, as well as the dwarf planet Pluto, all of which have subsurface water at some depth. Some of these bodies receive appreciable internal energy from tidal heating. One of the most intriguing cold outposts that might harbor life is Enceladus, the ice-covered moon of Saturn with a diameter of 500 km. Although common noon-time temperatures are ∼80 K, this tiny body has a warm subsurface region that is releasing water to space. It appears that there is liquid water at shallow depths, perhaps the result of tidally generated heat.

For planets, the conventional habitable zone moves outward with time as stars brighten. Typical stars brighten by a factor of ∼2.5 during their main-sequence (MS) lifetimes, the periods of their lives when they are stable stars fusing hydrogen to form helium. Main-sequence stars of all mass brighten by a similar fraction as the ratio of He/H in their cores increases with time. At present, the Sun is nearing half its MS lifetime and it is brightening at a rate of about 10% per billion years, and is currently about 30% more luminous than it was 4 billion years ago. More massive and less massive stars brighten at higher and lower rates proportionate to their total MS lifetimes (see Fig. 2.9).

As stars brighten, the inner and outer regions of the habitable zone move outward. Planets that form in the inner regions of the habitable zone eventually find themselves interior to the hot edge of the habitable zone and they lose their oceans. Some planets can remain in the habitable zone over a star's full MS lifetime and those that do are considered to be in the continuously habitable zone or CHZ.

The concept of a CHZ is actually a misnomer as it only applies to stars in their MS stage of evolution. All planets and moons that form in a planet's habitable zone eventually become uninhabitable due to the orders of magnitude increases in luminosity that occur as stars evolve beyond their MS stage and become red giants (see Fig. 2.9). There is, however, a very interesting phase in solar system history that occurs when the Sun becomes a red giant star but before it becomes "too bright". During this period the habitable zone reaches the outer solar system and solar heating of formerly frigid bodies may provide a substantial period of habitability. It has been estimated that Saturn's large moon Titan may become habitable in about 6 billion years and remain so for over a hundred million years, comparable to the time it may have taken for life to develop on Earth (Lorenz *et al.*, 1997).

The habitable zone concept becomes more complex when the ability to have photosynthesis is considered. A more restrictive consideration of surface habitability by organisms similar to planets and animals is the photosynthetic habitable zone

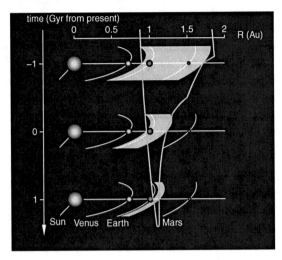

Fig. 4.1. The photosynthetic habitable zone (pHZ) over time, from 1 Gyr in the past to 1 Gyr in the future. The inner edge of the pHZ moves outwards as the Sun becomes brighter with age and the outer edge moves inwards as surface warming leads to decline of CO_2 in the atmosphere to the point where photosynthesis is not possible. (Figure from von Bloh *et al.*, 2002.)

or pHZ (von Bloh *et al.*, 2008). Photosynthesis requires atmospheric levels of CO_2 above some critical limit, approximately 10 ppm for known plants. The pHZ of a given star (see Fig. 4.1 for the case of the Sun) narrows over time as the star gets brighter. The inner edge moves outwards and the outer edge moves inwards. For a planet with land and surface water, weathering processes remove CO_2 from the atmosphere. The process involves sequestering CO_2 in carbonates and this becomes increasingly more effective as warming stars produce warmer planetary surfaces. This process can cause the pHZ to shrink to zero. Estimates by Franck *et al.* (2000) indicate that the Sun's pHZ will shrink to zero width when the Sun reaches an age of 6.5 billion years. The Earth, even now close to the inner edge of the habitable zone, will be left behind the moving pHZ, and lose most of its surface water, long before this time.

The general concept of habitable zones can be extended beyond individual stars to entire regions of galaxies. The concept of a galactic habitable zone (GHZ) considers regions of our Milky Way galaxy where habitable planets are more likely to form and prevail (Lineweaver *et al.*, 2004). Earth-like planets are composed of what astronomers refer to as heavy elements, primarily O, Mg, Si, and Fe. As is the case for all elements heavier than He, the fundamental building blocks needed to form planets are made inside stars and then are recycled back into space to form new generations of stars. From a planetary perspective, galaxies are factories for making heavy elements needed to make planets and life and the abundance

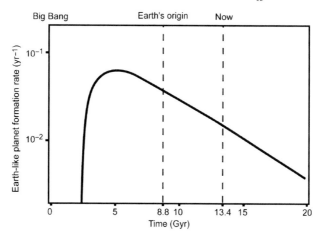

Fig. 4.2. The estimated formation rate of Earth-like planets in the Galaxy over time. The optimum time to form habitable planets like Earth peaked in the first half of the current age of the universe. (Figure adapted from Lineweaver, 2001.)

of planetary building blocks varies in galaxies due to differences of star formation history. The central region of the Milky Way contains higher abundances of planet-forming elements and it is probably the most efficient region for forming terrestrial planets. The center of the Galaxy, however, has a higher abundance of stellar explosions and other process that can harass life once formed. The GHZ, where habitable planets are most likely to form and support life, is a ring that does not include the Galaxy's center or its outer regions.

In a given Galactic locale, the formation rate of Earth-like planets is expected to change over time (Fig. 4.2). In the earliest history of the Galaxy there were no Earth-like planets and no habitable places because stars had not produced enough heavy elements to make them. As the abundance of newly synthesized heavy elements in a galaxy increases, the rate of formation of rocky planets should increase. In time, the planet formation rate reaches a peak and then begins a slow decline as the interstellar gas needed to form new stars becomes less abundant. The formation of new stars and planetary systems is initiated by gravitational collapse of interstellar gas and when the density of gas declines, the rate of star formation declines.

4.4 The habitable zone and other stellar effects

The long-term evolution of the Sun causes profound changes that limit the duration of a planet's habitability. Generally these effects can be considered to be migration of the boundaries of the habitable zone as stars brighten with age. In addition to the flux of visual light, other radiation from the Sun and other stars can also substantially affect the habitability of planets, even those that lie within habitable zones. Most of these processes are related to X-ray, ultraviolet, or strong particle fluxes. These processes are particularly important in the early stages of stellar evolution when higher rotation rates and higher magnetic fields result in higher levels of stellar activity observed in distant stars as flares, emission lines, and high X-ray emission (as discussed in Chapter 2). These processes can deplete atmospheres from some planets by both thermal and non-thermal processes when they produce loss rates that exceed a planet's ability to replenish lost atmospheric volatiles by outgassing from their interiors.

A prime example of the result of these processes is Mars. Mars may be in the Sun's habitable zone and it is also possible that Mars has or had life. However, compared to Earth, Mars appears to be poorly suited for life largely because of its thin atmosphere. Before an age of 3 Gyr the Martian atmosphere appears to have been denser, but for most of its existence the Martian atmosphere has been too thin to allow a stable presence of surface water. It has long been suspected that Mars has lost most of its atmosphere to space, a process that may have been exacerbated both by the planet's low escape velocity and by the lack of a strong dipolar magnetic field like that of Earth. Mars is locally magnetized but it does not have a coherent planetary dynamo field (see Chapter 7) and is thus not globally protected from the solar wind erosion. Observations by the Mars Global Surveyor spacecraft indicate that significant mass loss is occurring at the present time as ejected plasmoids. It has been suggested that this loss occurs due to the interaction of draped interplanetary magnetic field with localized magnetic anomalies in the Martian crust (Brain *et al.*, 2008). Earth is the only terrestrial planet with a strong magnetic field and it is the only such planet that does not show evidence for appreciable loss of atmospheric volatiles. Venus has an extensive atmosphere, but it lost essentially all the water that it formed with. The presence of a strong, long-lived magnetic field to shield a planet from solar or other stellar winds is also a factor to be considered for habitability.

The UV and X-ray emissions of stars can also ablate planetary atmospheres (Moore and Horwitz, 2007). In the case of the Sun, the strongest emission occurred early when it was a more active star (Chapter 2). In the case of lower mass stars, the period of increased activity can last for billions of years.

4.5 Earth before life

The Sun, the Earth, and the other planets formed over a brief period of time from a short-lived disk of gas and condensed matter called the solar nebula (Chapter 3). Earth's formation, like that of the other solid planets, occurred by accretion of solid materials. The processes began with particles of dust but collision and sticking processes rapidly led to the formation of larger and larger bodies. An important aspect of the growth of rocky planets is the amount of a planet's mass that is accreted in the form of large chunks. The accretional growth process yields a number of Moon to Mars-sized "embryos" in a given radial region of the nebula. The final assembly of a rocky planet involves both the accretion of numerous large embryos as well as gravitational ejection of some of them to other locales.

This formation mode that includes impacts of very large bodies is indicated both by the numerical simulation of accretion processes and by evidence that our Moon formed as the result of the impact of a Mars-sized body with the growing Earth. If planets had formed exclusively from small bodies and if they had formed slowly enough, they could have had cold interiors when fully formed. What is believed to normally occur is that the growth of planets involves at least a few collisions with embryos on the order of 10% of the mass of the final planet. Such impacts have severe consequences and bury accretion heat, the kinetic energy of the impactor, deep into the planet. As a result of this process, it is believed that the birth of Earth-like planets involves severe heating of interiors as well as exteriors. Planetary interiors were never cold and the violence of planet formation could radially mix materials, strip off early oceans, and send portions of early atmospheres back into space.

In spite of Earth's extremely violent birth, once the Moon-forming impact occurred it is possible that habitable conditions may have been reached quite early. Zahnle *et al.* (2007) describe a sequence of events where habitability could be reached in a time as short as 10–20 Myr after lunar formation, an event that is believed to have occurred $\sim$30 Myr after the initial formation of the solar nebula.

Following lunar formation, Earth's post-impact atmosphere of vaporized silicates may have condensed in $\sim$1000 yr (Fig. 4.3). The heat of the impact would have melted and partly vaporized Earth's mantle, but the resulting silicate magma ocean may have solidified in only a few million years. Once the magma ocean crystallized, cooling conditions would have allowed the large amount of water vapor injected into the atmosphere to condense and thus reduce the extreme greenhouse warming of the early Earth and allow surface temperatures to drop below 1000 K. Even with most of the water condensed, the atmosphere would still retain $\sim$100 bars of CO_2 whose greenhouse warming would keep the Earth's surface temperature at $\sim$500 K, even though the early Sun was $\sim$30% fainter that its present brightness. The final lowering of the Earth's surface temperature to habitable conditions requires transfer of most of the atmospheric CO_2 to the mantle and crust, a process that can happen over a time scale of 10–100 Myr. Although there is no evidence for it, it is possible that life could have existed on Earth only 10 Myr after formation of the Moon.

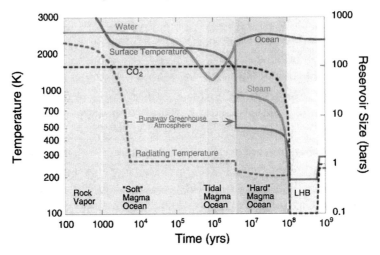

Fig. 4.3. The Earth's surface temperature and above-surface reservoirs of water and carbon dioxide after the Moon-forming collision. The water dip after $\sim 10^6$ yr occurs because of storage of water in Earth's short-lived magma ocean. The surface temperature drops below 1000 K after a few million years when Earth's steam atmosphere condenses, and it drops below 500 K to habitable conditions after about 100 million years when most of the atmospheric CO_2 is incorporated into the mantle. (Figure from Zahnle *et al.*, 2007.)

While the age of the Earth is 4.6 billion years, it is unfortunate that its earliest history is unknown because no rocks from that time have survived. Robust zircon crystals have survived for over 4 billion years, but Earth's oldest known rocks, whose properties could provide information about the early Earth, are just less than 3.9 billion years. This is a curious age: the Earth's oldest surviving rocks formed just after a rock-destroying time period known as the Late Heavy Bombardment or LHB. The existence of the LHB and information on its chronology was determined from the dating of rocks returned from the Moon by the Apollo program, many of which had ages close to 3.9 billion years. During its first 700 million years, the Moon collided with large bodies that created giant impact basins as large as the 2600 km diameter South Pole Aitkin basin. This period of large crater formation ended 3.9 billion years ago, but the bulk of the cratering may have happened over a relatively short period of time at ~ 3.9 billion years ago. Because the heavy cratering may have occurred as a spike, the LHB is also known as the Lunar Cataclysm. The episodes of large crater formation experienced by the Moon surely also happened on Earth, but with more severe consequences because of Earth's surface water, its higher surface gravity, and its larger collisional cross section.

The origin of the LHB has long remained a mystery. Solar system formation models as well as the observed crater record suggests that the LHB was not just the tail end of the planetary accretion process. The presence of heavily cratered regions on other bodies, including Mars, suggests that the LHB may have been a solar-system-wide process. An intriguing case has been made that the LHB may have been caused by a rearrangement of the outer planets. The "Nice hypothesis" (Gomes *et al.* 2005; see Section 3.6) suggests that a dramatic rearrangement of the outer planets gravitationally perturbed a large number of cometary bodies into orbits that penetrated the inner solar system and cratered the surfaces of all solar system bodies.

The LHB produced violent effects on Earth. Impacts of bodies of ~100 km or larger may have produced over 40 craters exceeding 1000 km in diameter. The heat from these events plus impressive greenhouse warming due to water vapor could vaporize oceans and heat the planet's surface to sterilization temperature. Until the LHB ended, Earth could not provide continually habitable conditions for life. It is possible that there was no life before the LHB ended but is also possible that life formed again and again only to be extinguished by the next giant impactor. Zahnle *et al.* (2007) even speculate that life in rocks ejected into space by impacts could later re-accrete on Earth and thus re-seed the planet after a global sterilizing impact. Even if the LHB was short-lived, the time between impacts would have been long, much longer than the time scale a planet needs to thermally recover to normal conditions before the next big one hits and much longer than the lifetimes of thousands of generations of organisms.

4.6 The early history of terrestrial life

The traditional fossil evidence for animals stretches back about 560 Myr when soft-bodied animals, called Ediacarans, are first seen in sedimentary rocks. Worm holes and tracks, so-called trace fossils, extend to somewhat earlier times. Before the Phanerozoic, the time when the fossil record is dominated by macroscopic plants and animals, evidence for life comes from microbial fossils, from physical structures such as stromatolites made by microbial communities, from characteristic chemical "biomarkers", and assorted isotopic signatures. The record becomes murkier and increasingly controversial with increasing age. The oldest "absolutely convincing" microfossils from the Archean (2.5–4 Gyr ago) are the 2.55 Gyr old microfossils from Transvaal chert from South Africa (Buick, 2007). Controversial microfossils or microfossil-like objects that resemble modern filamentous cyanobacteria date back to 3.5 Gyr. Assorted chemical and isotopic evidence is suggestive of the presence of life nearly back to the end of the Late Heavy Bombardment, 3.9 Gyr ago. It is clear that life developed on Earth within its first billion years and that it rapidly developed complex cellular and biochemical functions that allowed it to survive the evolving conditions on a young planet. The rapid emergence of microbial life, not long after the LHB, suggests that the formation of life on an Earth-like planet may not be difficult; at least it did not take long on Earth. The emergence of animals, at least those that formed fossils, took 4 billion years. The apparent difficulty of evolving animals, even on a hospitable planet like Earth, led to the so called Rare Earth hypothesis (Ward and Brownlee, 2000) suggesting that animal-supporting planets like Earth might be rare: it may be easy to make microbial-supporting planets but evolution of larger and more complex organisms is clearly more difficult and planets with animal-like life might be rare. Even Earth itself is only capable of supporting animals for a minor fraction of its life.

4.7 The rise of oxygen

One of the most important changes in our planet's history was the rise of atmospheric oxygen. This watershed event enabled the evolution of aerobic metabolism, a process needed for the development of larger and more complex organisms. The rise of oxygen involved a remarkable change in the infrared spectrum of Earth. This transition could have been seen by distant observers and detection of atmospheric oxygen is a key method used in the astronomical search for other worlds that might harbor life. Oxygen itself does not affect the IR spectrum but ozone does. Ozone is produced by ultraviolet photolysis of oxygen high in the atmosphere (see Chapter 16). Although ozone is a trace gas it is a strong infrared absorber.

The infrared spectrum of Earth shows strong absorption from water, carbon dioxide, and ozone, and this combination is believed to be a signature that can be used to detect Earth-like conditions and the presence of life on other planets. Free oxygen is not chemically stable in the atmosphere of a lifeless water-bearing planet and its presence must be maintained by biological processes. A major goal of proposed telescopic searches for habitable and even life-bearing planets is the detection of ozone, carbon dioxide, and water in the spectrum of an Earth-like planet.

Prior to 2.4 billion years ago, the Earth's atmosphere was essentially devoid of free oxygen. Although it was being produced by photosynthetic organisms such as cyanobacteria as well as the photolysis of water vapor, it was efficiently removed from the atmosphere as it oxidized compounds on the surface and in the atmosphere. Before this time, the atmosphere was dominated by nitrogen (N_2) but it contained appreciable amounts of CO_2, water vapor, and probably moderate amounts of methane (CH_4), possibly up to the percentage level. There is abundant evidence for low oxygen abundance on the early Earth including the oxidation state of various minerals, including iron oxide.

The most compelling evidence for a dramatic change in the atmospheric oxygen abundance and the best estimate of the time of the transition comes from isotopic composition of sulfur compounds (Fig. 4.4). As oxygen abruptly appeared 2.4 billion years ago, there was a fundamental change in isotopic composition

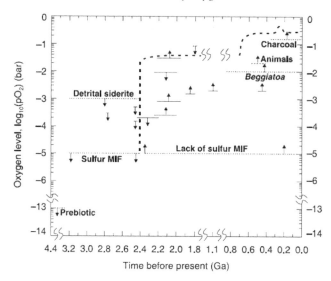

Fig. 4.4. Estimates of the evolution of Earth-atmospheric oxygen implied by various biological and mineralogical indicators. The presence of mass-independent fraction (MIF) of sulfur isotopes before 2.4 Gyr ago is the best determinant of the time marking the rise of oxygen and the decline of atmospheric methane. (Figure from Claire *et al.*, 2006.)

of sulfur compounds. In modern sulfur, the three isotopes (32, 33, and 34) are related in a mass-dependent manner. When normalized to average terrestrial values, deviations from the norm are mass dependent. For example, a process that increases the $^{34}S/^{32}S$ ratio will also increase the $^{33}S/^{32}S$ ratio but only by about half as much because the mass difference between 33 and 32 is only half the difference between 34 and 32. Most physical processes, such as evaporation or condensation, produce mass-dependent fractionation. Sulfur compounds deposited in marine environments before 2.4 billion years ago deviate from this pattern and show fractionation that is independent of mass. The change from mass-independent to mass-dependent fractionation is believed to be related to the rise of oxygen and the simultaneous decline of methane to a vanishingly low abundance (Zahnle *et al.*, 2006; Claire *et al.*, 2006). With this watershed change in the atmosphere, the chemical state of sulfur compounds near the surface changed and the presence of oxygen and ozone in the atmosphere blocked important wavelengths of ultraviolet light from reaching the lower atmosphere. Before oxygen was present, photolysis effects related to shelf-shielding by the most abundant sulfur isotope could result in mass-independent fractionation.

The crossover, i.e. the appearance of oxygen and simultaneous loss of methane, occurred 2.4 billion years ago. At this time the Earth entered a severe ice age, also called a "Snowball Earth" episode, during which the planet surface cooled to the point where ice formed at equatorial latitudes. It seems likely that this unusual cooling event was related to the rapid loss of significant greenhouse warming previously associated with the presence of methane.

4.8 The evolution and survival of animals

As viewed from an animal perspective, the evolution of animals is seen as perhaps the most important milestone in the history of the planet. It took ~4 billion years of biological and physical evolution for animals to first appear. The sudden rise of animals in the fossil record 540 million years ago is called the "Cambrian explosion" and is viewed as a wondrous time when many of the major groups of complex animals suddenly appeared in the fossil record.

If Earth's history could be described from the biased view of a microbe, the importance of animals would be likely to be underplayed. Animals are rather fragile creatures with limited ability to adapt to either extreme or changing environmental conditions. Their species easily become extinct and they are exceedingly rare compared to the concentrations of the microbial world where billions of organisms can live in a gram of host material. Microbes also live in almost every environment on Earth, while this is certainly not the case for animals.

Probably the most negative aspect of animals is that they are perhaps likely to be a somewhat short-lived phenomenon. While microorganisms may survive on Earth for most of the planet's full 12 billion-year lifetime, it is possible that animals will only survive a billion years or even less. Mass extinction events occur due to inevitable changes such as the ever-increasing brightness of the Sun. Extinctions of animal species also occur due to less predictable events that can be considered stochastic "planetary accidents". Since the Cambrian explosion there have been 15 major mass extinctions that have seriously altered the fate and evolution of animals on Earth (Ward, 2007). The causes of many of the extinctions remain controversial but there are numerous possible scenarios, including external events such as asteroid impacts and γ-ray bursts as well as internal effects due to instabilities in Earth's chemical and physical processes. Probably the most severe processes that are known to have occurred since the time of the late heavy bombardment are several Snowball-Earth episodes during which the Earth essentially froze over. This is an example of an event that could potentially cause extinction of all animals and even microbial life on a planet.

4.9 The decline of carbon dioxide

Photosynthesis is the primary means by which life on Earth derives energy from the Sun. The complex chemical processes involved with photosynthesis depend on the availability CO_2 in the atmosphere and CO_2 can be considered an essential "food of life" on our planet. Carbon dioxide on the present Earth is controlled by biogeo-chemical processes but in the future, as the Sun becomes brighter, the atmospheric CO_2 abundance will decline below the minimum ($\sim$10 ppm) amount needed to support plants. The end of CO_2 will mark the end of plants and animals that depend on direct contact with Earth's natural atmosphere. Unless artificial environments are developed on Earth, the end of CO_2 will mean a return to Earth's former state where all life was microbial.

We currently have major concerns with the CO_2 increase from burning fossil fuels and its global warming effects (see Chapter 12). However, this a short-term problem. Ultimately, all of the atmospheric CO_2 will become locked up in carbonates and removed from the atmosphere. Even now, most of the CO_2 that has ever been in the atmosphere is already in carbonates. Carbon dioxide is the dominant gas in the atmospheres of Venus and Mars and it must have been a major gas in the Earth's atmosphere before it declined due to carbonate formation. If Earth's total carbonate content were decomposed, it would yield over 20 atmospheres of CO_2, over 4×10^4 times the present CO_2 content of the atmosphere. As the Sun gets brighter, as all stars do as the hydrogen content of their cores is consumed, the Earth's surface temperature will increase and the CO_2 will decline as more and more is sequestered into carbonates. The removal process is related to weathering of rocks, a process whose rate increases with increasing temperature. The presence of silicates, water, and atmospheric CO_2 leads to the formation of carbonates. Presently, this process is dominated by biological processes such as the formation of shells, corals, and microscopic organisms such as foraminifera.

When atmospheric CO_2 is sufficiently depleted, Earth will have lost an important factor that has promoted the long-term stability of its surface temperature. Over Earth's history, the abundance of CO_2 and its greenhouse warming effects have varied in ways that have counteracted changes in atmospheric temperature. When Earth cools over long time periods, the CO_2 abundance can rise and promote greenhouse warming. When Earth warms, the CO_2 abundance can decline and promote cooling. This effect is called the carbonate–silicate cycle and it is a case of negative feedback, where change is resisted leading to stability. Cooling is promoted when CO_2 is removed by weathering processes that eventually form carbonates. This process is temperature dependent and becomes more effective with rising temperatures. Carbon is removed by weathering but is involved in a cycle because it is ultimately reintroduced back into the atmosphere. Carbonate deposits in the ocean floor are subducted beneath continents on a $\sim$100 Myr time scale where they are thermally decomposed and release CO_2 back into the atmosphere via volcanism. The CO_2 sink depends on weathering and carbonate deposition and the CO_2 source depends on subduction, an ongoing process associated with plate tectonics.

In the Earth's distant future, when the Sun becomes a red giant star, carbonates will be heated and CO_2 will again be released into the atmosphere. This may have no effects on life because life may have already been extinguished by other processes.

4.10 The loss of oceans and the end of plate tectonics

The loss of its oceans is one of the major events in the life of an Earth-like planet that forms in the habitable zone of a star. Ocean loss is inevitable and a natural consequence of stellar evolution. All stars brighten with age and oceans are lost when a critical threshold brightness is reached. Ocean loss is a drastic change for a planet, and for Earth it will mean a change to a seemingly "unearth-like" state, a planet more like Mars than the blue planet of its past. Ocean loss marks the end of what are often considered habitable conditions, but it does not really mean that the planet will actually be lifeless. If ocean loss does not involve globally lethal conditions, it is possible that a post-ocean Earth could harbor subterranean and even surface organisms for an extended period of time. Even without oceans, Earth will probably always have regional ponds or lakes fed by water derived from the mantle. The mantle is a reservoir that may contain several ocean-masses of water.

The most likely fate of Earth's oceans is loss by the "moist greenhouse" effect (Kasting, 1988), a process that occurs at present but at a very low rate. In this process, water is transported through the troposphere and stratosphere to heights where its hydrogen can be liberated by photolysis with solar UV photons. Near the exosphere the liberated hydrogen escapes to space, and forms Earth's geocorona. This process currently occurs at a rate of only a meter of ocean in a billion years due to the very low abundance of water vapor in the stratosphere. As the Sun warms, the partial pressure of water in the upper atmosphere rises and the time scale for water loss shortens. Modeling of this process indicates that the moist greenhouse effect will begin severely depleting the Earth's oceans in about a billion years or less. If surface water is not largely depleted by the rather gentle moist greenhouse process in roughly 3 billion years, a much more severe process will take over when the Sun is about 35% brighter that it is at present (Fig. 4.5, also Fig. 2.9). In a runaway, increasing temperatures introduce more greenhouse gas, providing positive feedback. This full runaway greenhouse advances to the critical point of water where density of water vapor equals the density of liquid water. In a runaway, the enormous amount of water vapor in the atmosphere produces greenhouse warming sufficient to melt surface rocks. Either the moist-greenhouse or the runaway-greenhouse process will result in the Earth's loss of its oceans to space and our planet will spend over half of its total life as an ocean-free planet, at least initially covered with salt and very oxidized rocks. If, as expected, the less severe

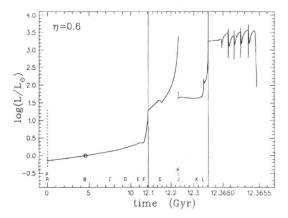

Fig. 4.5. Evolution of the luminosity of the Sun over its full life span. The first 12 billion years show the gradual brightness as hydrogen is depleted in the core of the the the Sun's main-sequence lifetime as a hydrogen-burning star (cf. Fig. 2.9). The large luminosity increases and pulses that follow this period include both the red giant and asymptotic giant branch (AGB) phases when the Sun swells in size and loses appreciable mass to space. (Figure from Sackmann *et al.*, 1993.)

moist-greenhouse loss mechanism occurs, microbial and perhaps even multicellular life might extend far into Earth's ocean-free era, perhaps even to the time when the Sun begins evolution into a red giant star.

The loss of oceans is likely to also lead to the end of plate tectonics. Hydrated minerals have lower melting points and in several ways the presence of water promotes the sinking of oceanic crust to subduct beneath continents. Without oceans it is expected that plate movement will stop and Earth – like all other planets in the solar system – will cease to have subduction and the drift of continents. Without subduction, the Earth's major mechanism for cycling CO_2 back into the atmosphere will be lost.

4.11 The red giant Sun and the fate of habitable zone planets

When the Sun evolves (Fig. 4.5) into its red giant (RG) phase, its core shrinks and becomes hotter but the diameter of the Sun increases and its effective surface temperature decreases (Fig. 2.9). As a giant, it becomes brighter, but with a cooler, redder photosphere. During the RG phase the Sun not only increases in brightness but it also loses about 30% of its mass. In its core, the red giant Sun fuses helium to form carbon. In the later portion of the giant phase, the depletion of helium in the core forces helium fusion to a thin shell surrounding a degenerate carbon/oxygen core. Above the helium-burning shell is a cooler shell where hydrogen continues to fuse into helium. This evolutionary stage is called the asymptotic giant branch (AGB) phase and it involves short-lived thermonuclear runaway pulses. As an AGB star, the Sun loses additional mass for a total mass loss as a giant of over 40%. During the RG and AGB evolutionary phases, planets that formed in the habitable zone are severely heated by greatly increased luminosity. In the case of the Sun, for most of its time in the RG and AGB phases its luminosity will be about 30 times brighter than at present and will extend over a period of several hundred million years. Near the ends of both the RG and AGB phases, pulses occur when the luminosity increases by much larger factors, over 2000 times its present brightness. These short-lived luminous phases will cause the melting of all rock landforms on all habitable zone planets and they also drive mass loss from the star.

As the luminosity of red giants increases, the effective surface temperature declines but the diameter increases by approximately two orders of magnitude. The physical expansion of giant stars results in the assimilation of many planets that may have formed in their formerly habitable zones. In the case of the Sun, current predictions indicate that the Sun will expand (Fig. 4.6) to nearly 1 AU, engulfing both Venus and Mercury. Because the Sun loses over 40% of its mass during the RG and AGB phases, Earth's orbit will actually expand to conserve angular momentum. This seemingly places Earth just beyond the presently modeled maximum diameter of the Sun, but detailed modeling indicates that Earth will be assimilated into the Sun because of tidal effects (Schroder and Smith, 2008).

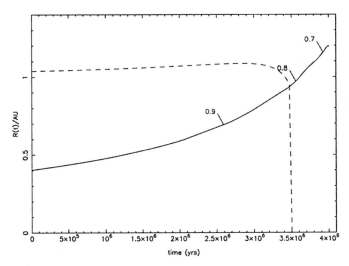

Fig. 4.6. The diameter of the red giant Sun (solid; in AU) and the size of Earth's orbit (dashed) during the 4 million years leading up to the phase when the Sun reaches maximum brightness. Earth's orbit expands slightly as the Sun loses mass but the Sun expands to the point where tidal drag causes Earth's orbit to decay and intersect the Sun's upper layers. These calculations by Schroder and Smith (2008) predict that Earth will be destroyed in the Sun's atmosphere 7.59 billion years from present.

Tidal forces raise a bulge in the Sun's upper layers that follows Earth and provide a retarding force that causes Earth's orbit to decay. Earth is totally vaporized by this process due the power generated by its ~ 25 km/s entry into the Sun's upper atmospheric layers. If Earth had formed 15% further from the Sun it would have escaped assimilation. Mars and all other planets are well beyond the effects of gas drag and tidal effects and are safe from total destruction although they are severely heated and rendered lifeless during the Sun's red giant phase.

It is the fate of all planets in the habitable zone to be made uninhabitable by stellar brightening and the fate of a significant fraction of these bodies to be assimilated into their stars during the red giant phase. No planet has a safe haven from its evolving star. The lifetimes of stars, and the potential lifetime of habitability of stars varies with stellar mass (e.g. Fig. 2.9). The Sun lasts about 12 billion years, while stars that are twice as massive become red giants in less than 4 billion years, the time it took animals to evolve on Earth and leave a fossil record. The lowest mass stars (M-type dwarf stars) can survive for over a trillion years before evolving into red giants that can toast planets that formed in their habitable zones.

In the far future, the universe will look quite different than it does at present. All massive bright stars will have evolved and become invisible. Only the slowly evolving and faint M stars will persevere. After several tens of billions of years of galactic evolution, questions about habitability will only concern the bodies that remain, these faint low-mass red stars and planets that orbit them in thin habitable zones close to their surfaces. There is considerable current interest in these stars both because they are the most numerous stars now in the universe and because in the long term they will be the only stars in the universe. Compared to the Sun these low-mass stars offer new challenges to understanding habitability. Although faint, they have pronounced flare activity which generates both UV and energetic particle fluxes capable of harassing life. Due to their faintness, their habitable zones are so close to the stars that planets can be tidally locked with one side always facing out to space. This can cause thin atmospheres to freeze out on the dark side of planets, although sufficiently thick atmospheres may be able to adequately distribute heat and prevent this calamity.

Understanding the subject of planetary habitability is a major challenge to modern science because of the wide range of scientific disciplines that must be involved to even try to understand the complexity of planetary and biological processes on billion-year time scales. Many of these processes are greatly affected by even very subtle changes in the stars that planets orbit. It is an important and timely scientific endeavor because we are just now at the threshold of evaluating extra-solar planets for potential habitability. Better understanding of habitability is a fascinating but daunting task that is a major factor stimulating the emerging interdisciplinary field of astrobiology (Lammer *et al.*, 2009; von Bloh *et al.*, 2009). A truly humbling aspect of this subject is that planets are vastly complex bodies with many interacting systems that, like weather, may fundamentally defy precise modeling. A major step towards understanding habitability is a more detailed understanding our own habitable planet, a body whose long-term history and even fundamental functions are remarkably poorly understood.

5

Solar internal flows and dynamo action

Mark S. Miesch

5.1 Magnetism with enthusiasm

As discussed in many chapters throughout these volumes, stars bristle with magnetic energy. They inherit some of their natural magnetism from their parent molecular clouds as they contract from protostellar cores (Chapter 3). However, stars are far from passive. After they ignite and enter the main sequence, much of their magnetism comes from within, bred by active hydromagnetic dynamos (Chapter 2; Vol. I, Chapter 3). Emerging magnetic flux influences the star's evolution and shapes its environment. It is the dissipation of magnetic energy in the solar corona that powers the solar wind, and the wind in turn carves out the heliosphere, a planetary cloister within the surrounding interstellar medium where the Sun holds sway. Magnetic fields originating in the solar interior permeate the heliosphere, weaving an intricate web with planetary magnetospheres and linking the Sun to the planets. The web changes continually as coronal mass ejections send sporadic bursts of magnetized plasma coursing through the heliosphere, restructuring fields and flows as they go.

Stars build magnetic fields by tapping the energy in their own corporeal constitution.[†] Thermonuclear fusion in their cores converts matter into thermal energy and electromagnetic radiation which, in the Sun, is transported outward via the diffusion of photons. In the solar envelope, the plasma becomes more opaque as the temperature drops, which inhibits radiative diffusion and steepens the temperature gradient relative to the adiabatic temperature gradient (Section 5.2). The stratification soon becomes superadiabatic and thermal convection takes over as the primary

---

[†] Pre-main-sequence stars that are still contracting from protostellar cores and post-main-sequence stars forming compact remnants also tap gravitational potential energy to help power convection and dynamo action and to heat the stellar plasma to temperatures where thermonuclear fusion can proceed. Meanwhile, compact remnants such as white dwarfs shine by emitting residual internal energy as they slowly cool. This too can support thin convection zones and possibly dynamo action.

Heliophysics: Evolving Solar Activity and the Climates of Space and Earth, eds. Carolus J. Schrijver and George L. Siscoe. Published by Cambridge University Press. © Cambridge University Press 2010.

mechanism for transporting energy to the solar photosphere where it is radiated into space.

The solar convection zone occupies approximately the outer 30% of the Sun by radius. It is here where internal energy of the plasma is converted to kinetic energy and then to magnetic energy, aided by radiation and gravity. Radiative heating in the lower convection zone and radiative cooling in the photosphere maintain a superadiabatic temperature gradient that sustains convective motions by means of buoyancy. In a rotating star, convection transports momentum as well as energy, establishing shearing flows and global circulations (Section 5.5). These mean flows work together with turbulent convection to amplify and organize magnetic fields through hydromagnetic dynamo action, giving rise to the rich display of magnetic activity so striking in modern solar observations.

Careful scrutiny of photospheric magnetograms[†] reveals that the amount of new magnetic flux appearing in bipolar active regions over the course of an 11-year sunspot cycle is of order $\Phi \sim 10^{24}$–10^{25} Mx (Schrijver and Harvey, 1994). If we attribute this to the emergence of toroidal flux rings with a characteristic field strength B_0, then this corresponds to a mean magnetic energy generation rate $\dot{E}_m \sim B_0 \Phi R_\odot/(4\tau)$ where $R_\odot$ is the solar radius and $\tau = 11$ yr. In the photosphere, bipolar active regions have a strength of order several kG but we expect submerged flux tubes to be stronger, expanding and weakening as they rise as a consequence of the density stratification. Taking $B_0 \sim 20$ kG yields $\dot{E}_m/L_\odot \sim 10^{-4}$–$10^{-3}$ where $L_\odot$ is the solar luminosity. So, as much as 0.1% of the solar luminosity may be converted to magnetic energy by the dynamo. This may be effortless for the Sun but many rapidly rotating stars exhibit intense magnetic activity that apparently pushes the limits of the available energy supply, posing significant modeling constraints (Section 2.3.2; also Rempel, 2008). Stellar dynamos can operate with impressive efficiency and enthusiasm!

Stellar observations confirm that convection breeds magnetism. Most solar-like stars with convective envelopes and fully convective lower-mass stars show signs of magnetic activity (Chapter 2). Even ostensibly inactive high-mass stars likely breed vigorous dynamos sequestered deep within their convective cores. Magnetic activity is also correlated with rotation and rotational shear, providing further clues to the nature of stellar dynamos.

Understanding solar convection is thus key to understanding the pervasive magnetism that underlies and unifies much of heliophysics. In Section 5.2 we discuss the nature of solar and stellar convection, highlighting the remarkable hierarchy of

---

[†] A *magnetogram* is a two-dimensional (latitude, longitude) map of the line-of-sight component of the photospheric magnetic field across the solar disk obtained by measuring the splitting of spectral emission or absorption lines due to the Zeeman effect (e.g. Figs. 2.1(left) and 5.3(left)). Similarly, a *Dopplergram* is a map of the line-of-sight velocity component obtained by means of the Doppler effect.

convective scales encountered as one conceptually travels from the photosphere to the deeper interior. We then consider how such flows generate magnetic fields in Sections 5.3 and 5.4, following the popular, albeit somewhat simplistic, paradigm that there are two solar dynamos rather than just one. One dynamo operates in the granulation layer near the surface and is responsible for the small-scale flux elements ubiquitous in photospheric magnetograms (Section 5.3). The other dynamo is global in nature and gives rise to patterns of magnetic activity such as the 11-year sunspot cycle (Section 5.4) and the longer-term variations reviewed in Chapter 2. This global dynamo relies critically on rotational shear and possibly also on mean meridional circulations; the origin of these mean flows is addressed in Section 5.5. The role of mean flows and turbulent convection in establishing long-term solar activity variations is explored further in Chapter 6. We conclude this chapter with a brief survey of the triumphs and tribulations that may be awaiting us in the coming years (Section 5.6) and an appendix that summarizes some physical characteristics of the solar interior.

5.2 The many faces of solar convection

Thermal convection is familar to most of us from our daily experience; warm air rises and cooler air sinks. When a fluid is heated from below it overturns, provided the temperature gradient is large enough, which here means that it must not only be greater than the adiabatic temperature gradient.[†] (the Schwarzschild criterion) but it must also overcome stabilizing influences such as thermal and viscous diffusion, rotation, compositional gradients (the Ledoux criterion), and magnetic flux.[‡] An intuitive way to think about convection (and to derive the Schwarzschild and Ledoux criteria) is to consider a small isolated volume, or *parcel*, of fluid that will buoyantly rise like a hot air balloon if its density is less than that of its surroundings or sink like a stone if its density is greater (the parcel is assumed to be in pressure equilibrium with its surroundings so density and temperature are anticorrelated). This is the conceptual framework behind mixing length theory, which goes on to say that the parcel will lose its identity, dispersing into the background, after traveling a vertical distance of order a pressure scale height H_p. For a discussion of mixing length theory see, for example, Hansen and Kawaler (1994).

[†] The adiabatic temperature gradient $dT/dr|_{ad}$, also called the adiabatic lapse rate, is that for which the specific entropy per unit mass S is constant with radius r: $\partial S/\partial r = 0$. For an ideal gas in hydrostatic equilibrium $dT/dr|_{ad} = -g/c_p$, where g is the gravitational acceleration and c_p is the specific heat per unit mass at constant pressure.

[‡] The literature on convective instability is vast; for an early exposition see Chandrasekhar (1961) and for more recent work concerning rotating spherical shells see Dormy *et al.* (2004).

With this intuitive picture in mind, we may expect that the vertical scale of solar convection should vary tremendously from the deep convection zone where the stratification is relatively gentle ($H_p \sim 35$ Mm; see Table 5.1 in the appendix) to the solar surface layers where the density and pressure drop precipitously ($H_p \sim 36$ km) as the star confronts the relative emptiness of the corona and the interplanetary space beyond. The associated drop in temperature near the surface triggers the recombination of hydrogen and other ions, which modifies the opacity,[†] decreases the particle number density, and releases latent heat, altering the thermodynamics (in particular the equation of state and the specific heats) and contributing to the convective enthalpy transport. Add in radiative energy transfer and the result is what we call solar granulation; the continually shifting pattern (lifetime $\tau \sim 5$ min) of small-scale convection cells (with a horizontal extent $L \sim 1$ Mm) that blankets the solar surface and accounts for the dappled appearance of the solar photosphere (Vol. I, Fig. 8.3).

5.2.1 Solar granulation

Solar granulation is driven by radiative cooling in a thin layer within about 100 km of the photosphere. As radiation streams away from a location on the photosphere the plasma cools, compresses, and sinks, pulled down by gravity. Surrounding fluid rushes in to fill the void, driven by horizontal pressure gradients. Upflows are thus a passive response to the negative buoyancy of downflows and are relatively isentropic in comparison to the distribution of low entropy in downflows (Stein and Nordlund, 1998). The nature of the induced circulation is such that the maximum upflow velocity occurs adjacent to the downflow region (Rast, 2003).

    To appreciate how this works in practice, consider a single convection cell, or granule, with an upflow region surrounded by narrow downflow lanes (see Vol. I, Fig. 8.3). We expect that the granule should be brightest around its edges, near the downflow lanes, where the recirculating vertical velocity and thus the convective enthalpy flux peaks. This is in fact what is seen in both photospheric observations and in numerical simulations (Stein and Nordlund, 1998). The radiative cooling-time scale in the photosphere is short (~ 20 s) relative to dynamical time scales so in order for the granule to survive, the vertical velocity at its center v_z must be large enough to balance the radiative losses at the surface. The convective enthalpy flux just under the photosphere is dominated by the latent heat flux associated with hydrogen ionization, so to get an idea of what value of v_z is needed we can set

---

[†] As the temperature, density, and pressure fall, the opacity first increases sharply ($0.94 < r/R_\odot < 0.99$) and then decreases even more sharply ($0.99 < r/R_\odot < 1$) as H^- ions begin to form near the photosphere. The H^- opacity is particularly sensitive to temperature, $\propto \rho^{1/2} T^9$ (Hansen and Kawaler, 1994).

$$\rho v_z y N_A \chi_H = \sigma T^4, \tag{5.1}$$

where y is the ionization fraction, N_A is the Avogadro constant, χ_H is the ionization energy for hydrogen, σ is the Stefan–Boltzmann constant, and ρ and T are the density and temperature. For photospheric conditions ($y \sim 0.1$, $T \sim 5800$, $\rho \sim 2 \times 10^{-7}$), this gives a minimum vertical velocity U_z of about $2\,\text{km s}^{-1}$ (Nordlund, 1985; Stein and Nordlund, 1998).

If v_z drops below U_z, a new downflow will form and the granule will split into two or more smaller granules. The process may be regarded as a thermal instability of the photospheric boundary layer, with positive feedback provided by radiation and ionization. As a location in the photosphere cools, its opacity and ionization fraction decrease (see footnote on p. 102) and its density increases. This increases the radiative cooling while simultaneously decreasing the convective enthalpy flux by decreasing both v_z and y. The imbalance cools the location further, accelerating the downflow.[*] The most dramatic manifestation of this is in so-called *exploding granules* (Rast, 1995). Boundary layer instabilities work together with the density stratification throughout the domain to generate a convection pattern characterized by an interconnected network of narrow downflow lanes surrounding broader, weaker, topologically disconnected upflows.

With this picture in mind, we can estimate the maximum horizontal size L of granules from mass conservation (Stein and Nordlund, 1989). If a granule simply overturns without altering the local density appreciably, then the continuity equation implies $L \sim D v_h / v_z$ where D is the vertical scale and v_h is the horizontal velocity (assuming a cylindrical geometry with upflow in the center and downflow around the periphery adds a factor of 1/2). The pressure-driven horizontal flows are unlikely to exceed the sound speed $c_s \sim 10\,\text{km s}^{-1}$ (shocks do form in numerical simulations of solar granulation but they are spatially and temporally intermittent), so if we set $v_z \geq U_z$ and $D \sim H_p \sim 400\,\text{km}$ we find that the horizontal scale of granules should not exceed about 1–2 Mm. This agrees well with photospheric observations and numerical simulations, although in practice the largest granules can span several pressure scale heights, with a correspondingly larger horizontal scale (Spruit *et al.*, 1990; Stein and Nordlund, 1998).

Radiation, ionization, and density stratification all promote the generation of narrow downflow lanes and plumes but they are not required. Indeed, sporadic plumes emanating from unstable thermal boundary layers are a hallmark of turbulent convection, ubiquitous in numerical simulations as well as laboratory experiments (Siggia, 1994). Related dynamics also occur in thermosolutal convection

---

[*] Partial ionization can further promote pluming instabilities in the thermal boundary layer by modifying the specific heat and thereby the thermal diffusivity (Rast and Toomre, 1993).

(e.g. salt fingers in the ocean) and in the Rayleigh–Taylor instability that occurs in many astrophysical flows including supernovae. Boussinesq convection simulations exhibit a downflow network reminiscent of solar granulation within the upper thermal boundary layer[†] (Vol. I, Chapter 2, Fig. 2.6).

5.2.2 *Mesogranulation and supergranulation*

With a little effort, one can discern other characteristic spatial and temporal scales in photospheric measurements over and above that of granulation. The most apparent of these is known as supergranulation ($L \sim 30\,\mathrm{Mm}$, $\tau \sim 20\,\mathrm{h}$). Although supergranulation is detectable in photospheric Dopplergrams (see footnote on p. 100; Hathaway *et al.*, 2000), in local helioseismic inversions (Hirzberger *et al.*, 2008), and in the correlation tracking of granular features (DeRosa and Toomre, 2004), it is most readily observed as the dynamical substrate underlying the quiet magnetic network, both figuratively and literally. The quiet magnetic network is a honeycomb-shaped pattern that is prominent in photospheric magnetograms and in chromospheric emission lines such as Ca II H and K (see Fig. 2.1; also Vol. I, Chapter 8). The chromospheric emission is attributed to enhanced heating connected in some way to vertical magnetic flux that is advected by converging horizontal flows into supergranule downflow lanes. The honeycomb pattern thus reveals that supergranulation has a horizontal structure similar in shape to that of granulation (e.g. Schrijver *et al.*, 1997) but on a larger scale, characterized by an interconnected network of narrow downflow lanes lining broader, gentler upwellings.

Motivated by this idea of convection advecting magnetic elements horizontally into converging downflows, you can conduct a fascinating experiment. Beginning with a time series of horizontal flow maps obtained by tracking of small-scale photospheric features (using, for example, cross-correlation methods for small sub-areas with a sequence of images), you may sprinkle the maps with passive tracer particles and follow their evolution, thus simulating the advection of buoyant corks floating across the solar surface (Spruit *et al.*, 1990). If your spatial and temporal resolution are good enough, you will see that within about ten minutes of elapsed solar time the corks will accumulate in the granulation downflow network. If you wait longer, say 30 minutes, the corks will trace out another distinct cellular pattern intermediate in scale between granulation and supergranulation. This phenomenon

---

[†] In the Boussinesq approximation, density variations are neglected in all but the buoyancy force so the fluid is essentially incompressible (Pedlosky, 1987). Boussinesq systems in a Cartesian geometry are symmetric about the mid-plane so a similar network of *upflows* exists in the lower thermal boundary layer. By contrast, many density-stratified granulation simulations have an open lower boundary where downflows pass through mainly as isolated, disconnected plumes and upflows respond to conserve mass (Stein and Nordlund, 1998).

is known as mesogranulation ($L \sim 5\,\mathrm{Mm}$, $\tau \sim 3\,\mathrm{h}$). As you continue to watch, the corks will gradually slip toward the vertices of the mesogranule network and, after about an hour, most will gather in isolated points of convergence, spaced nearly uniformly across the solar surface. If you are patient enough to watch for ten hours or more, the corks will eventually map out the supergranular network.

Mesogranulation is most easily seen as a mottled pattern in horizontal divergence maps obtained from local correlation tracking but it is also faintly detectable by other means, such as photospheric magnetograms (see Fig. 5.3). However, as demonstrated in Figure 5.1a, mesogranulation does not appear prominently in the velocity power spectrum obtained from photospheric Doppler measurements. By contrast, supergranulation is evident as a local maximum at spherical harmonic degree $\ell \sim 120$.

Observations are ambiguous on whether supergranulation and mesogranulation are convective phenomena. Thermal convection, by definition, requires correlations between vertical velocity and temperature (warm upflows, cool downflows) in order to extract buoyancy work. Supergranulation and mesogranulation are

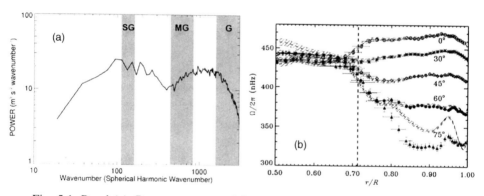

Fig. 5.1. Panel (a): Power spectrum of the convective velocity field in the solar photosphere obtained from Doppler measurements, plotted as a function of spherical harmonic degree ℓ (from Hathaway *et al.*, 2000). Mean flows and *p*-modes are filtered out. The falloff beyond $\ell \sim 1500$ reflects the resolution limit of the Michelson Doppler Imager (MDI) instrument onboard the SOHO spacecraft from which these data were obtained and is therefore artificial. Shaded areas indicate the approximate size ranges of supergranulation (SG), mesogranulation (MG), and granulation (G). Note that the expected granulation spectral peak at $\ell \sim 4400$, corresponding to $L \sim 1\,\mathrm{Mm}$, is not resolved. Panel (b): The solar internal rotation profile inferred from helioseismic inversions (from Thompson *et al.*, 2003). Angular velocity $\Omega/2\pi$ is shown as a function of fractional radius $r/R_{\odot}$ for several latitudes as indicated. Symbols and dashed lines denote different inversion methods, known as subtractive optimally localized averages (SOLA) and regularized least squares (RLS) respectively. Vertical 1-σ error bars (SOLA) and bands (RLS) are indicated and horizontal bars reflect the resolution of the inversion kernels. The vertical dashed line indicates the base of the convection zone.

phenomena most apparent in horizontal velocity measurements, with little or no detectable signatures in vertical velocity or intensity. Part of the problem is the tendency for vertical magnetic flux to enhance emission through magnetic reconnection and by depressing the photosphere (the hot-wall effect, see Spruit, 1976). Thus, supergranular lanes often appear bright even though the underyling plasma may be relatively cool.

The message from numerical simulations in Cartesian domains with large aspect ratios is clear: a macrocellular pattern analogous to mesogranulation is a general feature of turbulent convection (Cattaneo *et al.*, 2001; Rincon *et al.*, 2005; von Hardenberg *et al.*, 2007). To understand its origin in the solar context we begin with the granular downflow network in the photospheric boundary layer (Section 5.2.1). The strongest downflows occur at the interstices of the network so solar granulation may be loosely regarded as an ensemble of downflow plumes (Rast, 2003). The converging horizontal flow that feeds each plume advects and attracts other nearby plumes, promoting clustering and mergers. Plumes that are stronger or closer together on average will assimilate their neighbors, growing stronger still. Thus, plume interactions can establish larger-scale patterns that depend on the plume generation rate and effective cross section (Rast, 2003). The nature of the patterns is that of a lattice of convergence points as seen in empirical cork movies of mesogranulation (Spruit *et al.*, 1990). Vertical vorticity amplification in converging horizontal flows may provide dynamical stability (Cattaneo *et al.*, 2001).

This picture of interacting plumes is of course an idealization but it provides some insight into how granules may self-organize into larger-scale patterns. Further insight can be obtained by taking into account the density stratification. To see this, consider a horizontal reference level below the photosphere. Downflow lanes and plumes originating in the photospheric boundary layer above punch down through this reference level, inducing motions in the surrounding, relatively isentropic fluid by means of pressure fluctuations, entrainment, buoyancy, and turbulent mixing. By virtue of the high density at this level, the upward mass flux in these recirculating motions exceeds the mass flowing down from the relatively sparse boundary layer high above. Mass conservation therefore implies that most of the mass flowing up through our level does not make it to the photosphere (Stein and Nordlund, 1989). The low-entropy photospheric downflows themselves may extend for many density scale heights H_ρ, but the upflows turn over after a vertical distance comparable to H_ρ. As they descend, photospheric downflows are compressed by the density stratification and are pushed into progressively fewer, more widely spaced *superplumes* by the overturning upflows. The result is a heirarchical, fractal tree of branching downflows that increases in horizontal scale with increasing depth (Spruit *et al.*, 1990). Deep-seated pressure perturbations can imprint through to the photosphere, inducing large-scale horizontal flows that advect granules and other smaller-scale features as revealed by correlation tracking.

The idea of a heirarchical tree of convective structures suggests that there should be a continuous spectrum of cellular patterns, prompting some to question whether mesogranuluation is a distinct scale of motion after all. Its conspicuous absence from the photospheric power spectrum (Fig. 5.1a) does not help its existential case. Mesogranular patterns in flow maps obtained from correlation tracking may reflect the typical lifetime of magnetic elements more than the velocity field itself. Alternatively, the rapid falloff of the velocity amplitude with depth may create an edge in the spectrum that would highlight particular scales. Similar arguments also apply to supergranulation, although here other processes may also come into play, such as helium ionization and the influence of magnetism on radiative cooling. In any case, the formation of large-scale convective patterns through the self-organization of granules appears to be robust and stands in stark contrast to the down-scale (forward) cascade of energy characteristic of homogeneous, isotropic, hydrodynamic turbulence (Vol. I, Chapter 7).

5.2.3 Giant cells, helioseismology, and the solar internal rotation

In Section 5.2.1 we found that solar granulation is a boundary layer phenomenon and in Section 5.2.2 we found that larger-scale convective patterns such as mesogranulation and supergranulation may arise from the self-organization of granules. However, even supergranules are thought to be confined to the outer few percent of the solar interior by radius. With the outer 30% of the Sun convective, what lurks below? The bulk of the solar convection zone is believed to be occupied by larger-scale convective motions, evocatively known as giant cells.[†] Giant cells are the mysterious denizens of the deep in the solar convection zoo; we have compelling arguments for their existence but they are elusive in photospheric observations. Our most compelling insight into their nature comes not from measurements of the convective motions themselves but rather from viewing their handiwork, namely the solar internal rotation profile illustrated in Fig. 5.1b.

However, before discussing the solar internal rotation profile we must first address how it is obtained. How can one peer inside a star? Although they didn't quite appreciate it then, Leighton, Noyes, and Simon hit upon the answer in the early 1960s when they noticed that the surface of the Sun is vibrating with a period of about 5 minutes (Leighton *et al.*, 1962). Later it was realized that these are global acoustic oscillations; the Sun rings like a bell in millions of distinct tones

---

[†] We define giant cells in a loose sense here as convectively driven non-axisymmetric motions on scales larger than supergranulation, i.e. significantly larger than 20–30 Mm.

and the frequency of each oscillation mode provides information about the internal structure and dynamics of the star. A new science was born, that of *helioseismology*, whereby the internal conditions of the Sun are inferred from measuring acoustic or other wave fields at its surface (the extension to other stars is referred to as *asteroseismology*). For a thorough yet very readable review see Christensen-Dalsgaard (2002). Not all acoustic waves in the Sun are resonant modes of the entire sphere; small-wavelength modes are excited and damped faster than they can travel around to interfere with one another. These modes may be used to make detailed *solar subsurface weather* maps of horizontal velocities below the photosphere. For more on this vibrant and auspicious discipline of *local helioseismology* see Gizon and Birch (2005).

Most measurements of global solar acoustic oscillations, known as *p*-modes, are obtained from sequences of Dopplergrams that may be decomposed into spherical harmonic series in latitude and longitude and Fourier series in time. Distinct peaks in the frequency power spectrum for each spherical harmonic mode correspond to different radial eigenfunctions. In a non-rotating, spherically symmetric star, the mode frequencies are independent of the spherical harmonic order (longitudinal wave number) m, but rotation breaks this degeneracy, producing Doppler-induced *rotational frequency splittings*. Acoustic waves are most efficiently excited near the surface by granulation as a consequence of its relatively high Mach number (see Table 5.1 in the appendix) and each mode samples a limited range in depth, determined by a lower turning point where the vertical phase speed vanishes (Christensen-Dalsgaard, 2002). Thus, the observed rotational frequency splitting of a given mode is an integral over its sampling depth involving the local rotation rate Ω and an integration kernel that depends on the background stratification and the mode parameters. Various *inversion* techniques are then employed to infer the radial and latitudinal dependence of Ω based on the observed frequency splittings for a range of different modes. Similar inversion techniques may be employed to obtain structural information such as the radial profiles of sound speed, density, and adiabatic index $\Gamma_1 = (\partial \ln p / \partial \ln \rho)_s$.

We have known for nearly four centuries by tracking photospheric features such as sunspots and, more recently, by means of Doppler measurements that the surface of the Sun rotates differentially (Thompson *et al.*, 2003). According to Doppler measurements, the equatorial synodic rotation period is 25.6 d while polar latitudes rotate more slowly, with a period of approximately 33–36 d. Helioseismology now reveals that this monotonic decrease in angular velocity with increasing latitude persists throughout the convection zone, with an abrupt transition to nearly uniform rotation in the radiative interior (Fig. 5.1b). The transition region near the base of the convection zone is known as the solar tachocline and its far-reaching dynamical implications are discussed in Sections 5.4 and 5.5. There is also a less dramatic but no less significant *near-surface shear layer* in which the rotation rate systematically decreases by about 10–20 nHz from $r = 0.96 R_\odot$ to the photosphere. This is most apparent at low latitudes but may also occur at higher latitudes. Some dynamical implications of this layer are discussed in Section 5.5.1.

The striking difference in the rotation profile of the convective envelope and that of the radiative interior implicates convection as the primary source of the differential rotation. Furthermore, it tells us that giant cells are large enough and slow enough to be influenced by the rotation of the star. The magnitude of non-linear advection relative to the Coriolis force is quantified by the Rossby number R_o (Eq. 2.1 or 7.5). In the deep solar convection zone this is of order unity or less whereas it is much greater than unity in the solar surface layers (see Table 5.1 in the appendix). Coriolis-induced velocity correlations in the convection redistribute angular momentum via the Reynolds stress, generating a substantial rotational shear: $\Delta\Omega/\Omega \sim 30\%$ where $\Omega(r, \theta)$ is the angular velocity and $\Delta\Omega$ is the angular velocity difference between equator and pole (for more on how the solar differential rotation is maintained see Section 5.5). Furthermore, the nature of the redistribution is such that the angular velocity increases away from the rotation axis, $\partial\Omega/\partial\lambda > 0$ where $\lambda = r \sin\theta$ is the cylindrical radius. This is in stark contrast to the behavior one would expect from isotropic turbulent diffusion ($\Delta\Omega/\Omega \ll 1$) or from fluid parcels that tend to locally conserve their angular momentum as they move ($\partial\Omega/\partial\lambda < 0$). Giant cells must be a global phenomenon distinct from supergranulation.

5.2.4 *Properties of giant convection cells*

Some insight into the structure and evolution of giant cells can be obtained from global numerical simulations of solar convection as illustrated in Fig. 5.2. These simulations exclude the near-photospheric layers $r > 0.98 R_\odot$ because they do not have sufficient spatial resolution or physical realism (neglecting ionization, non-diffusive radiative transfer, and acoustic compression) to reliably capture granulation, mesogranulation, and supergranulation. For further details on the nature of these numerical models see Miesch (2005).

Near the upper boundary of the computational domain at $r = 0.98 R_\odot$ the convection pattern is reminiscent of granulation, albeit on a much larger scale ($L \sim 100$ Mm). As is typical with turbulent convective systems (see Sections 5.2.1 and 5.2.2), there is an interconnected network of downflow lanes emanating from the thermal boundary layer (Fig. 5.2a) that fragments into isolated lanes and plumes deeper down (Fig. 5.2b). As in granulation (Section 5.2.1), the horizontal scale of the network is determined by the cooling rate at the upper boundary together with the density scale height. Cooling at the boundary spawns

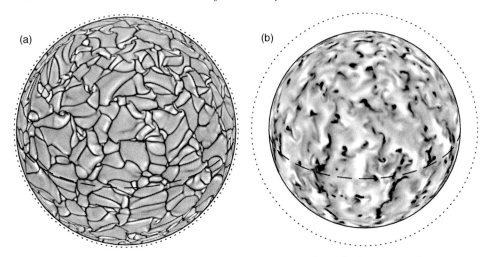

Fig. 5.2. Radial velocity patterns in a global simulation of solar convection are shown on two horizontal surfaces: Panel (a) lies within the thermal boundary layer near the outer boundary of the computational domain ($r = 0.98\,R_\odot$) and panel (b) lies in the middle of the convection zone ($r = 0.85\,R_\odot$). Both images are orthographic projections with the north pole tilted 30° toward the line of sight. The equator is indicated by a dashed line and the solar surface (outside the computational domain) is indicated by a dotted line. Bright tones denote upflow and dark tones downflow. The range of the grey scales is (*a*) −30 to +20 m s^{-2} and (*b*) −200 to +100 m s^{-1}. (From Miesch *et al.*, 2008.)

low-entropy downflows while upflows respond to the associated pressure gradients, with peak upflow velocities occurring adjacent to downflow lanes (Fig. 5.2a). When cells become too large to sustain the heat flux through the boundary they fragment.

The asymmetry between upflows and downflows is a consequence of the density stratification. Horizontally divergent upflows occupy more surface area so downflows must be stronger to balance the radial mass flux. This gives rise to a substantial inward flux of kinetic energy that can exceed 70% of the solar luminosity and that must be offset by an enhanced outward convective enthalpy flux (Miesch *et al.*, 2008). Downflows are also more turbulent than upflows, partly as a consequence of their high velocities and small horizontal scales which promote shear instabilities, and partly because of vorticity amplification by compression. In order to appreciate the latter, consider the enstrophy ω^2 which is defined as the square of the fluid vorticity $\boldsymbol{\omega} = \nabla \times \mathbf{v}$. An equation for the evolution of ω^2 can be derived by taking the curl of the momentum equation and projecting the result onto $\boldsymbol{\omega}$. This yields the following terms, among others:

$$\frac{\partial \omega^2}{\partial t} = (\nabla \cdot \mathbf{v})\,\omega^2 + \cdots \approx -\frac{v_r}{H_\rho}\omega^2 + \cdots, \qquad (5.2)$$

where H_ρ is the density scale height. Thus, enstrophy is enhanced in downflows and diminished in upflows, which is a general feature of turbulent compressible convection (Spruit *et al.*, 1990; Porter and Woodward, 2000).

Rotation further enhances vorticity, particularly in downflows. The converging flows that feed the downflow lanes in the upper boundary layer tend to conserve their angular momentum locally, acquiring cyclonic vorticity ($\boldsymbol{\omega}\cdot\boldsymbol{\Omega} > 0$) by means of the Coriolis force. This is evident in Fig. 5.2a as a distinct counter-clockwise swirl in the northern hemisphere (and clockwise in the southern hemisphere) within the downflow network at mid to high latitudes. Intense cyclonic downflows exist amid a weaker background of anticyclonic vorticity in the diverging upflows, implying a negative kinetic helicity $H_k = \boldsymbol{\omega}\cdot\mathbf{v} < 0$. This persists throughout most of the convection zone but switches sign in the overshoot region (or near the lower boundary if the simulation is non-penetrative) where downflows diverge and recirculate, inducing a weak positive helicity.

The strongest cyclonic vorticity occurs in highly intermittent, helical downflow plumes located at the interstices of the network. Some vortices can become intense enough to centrifugally evacuate their cores, reversing the buoyancy work and forming a new upflow in a phenomenon known as *dynamical buoyancy* (Brummell *et al.*, 1996). Several examples of this are evident in Fig. 5.2a as small, nearly circular upflows embedded in the downflow network. By deflecting motions perpendicular to $\boldsymbol{\Omega}$, the Coriolis force also tends to tilt helical downflow plumes away from the vertical and toward the rotation axis in a phenomenon known as *turbulent alignment* (Brummell *et al.*, 1996).

Close inspection of Fig. 5.2 reveals another type of rotational alignment at low latitudes manifested by extended downflow lanes with a preferential north–south orientation. This is most apparent in the mid-convection zone (Fig. 5.2b) but can also be discerned amid the more intricate near-surface network (Fig. 5.2a). These structures are closely related to the columnar convection cells prominent in simulations of convection and dynamo action in planetary interiors (Section 7.5). Variously referred to as "Busse columns" after the linear theory developed by Busse (1970), thermal Rossby waves after their longitudinal propagation mechanism (Busse, 2002), and banana cells after their sheared, warped appearance in more slowly rotating systems (Glatzmaier, 1985), these are approximately two-dimensional convective rolls that are oriented parallel to the rotation axis and that exist outside of the *tangent cylinder*, which is defined as the cylindrical surface aligned with the rotation axis and tangent to the base of the convective envelope. The preferred modes of convection inside and outside the tangent cylinder can be strikingly different in rapidly rotating systems. For a demonstration of this in a stellar context exhibiting spatially modulated convective patterns see Brown *et al.* (2008).

Although these rotationally aligned, columnar convective structures were originally identified and studied using linear analysis, their vestiges persist far into the
non-linear regime, changing their nature somewhat from steady columnar rolls to
smaller-scale episodic sheets that sprout from the inner boundary layer (Kageyama
et al., 2008). Even in the midst of intense turbulence, the most efficient way for
convection to transport heat outward at low latitudes is through extended radial
structures with minimal flow variation parallel to the rotation axis in accord with
the Taylor–Proudman theorem (see also Section 5.5.2; Pedlosky, 1987). This is
manifested in horizontal surfaces such as Fig. 5.2 as an apparent north–south
alignment at low latitudes.

Yet, two things set these solar convective structures apart from their planetary
analogs. First, although the Rossby number in the deep solar convection zone
is small relative to the solar surface layers, it is still orders of magnitude larger
than in planetary interiors (compare Table 5.1 in the appendix and Section 7.5
below Eq. (7.5)). In the Sun rotational influence is moderate enough that non-linear
advection still plays a considerable role, producing a downward cascade of energy
by means of vortex stretching as in canonical hydrodynamic turbulence (Vol. I,
Chapter 7). Although some self-organization likely occurs (see Section 5.2.2), the
rotationally aligned downflow lanes are embedded within a more isotropic sea of
smaller-scale turbulent fluctuations. The second difference is the density stratification, which is much larger in the Sun than in most planetary interiors and which
(along with the spherical geometry) breaks the symmetry between upflows and
downflows. The density stratification also influences the longitudinal propagation
of these rotationally aligned structures (which is generally prograde relative to
the local rotation rate) because it contributes to variations in potential vorticity
(Miesch, 2005; for a definition of potential vorticity see also Section 5.5.1).

5.3 Local dynamos and the magnetic carpet

5.3.1 Lagrangian chaos

Let's say we have a magnetized plasma that obeys the equations of magnetohydrodynamics (MHD) as discussed in Vol. I, Chapters 2 and 3. We can rewrite the
magnetic induction, Eq. (3.4) from Vol. I, as follows:

$$\frac{D\mathbf{B}}{Dt} \equiv \frac{\partial \mathbf{B}}{\partial t} + (\mathbf{v}\cdot\nabla)\,\mathbf{B} = (\mathbf{B}\cdot\nabla)\,\mathbf{v} - \mathbf{B}\,(\nabla\cdot\mathbf{v}) - \nabla \times (\eta\nabla \times \mathbf{B}), \qquad (5.3)$$

where D/Dt is the Lagrangian derivative, $\mathbf{B}$ is the magnetic field and η is the magnetic diffusivity. Now consider two fluid parcels in the flow that are infinitesimally
close together at some time t_0 and some location $\mathbf{x}_0$. We write the vector distance
between these two parcels as $\boldsymbol{\delta}(\mathbf{x}_0, t)$. At any instant, the time rate of change of $\boldsymbol{\delta}$ is
proportional to the velocity gradient in the direction of the displacement:

$$\frac{d\boldsymbol{\delta}}{dt} = (\boldsymbol{\delta}\cdot\nabla)\,\mathbf{v}, \tag{5.4}$$

where $\mathbf{v} = \mathbf{v}(\mathbf{x}(t), t)$ is the velocity at the parcels' current position $\mathbf{x}(t)$, defined such that $\mathbf{x}(t_0) = \mathbf{x}_0$. Comparing Eqs. (5.3) and (5.4) we see that they are identical if we replace $\mathbf{B}$ by $\boldsymbol{\delta}$, if we assume the fluid is incompressible $\nabla\cdot\mathbf{v}=0$, and if we neglect Ohmic dissipation, $\eta = 0$. This is true provided that we consider an ensemble of parcels at all locations $\mathbf{x}_0$ and if we interpret the Lagrangian derivative as the total time derivative in a reference frame moving with the flow.

We say that a flow exhibits *Lagrangian chaos* if the distance between parcels $\boldsymbol{\delta}(\mathbf{x}_0, t)$ increases exponentially with time in a globally averaged sense. The rate at which parcels exponentially diverge or converge at a particular location $\mathbf{x}_0$ is known at the *local Lyapunov exponent*. As discussed in Vol. I, Chapter 3, the kinematic dynamo problem is concerned with whether a magnetic field at some initial time $\mathbf{B}(\mathbf{x}, t_0)$ will exponentially grow or decay for a specified velocity field $\mathbf{v}(\mathbf{x}, t)$, neglecting Lorentz force feedbacks. Furthermore, *fast dynamos* are defined as flows for which the dynamo growth rate approaches a finite asymptotic value as $\eta \to 0$, or equivalently, as $R_\mathrm{m} \to \infty$, where $R_\mathrm{m} = vL/\eta$ is the magnetic Reynolds number. Given the large values of R_m in the solar convection zone (see Table 5.1), these are the dynamos of most interest to us. Thus, it is clear that the key to fast dynamo action (exponential growth of the magnetic field in the limit of large R_m) is Lagrangian chaos.

Although this is a profound insight, there is a catch. The evolution of $\boldsymbol{\delta}$ depends on the direction of the initial displacement; there are three dimensions so at each position $\mathbf{x}_0$ there are three local Lyapunov exponents.[†] In an incompressible flow, these must sum to zero; if a fluid parcel is stretched in one direction it must be squeezed in another to conserve mass. In a flow with a large but finite R_m, squeezing fields together with opposite polarities eventually leads to ohmic dissipation. Whether or not stretching can overcome dissipation depends on the global topology of the flow (including whether it is "rough" or "smooth"; see below). For more on this fascinating and subtle problem see the tutorial reviews by Childress and Gilbert (1995), Ott (1998), and Cattaneo and Hughes (2001).

5.3.2 Small-scale dynamos

Solar convection is highly turbulent, possessing extremely large Reynolds and magnetic Reynolds numbers (see Table 5.1 in the appendix to this chapter). Thus, we expect that it should exhibit Lagrangian chaos and probably dynamo action. Furthermore, because the time scale associated with granulation ($\tau \sim 5$ min;

[†] Local Lyapunov exponents are defined in terms of the eigenvalues of the Jacobian so the largest corresponds to the direction of maximum stretching.

see Section 5.2.1) is much shorter than that associated with giant cells ($\tau \sim L/U \sim 12\,\mathrm{d}$; see Table 5.1), we might expect that granulation should rapidly generate magnetic field with little regard to what may be occurring deeper in the convection zone.

Thus we arrive at the idea of a *local stellar dynamo* whereby magnetic fields are generated in a layer near the photosphere by small-scale convective motions including granulation, mesogranulation, and perhaps supergranulation. This concept is phenomenologically distinct from the small-scale dynamo introduced in Chapter 3 in Vol. I, which is defined as a dynamo in which the characteristic scale of the magnetic field is comparable to or smaller than that for the velocity field. However, as we will see, the two concepts are related in practice; the local dynamo, as it turns out, is a small-scale dynamo.

In the epic battle between field amplification by chaotic stretching and field dissipation by Ohmic diffusion, the easiest way to prevail is with a spatially smooth but temporally random flow.[†] For large R_m this generates an intricate web of folded magnetic sheets, ribbons, and filaments, with field lines pulled along the periphery of shearing eddies, parallel to the local flow and perpendicular to the local velocity gradient (Schekochihin *et al.*, 2004). Variations of **B** in the direction of **B** occur on the length scale of the eddies ℓ_e whereas variations in the direction perpendicular to **B** occur on a resistive scale $\ell_\mathrm{r} \sim R_\mathrm{m}^{-1/2}\ell_\mathrm{e}$. This develops as Lagrangian stretching and squeezing builds finer and finer transverse structure until diffusion eventually becomes comparable to advection in Eq. (5.3), such that $u_\mathrm{e}\ell_\mathrm{e}^{-1} \sim \eta\ell_\mathrm{r}^{-2}$ where u_e is the velocity amplitude. From a spectral perspective, the magnetic energy $E_\mathrm{m}(k)$ increases with wavenumber k, peaking near the resistive wavenumber $k_\mathrm{r} = 2\pi/\ell_\mathrm{r}$, and then dropping sharply thereafter, as the smallest scales are annihilated by diffusion.

Turbulent flows are not smooth. Rather, they contain motions on a vast range of dynamical scales that are typically self-similar over some intermediate inertial range where the kinetic energy spectrum follows a power law $E_k(k) \propto k^{-p}$ (Vol. I, Chapter 7). However, turbulent dynamos can still exhibit smooth, random stretching if the kinematic viscosity ν is larger than the magnetic diffusivity η, or in other words, if the magnetic Prandtl number $P_\mathrm{m} = \nu/\eta > 1$. Then the magnetic field on scales below the viscous dissipation scale ℓ_v is efficiently amplified by larger-scale, random shear as described above. For example, a large P_m dynamo with a Kolmogorov spectrum $p = 5/3$ will generate a magnetic energy spectrum $E_\mathrm{m}(k)$ that peaks in the subviscous range $k \sim k_\mathrm{r} > k_\mathrm{v} = 2\pi/\ell_\mathrm{v} \sim R_\mathrm{e}^{3/4}k_0$, where k_0 is the wavenumber corresponding to the energy injection scale, also called the outer scale

[†] A steady linear shear flow cannot be a dynamo. After an initial linear growth phase, the field strength will eventually tend toward a finite value or diffuse away entirely.

of the fluid motions, and $R_e = vL/\nu$ is the Reynolds number. An estimate for k_r can be obtained as in the previous paragraph by setting $\ell_e \sim \ell_\nu$ and $u_e \sim u_{\ell_\nu}$ where u_ℓ is the velocity on scale ℓ. At the viscous scale, $u_\ell \ell \sim \nu$ so the effective R_m is equal to P_m, which gives $k_r \sim P_m^{1/2} k_\nu$.

Most numerical simulations of small-scale turbulent dynamos published in the literature are concerned with the $P_m \geq 1$ regime. These work so well that for some time it was wondered whether turbulent dynamos could even exist for $P_m \ll 1$. However, this is the regime of most relevance to stars and planets.[*] In the Sun, P_m varies from 10^{-7} in the granulation layer to 10^{-5} in the lower convection zone (see Table 5.1). Only relatively recently have we verified from theoretical models (Boldyrev and Cattaneo, 2004), numerical simulations (Iskakov *et al.*, 2007), and laboratory experiments (Monchaux *et al.*, 2007) that small-scale dynamos can indeed operate for $P_m \ll 1$ but only with much more effort; the critical (minimum) value of R_m needed to sustain dynamo action increases sharply as P_m is decreased below unity. Although this poses a significant challenge for numerical simulations, stars need not be too concerned; their values of $R_m \sim 10^5$–10^8 (see Table 5.1) are safely supercritical.

In a low-P_m small-scale dynamo the resistive scale ℓ_r lies within the turbulent inertial range. Chaotic stretching by larger-scale motions must overcome turbulent mixing by smaller-scale motions that can enhance Ohmic dissipation. Yet, we may still expect the magnetic energy spectrum to peak near k_r because these are the scales that can most efficiently amplify field while at the same time avoiding Ohmic dissipation. To see this, consider the scale-dependent dynamo growth rate γ_ℓ. According to Eq. (5.4) this should be comparable to the velocity gradient, which scales as $k u_\ell \sim k^{(3-p)/2}$ within the inertial range. Thus, for a general rough[‡] velocity field with $p < 3$, γ_ℓ increases with k; the smallest eddies amplify field most rapidly because they have the fastest turnover time. However, Ohmic diffusion eventually dominates at large k so the spectral peak should lie near k_r. This is a plausible argument, but it is unclear how applicable it is to real flows even in the kinematic regime, let alone to the saturated regime where the Lorentz force must be taken into account. The dynamical behavior of non-helical turbulent dynamos in the limit $R_e \gg R_m \gg 1$ ($P_m \ll 1$) is still poorly understood (Iskakov *et al.*, 2007). Nevertheless, what is clear is that small-scale dynamos can efficiently

---

[*] Large-scale dynamos as discussed in Section 5.4 have an easier time operating at $P_m \ll 1$. For example, the critical value of R_m is lower for helical flows.

[‡] The roughness of a velocity field can be quantified by means of its second-order longitudinal structure function $\delta v^2(r) = \langle |(\mathbf{v}(\mathbf{x}+\mathbf{r}) - \mathbf{v}(\mathbf{x})) \cdot \mathbf{r}/r|^2 \rangle$, where the average $\langle \rangle$ is over space and time. If $\delta v^2(r) \propto r^{2\epsilon}$ either by construction or by fitting empirical or numerical data then ϵ is defined as the roughness exponent (Boldyrev and Cattaneo, 2004; Tobias and Cattaneo, 2008). For an isotropic, homogeneous, random field, $\epsilon = (p-1)/2$. Generally, ϵ will range between zero (bounded $E(k)$ integral) and unity (linear divergence/convergence), with larger values corresponding to smoother fields in an unfortunate twist of terminology.

generate magnetic energy on scales much smaller than the correlation length of the velocity field.

5.3.3 The Sun's magnetic carpet

In astrophysical and geophysical dynamos, coherent flow structures can provide efficient random stretching that can dominate turbulent field generation (Tobias and Cattaneo, 2008). This is particularly the case for solar and stellar convection, which is riddled with coherent downflow plumes and lanes (Section 5.2). However, these same coherent structures are also good at transporting magnetic flux, which can be bad news from the perspective of the local dynamo if flux is transported out of the granulation layer before it is recirculated and further amplified. The topological connectivity of the downflow network as well as the larger velocity of downflows relative to upflows gives rise to a preferential downward transport of magnetic flux known as *magnetic pumping*. For a discussion of this phenomenon within the context of the mean-field formulation see Vol. I, Chapter 3, and for a numerical demonstration in turbulent, compressible, penetrative convection see Tobias *et al.* (1998).

Magnetic pumping is a particular manifestation of the more general tendency for convective flows to expel magnetic flux from regions of high turbulent intensity. Other manifestations include flux expulsion by steady flows with closed stream-lines (Moffatt, 1978), flux separation in magnetoconvection (Tao *et al.*, 1998) and the mean-field concept of turbulent diamagnetism (Krause and Rädler, 1980). In stars, loss of magnetic flux may also occur by means of magnetic buoyancy (Fan, 2004). The solar granulation layer in particular has open boundaries both above and below where flux can escape. On the other hand, magnetic flux is contintually supplied to the solar surface layers by the global dynamo seething below (Section 5.4). This is most evident in the emergence of large bipolar active regions that fragment and disperse throughout the photosphere, accounting for a substantial fraction of the observed photospheric flux (Schrijver and Harvey, 1994).

Thus, it is a tricky and to some extent a moot question to ask whether a local dynamo can operate in the solar surface layers in isolation from the deeper convection zone. Small-scale fields generated by chaotic stretching must pervade the entire convection zone, linked by an intricate nexus of entangled field lines (see Section 5.4.4). Yet, observations and numerical simulations do suggest that much of the small-scale magnetic flux in the quiet Sun, away from active regions, plages, and the magnetic network (i.e. intranetwork flux) may indeed be attributed to local dynamo action by the particularly vigorous convection in the photospheric boundary layer.

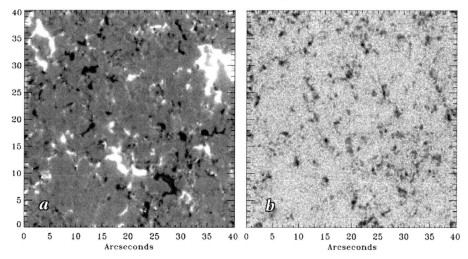

Fig. 5.3. The apparent flux density for the (a) vertical and (b) horizontal magnetic field components in the quiet Sun at disk center obtained from the solar optical telescope spectro-polarimeter (SOT/SP) onboard the Hinode spacecraft. (From Lites *et al.*, 2008.) In panel (a) black and white denote inward and outward flux respectively and the grey scale saturates at $\pm 50\,\text{Mx}\,\text{cm}^{-2}$, with an average amplitude of $11\,\text{Mx}\,\text{cm}^{-2}$. In panel (b) dark tones denote high values, with a saturation level of $200\,\text{Mx}\,\text{cm}^{-2}$ and an average value of $55\,\text{Mx}\,\text{cm}^{-2}$. For reference, the typical size of a granule is about $1.4''$ which corresponds to $1\,\text{Mm}$. Note that the values for the apparent flux density are related to the intrinsic field strength in gauss but the precise relationship is subject to uncertainty associated with the nature and sensitivity of the measurement, the calibration, and the spatial, temporal, and spectral averaging (Lites, 2008). (Reproduced by permission of the AAS.)

The solar photosphere is teeming with small-scale magnetic flux, blanketed by what Title and Schrijver (1998) have elegantly referred to as the Sun's *magnetic carpet*. With a spatial resolution of about $0.3''$ ($\approx 200\,\text{km}$), the Hinode observations shown in Fig. 5.3 are the highest-resolution spectro-polarimetric measurements yet achieved, but there are indications that even these are not capturing everything; there is likely magnetic structure on even smaller scales, at least an order of magnitude smaller than that of the granulation pattern ($\sim 1\,\text{Mm}$; Section 5.2.1). Unlike network fields, which are predominantly vertical, the intranetwork fields have a strong horizontal component as well (Lites *et al.*, 2008). Strong horizontal fields are generally found over the granules whereas vertical fields are found in the intergranular lanes, suggesting a local dynamic origin rather than a kinematic shredding and redistribution of larger-scale flux. This suggestion is further substantiated by the statistical properties of bipolar flux in small-scale ephemeral regions, which are approximately uniformly distributed across the solar surface, are randomly oriented, are almost independent of the phase of the solar activity cycle, and are globally replenished on a time scale of less than a day (Hagenaar *et al.*, 2003). Numerical simulations of local dynamo action and magnetoconvection in the photosphere exhibit properties that are qualitatively similar to what is observed (Stein and Nordlund, 2006; Abbett, 2007; Vögler and Schüssler, 2007; Schüssler and Vögler, 2008).

The local solar dynamo plays an important role in the dynamics and energetics of the quiet solar atmosphere, particularly with regard to coronal heating (Vol. I, Chapter 8). Photospheric convection and local dynamo processes may also be responsible for part or all of the basal emission observed in cool stars, which is a well-defined minimum level of chromospheric emission that varies with stellar type and that is often found in slowly rotating stars with little or no indication of cyclic activity and possibly, by implication, a weak or non-existent global dynamo (Schrijver and Zwaan, 2000). However, the chromospheric emission of the quiet Sun is larger than this basal emission level (Section 2.2.3.2), suggesting that rotation either enhances the local dynamo or, more likely, that much of the photospheric flux in ephemeral regions is connected with turbulent dynamo action lower in the convection zone where rotation plays a bigger role. In order to understand this, as well as patterns of magnetic activity on large spatial and temporal scales, we must probe deeper.

5.4 Global dynamos or "How to build a sunspot"

We saw in Section 5.3 that turbulent flows tend to beget turbulent magnetic fields. We observe such fields on the Sun but we also observe something else, even more striking and enigmatic. Superposed on the turbulent field component is an organized large-scale field component characterized by a prominent dipole moment that reverses sign every 11 years (plus or minus a year or two), and the associated appearance of strong bipolar active regions (including sunspots) with well-defined rules for emergence latitudes, orientations, and tilt angles (discussed in Chapter 2). Global solar dynamo theory is concerned with how this organized large-scale field arises and how it evolves on time scales of decades and longer. In this section we focus on how turbulent solar and stellar convection might generate such a large-scale field. For a discussion of the origins and long-term modulations of cyclic activity in the Sun and other stars see Chapter 2.

In Chapter 3 of Vol. I a *large-scale dynamo* was defined as a dynamo that can generate magnetic structure on scales larger than the correlation length of the velocity field. The central issue here is not how to generate *magnetic energy* but rather how to generate *magnetic flux*. Just as Lagrangian chaos is the key to small-scale turbulent dynamo action (Section 5.3), helicity and shear are the key additional ingredients that give rise to large-scale dynamo action.[*] However, as in Section 5.3, we will see that the issue is more subtle than it may first appear.

Throughout this section and also Section 5.5, angular brackets $\langle \rangle$ denote averages over longitude and time such that $\langle \mathbf{B} \rangle$ and $\langle \mathbf{v} \rangle$ are the mean fields and flows. Angular brackets with a subscript, $\langle \rangle_V$, denote averages over the volume of the domain in question.

5.4.1 Magnetism with a twist

For a clue as to how a large-scale dynamo might operate, consider the intuitive picture of a magnetic field as a collection of discrete lines of force as expressed in Vol. I, Chapter 4. The topololgy of how these magnetic field lines are linked, twisted, and writhed is reflected by the magnetic helicity $H_m = \langle \mathbf{A} \cdot \mathbf{B} \rangle_V$ (for magnetic field $\mathbf{B}$ and its vector potential $\mathbf{A}$; see Eq. (4.14) in Vol. I). Because H_m is an ideal invariant of the MHD equations, any local change in the helicity density $\mathbf{A} \cdot \mathbf{B}$ (at high R_m) must be accompanied by a global restructuring of the field. You might think that if you twist the field just right on small scales, the field lines may unfurl, forming braided loops on much larger scales. As it turns out, this is not a bad analogy.

In stars, magnetic helicity is generated by kinetic helicity $H_k = \langle \mathbf{v} \cdot \boldsymbol{\omega} \rangle_V$, which is in turn generated by rotation and density stratification. This is particularly the case for giant cells where the rotational influence is strong (Section 5.2.4). In mean-field dynamo theory, the lifting and twisting motions associated with kinetic helicity induce a turbulent electromotive force (EMF) known as the α-*effect* which has formed the backbone of solar dynamo theory for half a century (Vol. I, Chapter 3). The α-effect may be regarded more generally as a physical process that transfers magnetic helicity between the mean and fluctuating field components, thus sustaining the former against ohmic and turbulent diffusion (Seehafer, 1996).

There are two serious shortcomings to the theoretical underpinnings of the turbulent α-effect as applied to stars. The first is that it is essentially a kinematic concept that is not strictly valid for a physical, saturated dynamo. We return to this issue in Section 5.4.2. The second shortcoming is that the traditional derivation of the α-effect assumes a scale separation between mean and fluctuating fields that is generally not justified for a turbulent flow. However, a closely related process does occur in turbulent flows with a continuous spectrum of scales.

---

[*] Although helicity and shear are both important for the global solar dynamo, large-scale dynamo action can proceed with one or the other. For example, shear plays no role in the α^2 dynamos of the mean-field approximation (Vol. I, Chapter 3) and some astrophysical objects such as interstellar clouds may exhibit non-helical mean-field generation by means of shear alone (Rogachevskii and Kleeorin, 2003).

Helical MHD turbulence exhibits an *inverse cascade of magnetic helicity*. This is a self-similar transfer of magnetic helicity from small to large scales that extends beyond the outer scale of the motions k_0 to the global scales of the system. This phenomenon was first discovered by Pouquet *et al.* (1976) using a turbulence closure model and has since been verified by direct numerical simulation (Brandenburg, 2001; Alexakis *et al.*, 2006). The origin of the inverse cascade can be understood heuristically by considering non-linear interactions among triads of like-signed helical Fourier modes in the kinematic regime. If such interactions are to conserve both energy and magnetic helicity (as they must), then the latter must be transferred to lower k (Pouquet *et al.*, 1976; Brandenburg and Subramanian, 2005).

The α-effect and the inverse cascade both produce mean fields by means of an upscale (high k to low k) spectral transfer of magnetic helicity. The distinction between the two is that the upscale transfer is non-local in spectral space in the case of the α-effect (scale separation) and local for the inverse cascade. Numerical simulations suggest that the inverse cascade is more applicable in the kinematic regime, but saturated large-scale dynamos tend to transfer magnetic helicity directly from the forcing scale to the mean field (Brandenburg, 2001; Alexakis *et al.*, 2006). Although the latter scale separation is suggestive of an α-effect, many helical turbulent systems do not exhibit any clear correlation between the turbulent EMF and the mean field, calling into question the kinematic mean-field paradigm (Cattaneo and Hughes, 2006). An alternative interpretation for the non-local magnetic helicity transfer is in the topological properties of flux tubes; large-scale twisting can generate small-scale helicity associated with tightly spiraling field lines while small-scale reconnection can alter large-scale linkages (Alexakis *et al.*, 2006).

5.4.2 Dynamo saturation and open boundaries

In Section 5.4.1 we saw that large-scale dynamos can build mean fields through the upscale transfer of magnetic helicity. However, this comes at a price. The only way to sustain the upscale transfer (whether local or non-local) is by generating an equal amount of small-scale helicity of the opposite sign. The rapid accumulation of small-scale helical fields can feed back on the flow through the Lorentz force (specifically magnetic tension), suppressing the helicity transfer and shutting down the large-scale dynamo. In the context of mean-field theory this is known as dynamical α-quenching or, more ominously, as catastrophic α-quenching, and it has profound implications for solar and stellar dynamos. For further details see Vol. I, Section 3.5.2 and the reviews by Brandenburg and Subramanian (2005) and Miesch and Toomre (2009).

Catastrophic α-quenching may be avoided either by dissipating small-scale helicity or by transporting it out of the domain. A possible mechanism for dissipating small-scale magnetic helicity was found in recent simulations by Alexakis *et al.* (2006). They consider a periodic Cartesian domain with helical mechanical forcing and find that the inverse cascade of magnetic helicity for scales above the forcing scale $k < k_f$ is accompanied by a forward cascade of the oppositely signed helicity toward smaller scales $k > k_f$ where it is eventually dissipated. If this forward cascade at small scales holds up for stellar parameter regimes $R_e \gg R_m \gg 1$, it could play an important role in stellar dynamos. Alternatively, turbulent diffusion may help dissipate small-scale magnetic helicity (Brandenburg and Subramanian, 2005). However, it is difficult to see how small-scale motions might dissipate helicity of one sign through turbulent mixing while simultaneously transferring helicity of the opposite sign toward larger scales (unless aided by a forward cascade).

Helicity loss through the boundaries is perhaps a more plausible means by which stellar convection zones might sustain a large-scale dynamo. However, if such losses are to promote rather than inhibit mean-field generation, they must occur preferentially on small scales. Observations of magnetic structures in the photosphere and corona do confirm that emerging magnetic fields are helical in nature, with systematic sign preferences in the northern and southern hemisphere (Pevtsov and Balasubramaniam, 2003). Furthermore, ejection of magnetic flux and magnetic helicity in coronal mass ejections appears to play a crucial role in the global restructuring of the coronal and even the heliospheric magnetic field during the course of the solar activity cycle, culminating in periodic polarity reversals (Low, 2001; Zhang and Low, 2005). However, it is unclear whether the associated helicity flux may be construed as a small-scale helicity sink within the context of a turbulent large-scale dynamo. Magnetic helicity may also be generated through the shearing of partially emerged, quiescent coronal loops by photospheric differential rotation and the ensuing magnetic reconnection (van Ballegooijen and Martens, 1990).

In Section 5.3.3 we saw that turbulent convection tends to expel magnetic flux out of the convection zone, with a preferential downward transport known as magnetic pumping. Might it also expel helicity? This is possible but numerical simulations of rotating penetrative convection suggest that flux expulsion and magnetic pumping alone are not enough to promote efficient large-scale dynamo action (Tobias *et al.*, 2008; Käpylä *et al.*, 2008). However, it seems that rotational shear can do the trick, provided it is of the right orientation (Käpylä *et al.*, 2008).

Rotational shear induces a helicity flux directed along angular velocity contours (Brandenburg and Subramanian, 2005). Thus, latitudinal shear (as in the bulk of the solar convection zone; see Fig. 5.1b) can transport helicity radially, either outward through the surface or inward into the overshoot region and tachocline. This has been demonstrated in Cartesian domains by Käpylä *et al.* (2008) and may account for helical field generation in global simulations of convective dynamos that possess a tachocline (Browning *et al.*, 2006; Miesch *et al.*, 2009). In spherical systems, radial angular velocity shear may also induce helicity flux across the equator, enabling helical mean-field generation without the accumulation of small-scale helicity. This possiblity is in need of further exploration.

In rapidly rotating stars and planets, the saturation level of the dynamo (Section 2.3.2) depends primarily on the energy flux through the convection zone, which in stars is the luminosity (Section 7.6.5). If these objects are operating a turbulent large-scale dynamo in the canonical sense (Section 5.4.4), this suggests that helicity loss through the boundaries or through Ohmic dissipation is efficient enough that the buildup of small-scale helicity does not regulate dynamo saturation. Despite such constraints, rotating stars find a way to build strong magnetic fields.

5.4.3 Of sunspots and shear

Rotational shear not only promotes large-scale dynamo action by helping to dispose of small-scale helicity, it also generates non-helical mean toroidal field from mean poloidal field by virtue of the Ω-effect (Vol. I, Chapter 3). The linear nature of the Ω-effect contrasts sharply with the highly non-linear nature of turbulent field generation and appears to be an essential ingredient in establishing the regular, cyclic character of solar activity (Chapter 6).

One look at the solar internal rotation profile in Fig. 5.1b and we can guess where the Ω-effect might be most efficient. The strongest shear lies in the tachocline near the base of the convection zone. According to our current paradigm, axisymmetric fields generated in the convection zone are supplied to the tachocline by magnetic pumping and meridional circulation and there amplified into strong toroidal structures by the rotational shear. Thus, the generation mechanisms for mean poloidal and toroidal fields are spatially separated, which minimizes the back reaction of the Lorentz force and can thereby promote strong mean-field generation even in the presence of catastrophic α-quenching (Charbonneau and MacGregor, 1996). The subadiabatic stratification of the lower tachocline inhibits magnetic buoyancy instabilities, enabling long-term storage of toroidal flux (Fan, 2004). As the field strength grows, toroidal flux eventually destabilizes, fragmenting into localized tubes that buoyantly rise through the convection zone and emerge from the photosphere as bipolar active regions.

Bipolar active regions and sunspots are thus regarded as a manifestation of subsurface toroidal flux, most likely originating in the tachocline or lower convection zone (for an alternative view see Brandenburg, 2005). The prodigious strength of sunspot fields ($\sim$3000 G) relative to the mean poloidal field at the surface ($\sim$10 G) betrays the prominent influence of rotational shear. In the language of mean-field theory, the Sun is an $\alpha\Omega$ dynamo (Vol. I, Chapter 3).

Although there is little doubt that bipolar active regions originate from the emergence of toroidal flux tubes created by rotational shear, the details of how such tubes form remain poorly understood. A simple estimate based on flux freezing suggests that the magnetic field strength in a horizontal flux tube should scale as the density ρ. Because ρ increases by more than five orders of magnitude from the photosphere to the tachocline (see Table 5.1 in the appendix to this chapter), one would expect progenitor fields to be much stronger than sunspot fields. Numerical models of rising flux tubes (reviewed by Fan, 2004) suggest less extreme field strengths of 10^4–10^5 G but this still implies a magnetic energy in the tachocline that equals or exceeds the kinetic energy in the convection and the differential rotation (using the values in the appendix, the equipartition field strength is $B_{eq} = 2U\sqrt{\pi\rho} \sim$ 8000 G). Achieving such super-equipartition field strengths is a formidable challenge for dynamo models. Alternatively, flux tubes may somehow amplify as they rise through the convection zone (for possible mechanisms see Rempel and Schüssler, 2001, and Parker, 2008).

Global simulations of convective dynamos confirm that the presence of a tachocline promotes the generation of strong, relatively stable mean fields (Browning *et al.*, 2006; Miesch *et al.*, 2009). The magnetic structure in the tachocline is reminiscent of the paradigm outlined above, dominated by strong ($\sim$4 kG) toroidal flux sheets antisymmetric about the equator. However, there is as yet little indication for magnetic buoyancy instabilities or latitudinal propagation (as expected from the solar butterfly diagram; see Chapter 2). Whether this is a challenge to our dynamo paradigm or simply a consequence of insufficient resolution (which in turn implies insufficient stiffness and excessive diffusion) is currently unclear.

Faster rotation, Ω, promotes both stronger shear, $\Delta\Omega$, and stronger magnetic activity, $\langle B \rangle$, until a saturation point is reached ($\sim$10$\Omega_\odot$ where $\Omega_\odot$ is the solar rate) beyond which $\Delta\Omega$ and $\langle B \rangle$ become independent of Ω (Schrijver and Zwaan, 2000; Rempel, 2008; Brown *et al.*, 2008, 2009; see also Section 2.3.2). The increase in shear and activity is associated with a change in the topology of the magnetic field, favoring large-scale field patterns with an increasingly prominent toroidal component. This is supported by both observations (Petit *et al.*, 2008) and numerical models (Brown *et al.*, 2009). An example is shown in Fig. 5.4 for a solar-type star rotating at 3$\Omega_\odot$. In the simulation shown, rotational shear (Fig. 5.4a) generates strong ($>10^4$ G) toroidal bands (Fig. 5.4b) that are antisymmetric about the equator and that remain coherent in the midst of the turbulent convection zone for thousands of days, far longer than the rotation period ($\sim$10 d) and the convective turnover time ($\sim$25 d). Despite their coherence and persistence, these are not isolated flux surfaces; constituent field lines meander throughout the convection zone

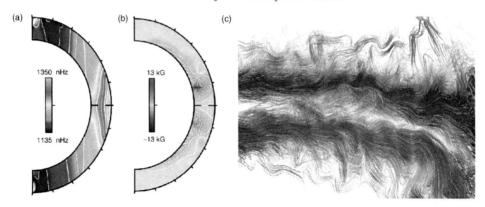

Fig. 5.4. Rotational shear and toroidal fields in a global 3D dynamo simulation of a solar-type star rotating at three times the solar rate. (a) Angular velocity Ω and (b) toroidal field $\langle B_\phi \rangle$, averaged over longitude and time. (c) A 3D rendering of magnetic field lines in a portion of the convection zone, spanning about 50° in longitude. Grey and black lines denote positive and negative B_ϕ respectively. The view is radially outward from a vantage point under the convection zone and slightly above (north of) the equator. (From Brown *et al.*, 2009.) See Color Plate 3.

(Fig. 5.4c). This is consistent with MHD simulations of toroidal flux generation by horizontal shear (Cattaneo *et al.*, 2006) but stands in stark contrast to the idealized flux tubes typically considered in numerical and theoretical models of flux emergence (reviewed by Fan, 2004). At somewhat higher rotation rates ($\Omega_\odot$) such convective dynamo simulations exhibit similar toroidal structures with quasi-cyclic polarity reversals, punctuated by sporadic symmetric states with a single toroidal band (Brown *et al.*, 2009).

5.4.4 Are global dynamos large-scale dynamos?

The canonical picture of a large-scale turbulent dynamo as applied to stars is that painted in Sections 5.4.1 to 5.4.3: turbulent convection on relatively small scales generates mean fields via the upscale transfer of magnetic helicity (whether local or non-local). The mean fields thus generated are helical in nature and the magnetic energy peaks at the largest scales in the system. Rotational shear then converts mean poloidal field to mean toroidal field, pumping more energy into the latter and potentially promoting cyclic reversals. This is a reasonable scenario but is it the way stellar dynamos really operate?

Rotating turbulent flows exhibit a wide range of dynamically active scales and the effective Rossby number on each scale $R_o(\ell) = v/\Omega\ell$ will generally increase with decreasing ℓ, inversely proportional to the the eddy turnover time (see Section 5.3.2). For flows with high R_e, large scales may feel a strong rotation influence ($R_o(\ell) < 1$) while small scales may be dominated by inertia ($R_o(\ell) > 1$). Here the distinction between a large-scale and a small-scale dynamo blurs and the magnetic energy spectrum may be multimodal, particularly in anisotropic, inhomogeneous flows with coherent structures.

Global simulations of convective dynamos in a solar context exhibit mean-field amplitudes comparable to observed photospheric values, $\sim$10–100 G for the poloidal component and several kG for the toroidal component (Brun *et al.*, 2004; Browning *et al.*, 2006; Miesch *et al.*, 2009). However, this represents the low-wavenumber tail of the spectrum; smaller-scale fields are even stronger, accounting for over 95% of the magnetic energy in the convection zone. The spectral peak of the magnetic energy occurs on scales comparable to or less than that of the kinetic energy, making these global simulations technically small-scale dynamos as defined in Section 5.3.2 (see also Vol. I, Chapter 3).

It is possible that at much higher values of R_e and R_m the peak in the velocity spectrum will shift toward smaller scales and the greater scale separation between the outer scale of the turbulence k_0 and the global scales of the system will enhance upscale magnetic helicity transfer (Käpylä *et al.*, 2008). However, small scales would still be less influenced by rotation, less helical in nature, and perhaps more inclined toward small-scale dynamo action. Mean-field generation may still be dominated by large-scale structures such as the rotationally aligned downflow lanes discussed in Section 5.2.4.

Another possibility is that physical mechanisms other than turbulent helical convection and rotational shear contribute to mean-field generation in the Sun and other stars, particularly in solar-like stars that possess a tachocline. What is needed for mean-field generation is a systematic correlation between the non-axisymmetric velocity and magnetic field components $\mathbf{v}'$ and $\mathbf{B}'$ such that there is a substantial mean EMF $\langle \mathbf{v}' \times \mathbf{B}' \rangle$. Essentially laminar processes may turn out to be more efficient at inducing such correlations than turbulent processes. Possible candidates include instabilities associated with magnetic buoyancy and magnetic tension in flux tubes, joint instabilities arising from toroidal fields and rotational shear in the tachocline (Section 5.5.5), and the fragmentation and dispersal of magnetic flux in photospheric active regions by granulation and supergranulation (Chapter 2).

5.5 Rotational shear and meridional circulation

The solar internal rotation profile $\Omega(r, \theta)$ was presented in Section 5.2.3 as a manifestation of giant cell convection. There a remarkable property of Ω was highlighted, namely, that the angular momentum increases with increasing distance from the rotation axis such that

$$\frac{\partial \mathcal{L}^2}{\partial \lambda} > 0, \tag{5.5}$$

where $\lambda = r \sin \theta$ is the cylindrical radius and

$$\mathcal{L} = \lambda^2 \Omega = \lambda \left(\lambda \Omega_0 + \langle v_\phi \rangle \right) \tag{5.6}$$

is the specific angular momentum per unit mass. The final expression in Eq. (5.6) corresponds to a coordinate system rotating at a rate of Ω_0 and $\langle v_\phi \rangle$ is the mean zonal (longitudinal) velocity relative to the rotating frame, defined as the *differential rotation* (as in Section 5.4, brackets denote averages over longitude and time).

In fact, we expect Eq. (5.5) to be a general property of rotating stars because the alternative situation, $\partial \mathcal{L}^2 / \partial \lambda \leq 0$, is unstable. This result was established in the 1930s and 1940s by Solberg and Høiland and is a stellar analog of the Rayleigh criterion for rotating fluids; for a thorough derivation and historical perspective see Tassoul (1978). Specifically, Høiland's criterion says that a rotating star is linearly unstable to axisymmetric, adiabatic perturbations if $\mathcal{L}$ decreases with increasing λ along a surface of constant specific entropy S. Although the analysis only strictly applies if we neglect convection, meridional circulation, magnetic fields, compositional gradients, and viscous diffusion, this is still a useful rule of thumb. An intuitive appreciation of Høiland's criterion can be obtained by considering a fluid parcel in a differentially rotating star with $\partial \mathcal{L}^2 / \partial \lambda < 0$. If the parcel is displaced outward, it will have more angular momentum than its surroundings and the excess centrifugal force will push it farther out. Likewise, a parcel displaced inward will continue inward due to a deficit of angular momentum.

Taking magnetism into account gives rise to an even stronger constraint on Ω:

$$\frac{\partial \Omega^2}{\partial \lambda} > 0. \tag{5.7}$$

If Eq. (5.7) is violated, again along isentropic surfaces, the star becomes susceptible to the magnetorotational instability, or MRI (Balbus, 1995). Poloidal field lines tether fluid parcels together at different λ, promoting angular momentum transport that feeds the instability.

Rotation profiles violating Eq. (5.7) or even Eq. (5.5) are not impossible, but in order to exist they must be maintained. In the Sun, for example, Eq. (5.7) is violated across the solar tachocline at high latitudes (Parfrey and Menou, 2007). Because the MRI operates on a shear time scale of order several months, some efficient mechanism is needed to oppose it. Even nominally stable rotation profiles ($\partial \Omega^2 / \partial \lambda > 0$) must be maintained against the potentially disruptive effects of turbulent mixing (in convection zones), magnetic tension, meridional circulations, and thermal diffusion (in radiative zones; see Section 5.5.2). We will see now how stars manage to sustain their rotational shear.

5.5.1 The Reynolds stress and gyroscopic pumping

An equation describing the temporal evolution of the specific angular momentum $\mathcal{L}$ follows straighforwardly by multiplying the zonal component of the momentum equation by λ and averaging over longitude and time. In light of the large Reynolds numbers in stellar interiors, we can safely neglect viscous diffusion. In this section we will also neglect the Lorentz force. Although this is not justified in general, magnetism typically acts to suppress shear rather than enhance it.[*] Thus, the question of how rotational shear is established and maintained can legitimately be approached from a hydrodynamic perspective, at least initially. If we assume that $\mathcal{L}$ is statistically steady over some intermediate time scale we then obtain

$$\nabla \cdot (\langle \rho \mathbf{v}_{\mathrm{m}} \rangle \, \mathcal{L}) = -\nabla \cdot \left(\lambda \left\langle \rho v_{\phi}' \mathbf{v}_{\mathrm{m}}' \right\rangle \right), \tag{5.8}$$

where $\mathbf{v}_{\mathrm{m}} = v_r \hat{r} + v_\theta \hat{\theta}$ is the velocity in the meridional plane and primes denote fluctuating components, e.g. $v_\phi' = v_\phi - \langle v_\phi \rangle$. The term on the left-hand side is the advection of angular momentum by the meridional circulation $\langle \rho \mathbf{v}_{\mathrm{m}} \rangle$ and the term on the right-hand side is the transport of angular momentum by the Reynolds stress. In a steady state, these must balance.

In stellar convection zones, the Reynolds stress arises from convective velocity correlations induced by the Coriolis force. In solar parameter regimes, the dominant contribution is from the rotationally aligned downflow lanes discussed in Section 5.2.4. The horizontally converging flow that feeds these lanes in the upper convection zone has an east–west orientation. The Coriolis force diverts the prograde flows equatorward and the retrograde flows poleward, yielding an equatorward angular momentum flux such that $\langle v_\theta' v_\phi' \rangle$ is positive in the northern hemisphere and negative in the southern.

As the rotation rate is increased, convective columns become more aligned with the rotation axis and acquire systematic tilts in the λ-ϕ plane as cylindrically inward and outward fluid motions tend to conserve their potential vorticity,[‡] $Q = (\omega_z + 2\Omega_0)/(H\rho)$, where ω_z is the axial component of the fluid vorticity in the rotating frame and H is the height of a fluid column (Aurnou *et al.*, 2007; Brown

---

[*] There are exceptions to the maxim that magnetism acts to suppress shear. For example, tipping instabilities of toroidal flux tubes in the tachocline can induce prograde zonal jets (Miesch *et al.*, 2007) and buoyantly rising flux tubes can induce retrogrograde jets (Fan, 2008). In both cases, the Maxwell stress works in conjunction with the Coriolis force to accelerate zonal flows.

[‡] Potential vorticity is a useful but somewwhat malleable concept that takes a slightly different form for different systems. For more on its properties and significance, see Müller (1995).

et al., 2008). The spherical geometry and the density stratification both contribute
to Q, with the former typically dominating for planets and the latter more important for stars. This generally leads to cylindrical rotation profiles $\partial\Omega/\partial z = 0$ with
$\partial\Omega/\partial\lambda > 0$ outside the tangent cylinder (Section 5.2.4). Slowly rotating stars
and planets (such as red giants and ice giants), on the other hand, tend to exhibit
more isotropic convective motions that mix angular momentum by means of the
Reynolds stress, producing anti-solar rotation profiles, $\partial\Omega/\partial\lambda < 0$ (Aurnou *et al.*,
2007; Miesch and Toomre, 2009).

Because the Mach number is small (and the flow is assumed to be steady) the
meridional circulation is divergence free and we can write Eq. (5.8) in the more
general, and instructive, form of

$$\langle\rho\mathbf{v}_{\mathrm{m}}\rangle\cdot\nabla\mathcal{L} = \mathcal{F}, \tag{5.9}$$

where $\mathcal{F}$ is some unspecified force that locally accelerates or decelerates the zonal
flow. In the case of Eq. (5.8), $\mathcal{F}$ is the convergence of the Reynolds stress but more
generally it can just as well incorporate the Lorentz force and viscous diffusion.
The latter is neglible in stars but not neccessarily in numerical simulations. Note
that because the mass flux is divergenceless, the net angular momentum flux due
to the meridional circulation across closed surfaces of constant $\mathcal{L}$ (or across some
closed intersection of $\mathcal{L}$ isosurfaces and impermeable boundaries) must vanish (this
is most readily appreciated using the divergence form in Eq. (5.8)).

It is clear from Eq. (5.9) that the effect of the zonal forcing $\mathcal{F}$ is to induce
a meridional circulation across surfaces of constant $\mathcal{L}$. Because $\nabla\mathcal{L}$ is generally
directed away from the rotation axis in stars (Eq. 5.5), then an acceleration of the
zonal flow ($\mathcal{F} > 0$) will induce a circulation away from the rotation axis and
a deceleration ($\mathcal{F} < 0$) will induce a circulation toward the rotation axis. We will
refer to this phenomenon as *gyroscopic pumping* after McIntyre (1998, 2007). Note
that gyroscopic pumping is a concept associated with the dynamical equilibrium
of Eq. (5.9) and is distinct from the Coriolis acceleration of the zonal flow. If a
specified forcing $\mathcal{F}$ is introduced at some time, the circulation and the rotation
profile $\mathcal{L}$ will adjust until (5.9) is satisfied. This adjustment phase is mediated by
the Coriolis force but may also involve other physical processes, as discussed in
Section 5.5.3.

A possible explanation for the near-surface shear layer evident in Fig. 5.1b and
discussed in Section 5.2.3 is that high-Rossby-number convection near the photospheric boundary layer, in particular supergranulation (Section 5.2.2), tends to
homogenize angular momentum, producing a divergence of the Reynolds stress.
Whatever the origin, Eq. (5.9) implies that the near-surface deceleration should be
accompanied by a poleward meridional circulation. Such a poleward circulation is
in fact observed in Doppler measurements, correlation tracking, and helioseismic

inversions (reviewed by Miesch, 2005). To what extent this poleward circulation is a boundary layer phenomenon or a more global consequence of convection zone dynamics remains an open question.

Another interesting application of Eq. (5.9) concerns the case where $\mathcal{F} = 0$. Here we see that in order for the meridional circulation to be steady, it must flow along surfaces of constant $\mathcal{L}$. Or, looking at it another way, the equilibrium rotation profile is such that $\mathcal{L}$ is constant along streamlines of the mass flux. Thus, a meridional circulation profile with large cells extending over a considerable range of latitudes (and λ) would tend to speed up the poles relative to the equator ($\partial \Omega^2 / \partial \lambda < 0$). We will return to this issue in Section 5.5.3.

5.5.2 Warm poles and thermal winds

In addition to Eq. (5.5), another striking feature of the solar internal rotation noted in Section 5.2.3 is the nearly radial angular velocity contours at mid-latitudes. In order to appreciate the significance of this, consider the zonal component of the vorticity equation, obtained by taking the curl of the momentum equation. Again we will average over longitude and time, assuming a statistically steady state, and we will neglect magnetism and viscous diffusion. Furthermore, we will assume that the Rossby number R_{o} is much less than unity so that we may neglect the Reynolds stress relative to the Coriolis force associated with the differential rotation. The result is

$$\Omega_0 \cdot \nabla \Omega = \Omega_0 \frac{\partial \Omega}{\partial z} \approx \left\langle \frac{\nabla \rho \times \nabla p}{2\lambda \rho^2} \right\rangle \approx \frac{g}{2\lambda r c_p} \frac{\partial \langle S \rangle}{\partial \theta} \qquad (R_{\mathrm{o}} \ll 1) \qquad (5.10)$$

where g is the gravitational acceleration and c_p is the specific heat per unit mass at constant pressure. Equation (5.10) reflects a balance between the Coriolis force, pressure gradients, and buoyancy. It is a consequence of geostrophic and hydrostatic balance, but in geophysical applications it often goes under the alternative name of *thermal wind balance*. The final expression applies under the additional assumptions (justified in stellar convection zones) of an ideal gas equation of state and a background stratification that is approximately adiabatic and hydrostatic.

We first consider the case where isosurfaces of p and ρ coincide, as would be the case for a barotropic equation of state $p = p(\rho)$. Here the right-hand side of Eq. (5.10) vanishes and the angular velocity contours are constant on cylinders aligned with the rotation axis; $\partial \Omega / \partial z = 0$. This may be regarded as a stellar manifestation of the Taylor–Proudman theorem. Conversely, we can say that a rotating star with a cylindrical rotation law $\partial \Omega / \partial z = 0$ satisfying Eq. (5.10) will also possess coincident p and ρ isosurfaces, *regardless of the equation of state*. Thus, stars with $\partial \Omega / \partial z = 0$ are referred to as *pseudo-barotropes* and they dominate the early literature on stellar rotation (Tassoul, 1978). Among the most well-known of the classical results is Von Zeipel's paradox, which states that a pseudo-barotrope in a state of steady rotation cannot be in radiative equilibrium. The paradox is usually resolved by admitting meridional circulations, but even here it is notoriously difficult to achieve a stable equilibrium in a differentially rotating star (Tassoul, 1978). Even stellar radiative zones are dynamic!

More generally, the *baroclinic vector* $\nabla p \times \nabla \rho$ does not vanish, the Taylor–Proudman constraint is broken, $\Omega = \Omega(\lambda, z)$, and the star is referred to as a *barocline*. This is clearly the case for the Sun, where the nearly radial angular velocity contours imply a poleward entropy gradient; $\partial S/\partial \theta < 0$ (> 0) in the northern (southern) hemisphere (assuming the balance in Eq. (5.10) holds). The amplitude of the expected thermal variation is small, corresponding to a temperature increase of about $10\,\mathrm{K}$ from equator to pole, compared to a background temperature of $2 \times 10^6\,\mathrm{K}$ in the lower convection zone (Miesch, 2005). This is currently beyond the sensitivity limits of helioseismic structure inversions (Gough *et al.*, 1996). A thermal variation of this amplitude would be detectable in photospheric observations, but Eq. (5.10) breaks down near the surface where R_o is not small and photospheric convection may mask deep-seated thermal variations. Still, recent observations suggest that the solar photosphere may be a few degrees warmer at the poles than the equator (Rast *et al.*, 2008). The photospheric thermal variation may be larger in more rapidly rotating stars with lower Rossby numbers and stronger shear and may warrant an observational search for warm poles (Brown *et al.*, 2008).

In radiative zones, thermal diffusion along isobaric (constant p) surfaces will suppress temperature and entropy variations, diminishing the baroclinic vector and eventually moving the system toward a cylindrical rotation profile (Tassoul, 1978). This occurs on a diffusive time scale, which is of order 10^6 yr for the Sun. In convection zones, however, latitudinal entropy gradients can be sustained indefinitely, as we now address.

5.5.3 Dynamical balances

The dynamical force balances expressed by Eqs. (5.8) and (5.10) provide useful insight into how mean flows are maintained, but they are not a closed system so they do not uniquely determine the mean flow profiles. Rather, they are part of a complex non-linear interplay between convection, differential rotation, meridional circulation, and thermal gradients as outlined in Fig. 5.5.

Convection redistributes linear and angular momentum by means of the Reynolds stress, establishing differential rotation ($a \rightarrow b$ in Fig. 5.5) and meridional

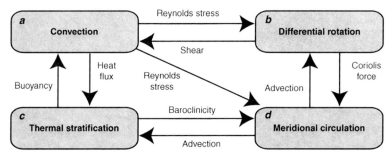

Fig. 5.5. Diagram illustrating the maintenance of mean flows in stellar convection zones. (From Miesch, 2007.)

circulations ($a \rightarrow d$). It also transports energy by means of the convective heat flux, which will in general depend on latitude as a consequence of the anisotropic nature of coherent structures in rotating convection (Section 5.2.4). This can establish latitudinal entropy gradients ($a \rightarrow c$) which can in turn induce meridional circulations by means of baroclinicity ($c \rightarrow d$) as discussed in Section 5.5.2. Meridional circulations are also established via the Coriolis acceleration of zonal flows ($b \rightarrow d$), and can feed back on the rotation profile ($d \rightarrow b$) and the thermal stratification ($d \rightarrow c$) through the advection of angular momentum and entropy. Differential rotation can shear out convection cells, inhibiting turbulent transport ($b \rightarrow a$) and changes in the stratification can affect the convection by altering the buoyancy driving ($c \rightarrow a$). Meridional circulations can in principle feed back directly on the convection as well ($d \rightarrow a$) but this is typically negligible in practice because they are relatively weak.

When considering how mean flows are established in the Sun, two potential scenarios come to mind, corresponding to a clockwise or counter-clockwise circuit around Fig. 5.5. The first begins with the establishment of rotational shear by the Reynolds stress followed by a Coriolis-induced meridional circulation that alters the thermal stratification ($a \rightarrow b \rightarrow d \rightarrow c$). However, if the Reynolds stress acts to accelerate the equator relative to higher latitudes as in the Sun, then the sense of the induced circulation in the convection zone tends to cool the poles rather than warm them (Rempel, 2005; Miesch *et al.*, 2006). The most likely outcome is for the circulation to redistribute angular momentum and entropy until it achieves a Taylor–Proudman state, minimizing the baroclinic torque and establishing a cylindrical rotation profile ($\partial\Omega/\partial z = 0$). However, as emphasized in Section 5.5.2, the solar rotation profile is not cylindrical. Rather, it is more conical, with approximately radial angular velocity contours at mid-latitudes. In short, a simple clockwise circuit will not work in the sense that it will not produce a solar-like rotation profile that satisfies Eqs. (5.8) and (5.10).

The second simple scenario is a counter-clockwise circuit whereby a latitude-dependent convective heat flux establishes warm poles, which in turn induces a baroclinic circulation that redistributes angular momentum ($a \to c \to d \to b$). However, it is clear from Eq. (5.9) and the discussion in Section 5.5.1 that this will not work either. Without a substantial Reynolds stress, a meridional circulation will tend to establish a rotation profile such that $\mathcal{L}$ is constant on streamlines ($\partial\Omega/\partial\lambda < 0$). Thermal wind balance (Eq. 5.10) will be achieved but at a heavy price; the poles will spin faster than the equator, in stark contrast to the Sun.

There are several ways out of this dilemma. The most obvious is to abandon our naive hope for a one-way street; the Reynolds stress and the convective heat flux both contribute to maintaining the solar differential rotation. The former establishes the equator–pole angular velocity contrast $\Delta\Omega$ while the latter shapes the Ω contours. Alternatively (or in addition), we can look to the boundaries for salvation. This is not as facetious as it sounds. A casual look at the solar rotation profile in Fig. 5.1b suggests where we might expect to find the largest thermal variations – in the tachocline (provided that strong toroidal fields do not disrupt thermal wind balance). It is here where the rotational shear $\partial\Omega/\partial z$ is largest. Furthermore, if a mechanical forcing tends to establish a solar-like rotational shear in the subadiabatic portion of the tachocline, the sense of the induced ciculation *does* tend to warm the poles, establishing an appropriate thermal wind balance. The convective heat flux may then transmit these thermal variations throughout the entire convection zone, promoting baroclinicity and a solar-like, conical rotation profile (Rempel, 2005; Miesch *et al.*, 2006).

The conceptual picture put forth in this and previous sections (Sections 5.5.1–5.5.3) is largely borne out by numerical simulations of solar and stellar convection (reviewed by Miesch, 2005). For solar parameter regimes, Eq. (5.10) is satisfied in the lower convection zone but breaks down in the upper convection zone where the Rossby number can exceed unity. Rotational shear is maintained by the Reynolds stress and the meridional circulation, which is in turn maintained by gyroscopic pumping, baroclinicity, and to a lesser extent, the meridional Reynolds stress. However, in most simulations, excessive viscous diffusion can throw off the delicate balance in Eq. (5.8), which may adversely affect the meridional circulation profile. More generally, the meridional circulation is a relatively weak flow, with an integrated kinetic energy roughly two orders of magnitude less than that in the convection and differential rotation, and its structure is sensitive to parameter regimes (particularly R_e), and to some extent, boundary conditions. It generally exhibits large temporal fluctuations and may possess multiple cells in latitude and radius. Still, some results are robust when averaged over at least several turnover time scales, including a poleward flow near the surface at low latitudes and an equatorward flow at mid-latitudes in the overshoot region below the base of the convection zone. For further details and discussion see Miesch (2005).

5.5.4 Stellar and planetary winds

The dynamical processes discussed in this section are not unique to stars. Indeed, much of the early theoretical work on rotating fluid systems that led to equations such as Eq. (5.5) and Eq. (5.10) was motivated not by stars but by planets, with particular emphasis on our own; these are offshoots of our continuing effort to understand the shape, weather, and ocean currents of the Earth (Tassoul, 1978; Pedlosky, 1987).

An important difference between the internal fluid dynamics of most stars and planets is that the latter often tend to be quasi-two-dimensional. This is largely a consequence of rapid rotation. Planets are smaller than stars and generally spin faster (with the exception of compact remnants such as pulsars). In the fluid cores and mantles of terrestrial planets and the extended atmospheres of many gas giant planets, the convective time scales are much longer than the rotation period, implying very low Rossby numbers (Section 5.2). As noted in Section 5.2.4 (see also Chapter 7), this gives rise to elongated, quasi-2D convective structures such that the flow is relatively invariant in the direction parallel to the rotation axis. In the atmospheres and oceans of terrestrial planets, on the other hand, quasi-2D dynamics arises simply by virtue of the geometry; global-scale horizontal motions are confined to thin spherical shells.

Quasi-2D hydrodynamic turbulence is significant with regard to mean flows because it exhibits self-organization via an *inverse cascade of kinetic energy*. As discussed in Section 5.4.1, an inverse cascade refers to local spectral transfer from small scales to large scales. In 2D hydrodynamic turbulence, small vortices merge to form larger vortices through non-linear advection. In the absence of rotation the inverse cascade can proceed to the largest scales of the system, but in rotating fluids large-scale vortices can propagate away as *Rossby waves* before they merge. Rossby waves are large-scale vortical waves that arise in rotating systems as a consequence of spherical curvature, topography, or density stratification; for further details see Chapter 15.

Suppression of the inverse cascade of kinetic energy in quasi-2D turbulence by Rossby-wave propagation is highly anisotropic; it only inhibits transfer to small latitudinal wavenumbers. Transfer to small longitudinal wavenumbers can proceed, producing alternating bands or jets of east–west (zonal) flows. The width of the bands is determined by what is known as the *Rhines scale*, after Rhines (1975) who first mapped out this transition between the 2D-turbulence and Rossby-wave regimes. This is the essential dynamics underlying a number of familiar phenomena, including oceanic jet streams on Earth and banded zonal winds on Jupiter and Saturn.[†] Uranus and Neptune rotate more slowly as quantified by a higher Rossby number. As noted in Section 5.5.1, this likely accounts for their relatively smooth wind profiles, with retrograde zonal flow at the equator (see also Aurnou *et al.*, 2007).

Another important difference between stellar interiors and planetary atmospheres is that the former are ionized and thus magnetized. Magnetism suppresses the inverse cascade of kinetic energy in 2D turbulence so zonal flows (i.e. differential rotation) in the convection zones of even the most rapidly-rotating stars likely bear little resemblance to Jovian winds.[‡]

Yet stars, like planets, likely do admit Rossby waves. Other significant commonalities between the two include internal gravity waves which transport momentum, heat, and chemical elements in stellar radiative zones as well as planetary atmospheres (Section 5.5.5; Chapter 15). In fact, angular momentum transport by breaking gravity waves in the Earth's stratosphere induces a meridional circulation by means of gyroscopic pumping in much the same manner as discussed in Section 5.5.1 (Holton *et al.*, 1995). Who would have thought that stellar plasmas and planetary atmospheres would share so much common ground?

5.5.5 The solar tachocline

Throughout this chapter we have focused on stellar convection zones because that is where most of the action is. However, convection is not the whole story when it comes to the fluid dynamics and magnetohydrodynamics of stellar interiors. Interesting dynamics also occur in stellar radiative zones, including waves, instabilities, and perhaps even dynamo processes. Nowhere is this more evident than in the solar tachocline, where the turmoil of the convective envelope meets the relative tranquility of the radiative interior.

The tachocline is an internal boundary layer of strong rotational shear $\partial\Omega/\partial r$ located near the base of the solar convection zone, identified by means of helioseismic rotational inversions (Section 5.2.3). The lower tachocline is stably stratified but the upper tachocline may overlap with the overshoot region at high latitudes (Thompson *et al.*, 2003). It is notable that the rotation rate of the radiative interior, Ω_i, is intermediate between the equatorial and polar rotation rates of the convective envelope (Fig. 5.1b). Indeed, Gilman *et al.* (1989) estimate that Ω_i is within one

---

[†] There is still a debate on whether the banded zonal flows in Jupiter and Saturn arise from deep convection or from shallow flows confined to the upper atmosphere. Either case may be interpreted in terms of the Rhines scale; see Vasavada and Showman (2005); Heimpel and Aurnou (2007).

[‡] The inverse cascade of kinetic energy in 2D turbulence relies on the local conservation of potential vorticity, which is broken in the presence of the Lorentz force; see Tobias *et al.* (2007).

percent of the value that would yield no net viscous torque across the tachocline. Yet the solar envelope loses angular momentum through the solar wind, slowly spinning down over evolutionary time scales ($\sim 10^9$ yr, Section 2.5). Thus, the intermediate value of Ω_i implies that there must be some coupling between the convection zone and the radiative interior on a time scale less than (perhaps much less than) a billion years.

A potential coupling mechanism may exist through the action of internal gravity waves. These waves are generated by penetrative convection and propagate into the radiative interior, inducing transport of angular momentum and chemical elements between regions of excitation and dissipation. As gravity waves propagate downward, rotational shear and toroidal fields in the tachocline can filter out modes with phase speeds comparable to the local rotation rate or Alfvén speed, setting up time-dependent zonal flows and residual circulations (Fritts *et al.*, 1998; Talon *et al.*, 2002; Miesch, 2005; Rogers *et al.*, 2008). However, wave transport tends to be non-local and non-diffusive so it alone cannot account for the nearly uniform rotation of the radiative interior (Gough and McIntyre, 1998).

This brings us to one of the most compelling mysteries about the tachocline, namely why it exists at all. The rotational shear in the tachocline implies thermal variations via Eq. (5.10), which will tend to spread inward due to radiative diffusion, inducing baroclinic circulations that redistribute angular momentum. This will act to suppress vertical shear and broaden the tachocline on time scales comparable to the age of the Sun (Spiegel and Zahn, 1992). Thus, the current thinness of the tachocline implies that it must be dynamically maintained. A promising candidate for keeping the tachocline thin is a fossil poloidal magnetic field in the radiative interior, that may be left over from the formation of the Sun, and that can in principle maintain uniform rotation through magnetic tension (Gough and McIntyre, 1998). However, this will only work if the field lines are confined to the radiative interior, otherwise magnetic torques will transmit the differential rotation of the envelope downward (MacGregor and Charbonneau, 1999). Furthermore, purely poloidal fields are unstable so the field in the radiative zone must also have a toroidal component (Braithwaite, 2009). Due to these and other outstanding issues, tachocline confinement remains an active area of research.

The solar tachocline is a rotating, stratified, magnetized shear layer and as such it is potentially susceptible to a host of fluid instabilities. As emphasized in Section 5.4.3, the storage, amplification, and buoyant destabilization of toroidal flux in the tachocline is an essential component of our current paradigm for how the global solar dynamo operates. The volatile mix of shear and magnetism can also trigger other instabilities, including the magnetorotational instability (Section 5.5), global tipping and clam-shell instabilities of toroidal loops, the poleward slip instability, and the Tayler instability which is a stratified magnetic pinch mode most efficient at high latitudes (Gilman and Fox, 1997; Dikpati and Gilman, 1999; Spruit, 1999; Parfrey and Menou, 2007). Such instabilities enhance the dynamical richness of the tachocline region and may influence mean flow profiles and dynamo processes in the convective envelope as well as the radiative interior. For a more thorough discussion of this and other aspects of tachocline dynamics see Miesch (2005) and Hughes *et al.* (2007).

5.6 Puzzles and prospects

The conceptual framework reviewed here has come about through a powerful synthesis of solar and stellar observations, numerical simulations, and laboratory experiments. Together these have shaped our theoretical paradigms and they will continue to do so in the years to come.

The most urgent and formidable challenge facing global convection and dynamo models is to better understand how the complex boundary layers straddling the solar convection zone influence the convection structure, mean flow profiles, and magnetic field generation within the bulk of the envelope. At the top we have the vigorous small-scale convective motions of the photospheric boundary layer (Sections 5.2.1 and 5.2.2) which may influence the buoyant driving of giant cells (Section 5.2.4), may induce a poleward circulation through gyroscopic pumping (Section 5.5.1), and may generate poloidal magnetic flux via the dispersal of bipolar active regions (Section 5.4.4). The loss of magnetic flux and helicity through the photosphere by means of magnetic buoyancy may also play an important role in large-scale dynamo action (Section 5.4.2). The concept of a local dynamo (Section 5.3) needs clarification within the context of the global dynamo below (Section 5.4) and the solar atmosphere above. One of the most fundamental touchstones to guide and challenge our understanding of global convection and its interaction with the photospheric boundary layer is the emergent radiation field itself, in particular the remarkably small variation of the solar photospheric intensity and luminosity with latitude and time (a few tenths of a percent, see Chapter 10).

At the bottom of the solar convection zone we have the solar tachocline and overshoot region, which act as a factory and repository for the toroidal magnetic flux that ultimately emerges to form photospheric active regions and sunspots (Section 5.4.3). The tachocline mediates the mechanical, thermal, and magnetic coupling between the convection and radiative zones and may influence mean flow profiles throughout the solar interior by means of baroclinicity (Section 5.5.3), magneto-shear instabilities, and internal gravity waves (Section 5.5.5). This relates to another important puzzle, namely that of tachocline confinement (Section 5.5.5). The disparate time scales compound the modeling challenge; the convective turnover time deep in the convection zone is of order a month while the radiative interior evolves on a Kelvin–Helmholtz time scale of order 30 million years.

Understanding the upper and lower boundary layers is a prerequisite for understanding other pressing questions regarding the global solar dynamo, including where the mean poloidal field is generated (different models place it either in the bulk of the convection zone, the surface layers, or the tachocline), how sunspot progentitor flux tubes are formed (Section 5.4.3), and how cyclic activity is established (see also Chapter 6). All of these questions are addressed with MHD simulations, but the latter in particular also requires ongoing observations of solar and stellar magnetic activity to make further progress. Solar observations will continue to characterize and quantify the properties of emerging flux, such as active longitudes, magnetic helicity, and temporal variability, while stellar observations will clarify how magnetic activity depends on stellar type and rotation rate, particularly with regard to dynamo saturation, magnetic topology, and cycle periods (where applicable).

Ongoing solar and stellar observations and 3D MHD simulations are also essential to understand the excitation and damping of p-modes (acoustic waves that pervade the Sun) and g-modes (internal gravity waves trapped in the solar radiative core and evanescent in the convective envelope), spectral line formation and abundance determinations (Asplund, 2005), and the complex coupling between photospheric convection and the overlying chromosphere and corona, which is largely responsible for coronal heating (Chapter 2 and Vol. I, Chapter 8). Yet another frontier is to elucidate the nature of turbulent dynamo action at low magnetic Prandtl number (Section 5.3.2) and the related issue of turbulent magnetic diffusion, which Parker (2008) has recently emphasized as a profound deficiency in our current understanding of solar dynamo theory. Future insights should arise from laboratory and numerical experiments and observations of turbulent, magnetized plasmas throughout the heliosphere.

Appendix: Physical characteristics of the solar convection zone

Table 5.1 summarizes some fundamental physical characteristics of the solar convection zone that are particularly relevant from a dynamical perspective. Fiducial radial levels are selected to represent the base of the convection zone ($r = 0.713\,R_\odot$; Christensen-Dalsgaard *et al.*, 1991), the mid convection zone ($r = 0.85\,R_\odot$), and the granulation layer near the surface ($r = 0.999\,R_\odot$). Here r is the radial coordinate and $R_\odot = 696$ Mm is the radius of the Sun. Quantities are also listed for the portion of the radiative interior immediately below the convective envelope ($r = 0.68\,R_\odot$).

Table 5.1. *Physical characteristics of the solar convection zone*

	Description	Units	Radiative interior	Convection zone base	Mid conv. zone	Granulation layer
$r/R_\odot$	Fractional radius	—	0.68	0.713	0.85	0.9995
ρ	Density	g cm^{-3}	0.25	0.19	0.054	5.1×10^{-7}
T	Temperature	K	2.5×10^{6}	2.2×10^{6}	9.6×10^{5}	1.1×10^{4}
p	Pressure	dyn cm^{-2}	8.4×10^{13}	5.7×10^{13}	7.0×10^{12}	4.0×10^{5}
$\partial S/\partial r$	Radial entropy gradient	erg g^{-1} K^{-1}cm^{-1}	7.7×10^{-3}	0	-7.0×10^{-8}	-3.2
H_p	Pressure scale height	Mm	59	57	35	0.39
H_ρ	Density scale height	Mm	81	95	58	0.29
κ	Thermal diffusivity	cm^{2} s^{-1}	1.3×10^{7}	1.2×10^{7}	3.2×10^{6}	7.8×10^{9}
ν	Kinematic viscosity	cm^{2} s^{-1}	16	15	6.7	10
η	Magnetic diffusivity	cm^{2} s^{-1}	2.0×10^{4}	2.5×10^{4}	8.5×10^{4}	6.9×10^{7}
U	Velocity scale	m s^{-1}	0.1	50	100	2000
M_a	Mach number	—	4×10^{-7}	2×10^{-4}	7×10^{-4}	0.2
R_e	Reynolds number	—	6×10^{8}	3×10^{11}	10^{13}	2×10^{12}
R_m	Magnetic Reynolds number	—	5×10^{5}	2×10^{8}	10^{9}	3×10^{5}
P_r	Prandtl number	—	10^{-6}	10^{-6}	2×10^{-6}	3×10^{-8}
P_m	Magnetic Prandtl number	—	8×10^{-4}	6×10^{-4}	8×10^{-5}	10^{-7}
R_o	Rossby number	—	2×10^{-3}	0.9	0.2	400

The density ρ, temperature T, pressure p, and radial gradient of the specific entropy per unit mass $\partial S/\partial r$ are obtained from a solar structure model, commonly referred to as Model S, which is described by Christensen-Dalsgaard *et al.* (1996) and available on the Internet.[†] The pressure and density scale heights, H_p and H_ρ are defined as $-(\partial \ln p/\partial r)^{-1}$ and $-(\partial \ln \rho/\partial r)^{-1}$ respectively.

The thermal diffusivity κ is obtained from Model S under the radiative diffusion approximation $\kappa = 4acT^3/(3\chi\rho^2 c_p)$, where $a = 7.57 \times 10^{-15}$ erg cm^{-3} K^{-4}

[†] URL: http://www.phys.au.dk/~jcd/solar_models

is the radiation pressure constant, c is the speed of light, χ is the opacity, and c_p is the specific heat per unit mass at constant pressure (Hansen and Kawaler, 1994). The kinematic viscosity η and the magnetic diffusivity are computed assuming a fully ionized hydrogen plasma where $\nu \approx 4 \times 10^{-16} T^{5/2} \rho^{-1}$ and $\eta \approx 8 \times 10^{13} T^{-3/2}$ (Spitzer, 1962). Both approximations break down near the photosphere but they will suffice as a rough estimate for illustrative purposes at $0.9995\,R_{\odot}$.

The velocity scale U listed in Table 5.1 was estimated by various means. Near the photosphere, the balance between convective and radiative energy fluxes (Eq. 5.1) yields a velocity scale $U \sim 2\,\mathrm{km\,s^{-1}}$, which is consistent with numerical simulations of solar granulation (Stein and Nordlund, 1998). According to mixing length theory, the vertical velocity in the mid convection zone should be of order $U \sim \sigma \lambda_{\mathrm{m}}$ where $\sigma^2 \equiv -g c_p^{-1} \partial S/\partial r$, λ_{m} is the mixing length and g is the gravitational acceleration. Taking $\lambda_{\mathrm{m}} \sim H_p$, and inserting values from Model S at $0.85\,R_{\odot}$ ($g = 3.8 \times 10^4\,\mathrm{cm\,s^{-2}}$, $c_p = 3.5 \times 10^8\,\mathrm{erg\,g^{-1}\,K^{-1}}$, other values from Table 5.1), we obtain $U \sim 100\,\mathrm{m\,s^{-1}}$. This is in good agreement with global convection simulations (Miesch *et al.*, 2008).

At the base of the convection zone, global convection simulations suggest $U \sim 50\,\mathrm{m\,s^{-1}}$ (Miesch *et al.*, 2008), which is comparable but somewhat less than the zonal velocity difference across the tachocline inferred from helioseismology $\Delta v_\phi \sim 100\,\mathrm{m^{-1}}$ (Thompson *et al.*, 2003). Because the rotational shear is thought to be a dynamical consequence of the convection (Section 5.5), we use former value. In the radiative interior at $0.68\,R_{\odot}$ the largest-amplitude motions are likely to be internal gravity waves excited by penetrative convection. Estimates of wave amplitudes vary widely; here we take $U \sim 10\,\mathrm{cm\,s^{-1}}$ as a fiducial value. Much weaker meridional circulations are also likely to be present, induced by rotation, rotational shear, baroclinicity, and wave damping (Tassoul, 1978; Gough and McIntyre, 1998; Fritts *et al.*, 1998).

The Mach number $M_{\mathrm{a}} = U/c_{\mathrm{s}}$ is calculated using the sound speed from Model S. The Reynolds and magnetic Reynolds numbers are defined as $R_{\mathrm{e}} = UL/\nu$ and $R_{\mathrm{m}} = UL/\eta$ where L is a characteristic length scale. In the photosphere we use the granulation scale $L \sim 1\,\mathrm{Mm}$ and in the mid convection zone we use $L \sim 100$ Mm, half the width of the convection zone. At the base of the convection zone we use $L \sim 10$ Mm, which is the approximate width of the overshoot region and perhaps the tachocline. We also use $L \sim 10\,\mathrm{Mm}$ in the radiative zone as a typical vertical wavelength for a gravity wave. The Prandtl and magnetic Prandtl numbers are defined as $P_{\mathrm{r}} = \nu/\kappa$ and $P_{\mathrm{m}} = \nu/\eta$ and the Rossby number is $R_{\mathrm{o}} = U/(2\Omega L)$ where $\Omega \sim 2.7 \times 10^{-6}$ is the rotation rate of the solar interior, which matches the surface rotation rate at mid-latitudes (Fig. 5.1).

6

Modeling solar and stellar dynamos

Paul Charbonneau

6.1 The dynamo problem

Magnetic fields are ubiquitous in stars, and in most cases they are believed to be generated contemporaneously by a dynamo mechanism powered by the inductive action of flows pervading stellar interiors. These astrophysical dynamo processes are expected to be well described by the magnetohydrodynamical (MHD) induction equation (see e.g. Davidson, 2001; Vol. I, Chapter 3):

$$\frac{\partial \mathbf{B}}{\partial t} = \nabla \times (\mathbf{v} \times \mathbf{B} - \eta \nabla \times \mathbf{B}), \qquad (6.1)$$

where $\mathbf{B}$ is the magnetic field, η is the magnetic diffusivity, inversely proportional to the plasma's electrical conductivity, and the flow field $\mathbf{v}$ includes large-scale flows such as differential rotation, as well as small-scale turbulence, if present.

In the solar/stellar dynamo context it is possible to numerically solve Eq. (6.1) together with suitable evolution equations for $\mathbf{v}$ describing turbulent, thermally driven convection in a rotating stratified sphere (or thick shell) of electrically conducting fluid. While improving steadily, such global MHD simulations have not yet reached the point where they can yield adequate models of the solar cycle (see Chapter 5). Consequently, the focus of this chapter is on simplified MHD models of dynamo action in the Sun and stars, simplified in the sense that Eq. (6.1) is solved for flows that are either prescribed, or that evolve in time in response to the time-varying magnetic field through simple non-linear response functions or through more complex equations that prescribe some dependence of $\mathbf{v}$ on $\mathbf{B}$. Such models still account for the bulk of solar and stellar dynamo models to be found in the literature, and are likely to continue to do so in the immediately foreseeable future. The following are a few technical reviews on this topic, each with a somewhat different angle on the subject: Ossendrijver (2003), Charbonneau (2005), Brandenburg (2009).

Heliophysics: Evolving Solar Activity and the Climates of Space and Earth, eds. Carolus J. Schrijver and George L. Siscoe. Published by Cambridge University Press. © Cambridge University Press 2010.

As discussed in Chapter 3 in Vol. I, the *existence* of solar and stellar magnetic fields is in itself not really surprising; any large-scale fossil field present at the time of stellar formation would still be there today at almost its initial strength, because the Ohmic dissipation time scale is extremely large for most astrophysical objects. The challenge is instead to reproduce the various observed spatiotemporal patterns reviewed in Chapter 2, most notably the cyclic polarity reversals on decadal time scales.

All solar and stellar dynamo models to be considered in this chapter operate within a sphere of electrically conducting fluid embedded in vacuum. We restrict ourselves here to axisymmetric mean-field-like models, in the sense that we will be setting and solving evolutionary equations for the large-scale magnetic field, and subsume the effects of small-scale fluid motions and magnetic fields into coefficients of these partial differential equations. Working in spherical polar coordinates (r, θ, ϕ), we begin by writing

$$\mathbf{v}(r, \theta) = \mathbf{u}_p(r, \theta) + \varpi \Omega (r, \theta)\hat{\mathbf{e}}_\phi, \tag{6.2}$$

$$\mathbf{B}(r, \theta, t) = \nabla \times (A(r, \theta, t)\hat{\mathbf{e}}_\phi) + B(r, \theta, t)\hat{\mathbf{e}}_\phi, \tag{6.3}$$

where $\varpi = r \sin \theta$, $\mathbf{u}_p$ is a notational shortcut for the component of the large-scale flow in meridional planes, and Ω is the angular velocity of rotation, which in the solar interior varies with both depth and latitude, and is now well constrained by helioseismology (e.g. Christensen-Dalsgaard, 2002). Note that in this prescription neither of these large-scale flow components is time dependent. This **kinematic approximation** is an assumption that is tolerably well supported observationally. Substituting these expressions in the MHD induction equation in Eq. (6.1) allows separation into two coupled 2D partial differential equations for the scalar functionals A and B defining respectively the poloidal and toroidal components of the magnetic field:

$$\frac{\partial A}{\partial t} = \eta \left(\nabla^2 - \frac{1}{\varpi^2} \right) A - \frac{\mathbf{u}_p}{\varpi} \cdot \nabla(\varpi A), \tag{6.4}$$

$$\frac{\partial B}{\partial t} = \eta \left(\nabla^2 - \frac{1}{\varpi^2} \right) B + \frac{1}{\varpi} \frac{\partial(\varpi B)}{\partial r} \frac{\partial \eta}{\partial r} - \varpi \nabla \cdot \left(\frac{B}{\varpi} \mathbf{u}_p \right)$$
$$+ \varpi (\nabla \times (A\hat{\mathbf{e}}_\phi)) \cdot \nabla \Omega, \tag{6.5}$$

where we retain the possibility that η varies with depth. The shearing term ($\propto \nabla\Omega$) on the right-hand side of Eq. (6.5) acts as a source of toroidal field. However, no such source term appears in Eq. (6.4). This is the essence of Cowling's theorem (see Section 3.3.8 in Vol. I), which in fact guarantees that an axisymmetric flow of the general form given by Eq. (6.2) *cannot* act as a dynamo for an axisymmetric magnetic field as described by Eq. (6.3). The construction of solar and stellar dynamo models therefore hinges critically on the addition of an extraneous source term in Eq. (6.4). The physical origin of this source term is what fundamentally distinguishes the various classes of solar and stellar dynamo models described in the following sections.

Shearing of the poloidal magnetic field into a strong toroidal component by differential rotation is an essential ingredient of all solar cycle models discussed below. The growing magnetic energy of the toroidal field is supplied by the kinetic energy of the rotational shearing motion, which makes for an attractive field amplification mechanism, because in the Sun and stars the available supply of rotational kinetic energy is immense (unless the dynamo is entirely confined to a very thin layer, for example the tachocline; on this see Steiner and Ferriz-Mas, 2005). Moreover, a strong, axisymmetric and temporally quasi-steady internal differential rotation is likely responsible for the observed high degree of axisymmetry observed in the Sun's magnetic field on spatial scales comparable to its radius. This situation is very different from that encountered in planetary core dynamos (see Chapter 7), where differential rotation is believed to be much weaker, and energetics pose a much stronger constraint on dynamo action. Lacking the large-scale organization provided by differential rotation, planetary core dynamos also tend to produce non-axisymmetric large-scale fields. The one outstanding exception appears to be Saturn, and indeed in this case the high axisymmetry of the observed surface field may well reflect the symmetrizing action of differential rotation in the envelope overlying the metallic-hydrogen core. The important point remains that, in the solar dynamo context, the assumption of an axisymmetric large-scale magnetic field is consistent with the observed and helioseismically inferred axisymmetry and quasi-steadiness of internal differential rotation.

6.2 Solar dynamo models

6.2.1 Mean-field dynamo models

Turbulence at a high magnetic Reynolds number (R_m) is known to be quite effective at producing a lot of small-scale magnetic fields, where "small-scale" is roughly $R_m^{-1/2}$ times the length scale of the flow (see Chapter 5). In addition, under certain conditions, solar/stellar convective turbulence can also produce magnetic fields with a mean component building up on large spatial scales. These **mean-field dynamo models** remain arguably the most "popular" descriptive models for dynamo action in the Sun and stars, and also in planetary metallic cores, stellar accretion disks, and even galactic disks.

Under the assumption that a good separation of scales exists between the large-scale "laminar" magnetic field **B** and the flow **v**, and the small-scale turbulent field $\mathbf{B}'$ and flow $\mathbf{v}'$, it becomes possible to express the inductive and diffusive action of the turbulence on $\mathbf{B}$ in terms of the statistical properties of the small-scale flow and field. The corresponding theory of mean-field electrodynamics is discussed in details in Chapter 3 in Vol. I. The turbulent flow introduces on the right-hand side of the induction equation (6.1) a term of the form $\nabla \times \boldsymbol{\xi}$, where $\boldsymbol{\xi} = \langle \mathbf{u} \times \mathbf{B}' \rangle$ is a mean electromotive force. For mildly inhomogeneous and near-isotropic turbulence, $\boldsymbol{\xi}$ can be expressed in terms of the large-scale field $\mathbf{B}$ as

$$\boldsymbol{\xi} = \alpha \mathbf{B} + \beta \nabla \times \mathbf{B}, \tag{6.6}$$

with

$$\alpha = -\frac{1}{3}\tau_{\mathrm{c}}\langle \mathbf{v}' \cdot (\nabla \times \mathbf{w})\rangle \ \ [\mathrm{m\,s^{-1}}], \qquad \beta = \frac{1}{3}\tau_{\mathrm{c}}\langle \mathbf{v}'^2\rangle \ \ [\mathrm{m^2\,s^{-1}}], \tag{6.7}$$

where τ_{c} is the correlation time for the turbulent flow. Note that the α-term is proportional to the (negative) kinetic helicity of the turbulence, which requires a break of reflectional symmetry. In stellar interiors and planetary metallic cores alike, this anisotropy is provided by the Coriolis force. Small-scale turbulence thus impacts the induction equation for the mean-field in two ways: it introduces a field-aligned electromotive force (the α-term), which acts as a source term and is called the "α-effect", and an enhanced "turbulent diffusion" (the β-term), associated with the folding action of the turbulent flow. In principle, the α and β coefficients can be calculated from the lowest-order statistics of the turbulent flow. In practice, more often than not they are chosen a priori, although with care taken to embody in these choices what can be learned from mean-field theory.

Under mean-field dynamo theory, Eqs. (6.4) and (6.5) are now taken to apply to an axisymmetric large-scale mean magnetic field. With the inclusion of the mean-field α-effect and turbulent diffusivity, scaling all lengths in terms of the radius R of star or planet, and time in terms of the diffusion time

$$\tau = R^2/\eta_{\mathrm{e}} \tag{6.8}$$

based on the (turbulent) diffusivity in the convective envelope, these expressions become

$$\frac{\partial A}{\partial t} = \eta \left(\nabla^2 - \frac{1}{\varpi^2} \right) A - \frac{R_m}{\varpi}\mathbf{u}_{\mathrm{p}} \cdot \nabla(\varpi A) + C_\alpha \alpha B, \tag{6.9}$$

$$\frac{\partial B}{\partial t} = \eta \left(\nabla^2 - \frac{1}{\varpi^2} \right) B + \frac{1}{\varpi}\frac{\partial(\varpi B)}{\partial r}\frac{\partial \eta}{\partial r} - R_{\mathrm{m}}\varpi \nabla \cdot \left(\frac{B}{\varpi}\mathbf{u}_{\mathrm{p}} \right)$$
$$+ C_\Omega \varpi (\nabla \times (A\hat{\mathbf{e}}_\phi)) \cdot (\nabla \Omega) + C_\alpha \hat{\mathbf{e}}_\phi \cdot \nabla \times [\alpha \nabla \times (A\hat{\mathbf{e}}_\phi)]. \tag{6.10}$$

We continue to use the symbol η for the total diffusivity, with the understanding that within the convective envelope this now includes the (dominant) contribution from the β-term of mean-field theory. Three non-dimensional numbers have materialized:

$$C_\alpha = \frac{\alpha_0 R}{\eta_e}, \qquad C_\Omega = \frac{\Omega_0 R^2}{\eta_e}, \qquad R_m = \frac{u_0 R}{\eta_e}, \qquad (6.11)$$

with α_0, u_0, and Ω_0 as reference values for the α-effect, meridional flow, and envelope rotation, respectively. The quantities C_α and C_Ω are **dynamo numbers**, measuring the importance of inductive versus diffusive effects on the right-hand side of Eqs. (6.9) and (6.10). The magnetic Reynolds number R_m here measures the relative importance of advection versus diffusion in the transport of A and B in meridional planes. Structurally, Eqs. (6.9) and (6.10) only differ from Eqs. (6.4) and (6.5) by the presence of two new source terms on the right-hand side, both associated with the α-effect. The appearance of this term in Eq. (6.9) is crucial for evading Cowling's theorem.

6.2.1.1 Linear $\alpha\Omega$ dynamo solutions

In constructing mean-field dynamos for the Sun, it has been a common procedure to neglect meridional circulation, because it is a very weak flow. It is also customary to drop the α-effect term on the right-hand side of Eq. (6.10) on the grounds that with $R \simeq 7 \times 10^8$ m, $\Omega_0 \sim 10^{-6}$ rad s^{-1}, and $\alpha_0 \sim 1$ m s^{-1}, one finds $C_\alpha/C_\Omega \sim 10^{-3}$, independently of the assumed (and poorly constrained) value for η_e. Equations (6.9) and (6.10) then reduce to the so-called $\alpha\Omega$ dynamo equations. In the spirit of producing a model that is solar-like we use a fixed value $C_\Omega = 2.5 \times 10^4$, obtained by assuming $\Omega_e \equiv \Omega_{Eq} \simeq 10^{-6}$ rad s^{-1} and $\eta_e = 5 \times 10^7$ m^2 s^{-1}, which leads to a diffusion time $\tau = R^2/\eta_e \simeq 300$ yr.

For the total magnetic diffusivity, we use a steep but smooth variation of η from a high value (η_e) in the convective envelope to a low value (η_c) in the underlying core, as described by the following convenient parameterization:

$$\frac{\eta(r)}{\eta_e} = \Delta\eta + \frac{1 - \Delta\eta}{2}\left[1 + \mathrm{erf}\left(\frac{r - r_c}{w}\right)\right], \qquad \Delta\eta = \frac{\eta_c}{\eta_e}, \qquad (6.12)$$

where $\mathrm{erf}(x)$ is the error function. A typical profile is shown in Fig. 6.1A (dash-dotted line). In practice, the core-to-envelope diffusivity ratio $\Delta\eta \equiv \eta_c/\eta_e$ is treated as a model parameter, with of course $\Delta\eta \ll 1$, since we associate η_c with the microscopic magnetic diffusivity, and η_e with the presumably much larger mean-field turbulent diffusivity. Taking at face values estimates from mean-field theory, one should have $\Delta\eta \sim 10^{-9}$ to 10^{-6}. The solutions discussed below have $\Delta\eta = 10^{-3}$ to 10^{-1}, which is still small enough to illustrate important effects of radial gradients in total magnetic diffusivity.

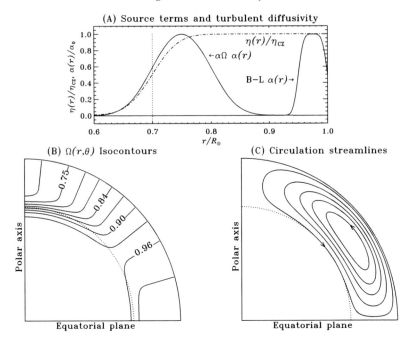

Fig. 6.1. Various "ingredients" for the dynamo models constructed in this chapter. Part (A) shows radial profiles of the total magnetic diffusivity and poloidal source terms. Part (B) shows contour levels of the rotation rate $\Omega(r, \theta)$ normalized to its surface equatorial value. The dotted line is the core–envelope interface at $r/R = 0.7$. Part (C) shows streamlines of meridional circulation, included in some of the dynamo models discussed here.

All solar dynamo models discussed in this chapter utilize the helioseismically calibrated solar-like parameterization of solar differential rotation described in Charbonneau *et al.* (1999), defined as

$$\Omega(r, \theta) = \Omega_c + \frac{\Omega_S(\theta) - \Omega_c}{2}\left[1 + \mathrm{erf}\left(\frac{r - r_c}{w}\right)\right], \qquad (6.13)$$

where

$$\Omega_S(\theta) = (1 - a_2 \cos^2\theta - a_4 \cos^4\theta), \qquad (6.14)$$

and with parameter values $\Omega_c = 0.939$, $a_2 = 0.1264$, $a_4 = 0.1591$, $r_c/R = 0.7$, and $w/R = 0.05$. The corresponding angular velocity contour levels are plotted in Fig. 6.1B. Such a solar-like differential rotation profile is quite complex from the point of view of dynamo modeling, in that it is characterized by multiple partially overlapping shear regions: a rotational shear layer, straddling the core–envelope interface, known as the **tachocline**, with a strong positive radial shear in its equatorial regions and an even stronger negative radial shear in its polar regions, as well as a significant latitudinal shear throughout the convective envelope and extending partway into the tachocline. For a tachocline of half-thickness $w/R_\odot = 0.05$, the mid-latitude latitudinal shear at $r/R_\odot = 0.7$ is comparable in magnitude to the equatorial radial shear, and its potential contribution to toroidal field production cannot be casually dismissed.

For the dimensionless functional $\alpha(r, \theta)$ we use an expression of the form

$$\alpha(r, \theta) = \frac{1}{4}\left[1 + \mathrm{erf}\left(\frac{r - r_c}{w}\right)\right]\left[1 - \mathrm{erf}\left(\frac{r - 0.8}{w}\right)\right]\cos\theta. \qquad (6.15)$$

This concentrates the α-effect in the bottom half of the envelope, and lets it vanish smoothly below, just as the net magnetic diffusivity does (see Fig. 6.1A). Various lines of argument point to an α-effect peaking in the bottom half of the convective envelope, because there the convective turnover time is commensurate with the solar rotation period, a most favorable setup for the type of toroidal field twisting at the root of the α-effect (see Fig. 3.5 in Vol. I). Likewise, the $\cos\theta$ dependency reflects the hemispheric dependence of the Coriolis force, which also suggests that the α-effect should be positive in the northern hemisphere.

The dimensionless number C_α, which measures the strength of the α-effect, is treated as a free parameter of the model.

With α, η and the large-scale flows given, the mean-field dynamo equations become linear in A and B. With none of the coefficients in these partial differential equations depending explicitly on time, one can seek eigensolutions of the form

$$\begin{bmatrix} A(r, \theta, t) \\ B(r, \theta, t) \end{bmatrix} = \begin{bmatrix} a(r, \theta) \\ b(r, \theta) \end{bmatrix} e^{\lambda t}, \qquad (6.16)$$

where $\lambda = \sigma + i\omega$. Solutions are sought in a meridional plane of a sphere of radius R, and are matched, for convenience, to a potential field in the exterior ($r/R > 1$). Regularity requires $A(r, 0) = A(r, \pi) = 0$ and $B(r, 0) = B(r, \pi) = 0$ on the symmetry axis. The boundary condition on the equatorial plane then sets the parity of the solution.

Substituting Eq. (6.16) into the $\alpha\Omega$ dynamo equations yields a classical linear eigenvalue problem, with σ interpreted as the growth rate and ω as the cyclic frequency of the corresponding eigenmodes, both expressed in units of the inverse diffusion time $\tau^{-1} = \eta_e/R^2$. In a model for the (oscillatory) solar dynamo, we are looking for solutions where $\sigma > 0$ and $\omega \neq 0$. In such linear $\alpha\Omega$ models the onset of dynamo activity ($\sigma > 0$) turns out to be controlled by the *product* of C_α and C_Ω:

$$D \equiv C_\alpha \times C_\Omega = \frac{\alpha_0 \Omega_0 R^3}{\eta_e^2}, \qquad (6.17)$$

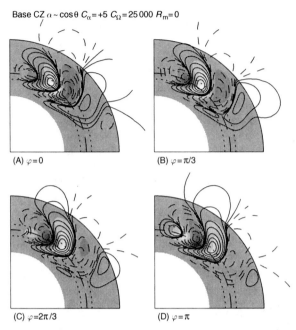

Fig. 6.2. Four snapshots in meridional planes of our minimal linear $\alpha\Omega$ dynamo solution with defining parameters $C_\Omega = 25\,000$ (see Eq. 6.11), $\Delta\eta = 0.1$ (see Eq. 6.12), $\eta_e = 5 \times 10^7$ m^2 s^{-1}. With $C_\alpha = +5$, this is a mildly supercritical solution, with oscillation frequency $\omega \simeq 300\,\tau^{-1}$. The toroidal field is plotted as filled contours (grey to black for negative B, grey to white for positive B, normalized to the peak strength and with increments $\Delta B = 0.2$), on which poloidal field lines are superimposed (solid for clockwise-oriented field lines, dashed for counter-clockwise orientation). The long-dashed line is the core–envelope interface at $r_c/R = 0.7$.

with positive growth rates materializing above a threshold value known as the **critical dynamo number**. The eigenvalue for the model defined above depends very little on the solution parity.

Figure 6.2 shows half a cycle of the dynamo solution, in the form of snapshots of the toroidal (grey scale) and poloidal eigenfunctions (field lines) in a meridional plane, with the symmetry axis defined by the stellar rotation oriented vertically. The four frames are separated by a phase interval $\varphi = \pi/3$, so that panel (D) is identical to panel (A) except for reversed magnetic polarities in both magnetic components. Such linear eigensolutions leave the absolute magnitude of the magnetic field undetermined, but the relative magnitude of the poloidal to toroidal components is found to scale approximately as $|C_\alpha/C_\Omega|$.

The eigenmode is concentrated in the vicinity of the core–envelope interface, and has very little amplitude in the underlying, low-diffusivity radiative core. This is due to the oscillatory nature of the solution, which restricts penetration into the core to a distance of the order of the electromagnetic skin depth $\ell = \sqrt{2\eta_c/\omega}$, where η_c is the core diffusivity. Having assumed $\eta_e = 5 \times 10^7 \, \mathrm{m^2 \, s^{-1}}$, with $\Delta\eta = 0.1$, a dimensionless dynamo frequency $\omega \simeq 300$ corresponds to $3 \times 10^{-8} \, \mathrm{s^{-1}}$, so that $\ell/R \simeq 0.026$, quite small indeed.

Careful examination of Fig. 6.2A–D also reveals that the toroidal/poloidal flux systems present in the shear layer first show up at high latitudes, and then *migrate equatorward* to finally disappear at mid-latitudes in the course of the half-cycle. These **dynamo waves** travel in a direction **s** given by

$$\mathbf{s} = \alpha \nabla\Omega \times \hat{\mathbf{e}}_\phi, \qquad (6.18)$$

i.e. along contours of equal angular velocity, a result known as the **Parker–Yoshimura sign rule** (Parker, 1955a; Yoshimura, 1975; Stix, 1976). Here with a negative $\partial\Omega/\partial r$ in the high-latitude region of the tachocline, a positive α-effect results in an equatorward propagation of the dynamo wave, in qualitative agreement with the observed equatorward drift of the latitudes of sunspot emergences as the solar cycle unfolds (see Fig. 2.2).

6.2.1.2 Kinematic $\alpha\Omega$ models with α-quenching

Obviously, the exponential growth characterizing supercritical ($\sigma > 0$) linear solutions must stop once the Lorentz force associated with the growing magnetic field becomes dynamically significant for the inductive flow. Because the solar surface and internal differential rotation show little variation with the phase of the solar cycle, it is usually assumed that magnetic back-reaction occurs at the level of the α-effect. In the mean-field spirit of *not* solving dynamical equations for the small scales, it has become common practice to introduce an ad hoc algebraic non-linear quenching of α directly on the mean toroidal field B by writing:

$$\alpha \rightarrow \alpha(B) = \frac{\alpha_0}{1 + (B/B_{eq})^2}, \qquad (6.19)$$

where $B_{eq} = (\mu_0 \rho u_t^2)^{1/2}$ is the equipartition field strength, of order $\sim 0.1\,\mathrm{T}$ (or $10^4 \, \mathrm{G}$) at the base of the solar convective envelope. Needless to say, this simple α-quenching formula is an *extreme* oversimplification of the complex interaction between flow and field that is known to characterize MHD turbulence, but its wide usage in solar dynamo modeling makes it the non-linearity of choice for the illustrative purpose of this chapter.

Introducing α-quenching in our model renders the $\alpha\Omega$ dynamo equations nonlinear, so that solutions are now obtained as initial-value problems starting from an arbitrary seed field of very low amplitude, in the sense that $B \ll B_{eq}$ everywhere in the domain. Figure 6.3 shows time series of the total magnetic energy in the simulation domain for a sequence of α-quenched solutions with increasing C_α. At

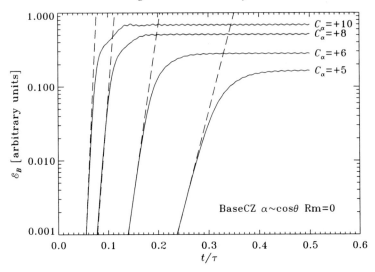

Fig. 6.3. Time series of magnetic energy for a set of $\alpha\Omega$ dynamo solutions using our minimal $\alpha\Omega$ model including algebraic α-quenching, and different values for dynamo number C_α (Eq. 6.11), as labeled. Magnetic energy is expressed in arbitrary units; time is expressed in units of the diffusion time scale (Eq. 6.8). The dashed line indicates the exponential growth phase characterizing the linear regime.

early times, $B \ll B_{\mathrm{eq}}$ and the equations are effectively linear, leading to exponential growth (dashed lines). Eventually however, B becomes comparable to B_{eq} in the region where the α-effect operates, leading to a break in exponential growth, and eventual saturation. The saturation energy level increases with increasing C_α, an intuitively satisfying behavior because solutions with larger C_α have a more vigorous poloidal source term. The cycle frequency for these solutions is very nearly independent of the dynamo number, and is slightly *smaller* than the frequency of the linear critical mode (here by some 10–15%), a behavior that is typical of kinematic α-quenched mean-field dynamo models. Yet the overall form of the dynamo solutions very closely resembles that of the linear eigenfunctions plotted in Fig. 6.2.

6.2.1.3 Interface dynamos

The α-quenching expression in Eq. (6.19) implies that dynamo action saturates once the mean, dynamo-generated large-scale magnetic field reaches an energy density comparable to that of the driving small-scale turbulent fluid motions. However, various calculations and numerical simulations have indicated that long before the mean toroidal field B reaches this strength, the helical turbulence reaches equipartition with the *small-scale*, turbulent component of the magnetic field. Such calculations also suggest that the ratio between the small-scale and mean magnetic components should itself scale as $R_{\mathrm{m}}^{1/2}$, where $R_{\mathrm{m}} = v\ell/\eta$ is a magnetic Reynolds number based on the turbulent speed but *microscopic* magnetic diffusivity. This then leads to the alternative quenching expression

$$\alpha \rightarrow \alpha(B) = \frac{\alpha_0}{1 + R_{\mathrm{m}}(B/B_{\mathrm{eq}})^2}, \tag{6.20}$$

known in the literature as *strong α-quenching* or *catastrophic quenching* (see Chapter 3 in Vol. I, and Section 5.4.2). Because $R_{\mathrm{m}} \sim 10^8$ in the solar convection zone, this leads to quenching of the α-effect for very low amplitudes of the mean magnetic field, of order 10^{-5} T. Even though significant field amplification is likely in the formation of a toroidal flux rope from the dynamo-generated magnetic field, we are now a very long way from the 1–10 T demanded by simulations of buoyantly rising flux ropes and sunspot formation.

A way out of this difficulty exists in the form of **interface dynamos** (Parker, 1993; MacGregor and Charbonneau, 1997; Tobias, 1997; Petrovay and Kerekes, 2004; Mason *et al.*, 2008). The idea is beautifully simple: to produce and store the toroidal field away from where the α-effect is operating. Parker (1993) showed that in a situation where a radial shear and α-effect are segregated on either side of a discontinuity in magnetic diffusivity taken to coincide with the core–envelope interface, the constant coefficient, Cartesian forms of the $\alpha\Omega$ dynamo equations support solutions in the form of traveling surface waves localized on the discontinuity in diffusivity. For supercritical dynamo waves, the ratio of peak toroidal field strength on either side of the discontinuity surface is found to scale as $(\eta_e/\eta_c)^{-1/2}$. With the core diffusivity η_c equal to the microscopic value, and if the envelope diffusivity is of turbulent origin so that $\eta_e \sim \ell v$, then the toroidal field strength ratio scales as $\sim (v\ell/\eta_c)^{1/2} \equiv R_{\mathrm{m}}^{1/2}$. This is precisely the factor needed to bypass strong α-quenching, at least as embodied in Eq. (6.20).

As an illustrative example, Fig. 6.4A shows a series radial cuts of the toroidal magnetic component at 15° latitude (grey curves), spanning half a cycle in a numerical spherical $\alpha\Omega$ interface-like solution. This is the same model as before, except that the differential rotation and α-effect profiles have been slightly altered so that they each peak at a finite distance on either side of the core–envelope interface, and the α-effect has been concentrated in the equatorial regions.

This model does achieve the kind of toroidal field amplification one would anticipate for interface dynamos. Notice how the toroidal field peaks below the core–envelope interface (vertical dotted line), well below the α-effect region and near the peak in radial shear. Figure 6.4B shows how the ratio of peak toroidal field below and above r_c varies with the imposed diffusivity contrast $\Delta\eta$. The dashed line shows the expected $(\eta_e/\eta_c)^{-1/2}$ dependency. For relatively low diffusivity contrast, $-1.5 \lesssim \log(\Delta\eta) \lesssim 0$, both the toroidal field ratio and dynamo period increase as $\sim (\Delta\eta)^{-1/2}$. Below $\log(\Delta\eta) \sim -1.5$, the max($B$)-ratio increases more slowly,

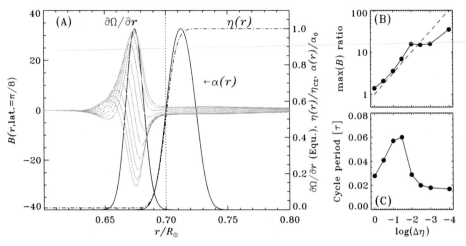

Fig. 6.4. A representative interface dynamo model in spherical geometry. This solution has dynamo numbers $C_\Omega = 2.5 \times 10^5$ and $C_\alpha = +10$ (see Eq. 6.11), and a core-to-envelope diffusivity contrast of 10^{-2}. Panel (A) shows a series of radial cuts of the toroidal field at latitude 15° (grey lines). The (normalized) radial profiles of magnetic diffusivity, α-effect, and radial shear are also shown, again at latitude 15°. The core–envelope interface is at $r/R = 0.7$ (dotted line), where the magnetic diffusivity varies near-discontinuously. Panels (B) and (C) show the variations of the core-to-envelope peak toroidal field strength and dynamo period with the diffusivity contrast, for a sequence of otherwise identical dynamo solutions.

and the cycle period falls, as can be seen in Fig. 6.4C. This is an electromagnetic skin-depth effect; unlike in the original picture proposed by Parker, here the poloidal field must diffuse down a finite distance into the tachocline before shearing into a toroidal component can commence. With this distance set by our adopted profile of $\Omega(r, \theta)$, as $\Delta\eta$ becomes very small there comes a point where the dynamo period is such that the poloidal field cannot diffuse as deep as the peak in radial shear in the course of a half cycle. The dynamo then runs on a weaker shear, thus yielding a smaller field strength ratio and weaker overall cycle.

6.2.1.4 $\alpha\Omega$ models with meridional circulation

Meridional circulation is unavoidable in turbulent, rotating convective shells (see Section 5.5). The $\sim 15\,\mathrm{m\,s^{-1}}$ poleward flow observed at the surface (Hathaway, 1996) has been detected helioseismically, down to $r/R_\odot \simeq 0.85$ without significant departure from the poleward direction, except locally and very close to the surface, in the vicinity of active region belts (see Gizon, 2004, and references therein). Mass conservation evidently requires an equatorward flow deeper down.

Meridional circulation can bodily transport the dynamo-generated magnetic field (terms $\propto \mathbf{u}_p \cdot \nabla$ in Eqs. (6.4) and (6.5)). At low circulation speeds, the primary effect is a Doppler shift of the dynamo wave, leading to a small change in the cycle period and equatorward concentration of the activity belts (Roberts and Stix, 1972). However, for a (presumably) solar-like equatorward return flow that is vigorous enough, it can overpower the Parker–Yoshimura propagation rule and produce equatorward propagation no matter what the sign of the α-effect is (Choudhuri *et al.*, 1995; Küker *et al.*, 2001). The behavioral turnover from dynamo wave-like solutions sets in when the circulation speed in the dynamo region becomes comparable to the propagation speed of the dynamo wave. In this advection-dominated regime, the cycle period loses sensitivity to the assumed turbulent diffusivity value, and becomes determined primarily by the circulation's turnover time. Solar cycle models achieving equatorward migration of activity belts in this manner are often called **flux transport dynamos**.

Panels A through F in Fig. 6.5 show meridional-plane snapshots spanning half a cycle of our $\alpha\Omega$ model, now including a meridional flow within the convective envelope which is poleward at the surface and equatorward at the core–envelope interface (see Fig. 6.1C), as defined by the convenient parametric form designed by van Ballegooijen and Choudhuri (1988). Advective transport of the magnetic field by meridional circulation is clearly apparent, and carries the toroidal field to low latitudes. Note also how poloidal field experiences strong stretching in the latitudinal direction within the tachocline (panels C through F) as a direct consequence of shearing – in addition to plain transport – by the equatorward flow. One interesting consequence is that induction of the toroidal field is now effected primarily by the *latitudinal* shear within the tachocline, with the radial shear, although larger in magnitude, playing a lesser role because $B_r/B_\theta \ll 1$. The meridional flow also has a profound impact on the magnetic field evolution at $r = R$, as it concentrates the poloidal field in the polar regions. This leads to a large amplification factor through magnetic flux conservation, so that dynamo solutions such as shown in Fig. 6.5 are typically characterized by very large polar field strengths, here some 20% of the toroidal field magnitude in the tachocline, even though we have here $C_\alpha/C_\Omega = 10^{-6}$. This concentrated poloidal field, when advected downwards to the polar regions of the tachocline, is responsible for the strong polar branch often seen in the time–latitude diagram of dynamo solutions including a rapid meridional flow. This difficulty can be alleviated, at least in part, by a number of relatively minor modifications to the model, such as the addition of a high-η subsurface layer, or displacement of the meridional flow cell towards lower latitudes, thus reducing the degree of polar convergence, as done for example in the "calibrated" solar cycle model of Dikpati *et al.* (2004).

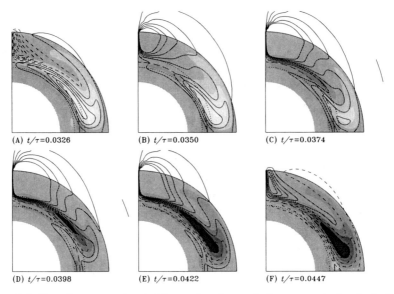

(A) t/τ=0.0326   (B) t/τ=0.0350   (C) t/τ=0.0374

(D) t/τ=0.0398   (E) t/τ=0.0422   (F) t/τ=0.0447

Fig. 6.5. Snapshots covering half a cycle of an $\alpha\Omega$ dynamo solution including meridional circulation, starting at the time of polarity reversal in the polar surface field (time in units of the diffusion time scale in Eq. 6.8). The grey-scale coding of the toroidal field and poloidal field lines is as in Fig. 6.2. This α-quenched solution uses the same differential rotation, diffusivity, and α-effect profiles as in Fig. 6.2, with parameter values $C_\alpha = 0.5$, $C_\Omega = 5 \times 10^5$, $\Delta\eta = 0.1$, $R_m = 2500$. Note the strong amplification of the surface polar fields and the latitudinal stretching of poloidal field lines by the meridional flow at the core–envelope interface.

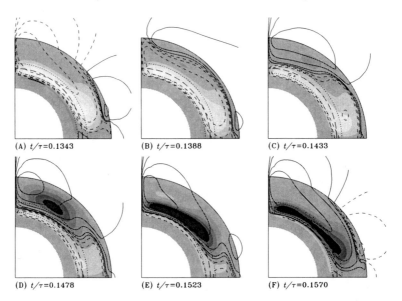

(A) t/τ=0.1343   (B) t/τ=0.1388   (C) t/τ=0.1433

(D) t/τ=0.1478   (E) t/τ=0.1523   (F) t/τ=0.1570

Fig. 6.6. As Fig. 6.5 for a Babcock–Leighton dynamo solution. This solution uses the same differential rotation, magnetic diffusivity and meridional circulation profile as for the advection-dominated $\alpha\Omega$ solution of Section 6.2.1.4, but now with the non-local surface source term defined through Eq. (6.21), with parameter values $C_\alpha = 5$, $C_\Omega = 5 \times 10^4$, $\Delta\eta = 0.003$, $R_m = 840$. Note again the strong amplification of the surface polar fields, and the latitudinal stretching of poloidal field lines by the meridional flow at the core–envelope interface.

It is noteworthy that to produce a butterfly-like time–latitude diagram of the toroidal field at the core–envelope interface, the required value of R_m in conjunction with the observed surface meridional flow speed and reasonable profile for the internal return flow, ends up requiring a rather low envelope magnetic diffusivity, $\lesssim 10^7 \, \mathrm{m^2 \, s^{-1}}$, which stands at the very low end of the range suggested by mean-field estimates such as provided by Eq. (6.7). Nonetheless, kinematic $\alpha\Omega$ mean-field models including meridional circulation and simple algebraic α-quenching can produce equatorially concentrated and equatorward propagating dynamo modes with a decadal period resembling that of the solar cycle for realistic, solar-like differential rotation and circulation profiles.

6.2.2 Solar cycle models based on active region decay

On the basis of his groundbreaking magnetographic observations of the Sun's magnetic field outside of sunspots, Babcock (1961) suggested that the polarity reversals of the high-latitude surface magnetic field are driven by the accumulation of magnetic fields released at low latitudes by the decay of bipolar magnetic regions. Figure 2.3 shows a numerical simulation illustrating this process, which leads to the buildup of a net poloidal hemispheric flux because the trailing member of the pair tends to be located at higher latitudes than the leading component, a pattern known as Joy's law (see also Sections 2.2.1.1 and 2.2.1.4), and therefore are subjected to less transequatorial dissipative flux cancellation than the leading members of the bipolar pair. Babcock went on to argue that, in conjunction with shearing by differential rotation, this could explain the observed patterns of solar cycle polarity reversals. In subsequent years Leighton (1964, 1969) turned this idea into a *bona fide* solar cycle model, known since as the Babcock–Leighton model.

That polarity reversals of the solar polar magnetic field occur in this manner has received strong support from later magnetographic observations. The crucial question is whether this is a mere side-effect of dynamo action taking place independently somewhere in the solar interior, or a dominant contribution to the dynamo process itself. The key point, from the dynamo perspective, is that the Babcock–Leighton mechanism taps into the (formerly) toroidal flux in the bipolar magnetic region to produce a poloidal magnetic component, and so can act as a source term on the right-hand side of Eq. (6.4). In the solar case, a $\sim 10^{-3} \, \mathrm{T}$ field pervading a polar cap of $\sim 30°$ angular width adds up to a poloidal magnetic flux of $\sim 10^{14} \, \mathrm{Wb}$. The total unsigned flux emerging in active regions, taken to be representative of the solar internal toroidal magnetic component, adds up to a few $10^{17} \, \mathrm{Wb}$ over a full sunspot cycle. The $\sim 0.1\%$ "efficiency" required in converting active region magnetic flux into a dipole moment is therefore quite modest.

To the degree that a positive dipole moment is being produced from a toroidal field that is positive in the northern hemisphere, this is operationally equivalent to a positive α-effect in mean-field theory. In both cases the Coriolis force is the agent imparting a twist on a magnetic field; with the α-effect this process occurs on the small spatial scales and operates on individual magnetic field lines. In contrast, the Babcock–Leighton mechanism operates on the large scales, the twist being imparted via the Coriolis force acting on the flow generated along the axis of a buoyantly rising magnetic flux tube that, upon emergence, gives rise to sunspot pairs.

Numerous dynamo models based on this mechanism of poloidal field regeneration have been constructed, based on the axisymmetric mean-field dynamo equations but with the α-effect replaced by a suitably designed source term on the right-hand side of Eq. (6.9) (e.g. Wang *et al.*, 1991; Durney, 1995, 2000; Dikpati and Charbonneau, 1999; Nandy and Choudhuri, 2001). One important difference from the mean-field $\alpha\Omega$ models considered earlier is that the two source regions are now spatially segregated: production of the toroidal field takes place in or near the tachocline, as before, but now production of the poloidal field is restricted to the surface layers. A transport mechanism is then required to link the two source regions for a dynamo loop to operate. Following Wang *et al.* (1989; see also Wang and Sheeley, 1991; Dikpati and Choudhuri, 1995), most incarnations of the Babcock–Leighton dynamo ascribe this key role to the meridional flow, which ends up acting as a form of conveyor belt, concentrating to high latitudes the surface magnetic field released by the decay of active regions, and dragging it down to the tachocline where shearing by differential rotation leads to the buildup of a new toroidal flux system, and thus to the onset of a new sunspot cycle.

The crux in formulating a mean-field-like Babcock–Leighton solar cycle model is to design a suitable source term to add to the right-hand side of Eq. (6.4). For the illustrative purposes of this section, we follow Charbonneau *et al.* (2005), and use a non-local α-effect-like source term with upper and lower operating threshold in toroidal field strength:

$$S(r, \theta, B(t)) = s_0 f(r) \sin\theta \cos\theta$$

$$\times \operatorname{erf}\left(\frac{B(r_c, \theta, t) - B_1}{w_1}\right)\left[1 - \operatorname{erf}\left(\frac{B(r_c, \theta, t) - B_2}{w_2}\right)\right] B(r_c, \theta, t). \quad (6.21)$$

The function $f(r)$ now concentrates the source term in the surface layers (see Fig. 6.1B), and the latitudinal dependency reflects the fact that the Babcock–Leighton mechanism is most efficient for flux ropes emerging at mid-latitudes.

The non-locality in B represents the fact that the strength of the source term is proportional to the field strength in the bipolar active region, itself presumably reflecting the strength of the diffuse toroidal field near the core–envelope interface, where the magnetic flux ropes originate that eventually give rise to the bipolar active regions. The combination of error functions restricts the operating range of the model to a finite interval in toroidal field strength. This is motivated by simulations of the stability and of the buoyant rise of thin flux tubes, which indicate that toroidal flux ropes require a minimal strength of $\sim$1 tesla for destabilization, and emerge at the surface without the tilt necessary for the Babcock–Leighton mechanism to operate if B exceeds a few tens of teslas (see Fan, 2004, and references therein).

Figure 6.6 shows a series of meridional-plane snapshots of one such Babcock–Leighton dynamo solution, covering one sunspot cycle and starting approximately at sunspot maximum (based on magnetic energy as a proxy for sunspot number). Surface poloidal flux from the current cycle has begun to build up at low latitudes, and is rapidly swept to the pole, with polarity reversal of the polar field taking place shortly thereafter (panel B). As with the advection-dominated $\alpha\Omega$ solution of the preceding section, this solution is characterized by strong surface polar fields resulting from the poleward transport by the meridional flow of the poloidal component produced at lower latitudes, and the equatorward propagation of the toroidal field in the tachocline is also driven by the meridional flow. The turnover time of the meridional flow is here again the primary determinant of the cycle period. With $\eta = 3 \times 10^7 \, \mathrm{m}^2 \, \mathrm{s}^{-1}$, this solution has a nicely solar-like half-period of 12.4 yr. All in all, this is once again a reasonable representation of the cyclic spatiotemporal evolution of the solar large-scale magnetic field.

Although it exhibits the desired equatorward propagation, the toroidal field in Fig. 6.6 peaks at higher latitude ($\sim$45°) than suggested by the sunspot butterfly diagram ($\sim$15°–20°). This occurs because this is a solution with high magnetic diffusivity contrast, where meridional circulation closes at the core–envelope interface, so that the *latitudinal* component of differential rotation dominates the production of the toroidal field. Note in Fig. 6.6D how the toroidal component starts building up already in the convective envelope, and later gets further amplified upon reaching the tachocline (going from panel D to E). This difficulty can be alleviated by letting the meridional circulation penetrate below the core–envelope interface, but this often leads to the production of a strong polar branch, again a consequence of both the strong radial shear present in the high-latitude portion of the tachocline, and of the concentration of the poloidal field taking place in the high-latitude surface layer prior to this field being advected down into the tachocline by meridional circulation (see Fig. 6.6B–D).

6.2.3 Models based on (magneto)hydrodynamical instabilities

In the presence of stratification and rotation, a number of hydrodynamical (HD) and magnetohydrodynamical (MHD) instabilities associated with the presence of a strong toroidal field in the stably stratified, radiative portion of the tachocline can lead to the growth of disturbances with a net helicity, which under suitable circumstances can produce a toroidal electromotive force, and therefore act as a source of poloidal field. Different types of solar cycle models have been constructed in this manner. In nearly all cases the resulting dynamo models end up being described by something closely resembling the axisymmetric mean-field dynamo equations, the novel poloidal field regeneration mechanisms being once again subsumed in an α-effect-like source term appearing of the right-hand side of Eq. (6.9).

Hydrodynamical stability analyses of the latitudinal shear profile in the solar tachocline indicate that the latter may be unstable to non-axisymmetric perturbations, with the instability planforms characterized by a net (negative) kinetic helicity (Dikpati and Gilman, 2001; Dikpati *et al.*, 2003). With inspiration provided by Eq. (6.7), this yields an azimuthally averaged (positive) α-effect-like source term that is directly proportional to the large-scale toroidal component, just as in mean-field electrodynamics. Dikpati and Gilman have presented various kinematic axisymmetric dynamo solutions relying on this poloidal field regeneration mechanism. Their model yields dynamo activity in the mid to low latitudes of the tachocline, and operates as a flux transport dynamo, with the equatorward meridional flow overpowering the dynamo wave that would otherwise propagate towards the pole, as per the Parker–Yoshimura sign rule.

The thin-flux-tube approximation can be used to study the stability of toroidal flux ropes stored immediately below the base of the convection zone, and determine the conditions under which they can be destabilized and give rise to sunspots (Fan, 2004, and references therein). Such calculations indicate that once the tube destabilizes, due to the influence of rotation the correlation between the non-axisymmetric flow and field perturbations is such as to yield a mean azimuthal electromotive force, which acts as a source of poloidal magnetic field (Ferriz-Mas *et al.*, 1994; Ossendrijver, 2000b). Although it has not yet been comprehensively studied, this dynamo mechanism has a number of very attractive properties. It operates without difficulty in the strong field regime (in fact in *requires* strong fields to operate). It also naturally yields dynamo action concentrated at low latitudes. Both analytic calculations and numerical simulations suggest a *positive* α-effect equivalent in the northern hemisphere, which should then produce *poleward* propagation of the dynamo wave at low latitudes. Meridional circulation could then perhaps produce equatorward propagation of the dynamo magnetic field even with a positive α-effect, as it does in true mean-field models (see Section 6.2.1.4).

6.3 Modeling the solar cycle

Given that the basic physical mechanism(s) underlying the operation of the solar cycle (see Section 2.3.2) are not yet agreed upon, attempting to understand the origin of the observed *fluctuations* of the solar cycle may appear to be a futile undertaking. Nonetheless, work along these lines continues at full steam, in part because of the high stakes involved: the Sun's radiative output and frequencies of all geo-effective eruptive phenomena relevant to space weather are strongly modulated by the amplitude of the solar cycle; varying levels of solar activity may contribute significantly to climate change; and certain aspects of the observed fluctuations may actually hold important clues as to the physical nature of the dynamo process.

6.3.1 Cycle modulation through stochastic forcing

Sources of stochastic "noise" abound in the solar interior; large-scale flows in the convective envelope, such as differential rotation and meridional circulation, are observed to fluctuate, an unavoidable consequence of dynamical forcing by the surrounding, vigorous turbulent flow. This convection is known to produce its own small-scale magnetic field (see Section 5.3), which amounts to a form of rapidly varying zero-mean additive "noise" superimposed on the slowly evolving mean magnetic field. In addition, the azimuthal averaging implicit in all models of the solar cycle considered earlier yields dynamo coefficients showing significant deviations about their mean values, as a consequence of the spatiotemporally discrete nature of the physical events (e.g. cyclonic updrafts, sunspot emergences, flux rope destabilizations, etc.) whose collective effects add up to produce a mean electromotive force. A straightforward way to model the impact of these fluctuations is to let the dynamo number fluctuate randomly in time about some pre-set mean value $\bar{C}_\alpha$:

$$C_\alpha \to \bar{C}_\alpha + \rho \times \delta C, \qquad \rho \in [-1, 1], \qquad \text{if}(t \bmod \tau) = 0. \qquad (6.22)$$

By most statistical estimates (see e.g. Hoyng, 1993), the expected magnitude of these fluctuations is quite large, i.e. many times the mean value, a conclusion also supported by numerical simulations (Ossendrijver *et al.*, 2001; Hoyng, 1993; Otmianowska-Mazur *et al.*, 1997). One typically also introduces a **coherence time** (τ) during which the dynamo number retains a fixed value. At the end of this time interval, this value is randomly readjusted. Depending on the dynamo model at hand, the coherence time can be related to the lifetime of convective eddies (α-effect-based mean-field models), to the decay time of sunspots (Babcock–Leighton models), or to the growth rate of instabilities (HD shear or buoyant MHD instability-based models).

The effect of stochastic forcing varies according to the type of dynamo model being forced, but some common trends nonetheless emerge (see e.g. Choudhuri, 1992; Moss *et al.*, 1992; Ossendrijver *et al.*, 1996). In most models stochastic forcing or noise increases both the average amplitude and duration of cycles. It also introduces long-time-scale modulations in the overall cycle amplitudes, "long" in the sense of being much longer than the coherence time for the noise and/or forcing, and often significantly longer than the cycle period itself. Often this can be traced to the production and storage of strong magnetic fields in the low-diffusivity

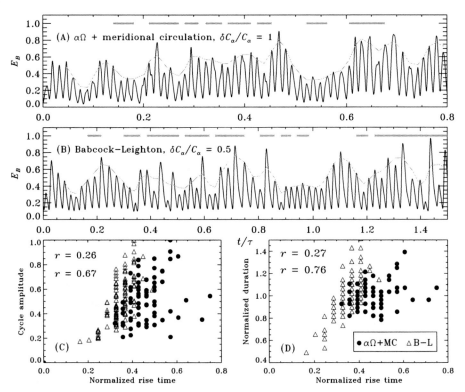

Fig. 6.7. Impact of stochastic fluctuations of the C_α dynamo number (Eq. 6.11) on the behavior of the advection-dominated mean-field $\alpha\Omega$ solution of Fig. 6.5, with 100% forcing of the poloidal dynamo number, and of the Babcock–Leighton solution of Fig. 6.6, with 50% fluctuation of the surface source term. In both cases the coherence time is ~5% of the cycle period. Panels (A) and (B) shows a time series for the magnetic energy (thick solid line), together with a 1-2-1 running mean of the peak amplitudes (grey line). The horizontal grey bars indicate epochs where a Gnevyshev–Ohl-like pattern holds. Panels (C) and (D) are Waldmeier-rule-like correlation plots between cycle peak, rise time and duration with cycle peak and duration normalized to the corresponding values in the parent non-fluctuating solution.

regions of the domain, below the core–envelope interface, where the resistive decay time can be quite long.

Figure 6.7 shows some representative results for the advection-dominated mean-field $\alpha\Omega$ model of Section 6.2.1.4, and for the Babcock–Leighton model of Section 6.2.2. In both cases the total magnetic energy (thick solid lines on panels A and B) is used as a proxy for sunspot number. These two specific stochastically forced solutions were selected because they exhibit a number of solar-like features, including relative ranges of variations in cycle amplitudes ($\pm$ ~40% of the mean) and duration ($\pm$ ~20% of the mean), amplitude modulation patterns spanning many cycles, and shorter-lived Dalton-minimum-like intervals of markedly reduced amplitude.

Both of these solutions do fairly well at reproducing Gnevyshev–Ohl-like alternating patterns of variations in cycle amplitude about their running mean. This is illustrated by the grey horizontal line segments, which flag epochs where cycle amplitudes alternate regularly above and below the 1-2-1 running mean of cycle amplitudes, plotted as grey lines on panels A and B. That such sequences should exist is in itself not surprising; purely random numbers would distribute themselves symmetrically about their mean, so that Gnevyshev–Ohl-like patterns can materialize only by chance. What is striking here is the distributions of durations for these epochs, which can greatly exceed (especially here in the Babcock–Leighton solution) what one could rightfully expect from Poisson statistics.

Such stochastically forced dynamo models, including the two shown in Fig. 6.7, do produce a moderate positive correlation between cycle rise time and duration (see Fig. 6.7D). The situation is not as good with regards to the observed anti-correlation between cycle amplitude and rise time (or duration) embodied in the Waldmeier rule. Whether forced stochastically through the dynamo number or via additive noise in the surface layers, most of the kinematic models considered here end up producing a positive correlation (rather than an anti-correlation) between these two cycle parameters.

6.3.2 Cycle modulation through the Lorentz force

The dynamo-generated magnetic field, in general, produces a Lorentz force that tends to oppose the driving fluid motions. This is a basic physical effect that should be included in any dynamo model. It is not at all trivial to do so, however, because in a turbulent environment both the fluctuating and mean components of the magnetic field can affect the flow on all spatial scales. Introducing magnetic back-reaction on differential rotation and/or meridional circulation rapidly becomes a tricky business, because one must then also, in principle, provide a model for the Reynolds stresses powering the large-scale flows in solar/stellar convective envelopes, as well as a procedure for computing magnetic back-reaction on these. This rapidly leads into the unyielding realm of MHD turbulence. A more practical approach consists in dividing the large-scale flow into two components, the first (**v**) corresponding to some prescribed, steady profile, and the second (**v**′) to a time-dependent flow field driven by the Lorentz force:

$$\mathbf{v} = \mathbf{v}_0(\mathbf{x}) + \mathbf{v}'(\mathbf{x}, t, \mathbf{B}),\qquad(6.23)$$

with the (non-dimensional) governing equation for **v**′ including only the Lorentz force and a viscous dissipation term on its right-hand side. If **v** amounts only to differential rotation, then **v**′ must obey a (non-dimensional) differential equation of the form

$$\frac{\partial \mathbf{v}'}{\partial t} = \frac{\Lambda}{4\pi\rho}(\nabla \times \mathbf{B}) \times \mathbf{B} + P_{\mathrm{m}}\nabla^2\mathbf{v}',\qquad(6.24)$$

where time has been scaled according to the magnetic diffusion time $\tau = R^2/\eta_{\mathrm{e}}$, as before (see e.g. Tobias, 1997; Bushby, 2006). Two dimensionless parameters appear in Eq. (6.24), the critical one being the **magnetic Prandtl number** $P_{\mathrm{m}} = \nu/\eta$, which measures the relative importance of viscous and Ohmic dissipation. When $P_{\mathrm{m}} \ll 1$, large velocity amplitudes in **v**′ can be produced by the dynamo-generated mean magnetic field. This effectively introduces an additional, long time scale in the model, associated with the evolution of the magnetically driven flow; for smaller P_{m}, the time scale increases. The resulting non-linear, non-kinematic dynamo solutions can exhibit a variety of behaviors, including amplitude and parity modulation, periodic or aperiodic, hemispheric asymmetries, as well as intermittency (see e.g. Sokoloff and Nesme-Ribes, 1994; Beer *et al.*, 1998; Küker *et al.*, 1999; Moss and Brooke, 2000). It has been argued that amplitude modulation in such models can be divided into two main classes: Type-I modulation corresponds to a non-linear interaction between modes of different parity, with the Lorentz force-mediated flow variations controlling the transition from one mode to another; Type-II modulation refers to an exchange of energy between a single dynamo mode (of some fixed parity) with the flow field. This leads to quasi-periodic modulation of the basic cycle, with the modulation period controlled by the magnetic Prandtl number. Both types of modulation can coexist in a given dynamo model, leading to a rich overall dynamical behavior (see Knobloch *et al.*, 1998).

The differential rotation can also be suppressed indirectly by magnetic back-reaction on the *small-scale* turbulent flows that produce the Reynolds stresses driving the large-scale mean flow. Inclusion of this so-called "Λ-quenching" in mean-field dynamo models, alone or in conjunction with other amplitude-limiting non-linearities, has also been shown to lead to a variety of periodic and aperiodic amplitude modulations, provided the magnetic Prandtl number is small (see e.g. Küker *et al.*, 1999). These models stand or fall with the turbulence model used to compute the various mean-field coefficients, and it is not yet clear which aspects of the results are truly generic to Λ-quenching.

The dynamical back-reaction of the large-scale magnetic field on meridional circulation has received comparatively little attention. The calculations of Rempel (2006) suggest that diffuse toroidal magnetic fields of strength up to 0.1 T can probably be advected equatorward at the core–envelope interface. That it can indeed do so is crucial in *any* flux transport model relying on the meridional flow to produce equatorward propagation of magnetic fields as the cycle unfolds.

6.3.3 Cycle modulation through time delays

In solar cycle models based on the Babcock–Leighton mechanism of poloidal field generation, meridional circulation effectively sets – and even regulates – the cycle period. In doing so, it also introduces a long time delay in the dynamo mechanism, "long" in the sense of being comparable to the cycle period. This delay originates with the time required for circulation to advect the surface poloidal field down to the core–envelope interface, where the toroidal component is produced. In contrast, the production of poloidal field from the deep-seated toroidal field is a "fast" process, with growth rates and buoyant rise times for sunspot-forming toroidal flux ropes being of the order of a month.

The long delay turns out to have important dynamical consequences. This was first explored using a very simple formulation of the Babcock–Leighton model, reducing the dynamo equation to a simple one-dimensional iterative map (Durney, 2000; Charbonneau, 2001), with closely analogous behavior later found in a kinematic axisymmetric Babcock–Leighton dynamo model essentially identical to that described in Section 6.2.2 (Charbonneau *et al.*, 2007, and references therein). It is indeed quite remarkable that a model including a simple algebraic amplitude-limiting non-linearity can produce the full range of non-linear behavior normally associated with truly non-linear, non-kinematic models, including period doubling, transition to chaos, and intermittency.

6.3.4 Intermittency

The term "intermittency" refers to systems undergoing apparently random, rapid switching from quiescent to bursting behaviors, as measured by the magnitude of some suitable system variable. In the context of solar cycle models, intermittency is invoked to explain the existence of Maunder minimum-like quiescent epochs of strongly suppressed activity randomly interspersed within periods of "normal" cyclic behavior.

Intermittency has been shown to occur through stochastic fluctuations of the dynamo number in mean-field dynamo models operating at or near criticality (see e.g. Ossendrijver and Hoyng, 1996). This mechanism for "on-off intermittency" works well, however there is no strong reason to believe that the solar dynamo is running just at criticality, so that is not clear how good an explanation this is of Maunder-type grand minima. Parity modulation driven by stochastic noise can also lead to a form of intermittency, by exciting the higher-order modes that perturb the normal operation of the otherwise dominant dynamo mode, producing strongly reduced cycle amplitudes (see e.g. Mininni and Gómez, 2002).

Another way to trigger intermittency in a dynamo model is to let nonlinear dynamical effects, for example a reduction of the differential rotation amplitude, push the effective dynamo number below its critical value; dynamo action then ceases during the subsequent time interval needed to re-establish differential rotation following the diffusive decay of the magnetic field; in the low P_m regime, this time interval can amount to many cycle periods, but P_m must not be too small, otherwise grand minima become too rare. Values of $P_m \sim 10^{-2}$ seem to work best. Such intermittency is again most readily produced when the dynamo is operating close to criticality (see e.g. Küker *et al.*, 1999; Brooke *et al.*, 2002). Similarly, intermittency can be observed in strongly supercritical models including α-quenching as the sole amplitude-limiting non-linearity (e.g. Tworkowski *et al.*, 1998). Such solutions can enter grand minima-like epochs of reduced activity when the dynamo-generated magnetic field completely quenches the α-effect. The dynamo cycle restarts when the magnetic field decays back resistively to the level where the α-effect becomes operational once again.

Intermittency can also arise naturally in dynamo models characterized by a lower operating threshold on the magnetic field. These include models where the regeneration of the poloidal field takes place via the MHD instability of toroidal flux tubes (Section 6.2.3), or the Babcock–Leighton mechanism (Section 6.2.2). In such models, the transition from quiescent to active phases requires an external mechanism to push the field strength back above threshold. This can be stochastic noise (e.g. Schmitt *et al.*, 1996; Charbonneau *et al.*, 2004), or a secondary large-scale dynamo process, such as an α-effect-based dynamo in the convective envelope, normally overpowered by the "primary" dynamo during active phases (Ossendrijver, 2000a).

Figure 6.8 shows a representative intermittent Babcock–Leighton solution, taken from Charbonneau *et al.* (2004). The top panel shows a sample trace of the tachocline toroidal field, and the bottom panel a time–latitude "butterfly" diagram of the toroidal magnetic field constructed at the core–envelope interface in the model. This is a dynamo solution which, in the absence of noise, operates in the

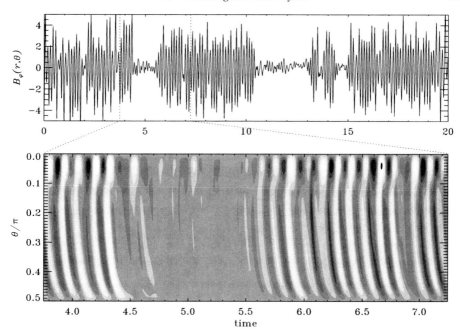

Fig. 6.8. Intermittency in a dynamo model based on the Babcock–Leighton mechanism (see Section 6.2.2). The top panel shows a trace of the toroidal field sampled at $(r, \theta) = (0.7, \pi/3)$. The bottom panel is a time–co-latitude diagram for the toroidal field at the core–envelope interface (with positive and negative toroidal field strengths shown lighter or darker, respectively, relative to a grey zero point).

singly periodic regime. Stochastic noise is added to the vector potential $A\hat{\mathbf{e}}_\phi$ in the surface layers, and the dynamo number is also allowed to fluctuate randomly about a pre-set mean value. The resulting solution exhibits both amplitude fluctuations and intermittency.

With its strong polar branch often characteristic of dynamo models with meridional circulation, Fig. 6.8 is not a particularly good fit to the sunspot butterfly diagram. Yet its fluctuating behavior is solar-like in a number of ways, including epochs of alternating higher-than-average and lower-than-average cycle amplitudes and residual pseudo-cyclic variations during quiescent phases, as suggested by [10]Be data through the Maunder minimum (Beer *et al.*, 1998). Here this feature is due at least in part to meridional circulation, which continues to advect the (diffusively decaying) magnetic field after the dynamo has fallen below threshold.

6.3.5 Model-based cycle forecasts

Because the Sun's magnetic cycle modulates its radiative output as well as the frequency of all geo-effective solar eruptive phenomena, predicting the characteristics of future sunspot cycles remains a very active research area (see Hathaway *et al.*, 1999 for a review). It is of course possible to treat this forecasting problem as a pure exercise in time series analysis and forecasting, without any physical input. The sunspot number time series is just a time series, and it can be extended using a number of techniques coming from statistics or dynamical system theory. To this day, forecasts based on such techniques have not fared significantly better than so-called "climatological" forecasts, which consist in simply "predicting", for instance, that the next cycle has the same amplitude as the current cycle, or an amplitude equal to the mean cycle amplitude over the length of the sunspot record, etc. In this section, we focus instead on forecast schemes that are based, in one form or another, on dynamo models. Towards this end, it is necessary to first answer the following basic questions:

(i) What type of dynamo powers the solar cycle: $\alpha\Omega$, $\alpha^2\Omega$, interface, Babcock–Leighton, etc.?
(ii) Which mechanism drives cycle duration and cycle-amplitude fluctuations: stochastic forcing, non-linear modulation due to the Lorentz force, time delay, etc.?
(iii) How do we "forecast" the sunspot number from a dynamo solution that describes the spatiotemporal evolution of a diffuse, large-scale magnetic field?

Remarkably, we are still not anywhere near being in a position to provide a reliable answer to any of these questions. However, we do expect the dynamo to operate on the existing magnetic field, therefore trying to forecast the next cycle using characteristics of the current cycle (and maybe recent past cycles as well) is definitely justified. This is the physical underpinning of all so-called "precursor methods".

6.3.5.1 The solar polar magnetic field as a precursor

Temporally extended synoptic magnetogram series suggest that the solar cycle can be divided into sequences of sub-steps whereby a poloidal field (P) produces a new toroidal component (T), which then leads to the buildup of a new poloidal component, with accompanying polarity reversals; schematically:

$$\cdots \rightarrow P(+) \rightarrow T(-) \rightarrow P(-) \rightarrow T(+) \rightarrow P(+) \rightarrow \cdots . \qquad (6.25)$$

This suggests that the optimal precursor for the amplitude of the sunspot-generating toroidal component should be sought by moving back up the causal chain by one sub-step, to the poloidal component produced in the previous sunspot cycle. This is the basis for the set of dynamo-inspired precursor schemes pioneered by Schatten *et al.* (1978), and elaborated in a variety of ways since (see e.g. Svalgaard *et al.*, 2005; Schatten, 2005).

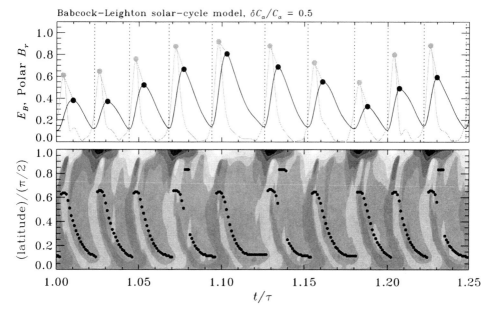

Fig. 6.9. Portion of a simulation run of a Babcock–Leighton model, with fluctuations at the ±50% level imposed in the magnitude of the surface source term. The unperturbed reference solution is that illustrated in Fig. 6.6. The top panel shows time series of the surface radial magnetic field sampled at the pole (grey), together with a time series of the total magnetic energy (black), used here as a proxy for the sunspot number. The dark grey line segment joins the peak poloidal field at (or near) "sunspot minimum" with the peak in the sunspot number proxy for the following cycle. The bottom panel is a time–latitude diagram of the the surface radial field, and the black dots trace the latitude of peak toroidal field strength at the core–envelope interface at the corresponding time.

The various dynamo models considered in this chapter can be used to test this idea, as illustrated in Fig. 6.9 for the stochastically forced Babcock–Leighton solution of Fig. 6.7B. The top panel shows a short segment of the time series for the magnetic energy, used – once more – as a proxy for the sunspot number, together with a time series of the surface polar field strength (in grey). The bottom panel shows a time–latitude diagram of the surface radial magnetic component, together with the latitudes of peak toroidal field strength at the core–envelope interface, where sunspots are presumed to originate (filled disks). The overall spatiotemporal evolution of the surface field, and its phase relationship to the deep-seated toroidal field, are both remarkably solar-like. Examination of the two proxies on the top panels of Fig. 6.9 reveals that the surface radial field peaks shortly following what one would identify with solar minimum on the basis of our proxy for the sunspot number. It is then a simple matter to pair the peak polar field at solar minimum with the sunspot number proxy of the following cycle, as indicated in Fig. 6.9 by the

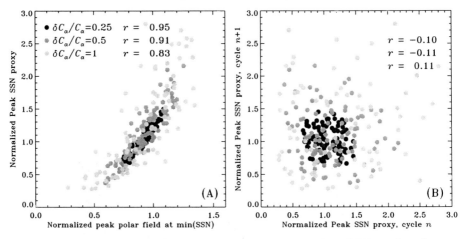

Fig. 6.10. (A) Correlations between peak sunspot number (SSN) and surface poloidal field strength in the stochastically forced Babcock–Leighton solutions of Fig. 6.6, for three different levels $\delta C_\alpha / C_\alpha$ of forced stochastic fluctuations in the surface source term, grey-coded as indicated and listing the corresponding linear correlation coefficient r. The red dots corresponds to the simulation run illustrated in Fig. 6.9. (B) Scatter diagram between the peak in sunspot number proxy for successive pairs of cycles. All amplitudes are normalized to those characterizing the non-fluctuating parent simulation.

connecting grey line segments. The resulting correlation plot is shown in Fig. 6.10, for three different levels of stochastic forcing, as grey-coded dots, with the solution of Fig. 6.9 in the darker shade of grey. In all cases, the time series for the sunspot number proxies and surface poloidal field strength have been normalized to the peak values characterizing a parent run without stochastic forcing.

The peak polar field at solar minimum clearly has a precursor value, but stochastic forcing rapidly degrades the forecasting accuracy. Consider, for example, the solution with 100% fluctuation of the dynamo number C_α (light grey); while a linear correlation coefficient of 0.83 may sound pretty good, the fact remains that a polar field of 0.8 (say) in the normalized units of Fig. 6.10 would lead to a sunspot number forecast covering a very broad range, namely 0.60–1.2 in normalized units, which is not a particularly accurate forecast.

Performing the same analysis on our other solar cycle model reveals that the polar surface field is also a good precursor of cycle amplitude for the advection-dominated mean-field model with meridional circulation of Section 6.2.1.4, but no precursor at all for the classical $\alpha\Omega$ model of Section 6.2.1.2. This curious situation can be traced to the fact that, in the former, the surface polar field does feed back into the dynamo loop, as circulation drags it down back into the tachocline, where it merges with the poloidal field produced there by the α-effect (see Fig. 6.5). In the circulation-free models, on the other hand, the poloidal field diffuses more or less radially outwards to the surface, with poloidal field of the subsequent cycle being generated completely independently at the base of the envelope (see Fig. 6.2). In retrospect, the logic between Schatten *et al.*'s precursor argument can be understood to hold only for a subset of dynamo models, namely those where some "feedback" of the surface polar field on the dynamo loop takes place. If the surface field is only a passive relic of dynamo action taking place independently in the deep interior, then it has no precursor value.

It is noteworthy that in all stochastically forced solar cycle models considered here, the value for the peak sunspot number proxy has little or no precursor value in forecasting the next sunspot number proxy peak. This is shown in Fig. 6.10B, and can be traced to the manner in which stochasticity is introduced in the model. In the case of imposed stochastic fluctuations in the poloidal source term, the scheme given by Eq. (6.25) must be replaced by something like

$$\cdots \longrightarrow P(+) \longrightarrow T(-) \xrightarrow{\text{stoch}} P(-) \longrightarrow T(+) \xrightarrow{\text{stoch}} P(+) \longrightarrow \cdots . \quad (6.26)$$

Precursor forecasting based on either component is only possible if the forecast does not traverse a "stochastic" dynamo sub-step.

6.3.5.2 Model-based forecasting using solar data

Some solar cycle amplitude forecasts have been made using solar cycle models of the Babcock–Leighton variety (Section 6.2.2), in conjunction with input of solar magnetic field observations. It is particularly instructive to compare and contrast the forecast schemes (and cycle 24 forecasts) of Dikpati *et al.* (2006) on the one hand, and Choudhuri *et al.* (2007) on the other, because these two schemes are so remarkably similar. They both use kinematic axisymmetric Babcock–Leighton-type models, with essentially the same differential rotation, magnetic diffusivity, and meridional circulation profiles. They do differ in their formulation of the poloidal source term, solar data used to drive the model, and manner in which this driving is implemented. Yet these two forecasting schemes end up producing cycle 24 amplitude forecasts that stand at opposite ends of the very wide range of cycle 24 forecasts produced by other techniques, as well as opposite ends of the range of past cycle amplitudes. The SSN = 150–180 cycle 24 forecast of Dikpati *et al.* (2006) would put it nearly on par with the highest cycle amplitudes on record, while SSN = 80, as forecasted by Choudhuri *et al.* (2007), would place it amongst the weakest of the past century. In the context of Babcock–Leighton models, the overall approach is definitely viable in principle, because the solar surface magnetic field is that which serves as seed to produce the sunspot-generating toroidal component of the next cycle. The one thing that these two model-based forecasting attempts have demonstrated, beyond any doubts, is that modeling details matter (see also Bushby and Tobias, 2007; Cameron and Schüssler, 2007, 2008; Yeates *et al.*, 2008).

6.4 Stellar dynamos

In going from the solar cycle to stellar dynamos, observational constraints diminish precipitously, and theoretical considerations take on an enlarged role. Theory does inform us that in the high-R_m regime pertaining to most astrophysical situations, the inductive action of convective turbulence is very good at producing magnetic fields. It also offers strong indications that in conjunction with stratification and rotation, and especially differential rotation, convection can also produce large-scale magnetic fields. So, offhand we are not in too bad a shape with regards to stellar dynamos. Stars certainly are stratified, and certainly rotate. Moreover, thermally driven convection is also present across a large part of the H-R diagram, but here we start to encounter complications that restrict the use of the solar exemplar.

Figure 2.10 illustrates, in schematic form, the internal structure of main-sequence stars, more specifically the presence or absence of convection zones (see also Section 2.3.1). A G-star like the Sun has a thick outer convection zone, spanning the outer 30% in radius. As one moves to lower masses, the relative thickness of the convective envelope increases until, somewhere around spectral type M5, stars become fully convective. Moving from the Sun to higher masses, the convective envelope becomes ever thinner, until somewhere around spectral type A0 it essentially vanishes. However, at around the same spectral type hydrogen burning switches from the p-p chain to the CNO cycle, for which nuclear reaction rates are much more sensitively dependent on temperature. Core energy release becomes strongly depth dependent, leading to convectively unstable temperature gradients. The resulting small convective core grows in size as one moves up to larger masses. In an early B-star of solar metallicity, the convective core spans the inner 25% or so in radius of the star.

6.4.1 Early-type stars

Main-sequence stars of the *O* and *B* types combine vigorous core convection and high rotation rates, which makes dynamo action more than likely. This expectation has been amply confirmed by 3D MHD numerical simulations of dynamo action in the convective cores of massive stars (see Brun *et al.*, 2005). For illustrative purposes we first consider kinematic axisymmetric core-dynamo models relying exclusively on the α-effect acting as a source term on the right-hand side of both Eqs. (6.4) and (6.5). Such mean-field models are known as α^2 dynamos. In the spirit of the other dynamo models discussed in this chapter, we adopt simple parametric profiles for α and η:

$$\alpha(r, \theta) = \frac{1}{2}\left[1 + \mathrm{erf}\left(\frac{r - r_{\mathrm{c}}}{w}\right)\right]\cos(\theta),\qquad(6.27)$$

$$\eta(r) = \eta_{\mathrm{e}} + \frac{\eta_{\mathrm{c}} - \eta_{\mathrm{e}}}{2}\left[1 - \mathrm{erf}\left(\frac{r - r_{\mathrm{c}}}{w}\right)\right].\qquad(6.28)$$

Equations (6.27) represent once again "minimal" assumptions on the spatial dependency of the α-effect: it changes sign across the equator ($\theta = \pi/2$), and vanishes rapidly outside of the convective core ($r \gtrsim r_{\mathrm{c}}$), the transition occurring across a spherical layer of thickness $\sim 2w$. With convective turbulence vanishing outside the nuclear burning core, we consider a situation where the core-to-envelope diffusivity contrast $\Delta\eta = \eta_{\mathrm{c}}/\eta_{\mathrm{e}}$ is now much larger than unity.

Linear kinematic dynamo models of the α^2 type usually produce *steady* magnetic fields, i.e. the solution eigenvalue is purely real ($\omega = 0$ in Eq. (6.16); but do see Rüdiger *et al.*, 2003). Figure 6.11 shows a sequence of three typical solutions with increasing diffusivity contrasts between the core and envelope, all with a negative dynamo number C_α; changing the latter's sign yields morphologically identical eigenmodes, except for a change in the relative signs of the poloidal and toroidal components. Growth rates are here of order 10 yr in dimensional units, leaving no doubt that ample time is available to amplify a weak seed magnetic field in the core of a massive star. The introduction of α-quenching leaves these eigenmodes essentially unaltered.

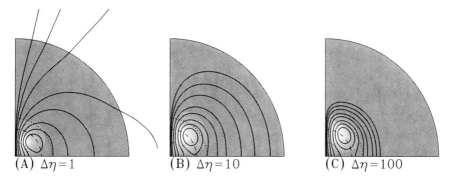

(A) $\Delta\eta = 1$ \qquad (B) $\Delta\eta = 10$ \qquad (C) $\Delta\eta = 100$

Fig. 6.11. Three antisymmetric steady α^2 dynamo solutions, computed using increasing magnetic diffusivity ratios between the core and envelope. The solutions are plotted in a meridional quadrant, with the symmetry axis coinciding with the left quadrant boundary. Poloidal field lines are plotted superimposed on a grey scale representation for the toroidal field (dark to light is weaker to stronger field). The dashed line marks the core–envelope interface depth r_{c}, and the two dotted lines indicate the depths $r_{\mathrm{c}} \pm w$ corresponding to the width of the transition layer between core and envelope.

If differential rotation exists within the convective core of massive stars, or at the core–envelope boundary, then additional modes of mean-field dynamo action become possible, including oscillating solutions of the $\alpha\Omega$ or $\alpha^2\Omega$ type. In such models the availability of an additional energy source leads to dynamo solutions with stronger magnetic field, where the toroidal field strength in general exceeds that of the poloidal field, unlike in the α^2 case where both poloidal and toroidal components have similar magnitudes.

All these core dynamo models have one thing in common: the large η_c/η_e diffusivity contrast leads to a "trapping" of the magnetic field in the lower part of the radiative envelope, a direct consequence of the difficulty experienced by an externally imposed magnetic field in diffusively penetrating a good electrical conductor. For the α^2 solutions plotted in Fig. 6.11, the surface-to-core poloidal field strength ratio falls rapidly from $\sim 10^{-2}$ at $\Delta\eta = 1$ to 3×10^{-4} at $\Delta\eta = 10^{-1}$, and way down to $\lesssim 10^{-8}$ at $\Delta\eta = 10^{-3}$. With the electromagnetic skin depth restricting even more the penetration of oscillating magnetic fields generated in the core into the overlying envelope, $\alpha\Omega$ and $\alpha^2\Omega$ dynamo models yield vanishingly small surface fields (see Charbonneau and MacGregor, 2001; also Brun *et al.*, 2005).

This long-recognized property of stellar core dynamos represents a rather formidable obstacle to be bypassed if the magnetic fields generated by dynamo action in convective cores are to become observable at the stellar surface (Levy and Rose, 1974). As discussed in Schuessler and Paehler (1978), the situation is in fact even worse than the eigensolutions of Fig. 6.11 may suggest. In a time-dependent situation where the core dynamo "turns on" at or shortly before the arrival on the zero-age main sequence, the time needed for the magnetic field to resistively diffuse to the surface can become larger than the star's main-sequence lifetime, for masses in excess of about 5 $M_\odot$.

An additional magnetic flux transport mechanism linking the interior to the surface is evidently needed. An appealing possibility is that the dynamo-generated magnetic field manages to produce toroidal flux ropes that then rise buoyantly to the surface. The analogy with the Sun becomes even more compelling if a tachocline-like rotational shear layer were to exist at the boundary between the inner convective core and outer radiative envelope. However, and unlike in the solar case, here the toroidal flux ropes are rising through a *stably* stratified environment, and so lose their buoyant force as they rise, because as they rise they cool faster than the surrounding stratification. Calculations performed in the thin flux tube approximation by MacGregor and Cassinelli (2003) suggest that such toroidal flux ropes, assuming they do form, could rise perhaps halfway across the radiative envelope, but are unlikely to make it all the way to the surface through buoyancy alone.

6.4.2 A-type stars.

Stars with spectral types ranging from late-B to early-F stand out as the least likely to support dynamo action, because they lack a convective region of substantial size. This squares well with various lines of observations; in particular, main-sequence A-stars are amongst the most "magnetically quiet" stars in the H-R diagram. A sub-set of late-B and A stars, namely the slowly rotating, chemically peculiar Ap/Bp stars, do show strong magnetic fields, but even those show no sign of anything even mildly analogous to solar activity. The single pattern of temporal evolution noted is a decrease, by factors of two to three, in the overall strength of the surface field, most prominent in the early stages of main-sequence evolution. This seems compatible with the idea of diffusive decay of residual higher-degree eigenmodes, and slow decreases associated with flux conservation as the stars slowly expand in the course of their main-sequence evolution. For these reasons, the fossil field hypothesis remains the favored explanatory model for the magnetic field of Ap stars. It is also quite striking that the high field strength observed in Ap stars (a few T), in magnetized white dwarfs ($\sim 10^5$ T), and in the most intensely magnetized neutron stars ($\sim 10^{11}$ T) all amount to approximately the same total surface magnetic flux, $\sim 10^{19}$ Wb, lending support to the idea that these high fields can be understood from simple flux-freezing arguments (see Chapter 3 in Vol. I).

Dynamo-based explanations for the magnetic fields of Ap stars certainly exist. Recent years have witnessed renewed interest in the possibility that dynamo action could take place in the radiative envelope of massive stars, through turbulence associated with the so-called Tayler–Spruit global instability of the magnetic field (see Spruit, 2002; Braithwaite, 2006; also Arlt *et al.*, 2003). This idea has attracted attention outside of the dynamo circles because the associated turbulent transport would also cause enhanced chemical mixing, known to be required to properly fit evolutionary tracks of massive stars, but whose origin remains mysterious (see Maeder and Meynet, 2003, 2005, and references therein). However, it appears that sustained dynamo action by this instability requires an external mechanism to maintain significant differential rotation in the radiative envelope, and this is certainly unlikely in the very slowly rotating Ap/Bp stars.

6.4.3 Solar-type stars

Until strong evidence to the contrary is brought to the fore, we are allowed to assume that late-type stars with a thick convective envelope overlying a radiative core host a solar-type dynamo. Observationally, a lot of what we know regarding dynamo activity in solar-type stars comes from the Mount Wilson Ca H+K survey. Two important pieces of information can be extracted from these data, as constraints on dynamo models. The first is the overall level of Ca H+K emission, which is taken as a measure of overall photospheric magnetic field strength, consistent with what one observes on the Sun at various phases of its activity cycle. As discussed in Section 2.3.2 (Fig. 2.11), this is found to increase with rotation up to 5–10 times the solar rotation rate, after which saturation sets in. The second is of course the cycle period, for the subset of stars in which a regular cycle can be detected (see Fig. 2.12). Cycle periods for stars of varying spectral types can be uniquely related not to the rotation rate per se, but rather to the Rossby number R_o, namely the ratio of the rotation period to the convective turnover time, through a power-law relation of the form $P_{cyc} \propto R_o{}^n$, with power-law index $n = 1.25 \pm 0.5$ (Noyes *et al.*, 1984b). In essence, the Rossby number measures the efficiency of the Coriolis force in breaking the mirror symmetry of convective turbulence, and thus producing a non-zero α-effect. Recall also that the larger the dynamo number, the more magnetic energy mean-field models can produce (see Fig. 6.3). So, in a rough qualitative sense, observations do fit our (naive) theoretical expectations.

At the modeling level, we are facing a number of difficulties in extrapolating solar dynamo models to stars other than the Sun. At the very least, we need to be able to specify:

  (i) how the form and magnitude of differential rotation and meridional circulation change with rotation rate and luminosity, the latter determining the magnitude of convective velocities, and thus the magnitude of the turbulent Reynolds stresses powering the large-scale flows important for dynamo action;

 (ii) how the α-effect and turbulent diffusivities vary in stars with rotation rate and convection zone properties;

(iii) how the process of sunspot formation (Babcock–Leighton models) varies with changing convection zone depth, rotation, etc.

Even in the Sun, we do not really know for sure what mechanism is responsible for the regeneration of the poloidal magnetic component. How then can we hope to go about modeling stellar dynamos with anything resembling confidence? The problem can be turned around, in that stellar observations can perhaps be used to distinguish between different classes of dynamo models. The possibility hinges, for instance, on the distinct dependency of the cycle period on model parameters in various models. For example, in the simple α-quenched mean-field solutions discussed in Section 6.2.1.2, the (dimensionless) cycle period is, to a first approximation, independent of the C_α dynamo number, so that in dimensional units the cycle period scales as

$$P_{cyc} \propto \eta^{-1}, \qquad (\alpha\text{-quenched basic } \alpha\Omega \text{ model}), \qquad (6.29)$$

where η is the (assumed) turbulent diffusivity; in other types of mean-field models dependencies on the strength of the α-effect and differential rotation can appear, and these dependencies can be sensitively dependent on the type of non-linearity saturating the dynamo amplitude (Tobias, 1998). Nonetheless, the primary parameter dependency usually remains on the turbulent diffusivity (see e.g. Saar and Brandenburg, 1999). On the other hand, in Babcock–Leighton dynamo models the cycle period is set primarily by the turnover time of the meridional flow. For example, in the Dikpati and Charbonneau (1999) model, the cycle period is found to vary as

$$P_{\text{cyc}} \propto u_0^{-0.89} s_0^{-0.13} \eta^{-0.22}, \qquad \text{(Babcock–Leighton)}, \qquad (6.30)$$

where u_0 is the surface meridional flow speed, and s_0 is the parameter measuring the magnitude of the Babcock–Leighton source term. Unfortunately, using this relationship in conjunction with observed stellar cycle data requires one to specify how the meridional flow speed varies with rotation, which currently remains highly uncertain on the theoretical front. But this offers at least a path ahead.

The preponderance of strong magnetic field concentrated at high latitude in rapidly rotating solar-type stars is also a potentially interesting discriminant. This can arise through channeling of buoyantly rising toroidal flux ropes along the polar axis (Schuessler *et al.*, 1996), or efficient poleward transport of surface magnetic flux (see Section 2.6). The rapid improvements in surface magnetic field data and reconstruction techniques may allow us to distinguish between these two possibilities, and may thus provide much needed additional constraints on flux transport dynamo models and/or magnetic flux rope formation.

6.4.4 Fully convective stars

With fully convective stars we encounter potential deviations from a solar-type dynamo mechanism; without a stably stratified tachocline and radiative core to store and amplify toroidal flux ropes, the Babcock–Leighton mechanism (Section 6.2.2), the tachocline α-effect, and the flux-tube α-effect (Section 6.2.3) all become problematic. Mean-field models based on the turbulent α-effect remain viable (e.g. Durney *et al.*, 1993), but the dynamo behavior becomes dependent on the presence and strength of internal differential rotation, about which we really don't know very much in stars other than the Sun. The full-sphere MHD simulations of an "M-star in a box" by Dobler *et al.* (2006) are particularly interesting in this respect, as they indicate that fully convective stars do produce significant internal differential rotation and well-defined patterns of hemispheric kinetic helicity, both supporting the growth of a spatially well-organized large-scale magnetic component.

Moving to even cooler stars, as the luminosity drops and surface temperature falls below a few thousand kelvin, the magnetic Reynolds number in the surface layers is expected to eventually fall back towards values approaching unity. Small-scale turbulent dynamo action may shut down, with magnetic activity then reflecting only the operation of a deep-seated, large-scale dynamo. Whether this transition is sharp or gradual, and whether it leads to well-defined observational signatures, remain open questions. There is certainly no a-priori reason to presume that dynamo action should cease. Indeed, in some ways, rapidly rotating very low-mass stars are getting closer to the physical parameter regime characterizing the geodynamo (see Section 7.6.5).

6.5 Outlook

The solar and stellar dynamo models discussed in this chapter all rely on severe geometrical and conceptual simplifications of the governing equations of magnetohydrodynamics. Even though they have foregone mathematical and physical determinism, these dynamo models do have a descriptive value. More important perhaps, they are thinking tools, traction aids for the brain. Almost surprisingly, many of the insights they have provided have been vindicated by large-scale numerical MHD simulations. The numerical verification of mean-field theory expressions for the α-effect and turbulent pumping by Käpylä *et al.* (2006) stand as a good case in point; so do the fully convective stellar dynamo simulations of Dobler *et al.* (2006).

Yet, even as descriptive or thinking tools, these simple models also have their limitations, as the solar case illustrates all too well. Despite decades of sustained modeling work, our state of understanding of the dynamo mechanism underlying the solar magnetic activity cycle is still rather minimal. No consensus currently exists regarding the primary means of regeneration for the solar poloidal magnetic component, or the non-linear mechanism regulating the strength of the dynamo-generated magnetic field. Likewise, the mechanism(s) responsible for the cycle's fluctuations in amplitude and duration, including Maunder-like grand minima, still cannot be pinned down with confidence on the basis of extant models and data.

Moving from the Sun to the stars, from the modeling point of view the big unknowns are at least identified: how do large-scale flows such as differential rotation vary with rotation rate and structural parameters? How far in the rapid rotation regime can we trust mean-field-based theories of the α-effect, already overstretched even in the solar case? Is a tachocline essential to solar-like magnetic activity? Are starspots key components of stellar dynamos? Some of these questions may be addressed via 3D hydrodynamical simulations, others require full MHD treatment (see Chapter 5). Such simulations, even though severely constrained by available computational resources, are likely to play an increasingly important role as alternate reality checks on mean-field-like dynamo models. Observations also provide critical insight; the variations (or lack thereof) in the observed mode(s) of magnetic activity at the fully convective boundary in M-type dwarf stars, and as surface convection disappears moving up the main sequence from late- to mid-F-type dwarf stars, both merit a lot of attention, the latter far more than it has received to date.

In closing, it is worth remarking that some of the field's most basic tenets continue to be challenged. One of the great success story of solar cycle modeling of the 1990s is the quantitative development of the rising flux rope model for bipolar active regions. Such models can reproduce the observed patterns of emergence latitudes and E–W tilts, and, in conjunction with corresponding stability analyses, give a unified, coherent picture of the storage, destabilization, and buoyant rise of magnetic fields from the overshoot region of the tachocline essentially all the way to the surface. Yet, some observations of evolving bipolar active regions immediately following emergence appear at odds with this grand unifying picture (see Kosovichev and Stenflo, 2008). Likewise, the existence of active longitudes could imply a significant non-axisymmetric component for the solar internal magnetic field. This component could be deep-seated and "enslaved" to the primary axisymmetric dynamo mode (see Bigazzi and Ruzmaikin, 2004, for an interesting example), but it is also possible that even the primary dynamo is non-axisymmetric, and the approximate axisymmetry of the surface large-scale solar magnetic field arises through axisymmetrization by differential rotation across the convective envelope. At an even more fundamental level, the idea of a deep-seated dynamo powered at least in part by the inductive action of internal large-scale flows now goes back a little over half a century to the seminal work of Parker (1955a, b), yet credible alternatives, whereby the solar cycle is seen as a purely surface or near-surface process, continue to be proposed (see e.g. Brandenburg, 2005; Schatten, 2009, for examples). Could we be completely off? Probably not, but such challenges act as a stark reminder that the road ahead is still long and hard, and quite possibly hiding some spectacular surprises.

7

Planetary fields and dynamos

Ulrich R. Christensen

7.1 Introduction

Over four centuries ago it was realized that the time-averaged direction of a compass needle is not affected by a force emanating from the sky, but by a magnetic field that is intrinsic to the Earth. The basic structure of the geomagnetic field and its slow variation with time was characterized long before magnetic fields were detected on other celestial bodies. By the middle of the twentieth century, the study of remanent magnetization of natural rocks had firmly established that the principal dipole component of the Earth's magnetic field had reversed its direction many times in the past.

Our understanding of the origin of the field by a dynamo process in the Earth's core has developed at a much slower pace, basically in parallel with that of astrophysical dynamos in general. Aside from understanding the intricate details of how a magnetic field is generated by a dynamo, we must ascertain that some fundamental requirements are fulfilled inside our planet. Geophysical observations have shown that one condition, namely the existence of an electrically conducting fluid region, is met inside the Earth, which has an outer core consisting of a liquid iron alloy. It is likely, but not completely certain, that all big planets have conducting fluid cores (see Fig. 7.5). However, some planets may not conform with another basic condition for a dynamo, namely sufficiently fast motion in the fluid layer. Convection is envisaged as the most likely source of a flow that can sustain a dynamo, but in some planets the fluid core may be stably stratified.

Since 1995, numerical modeling of the geodynamo has been thriving. Global models of convection-driven dynamos in a rotating spherical shell show magnetic fields that resemble the geomagnetic field in many respects – they are dominated by the axial dipole of approximately the right strength, they show spatial power spectra similar to that of Earth's magnetic field, and the magnetic field morphology and the temporal variation of the field resembles that of the geomagnetic field. While

Heliophysics: Evolving Solar Activity and the Climates of Space and Earth, eds. Carolus J. Schrijver and George L. Siscoe. Published by Cambridge University Press. © Cambridge University Press 2010.

these models represent direct numerical simulations of the fundamental magneto-hydrodynamic equations without parameterized induction effects, they do not match actual planetary conditions in a number of respects and their success appears somewhat surprising.

Space missions revealed that most planets in the solar system have internal magnetic fields (see Vol. I, Chapter 13), but there are exceptions (Venus, Mars). Some planets seem to have had a field that is now extinguished (e.g. Mars). In many cases with an active dynamo the axial dipole dominates the field at the planetary surface (Fig. 13.2 in Vol. I), but Uranus and Neptune are exceptions. Saturn is special because its field is extremely symmetric with respect to the planet's rotation axis. The field strengths at the planetary surfaces differ by a factor of 1000 between Mercury and Jupiter. A full understanding of this diversity in the morphology and strength of planetary magnetic fields is still lacking, but a number of promising ideas have been suggested and backed up by dynamo simulations. Some of the differences can be explained by a systematic dependence of the dynamo behavior on parameters such as rotation rate or energy flux, whereas others seem to require qualitative differences in the structure and dynamics of the planetary dynamos.

This chapter summarizes our state of knowledge about the structure and time dependence of the geomagnetic field and the more limited knowledge on the fields of other planets. The internal constitution and the thermal budget of the planets is discussed as far as it is essential for the understanding of planetary dynamos. The fundamentals of astrophysical dynamos have been described in Chapters 5 and 6, and in Vol. I, Chapter 3; here we discuss conditions for fluid flow and magnetic field generation that are particular to planetary cores and we contrast them with those in the Sun. We give special consideration to numerical simulations that have played a major role in our understanding of the generation of planetary magnetic fields.

7.2 Geomagnetic field

7.2.1 Field structure

For the last four hundred years the Earth's magnetic field has been mapped sufficiently well to determine its global structure. Most of the early measurements were taken routinely by mariners (Jackson *et al.*, 2000). Usually only the declination was recorded, i.e. the deviation of the horizontal component of the magnetic field from true north. Other measurements have been taken for scientific reasons and included the inclination, i.e. the angle between the field direction and Earth's surface. In 1832, Carl Friedrich Gauss developed a method that allowed the intensity of the field to be measured in absolute terms for the first time. Not much later the first permanent magnetic observatories were established.

Gauss was also the first to realize that the magnetic field **B** near the Earth's surface (and in general in a source-free region) can be represented as the gradient of a scalar potential Φ. He introduced the presentation of the field in terms of spherical harmonic functions and Gauss coefficients g_n^m and h_n^m, as they are now called:

$$\Phi = R_p \sum_{n=1}^{\infty} \sum_{m=0}^{n} \left(\frac{R_p}{r}\right)^{n+1} P_n^m(\cos\theta) \left(g_n^m \cos m\lambda + h_n^m \sin m\lambda\right), \quad (7.1)$$

where r is the distance from the planet's center, R_p is the (equatorial) radius of the planet, θ is the co-latitude, λ longitude, n and m are spherical harmonic degree and order, respectively, and P_n^m are the associated Legendre functions in the so-called Schmidt normalization. g_1^0 describes the axial dipole ("axial" means aligned with the planet's rotation axis), g_1^1 and h_1^1 the equatorial dipole, terms with $n = 2$ the quadrupole, those with $n = 3$ the octupole, and so on. Equation (7.1) is formulated such that the coefficients g and h have the unit of magnetic induction (here also called magnetic field strength). Usually the sub-unit nanotesla (nT) is used in geophysics and planetary sciences ($100\,000$ nT = 1 gauss).

Often a characterization of the magnetic field of planets other than Earth in terms of a dipole that is offset from the planet's center can be found in the literature. This is an outdated description, because it is very implausible that the dynamo region is not spherically symmetric with respect to the planet's center of mass. A combination of a planetocentric dipole and higher multipoles is equivalent to an off-center dipole.

The complete description of a potential field requires additional terms in Eq. (7.1) that vary with radius as $(r/R_p)^{n-1}$. These terms describe a field component of external (ionospheric or magnetospheric) origin. Gauss found that for the Earth they are small in comparison to those describing an internal field (see Section 7.3 for the impact on measurements of fields from other planets using spacecraft flybys).

The properties of the recent geomagnetic field have been mapped with high spatial resolution by dedicated satellite missions carrying magnetometers in a low-Earth orbit, namely MAGSAT in 1980, and ØRSTED and CHAMP since 1999 and 2000, respectively (Olsen *et al.*, 2007). Spherical harmonic representations of the Earth's internal magnetic field up to degree and order 100 are available. When aiming at an understanding of the geodynamo, it is more meaningful to consider the magnetic field structure at the surface of the core, within which the dynamo process operates, rather than at the Earth's surface. To the extent that there are no significant sources of the magnetic field in the Earth's crust and mantle (made of silicate rock), Eq. (7.1) can be used to downward continue the magnetic field from $r \approx R_p$, where it is observed, to the core surface at R_c. In Fig. 7.1 spatial power

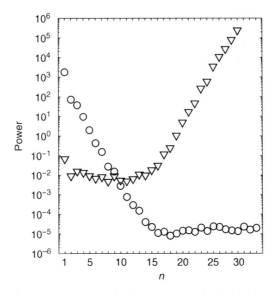

Fig. 7.1. Spatial power spectra of the geomagnetic field in 2004 according
to the POMME model (Maus *et al.*, 2006) as function of spherical harmonic
degree n at Earth's surface (circles) and at the core–mantle boundary (triangles;
offset in amplitude). Note that structures of the core field corresponding to
$n > 13$ are veiled by the crustal magnetic field, and that the apparent rise in the
power spectrum does not reveal the properties of the deep geomagnetic field.
Units are μT^2 for the surface field and mT^2 for the core field.

spectra of the magnetic field are compared for the Earth's surface (circles) and the
core–mantle boundary (triangles). The degree power at radius r is given by

$$P_n = (n+1)\left(\frac{R_p}{r}\right)^{2n+4}\sum_{m=0}^{n}\left(\left(g_n^m\right)^2 + \left(h_n^m\right)^2\right). \qquad (7.2)$$

The spectrum at Earth's surface drops sharply up to spherical harmonic degree 13,
and is nearly white beyond that. The spectrum of the field projected onto the core–
mantle boundary is almost white up to $n = 13$, except for the dipole term, which
stands out by a factor between five and ten. For $n > 13$ the spectrum rises steeply,
which is considered to be a very unlikely property of the core field. The generally
accepted interpretation of these spectra is that the field at the Earth's surface is dominated by the core field at large scales up to $n \approx 13$. At shorter scales the geometric
attenuation of the core field with radius is very strong (Eq. 7.1). The relatively
weak magnetic field due to the inhomogeneous remanent and induced magnetization of small amounts of ferromagnetic minerals in the Earth's crust takes over
and dominates the observed surface field. Projecting this small-scale field onto the
core–mantle boundary is unphysical and leads to the blue spectrum for $n > 13$. As
a consequence, we know the magnetic field at the surface of Earth's core only at

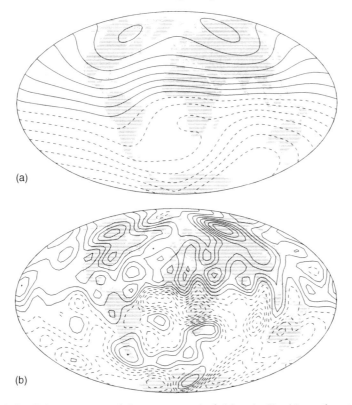

(a)

(b)

Fig. 7.2. Radial component of the geomagnetic field at the Earth's surface (a) and at the core–mantle boundary (b). Full lines for inward magnetic flux and dashed lines for outward flux. Contour intervals are arbitrary and different in the two panels.

large and intermediate wavelengths. Structures of the core field corresponding to $n > 13$ are veiled by the crustal magnetic field. The fine structure at the top of the solar dynamo is much better resolved than that of the geodynamo.

Figure 7.2 shows the radial component of the geomagnetic field at the Earth's surface (panel a) and, truncated at $n = 13$, at the core–mantle boundary (panel b). At the surface, the dipole part is very dominant. At the core–mantle boundary, in contrast, the dipole dominance is still visible, but there is significant structure at smaller scales. Most of the dipole field is formed by strong concentrations of magnetic flux into four lobes, two in each hemisphere, centered at $\pm(60°-70°)$ latitude. The prominent flux lobes in the Northern Hemisphere, under North America and Siberia, have counterparts in the Southern Hemisphere that lie at approximately the same longitudes. Close to the rotation poles, the flux is weak or even inverse with respect to the dominant polarity of the respective hemisphere. Patches of magnetic flux of both polarities are found at low and mid-latitudes.

   The rms magnetic field strength at the core–mantle boundary is 0.39 mT in harmonic degrees from 1 to 13. It is uncertain how much components with $n > 13$ add; possibly they might double the mean strength. The mean field strength inside the dynamo is even more difficult to estimate. Speculations that the toroidal magnetic field in the Earth's core would be much stronger than the poloidal magnetic field, as it likely is in the Sun, are not supported by geodynamo models. A range of 1–4 mT (10–40 G) seems plausible for the internal field strength of the geodynamo.

   Figure 7.2 represents a snapshot of a time-dependent magnetic field. Maps of the core field based on the historical record of observations have been constructed back until the year 1590, although with degrading spatial resolution (Jackson *et al.*, 2000, 2007). Although the details of the field structure change, some general traits seem to remain the same. The Northern Hemisphere flux lobes, in particular, are persistent and stay more or less in place.

7.2.2 Time-variability of Earth's field and the paleofield

The Earth's internal magnetic field changes on various time scales ranging from one year to a hundred million years. The changes that occurred during the past 400 years are documented by direct measurements. Going further back in time is possible by accessing the huge archive of magnetized rocks, which date back to various epochs of geological time and which recorded the magnetic field at the time of their formation.

7.2.2.1 Secular variation

The non-axial dipole part of the Earth's field, comprising the equatorial dipole and higher multipoles, changes significantly over a century. This is called the geomagnetic secular variation. Even in the eighteenth century it was noticed that part of the variation can be described as a westward drift of magnetic structures. The axial dipole changes more slowly; since 1840 the dipole moment has decreased by about 9%.

   Much more recently, the magnetic field changes at the core–mantle boundary have been used to infer the flow of liquid iron at the top of Earth's core. This is based on the assumption that on the decadal time scale the magnetic flux is approximately frozen into the moving fluid (Alfvén's theorem, see Vol. I, Section 3.2.3.1). This alone is not sufficient to invert the field changes uniquely for the pattern of the large-scale flow and additional assumptions must be made (Holme, 2007). Figure 7.3 shows an example for a map of the core flow; other maps are broadly similar. The predominantly westward flow associated with the westward magnetic drift is restricted to the Atlantic hemisphere of the globe (where the westward drift

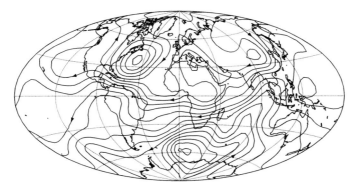

Fig. 7.3. Streamlines of the flow at the surface of the Earth's core inferred from the geomagnetic secular variation under the frozen flux assumption for the year 1980. (Adapted from Amit and Christensen, 2008.)

was discovered originally), but is not found globally. A typical flow velocity is $15 \, \text{km yr}^{-1} \, (0.5 \, \text{mm s}^{-1})$.

Using this velocity estimate and the estimate for the internal field strength of the geodynamo from above, the magnetic energy density in the core is roughly three orders of magnitude larger than the kinetic energy density (although there is some uncertainty about the energy in small-scale components, in particular of the velocity field). This is in contrast to conditions in the convection zone of the Sun, where these two energy densities are comparable on average.

7.2.2.2 The Earth's paleofield

Most rocks contain small (sometimes minute) amounts of ferromagnetic minerals. A remanent magnetization can be acquired in various ways when a rock is formed. Of particular importance is the thermoremanence of a magmatic rock that cools in an ambient magnetic field below the Curie temperature (where a mineral becomes magnetic) and the blocking temperature (where the acquired magnetization becomes insensitive to later changes in field direction). From oriented and dated rock samples the field direction and sometimes the magnetic field strength at the time of their formation can be determined. This is not straightforward, because alterations of the rock at some later time may involve the formation of new ferromagnetic grains and lead to a "magnetic overprint". To unravel the magnetic palimpsest, each rock sample is stepwise demagnetized in the laboratory, by heating it up or by the application of an AC magnetic field. The resulting changes in the direction and intensity of the remanence signal are measured in order to retrieve the primary magnetization.

This technique and its refinements have been applied to rock samples of all ages and also to artifacts, such as potsherds, which provide more detailed information for the past couple of thousand years. The oldest rocks that have been used for robust paleointensity measurements date back 3.2 billion years (Tarduno *et al.*, 2007). Although the intensity of the geomagnetic field fluctuates on various time scales, there is no long-term trend. For most of the time the intensity is found to be within a factor of two or three of the present field strength.

The detailed geometry of the field is more difficult to determine from paleomagnetic data, because the times of magnetization of samples from different locations are not synchronous. Furthermore, for rocks older than 5–10 million years continental drift becomes important, i.e. the location of the rock at the time when it was formed is not the same as it is today. In fact, the movements of the continents are calculated from paleomagnetic data under the assumption that the geomagnetic field is a geocentric axial dipole when averaged over long time intervals. Paleomagnetic data from the past 5 million years (for which the effects of continental drift are small) strongly support this hypothesis. The scatter found in these data, which is due to the combined influence of dipole tilt and of higher multipole contributions to the magnetic field, suggests that most of the time the amplitude of these two has been similar to what it is in the recent geomagnetic field. The dipole dominance is more difficult to prove for earlier times, but the available evidence is in support of it.

In summary, the Earth's magnetic field has not changed dramatically over the past three billion years in geometry or strength. A detailed account of our knowledge of the paleofield can be found in the book by Merrill *et al.* (1998).

7.2.2.3 Dipole reversals

One of the earliest findings by paleomagnetism is the occurrence of reversals of the dipole field. Today a detailed chronology of the geomagnetic polarity during the past couple of hundred million years has been established (Fig. 7.4 shows the past 120 million years). Compared to the length of periods with stable dipole polarity of some hundred thousand years, reversals are fairly rapid. The time interval during which the dipole axis is strongly tilted may last several thousand years. During reversals the dipole does not simply tip over, but also becomes much weaker, whereas the strength of higher multipole components does not seem to change much. Hence, the field at the Earth's surface becomes multipolar during a reversal. Aside from complete reversals, so-called geomagnetic excursions are also found in the paleomagnetic data. During these short events, the dipole axis becomes strongly tilted, often by more than 90°, but swings back to its original orientation.

On average, the geomagnetic field has reversed a few times in a million years during the recent geological past. In contrast to the cyclic behavior of the solar magnetic field, the timing of geomagnetic reversals is random: the probability of a reversal to occur is independent of the time that has passed since the last reversal. However, on time scales of 100 million years the reversal frequency itself

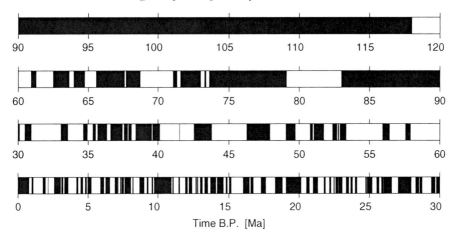

Fig. 7.4. Polarity of the geomagnetic field for the past 120 million years, with time running backward from left to right in each row (before present – BP, i.e. 1950 – in units of millions of years). Dark regions indicate times when the dipole polarity was the same as today, in white regions it has been opposite.

changes drastically. In a 35 million year time interval that ended 83 million years ago, the so-called Cretaceous superchron, no reversals occured at all. The reversal frequency increased gradually after the superchron, and decreased before the superchron. Other superchrons existed earlier in Earth's history, but are less well documented. The reversal frequency changes on a time scale that is comparable to the overturn time of the sluggish convection in the Earth's silicate mantle. For this reason it is assumed that the reversal frequency may be controlled by the slowly changing conditions in the lowermost mantle, for example its thermal structure, which would affect convection in the liquid core. Further details on reversals are found in Merrill *et al.* (1998) or Glatzmaier and Coe (2007).

7.3 Magnetic fields of other planets and satellites

The magnetic fields of all major planets in the solar system have been characterized by space missions during flybys or from orbiting spacecraft carrying vector magnetometers (Connerney, 2007; see Vol. I, Chapter 13 for a discussion of planetary magnetic fields and their associated magnetospheres). So far, this has provided only relatively crude snapshots in comparison to our knowledge of the geomagnetic field. Next to nothing is known about the time variability of the magnetic fields of planets other than Earth. In some cases the separation of internal (dynamo) and external (magnetospheric) contributions to the field observations is a significant source of uncertainty. Table 7.1 gives an overview on the field properties of the planets and some of their satellites.

Table 7.1. *Properties of magnetic fields of planets and satellites.*

Object	Active dynamo	R_c/R_p	B_{rms}[nT]	Dipole tilt	P_2/P_1	P_3/P_1
Mercury	Yes?	0.75	300	<5°?	0.1–0.5?	
Venus	No	0.55				
Earth	Yes	0.55	44 000	10.4°	0.04	0.24
Moon	No; yes in past?	0.2?				
Mars	No; yes in past	0.5				
Jupiter	Yes	0.84	640 000	9.4°	0.10	0.09
Ganymede	Yes	0.3	1 000	4°		
Saturn	Yes	0.6?	31 000	0°	0.02	0.22
Uranus	Yes	0.75	48 000	59°	1.3	1.5?
Neptune	Yes	0.75	47 000	45°	2.7	6?

Listed are the ratio R_c/R_p of the core to the planetary/satellite radius, the rms value B_{rms} of the surface magnetic field, the dipole tilt angle relative to the spin axis, and the ratios of quadrupole power or octupole power to the dipole power (P_2/P_1 and P_3/P_1) at the outer boundary of the dynamo at $r = R_c$. See Fig. 13.2 in Volume I for a graphical respresentation of the magnetic fields of the Earth and the giant planets.

Mercury   The discovery of Mercury's internal magnetic field during a flyby of Mariner 10 in 1975 came as a surprise. Before then, it was believed that internal activity had ceased in the small planet. The flybys of the MESSENGER spacecraft in 2008 confirmed that the field is dominated by a dipole slightly tilted relative to the rotation axis. The relative importance of higher multipole contributions remains uncertain. The field strength at the planetary surface is $B_{rms} \approx 300$ nT. Finding an explanation for this very low value, compared to those of other planets with a dynamo, is a challenge for the theory of planetary dynamos.

Venus   No intrinsic magnetic field has been observed at Venus. The upper limit for the dipole moment is 10^{-5} of Earth's value. Unlike in the case of Mars (see below), a small-scale magnetic field due to remanent magnetization of crustal rocks that could be indicative for an ancient dynamo has not been observed either. However, the Venusian surface temperature of $\sim$735 K is close to, or above, the Curie temperature of ferromagnetic minerals. Also, there is evidence that Venus' entire crust was renewed some 500 million years ago, which would have erased any magnetization that might have existed before. The answer to the question whether Venus once had an operating dynamo therefore remains elusive.

Moon   The Earth's satellite has no global field at present. Small-scale magnetic fields that locally reach a strength of several tens of nT have been observed. Lunar rock samples brought to Earth by the Apollo missions show remanent magnetizations. The origin of the magnetization could be the field of an ancient dynamo, but the small size of the lunar core and the associated geometric decrease of the field strength may be a problem. An alternative hypothesis for the acquisition of the magnetization involves strong local magnetic fields in the plasma clouds generated for a short time by big meteor impacts.

Mars   Mars has no global magnetic field, but strong fields of crustal origin exist at the local or regional scale. Their amplitude is several hundred nT at a spacecraft altitude of 200 km, corresponding to probably several thousand nT at the Martian surface. This is considerably stronger than the magnetic field contribution from crustal magnetization on Earth. The only plausible cause for its acquisition is the existence of a strong global field generated by an early dynamo. Pronounced local fields are found in the very old southern highlands on Mars and are nearly absent in the younger northern lowlands. From the magnetization (or its absence) associated with large dated impact basins it has been estimated that the dynamo ceased to operate 4.1 billion years ago (i.e. around the time of the Late Heavy Bombardment, see Chapter 4).

Jupiter   The detection of Jupiter's global magnetic field pre-dated the planet's exploration by spacecraft. It was inferred from the observation of strong emissions of radiowaves in the decameter wavelength range. These are generated by energetic electrons that gyrate around magnetic field lines close to Jupiter's surface (Barrow and Carr, 1992). Jupiter's field is about ten times stronger at the surface than the geomagnetic field, but the morphology is fairly similar: the dipole tilt is around ten degrees, and the ratios of the quadrupole and octupole components to the dominant dipole component are similar for both planets.

Saturn   Saturn's field is slightly weaker at the surface than Earth's field. The dipole tilt is indistinguishable from zero. Furthermore, only zonal quadrupole and octupole components are needed in addition to the axial dipole to fit the field measurements by passing spacecrafts and the Cassini orbiter. This creates a problem for dynamo theory, because a strictly axisymmetric magnetic field cannot be generated by a dynamo according to Cowling's theorem (Section 4.1.5 in Vol. I).

Uranus and Neptune   Uranus and Neptune can be dealt with jointly: their magnetic fields are similar to each other, yet distinct from those of other planets. So far, Uranus' and Neptune's fields have been characterised during a single flyby by Voyager 2 at each of these planets and uncertainties remain concerning details of the field structure. However, while the surface field strength is comparable to that at Earth, the geometry is clearly different. The dipole axis is strongly inclined with respect to the rotation axis and quadrupole and probably octupole contributions are comparable to the dipole magnitude at the surface. At the probable radius of the top of the dynamo region, the quadrupole and octupole field are stronger than the dipole field (Table 7.1). While all other dynamo-generated planetary magnetic fields in the solar system are dipole-dominated, those of Uranus and Neptune must properly be termed multipolar.

Ganymede Jupiter's largest moon Ganymede is the only satellite in the solar system for which a global field with a probable dynamo origin has been found. Ganymede orbits inside Jupiter's magnetosphere and the strength of the Jovian field at Ganymede is about 120 nT, or one-eighth of Ganymede's intrinsic field. Other Jovian satellites have weak induced fields. The temporal change of Jupiter's field at the satellite position due to the rotation of the planet with its tilted dipole induces currents in the electrically conducting interior of the satellite. The strength of the induced field is at most comparable to that of the inducing field. Ganymede's field also has this component, but the surface field strength of 1000 nT is much larger than that of the Jovian field at this distance. The observations taken during repeated flybys of the Galileo spacecraft require a nearly axial dipole field that is intrinsic to Ganymede.

7.4 Structure and energy budget of planetary interiors

In this section the internal structure and the energy budget of planetary interiors are discussed as far as they are relevant for the operation of a dynamo. We distinguish between the rocky (terrestrial) planets of the inner solar system and the gas planets in the outer solar system. Both types of planets can host dynamos, although their structure and their energetics are different. A schematic overview of planetary internal structure is given in Fig. 7.5.

7.4.1 Earth

Earth serves as the prototype for the terrestrial planets. Its interior structure is known in some detail from seismology. Observations of the travel times of compressional waves and shear waves, and of frequencies of free oscillations of the Earth (which are excited by big earthquakes), can be inverted for the distribution of elastic properties and density inside our planet. There is a core with radius $R_c \approx 0.55 R_p$. Its outer part does not support the propagation of shear waves and hence is liquid. The small inner core, with a radius $R_{ic} = 0.35 R_c$, is clearly distinct. Since its discovery in 1935, it has been assumed to be solid. This is not

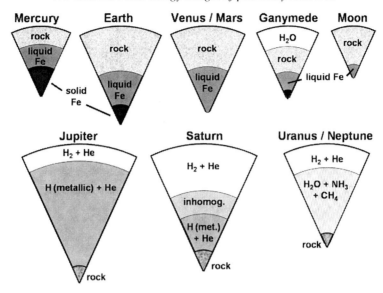

Fig. 7.5. Interior structure of planets with active or extinct dynamos. The top row shows the rocky (terrestrial) planets, and the bottom row the much larger gas planets. Larger planets are shown slightly larger, but relative sizes are not drawn to scale.

easy to prove, but from the observed frequencies of free oscillation modes that are particularly sensitive to the shear strength in the inner core, it has been made certain.

The core appears to consist predominantly of iron. Iron is the only element with sufficient cosmochemical abundance for which density and compressibility at the appropriate pressures and temperatures match the seismologically determined values of the core. Nickel also contributes, but is very similar in properties to iron. However, the density in the outer core is slightly less than that of pure iron–nickel and ~10% of a light chemical element must also be present. Silicon and oxygen are the top candidates, although others such as sulfur are likely to contribute. The composition of the solid inner core is closer to pure iron–nickel.

The total internal heat flow at the Earth's surface is 46 TW (although a large number, it is only 0.03% of the total power coming into the Earth's atmosphere by insolation). Roughly one-half of it is balanced by the heat generated by the decay of uranium, thorium, and the potassium isotope ^{40}K inside the Earth. The remainder of the heat flow is due to the cooling of the Earth. The loss of gravitational potential energy associated with the contraction of the Earth contributes a modest amount, but is much less important than it is in young stars or in gas planets. How much of the Earth's heat flow comes from the core is rather uncertain. Recent estimates that are based on different lines of evidence mostly fall into the range 5–15 TW

(Nimmo, 2007, Lay *et al.*, 2008), although values as low as 3–4 TW have also been discussed. Most of the radioactive elements reside in the silicate crust and mantle. Some potassium may be present in the core, but the majority of the core heat must be due to cooling. It is important to note that the heat loss from the core is regulated by the slow solid-state convection in the mantle. The core, which convects vigorously in comparison to the mantle and which is thermally well-mixed, delivers as much heat as the mantle is able to carry away.

Radiative heat transfer is not an issue in planetary cores, but liquid metal is a good thermal conductor. The heat flux that can be transported by conduction along an adiabatic temperature gradient, $(dT/dr)_{ad} = T/H_T$, is sometimes called the "adiabatic heat flow" (T is absolute temperature, $H_T = c_p/(\alpha g)$ is the temperature scale height with c_p the heat capacity, α the thermal expansivity and g the gravitational acceleration). In terrestrial planets, the adiabatic heat flow can be a large fraction of the actual heat flow, or it may exceed the actual heat flow, in which case at least the top layers of the core would be thermally stable. Near the top of Earth's core approximately 3–4 TW can be conducted along the adiabat (Lay *et al.*, 2008), i.e. close to the minimum estimates for the entire core heat flow. But even if all heat flux near the core–mantle boundary were carried by conduction, a convective dynamo can exist thanks to the inner core. At the inner core boundary, the adiabatic temperature profile of the convecting outer core crosses the melting point of iron. The latter increases with pressure more steeply than the adiabatic gradient, which is the reason why the Earth's core freezes from the center rather than from above. As the core cools, the inner core grows with time by freezing iron onto its outer boundary. This has two important implications for driving the dynamo. The latent heat that is released upon solidification is an effective heat source, which contributes to the heat budget approximately the same amount as the bulk cooling of the core. The heat flux that originates at the inner core decreases with radius as r^{-2} in the spherical geometry of the fluid core. The adiabatic temperature gradient is roughly proportional to r, because gravity decreases towards the center. Therefore, even if the actual heat flux were slightly less than the adiabatic heat flux near the core–mantle boundary, it must be superadiabatic deeper down. A second, perhaps more important effect is that the light elements in the outer core are preferentially rejected when iron freezes onto the inner core. Hence, they become concentrated in the residual fluid near the inner core boundary. This layering is gravitationally unstable because of the reduced density, which leads to compositional convection that homogenizes the light elements in the bulk of the fluid core. Compositional convection contributes as much as, or more than, thermal convection to the driving of the geodynamo in recent geological times.

Most models for the inner core growth rate imply that the inner core did not exist for most of the history of the Earth. Rather, it would have nucleated between

0.5 and 2 billion years ago. In the absence of an inner core, only thermal convection by secular cooling of the fluid core (and perhaps radioactive heating) can drive a dynamo, which is less efficient than the present-day setting. A change in the geomagnetic field properties might be expected upon the nucleation of the inner core, but no clear indication for such an event has been found in the paleomagnetic record.

7.4.2 Other terrestrial planets

Few data are available to constrain the internal structure and thermal budget of terrestrial planets other than Earth (Sohl and Schubert, 2007; Breuer *et al.*, 2007). The mean density and the composition of surface rocks strongly suggest that they are differentiated into crust, mantle, and core, as is the case for Earth. The moment-of-inertia factor $C/M R_p^2$, where C is the polar moment of inertia and M the planetary mass, is sensitive to the radial variation of density inside the planet. Aside from Earth, the only other terrestrial planet for which it has been constrained so far is Mars, where it confirms the existence of a core. The observed reaction of Mars to solar tides shows that the core must be at least partially liquid. The observation of forced librations, i.e. slightly uneven rotation under the influence of a solar torque, suggests the same for Mercury. Because of Mercury's high mean density its core must be very large in relation to the size of the planet. However, the core radius cannot be constrained precisely in the cases of Mercury, Venus, Mars, and the Moon. A major source of uncertainty is the amount of light elements in the cores of these bodies.

No direct evidence on the existence or non-existence of a solid inner core is available for any planet other than Earth. But the possible absence of an inner core could explain why Venus and Mars do not have an active dynamo. On Earth mantle convection reaches the surface in the form of plate tectonics, which is a fairly efficient mode of removing heat from the interior. None of the other terrestrial planets have plate tectonics. In their cases, mantle convection is confined to the region below the lithosphere, a rigid lid of some 100−300 km thickness through which heat must be transported by conduction. Without plate tectonics, the heat flow is expected to be significantly lower not only at the surface, but also at the top of the core, where it is very probably subadiabatic. If no inner core exists to provide latent heat, it is then subadiabatic throughout the core. Furthermore, compositional convection is also unavailable to drive a dynamo. The slower cooling of the planetary interior in the absence of plate tectonics concurs with the idea that an inner core has not (yet) nucleated in the cases of Mars and Venus. Early in the planets' history the cooling rate was probably much higher and the associated core heat flow large enough for thermal convection. The demise of the dynamo must have occurred when the declining heat flow dropped below the conductive threshold.

7.4.3 Gas planets

Jupiter and Saturn are similar in composition to the Sun (Guillot and Gautier, 2007). Shells of a hydrogen–helium mixture surround a small rocky core. In the outer envelope, where hydrogen forms H_2 molecules, the electrical conductivity is poor. At high pressure, hydrogen becomes a metallic liquid with free electrons (density is too high while temperatures are not high enough to call it a plasma). Shock-wave experiments show that there is no first-order phase transition, but the electrical conductivity rises gradually and reaches metallic values at around 1.3 Mbar pressure (Nellis *et al.*, 1999). This is reached at a depth corresponding to 84% of Jupiter's radius and 62% of Saturn's.

Uranus and Neptune also have an envelope rich in hydrogen and helium, but the bulk of their mass consists of a water-rich mixture of water, ammonia and methane, termed "ices" in planetology, even if in a fluid state (Guillot and Gautier, 2007). The ice layer extends to approximately 75% of the radius. It has ionic electrical conductivity, which is two orders of magnitude lower than the metallic conductivity in the cores of terrestrial planets and the large hydrogen planets, but probably sufficiently high to sustain a dynamo.

The internal heat flow of the gas planets has been determined by monitoring their infrared luminosity in excess of the re-emission of absorbed sunlight. The source of internal heat is mostly the potential gravitational energy lost upon contraction. The results of simple evolution models of the planetary interior agree with the observed luminosity of Jupiter, but underpredict it in the case of Saturn and overpredict it for Neptune and in particular for Uranus. The He/H-ratio in Saturn's atmosphere seems to be less than the solar ratio. Stevenson (1980) proposed that helium becomes immiscible with hydrogen in the upper part of the metallic layer in Saturn, resulting in a downward segregation in the form of a "helium rain". The gravitational energy of the ongoing internal differentiation boosts the luminosity to the observed value. The radial dependence of the helium depletion in the upper part of the metallic shell results in a stable compositional stratification which suppresses convection. For Uranus and Neptune it has been suggested that stratification in deeper parts of the ice layers inhibits convection and explains the reduced ability of these planets to lose internal heat.

The possible compositional stratification may impede convection in the electrically conducting cores of the outer planets, but thermal conduction along an adiabat is insufficient to transport the observed amounts of internal heat. In compositionally stratified regions the thermal gradient must be superadiabatic and unstratified layers should convect vigorously.

7.5 Some basics of planetary dynamos

Planetary dynamos share with stellar dynamos that the basic physical concept for their description is that of convection-driven magnetohydrodynamic flow in a rotating spherical shell combined with the associated magnetic induction effects. The principles of such dynamos have been discussed in detail in Chapter 6 and in Vol. I, Chapter 3, but here it is useful to recall some requirements and assumptions for planetary dynamos. Next, specific conditions for the magnetohydrodynamic flow in planetary cores are discussed; these flows are, for example, more strongly influenced by rotational forces than the flow in the solar convection zone.

Inside a shell of depth d with an electrical conductivity σ the fluid must move with a sufficiently large characteristic velocity v, so that the magnetic Reynolds number

$$R_\mathrm{m} = \frac{vd}{\lambda} \tag{7.3}$$

exceeds a critical value $R_\mathrm{m,crit}$ in order to have a self-sustained dynamo ($\lambda = 1/\mu_0\sigma$ is the magnetic diffusivity, with μ_0 magnetic permeability). The flow pattern must also be favorable for dynamo action, which requires a certain complexity. In particular, helical (corkscrew-type) motion with a large-scale order in the distribution of right-handed and left-handed helices is suitable. The Coriolis force plays a significant part in the force balance of the fluid motion and influences the pattern of convection. With this, the requirement for "flow complexity" seems to be satisfied and self-sustained dynamo action is possible above $R_\mathrm{m,crit} \approx 40$–$50$ (Christensen and Aubert, 2006).

At greater depth in the solar convection zone, the magnetic Reynolds number reaches values of order 10^9 for molecular values of the magnetic diffusivity (see appendix to Chapter 5). In the geodynamo R_m is approximately 1000. This fairly moderate value allows for the direct numerical simulation of the magnetic field evolution without the need to use an "effective diffusivity" or a parameterization of the induction process through a turbulent α-effect (Section 3.4.6 in Vol. I and Section 6.2.1). The ability to run simulations at the relevant value of R_m may be the most important cause for the success of geodynamo models.

The density in the Sun varies by many orders of magnitude with depth and the convection region spans many density scale heights. The density changes associated with radial motion are thought to be important. Flow helicity arises in the Sun because of the action of the Coriolis force on rising expanding and sinking contracting parcels of plasma (Section 5.2.4). Strong magnetic flux tubes have their own dynamics, because the reduction of fluid pressure that compensates magnetic pressure reduces their density and makes them buoyant (Section 5.4.3). In contrast, the dynamo region in Jupiter covers approximately one density scale height and it covers much less in terrestrial planets. The two compressibility effects mentioned before probably do not play a significant role in planetary dynamos. Present geodynamo models usually neglect the small density variation and assume incompressible flow in the Boussinesq approximation (where density differences are only taken into account for the calculation of buoyancy forces).

Many models of the solar dynamo assume that most of magnetic field generation occurs at the tachocline, the shear layer between the radiative deep interior and the convection zone of the Sun (Section 5.5.5). For planetary dynamos the process of magnetic field generation is thought to occur in the bulk of the convecting layer.

The thinking on planetary dynamos has been shaped by the theory for the onset of rotating convection and by theoretical arguments on the dominant force balance for the flow in planetary cores (Jones, 2007). The relevant equation of motion for an incompressible fluid is

$$\rho \left(\frac{\partial \mathbf{v}}{\partial t} + (\mathbf{v} \cdot \nabla)\mathbf{v} \right) + 2\rho\Omega\, \mathbf{e}_z \times \mathbf{v} + \nabla P = \rho\nu\nabla^2\mathbf{v} + \rho\alpha g T \mathbf{e_r} + \mathbf{j} \times \mathbf{B}, \quad (7.4)$$

where $\mathbf{v}$ is velocity, $\mathbf{e}$ a unit vector, Ω rotation rate, ρ density, P non-hydrostatic pressure, ν kinematic viscosity, α thermal expansivity, g gravity, T temperature, $\mathbf{B}$ magnetic field, $\mathbf{j} = \mu_0^{-1}\nabla \times \mathbf{B}$ current density, r radius, and z the direction parallel to the rotation axis. The terms in Eq. (7.4) describe, in order, the linear and non-linear parts of inertial forces, Coriolis force, pressure gradient force, viscous force, buoyancy force, and Lorentz force.

In the non-magnetic and rapidly rotating case, the primary force balance is between the pressure gradient force and the Coriolis force (geostrophic balance), similar to large-scale weather systems in the Earth's atmosphere. Ignoring all other terms in Eq. (7.4) and taking the curl, we arrive at the Taylor–Proudman theorem, which predicts the flow to be two-dimensional with $\partial\mathbf{v}/\partial z = 0$. The only type of perfectly geostrophic flow in a sphere, i.e. a flow that satisfies this condition, is the differential rotation of cylinders that are co-aligned with the rotation axis (geostrophic cylinders). Such flow can neither transport heat in the radial direction, nor can it act as a dynamo. Convection requires motion away from and towards the rotation axis. This must violate the Taylor–Proudman theorem, because a column of fluid that is aligned with the z-direction will then stretch or shrink because it is bounded by the outer surface of the sphere. Hence the velocity cannot be independent of z. The necessity to violate the Taylor–Proudman theorem inhibits convection and requires that some other force, such as viscous friction, must enter the force balance. In order for viscosity to do so, the length scale of the flow must become small, at least in one direction. But the flow maintains a nearly geostrophic structure as far as possible. At the onset of convection it takes the form of columns aligned with the rotation axis (Fig. 7.6; see

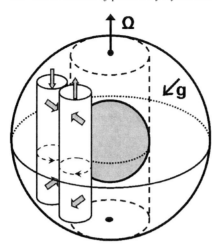

Fig. 7.6. Columnar convection in a rotating spherical shell near onset. The inner core tangent cylinder is shown by broken lines. Under Earth's core conditions the columns would be much thinner and very numerous.

also Section 5.2.4). They surround the inner core tangent cylinder like pins in a roller bearing. The tangent cylinder is parallel to the z-axis and touches the inner core at the equator. It separates the fluid core into dynamically distinct regions.

The primary circulation is around the axes of these columns. However, in addition there is a net flow along the column axes which diverges from the equatorial plane in anticyclonic vortices and converges towards the equatorial plane in columns with a cyclonic sense of rotation. The combination implies a coherently negative flow helicity in the northern hemisphere and positive helicity in the southern hemisphere. Busse (1975) demonstrated that this flow can serve as an efficient dynamo of the α^2-type (Section 3.4.6.2 in Vol. I).

When the motion becomes more vigorous at highly supercritical convection and when a strong magnetic field is generated, other forces such as inertia (advection of momentum) and the Lorentz force can affect the flow. However, one difference between the solar dynamo and planetary dynamos is the different role of inertial forces versus the Coriolis force. Their ratio is measured by the Rossby number

$$R_{\mathrm{o}} = \frac{v}{\Omega \ell}, \tag{7.5}$$

where v and ℓ are characteristic velocity and length scales, respectively. Deep in the solar convection zone $R_{\mathrm{o}} \approx 1$ when the pressure scale height is taken as ℓ. With typical estimates for the flow velocity in the Earth's core (1 mm s^{-1}), the Rossby number is of order 10^{-6} when a global scale such as the core radius or shell thickness is used for ℓ. Therefore, fluid motion in the geodynamo is often considered to be largely unaffected by inertial forces. The general force balance is believed to be that between Coriolis force, pressure gradient force, Lorentz forces, and buoyancy forces (MAC balance = Magnetic, Archimedean, Coriolis; Roberts, 1987). However, at small scales inertial forces may become important also in planetary dynamos and can potentially feed back on the large-scale flow (see Section 7.6.2).

Like rotation, the presence of an imposed uniform magnetic field inhibits convection in an electrically conducting fluid. However, the combination of a magnetic field and rotation reduces the impeding influence that either effect has separately. This constructive interference is most efficient when the Coriolis force and the Lorentz force are in balance. For an imposed uniform field this is the case when the Elsasser number,

$$\Lambda = \frac{\sigma B^2}{2\rho\Omega},\tag{7.6}$$

is of order one. For this force balance, called magnetostrophic, the flow pattern becomes large scaled. Applied to dynamos, it is argued that as long as the magnetic field is weak ($\Lambda \ll 1$), any field growth will intensify convection, meaning more efficient dynamo action and further increase of the field. Field growth at $\Lambda \gg 1$ would weaken convection, hence it is assumed that the field equilibrates at an Elsasser number of one. The field strength inside the geodynamo or in Jupiter's dynamo seems to agree with the Elsasser number rule (Stevenson, 2003). However, numerical dynamo simulations put some doubt on its validity (see Section 7.6.5).

A special condition applies to the integral force acting on geostrophic cylinders in the azimuthal direction. Buoyancy has no azimuthal component and the Coriolis force and the pressure gradient force are zero when averaged over the surface of these cylinders. If both viscous and inertial forces make a negligible contribution, the Lorentz force must also vanish, meaning that the magnetic field must maintain a special configuration. When this condition is satisfied, a dynamo is said to be in a "Taylor state". Disturbing the Taylor state will accelerate the cylinders, but the shearing of magnetic field lines penetrating neighboring cylinders, which is associated with the differential rotation, provides a restoring Lorentz force. The results are so-called torsional oscillations around the Taylor state, which in the Earth's core should have periods of some decades. Evidence for torsional oscillations has been claimed from fast secular variations of the geomagnetic field (Zatman and Bloxham, 1997). However, inertial effects of small-scale turbulent motion may contribute significantly to the acceleration of the cylinders, in which case the concept of a Taylor state is of limited value.

7.6 Numerical geodynamo models

7.6.1 Setup and parameters for geodynamo models

Most modern geodynamo models are direct numerical simulations of the equations for convective flow in the Boussinesq limit for a rotating spherical shell and of the magnetic induction equation (Christensen and Wicht, 2007). The equations are usually written in terms of non-dimensional variables and dimensionless control parameters. Here, we give them in the form used by Christensen and Aubert (2006):

$$\frac{\partial \mathbf{v}}{\partial t} + \mathbf{u} \cdot \nabla \mathbf{v} + 2\hat{\mathbf{z}} \times \mathbf{v} + \nabla P = E\nabla^2 \mathbf{v} + R_a^* \frac{\mathbf{r}}{r_o} T + (\nabla \times \mathbf{B}) \times \mathbf{B}, \qquad (7.7)$$

$$\frac{\partial \mathbf{B}}{\partial t} - \nabla \times (\mathbf{v} \times \mathbf{B}) = \frac{E}{P_m} \nabla^2 \mathbf{B}, \qquad (7.8)$$

$$\frac{\partial T}{\partial t} + \mathbf{v} \cdot \nabla T = \frac{E}{P_r} \nabla^2 T, \qquad (7.9)$$

$$\nabla \cdot \mathbf{v} = 0, \quad \nabla \cdot \mathbf{B} = 0. \qquad (7.10)$$

Equation (7.8) is the magnetic induction equation, which results from Maxwell's equations and Ohm's law for a moving fluid (e.g. Roberts, 2007), and Eq. (7.9) describes advection and diffusion of thermal energy, as measured by temperature T. The four non-dimensional control parameters are the Ekman number, measuring the ratio of viscous forces to the Coriolis force

$$E = \frac{\nu}{\Omega d^2}, \qquad (7.11)$$

a modified Rayleigh number, measuring the ratio of buoyancy forces to the impeding rotational forces

$$R_a^* = \frac{\alpha g_o \Delta T}{\Omega^2 d}, \qquad (7.12)$$

and the two diffusivity ratios: the Prandtl number

$$P_r = \frac{\nu}{\kappa}, \qquad (7.13)$$

and the magnetic Prandtl number

$$P_m = \frac{\nu}{\lambda}. \qquad (7.14)$$

Here, g_o is gravity at the outer boundary, ΔT the (superadiabatic) temperature contrast and κ the thermal diffusivity. $\left(R_a^*\right)^{1/2}$ is often called the convected Rossby number in the astrophysical literature. R_a^* is related to the conventional Rayleigh

Table 7.2. *Order of magnitude of dynamo control parameters and diagnostic numbers in the Earth's core and in planetary dynamo models.*

	Control parameters			
	$R_a^*/R_{a,c}^*$	E	P_m	P_r
Earth's core	~5000	10^{-15}–10^{-14}	10^{-6}–10^{-5}	0.1–1
Models	1–1000	10^{-3}–10^{-6}	0.1–10^3	0.1–10^3

	Diagnostic numbers			
	R_m	R_e	R_o	Λ
Earth's core	~10^3	~10^9	~10^{-5}	0.1–10
Models	50–3000	10–2000	3×10^{-4}–10^{-1}	0.1–100

Listed are the modified Rayleigh number (Eq. 7.12), relative to the critical value for the onset of convection in the absence of a magnetic field, the Ekman number E (Eq. 7.11), the magnetic Prandtl number P_m (Eq. 7.14), the Prandtl number P_r (Eq. 7.13), the magnetic Reynolds number R_m (Eq. 7.3), the Reynolds number $R_e \equiv vd/v$, the Rossby number R_o (Eq. 7.5), and the Elsasser number Λ (Eq. 7.6). For comparison with values of the Prandtl, Reynolds, and Rossby numbers in the solar interior, see Table 5.1.

number $R_a = \alpha g_0 \Delta T d^3/(v\kappa)$ by $R_a^* = R_a E^2 P_r^{-1}$. Equations (7.7)–(7.10) must be completed by appropriate boundary conditions. For Earth's core, impenetrable rigid boundaries with imposed constant temperatures or heat flux are usually taken. The magnetic field must match with an appropiate potential field outside the dynamo region.

In Table 7.2 we compare control parameter values used in geodynamo models with those for the Earth's core. The Rayleigh number has been normalized to its critical value $R_{a,c}^*$ for the onset of convection in the absence of a magnetic field. We also list several other non-dimensional numbers that are diagnostic for the dynamo and result from the model solution.

While the Prandtl number in the models is of the right order, the values of the other control parameters are far off. The Ekman number and the magnetic Prandtl number are too large in the models by factors of 10^{10} and 10^6, respectively. The modified Rayleigh number is too small with respect to supercriticality, but its absolute value is larger than the core value. In terms of physical parameters, the viscosity and thermal diffusivity are too large by a factor of order 10^6 compared to the magnetic diffusivity, which is about right. In addition, the rotation rate is too small by a factor of ~10^4 in most models. The magnetic Reynolds number R_m agrees with Earth values at least in the more advanced models, whereas the hydrodynamic Reynolds number $R_e = vd/v$ is far too small and the Rossby number R_o

is too large. The Elsasser number Λ can be taken as a non-dimensional measure for the magnetic field strength. The claim that a model reproduces the geomagnetic field strength actually means that it has an Elsasser number of order one.

Because of the large discrepancies in some of the control parameters it has been suspected that the dynamical regime in the models is different from that in planetary dynamos and that the agreement in the magnetic field properties found in some of them is fortuitous. In particular, viscosity might play an important role in the models whereas it is insignificant in the Earth's core.

7.6.2 Types of dynamo solutions

Many published geodynamo models have a magnetic field on the outer boundary that is strongly dominated by the axial dipole. Often the dipole in such models shows no tendency to ever reverse, although the model run time may not have been long enough to capture one of these rare events. Other numerical dynamos have a multipolar field, which is in many cases spatially complex without any obvious symmetries and is rapidly varying in time. The two classes of solution are rather distinct and few in-between cases have been found. Figure 7.7 shows examples of the two types. The underlying models are "advanced" in the sense that the Ekman number is decently small (from the point of numerical feasibility) and the magnetic Reynolds number is Earth-like in both cases. The spatial power spectrum at the outer boundary of the dynamo models is typically fairly white for harmonic degrees from 3 to 13, but in one class of solutions the dipole stands above the higher multipoles, as it does in the power spectrum of the geomagnetic field at the core–mantle boundary (Fig. 7.1), whereas it falls below the multipole level in the other class. In the fully developed multipolar regime the weak dipole component changes its polarity continuously in an erratic way.

Systematic model studies suggest that the ratio of inertial forces relative to the Coriolis force plays a key role for the selection of the magnetic field geometry. When inertia is weak, the field is very dipolar. When inertia becomes relevant, the dynamo switches to generating a multipolar field. The Rossby number (Eq. 7.5) calculated with the shell thickness is still significantly smaller than unity in the multipolar cases; for example in the models shown in Fig. 7.7 it is 0.01 and 0.02, respectively. Christensen and Aubert (2006) suggested that a "local" Rossby number is a more appropriate measure for the ratio between inertial forces and the Coriolis force:

$$R_{\text{ol}} = \frac{v}{\Omega \ell}. \tag{7.15}$$

The mean flow length scale ℓ is taken from the kinetic energy spectrum as function of wavelength. Analyzing a large number of model results, Christensen and

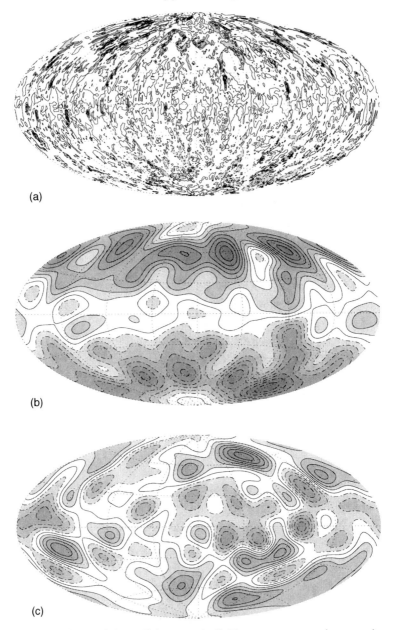

(a)

(b)

(c)

Fig. 7.7. Snapshots of the radial magnetic field component on the outer boundary from numerical dynamo models. Solid lines are used for outward flux and dashed lines for inward flux (arbitrary contour steps in each panel). Grey scale indicates absolute amplitude. (a) Model parameters $E = 10^{-5}$, $R_a^* = 0.12$, $P_m = 0.8$, $P_r = 1$. (b) Same field low-pass filtered to harmonic degrees $n < 14$. (c) Model parameters $E = 10^{-5}$, $R_a^* = 0.17$, $P_m = 0.5$, $P_r = 1$, low-pass filtered. R_m is approximately 900 in both cases; R_{ol} is 0.125 in (a) and (b), and 0.19 in (c). See Color Plate 4.

Aubert (2006) found that a transition from dipolar to multipolar magnetic field occurs when R_{ol} exceeds a critical value of approximately 0.12, irrespective of what the values of control parameters such as E, P_r, and P_m are. Dipolar dynamos that show occasional reversals have a local Rossby number near the transitional value. Hence a reversal may represent an accidental brief lapse of the basically dipolar dynamo into the multipolar regime. When the dipole recovers, it can then take either polarity.

Olson and Christensen (2006) derived an empirical rule based on numerical model data for relating the local Rossby number to the fundamental control parameters of the dynamo. It involves powers of all four control parameters and requires an extrapolation over a large range to apply it to the planets. Nonetheless, using appropriate parameter values for the Earth, $R_{ol} \approx 0.1$ is predicted for the geodynamo, which puts it close to the transition point between the dipolar and the multipolar class, in agreement with the occasional occurrence of reversals. One problem is that the flow length scale associated with this value of R_{ol} is only $\ell \approx 100$ m. Even if eddies of such size have significant energy in the core, at this scale the magnetic field is diffusion-dominated and cannot be affected directly by the flow. An indirect effect is conceivable. For two-dimensional turbulence it is well known that small scales transport energy into large flow scales ("inverse cascade", e.g. Davidson, 2004). The nearly geostrophic flow at the onset of rotating convection is quasi-two-dimensional. In the dynamo models the flow is still preferentially aligned with the axis of rotation. If this is also the case in the Earth's core, small eddies may affect the circulation at large scales and play a direct role in the induction process.

7.6.3 Flow structure and field generation mechanism

The stretching of magnetic field lines by differential rotation in the case of the solar dynamo, particularly at the tachocline, is thought to be of major importance for the generation of a toroidal magnetic field that is much stronger than the poloidal field (Section 5.4.3). In most geodynamo models, in contrast, differential rotation does not contribute much to the total kinetic energy and the toroidal and poloidal magnetic field components have similar strength. As mentioned before, the flow is strongly organized by rotational forces and the vortices are elongated in the z-direction. Even at a highly supercritical Rayleigh number and in the presence of a strong magnetic field, the flow outside the inner-core tangent cylinder is reminiscent of the helical convection columns found at onset. Inside the tangent cylinder, the flow pattern is different and often exhibits a rising plume near the polar axis (Fig. 7.8b). The plume is accompanied by a strong vortex motion (called a "thermal wind") with a retrograde sense of rotation near the outer surface changing to

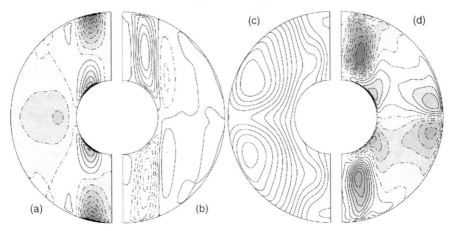

Fig. 7.8. Time-averaged axisymmetric components of velocity and magnetic-field components for a planetary dynamo model with $R_a^* = 0.225$, $E = 3 \times 10^{-4}$, $P_r = 1$, $P_m = 3$. $R_m \approx 250$ and $R_{ol} \approx 0.1$. The grey scale indicates absolute intensity. (a) Azimuthal velocity, broken lines are for retrograde flow, (b) streamlines of meridional velocity, full lines for clockwise circulation, (c) poloidal magnetic field lines, (d) azimuthal (toroidal) magnetic field, broken lines westward directed field.

prograde rotation at depth (Fig. 7.8a), because the Coriolis force acts on the associated converging flow near the inner core boundary and diverging flow near the outer boundary.

Several authors have analyzed their numerical dynamo solutions in order to understand the basic mechanism by which the magnetic field is maintained. In the tradition of mean-field dynamo theory it is considered how large-scale (e.g. axisymmetric) poloidal field is converted to large-scale toroidal field and vice versa. There is general agreement that the axial dipole field is generated from the axisymmetric toroidal field by an α-effect associated with the helical flow in the convection columns outside the tangent cylinder. In mean-field theory as it is used in astrophysics, the α-effect is associated with unresolved turbulent eddies (Vol. I, Section 3.4.3). In the geodynamo models a "macroscopic" α-effect is observed.

The mechanism for generating the axisymmetric toroidal field is less clear and both an α-effect and differential rotation (Vol. I, Sections 3.3.7 and 3.4) seem to play a role. Often two flux bundles in the azimuthal direction are found outside the tangent cylinder, with opposite polarity north and south of the equatorial plane (Fig. 7.8d). Olson *et al.* (1999) show that they are generated from the axisymmetric poloidal field by a similar macroscopic α-affect associated with the helical convection columns (α^2 dynamo). Other authors show that the Ω-effect (the shearing of poloidal field by differential rotation) contributes strongly to the generation

of axisymmetric toroidal field, even though the kinetic energy in the differential rotation is rather limited. While in weakly driven numerical dynamo models
the regions inside the tangent cylinder, north and south of the inner core, are
nearly quiescent, vigorous flow is found here in more strongly driven models. In
these cases a strong axisymmetric toroidal field is found inside the tangent cylinder region, produced by the shearing of poloidal field lines in the polar vortex
(Fig. 7.8a, c, d).

7.6.4 Comparison of geodynamo models with Earth's field

Some criteria to judge the similarity between the magnetic field of a dynamo model
and the geomagnetic field are (1) the agreement in field strength, (2) the agreement
in the shape of the spatial power spectrum, (3) qualitative agreement in the magnetic field morphology, (4) an agreement in the time scales of secular variation,
and (5) agreement in the frequency and characteristic properties of dipole reversals. Many published models do well with respect to some of these criteria and
a few satisfy most of them to a fair degree. A good guide for a dynamo model
to generate an Earth-like magnetic field is probably that the magnetic Reynolds
number and the local Rossby number must assume the appropriate values. Other
parameters may be less critical. We defer the discussion of the field strength to
Section 7.6.5 and address the other criteria below.

The shape of the geomagnetic power spectrum up to degree 13 at the core–
mantle boundary (Fig. 7.1) is reproduced rather well by several models, although
often the dipole is somewhat stronger or weaker relative to higher multipoles
than in the present geomagnetic field (see Christensen and Wicht, 2007, for more
details). Comparing Figs. 7.2b and 7.7b, similar morphological structures are
found. The model reproduces flux lobes at high latitudes, which are roughly aligned
on similar longitudes in both hemispheres, although they may be more numerous
than they are in the geomagnetic field. The model also shows weak flux at the poles
and scattered flux spots of both polarities at low latitudes.

The cause for these various magnetic structures in the core field has tentatively
been inferred from the flow structures that are predicted by theory and seen in the
dynamo models (Gubbins and Bloxham, 1987; Christensen *et al.*, 1998). The high-
latitude flux concentrations are related to the helical convection columns outside
of the inner core tangent cylinder. Cyclonic vortices are associated with downflow near the surface (Fig. 7.6) that concentrates magnetic flux. Low flux at the
poles can be related to the upwelling plume near the rotation axis which dis-
perses magnetic field lines. The variation of the geomagnetic field in the north
polar region of the core–mantle boundary, assuming that it is frozen into the fluid,
also supports the existence of an anticyclonic vortex motion near the core–mantle

boundary, which should accompany the rising plume (Olson and Aurnou, 1999; Hulot *et al.*, 2002).

Finally, bipolar pairs of flux spots at low latitudes are found in many dynamo models. They have been associated with the emergence of toroidal magnetic flux tubes through the core–mantle boundary, analogous to the mechanism for the formation of bipolar active regions on the Sun (Christensen *et al.*, 1998, Christensen and Olson, 2003). The pairs are often north–south rather than east–west aligned and their polarity is opposite to the global dipole polarity. Such a configuration arises in the models because strong toroidal fields of opposite polarity are found at close distance north and south of the equator (Fig. 7.8d) and because the columnar flow is north–south aligned and acts on both toroidal tubes in the same way. Comparable structures exist in the Earth's field at the core–mantle boundary (see Fig. 7.2b, beneath Africa and the Atlantic Ocean) and have been explained by flux expulsion (Bloxham, 1989). However, in the geomagnetic field they are more strongly offset from the equator than they are in dynamo models and overall the semblance between model field and geomagnetic field is less convincing regarding the low-latitude flux spots than it is for other magnetic structures.

Matching the time scales of secular variation is a matter of magnetic Reynolds number. For Earth-like values of R_m the model magnetic field changes at the observed rates, provided model time is scaled to real time using the magnetic diffusion time scale d^2/λ. There is a certain circularity in this argument, because the magnetic Reynolds number of Earth's core is estimated under the assumption that most of the observed secular variation is due to the frozen-flux advection of magnetic structures.

Geodynamo models that are in the right regime for dipole reversals often show a degree of agreement with the paleomagnetic record that goes beyond the simple occurrence of reversals, even in cases with very modest parameter values such as a relatively large Ekman number. Figure 7.9 shows time series of the dipole tilt, dipole moment and relative dipole field strength in such a model. Some of these properties resemble traits of the geomagnetic field: (1) the directional change of the dipole field is a relatively brief event compared to the length of the period in which the dipole is nearly aligned with the rotation axis; (2) the dipole moment starts to drop before the directional change occurs, and during the reversal the magnetic field is multipolar; and (3) apart from complete reversals, strong changes occur in the dipole direction that are brief and non-persistent (geomagnetic excursions). The actual frequency of reversals in geodynamo models seems to depend on the fine tuning of parameters. The search for a clearly defined "mechanism" for reversals in the dynamo models has not yet come up with a unique answer.

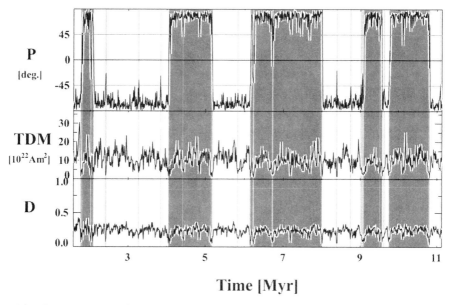

Time [Myr]

Fig. 7.9. Evolution of the dipole tilt for a modeled planetary magnetic field with respect to the equator (P), the true dipole moment (TDM), and the dipole strength relative to the total field strength at the core–mantle boundary (D) for a dynamo model with $E = 10^{-3}$, $R_a^* = 0.5$, $P_m = 10$, $P_r = 1$. The present TDM of the geomagnetic field is 8×10^{22} A m^2. Dark and light bands indicate polarity intervals. (From Christensen, 2009, courtesy of Johannes Wicht.)

7.6.5 Scaling of magnetic field strength

Dynamo scaling laws relate characteristic properties, for example the mean magnetic field strength, to fundamental quantities of the planet and its core, such as core radius R_c, density ρ, conductivity σ, rotation rate Ω, and convected energy flux q_c. As explained in Section 7.5, it has commonly been assumed that the magnetic field strength in a planetary dynamo is determined by a magnetostrophic force balance. Magnetostrophic balance is often associated with an Elsasser number of order one, which means that the magnetic field strength scales as $B \propto (\rho\Omega/\sigma)^{1/2}$. Notably, the Elsasser number is independent of the energy flux. Stevenson (1983) pointed out that this scaling is unlikely to be universally applicable, because it ignores the requirement that sufficient energy must be available for balancing Ohmic dissipation. Alternative scaling laws based on a magnetostrophic balance, which make different assumptions on how the Lorentz force depends on characteristic properties than those used for deriving the Elsasser number, have also been proposed (e.g. Starchenko and Jones, 2002).

Christensen and Aubert (2006) and Christensen *et al.* (2009) suggest that the magnetic field strength is not determined by a force balance, but solely by the energetics of the dynamo, at least in rapidly rotating cases with a dipole-dominated magnetic field. For a thermally driven dynamo, the heat flux that is available per unit volume for conversion to other forms of energy is given by $H_T^{-1} q_c$, where q_c is the convected part of the heat flux and H_T is temperature scale height. The rate at which magnetic energy is dissipated scales as $\lambda B^2 / \ell_B^2$, where ℓ_B is the length scale of the field. Equating energy generation and dissipation, the following scaling for the magnetic energy density is obtained:

$$\frac{B^2}{2\mu_o} \propto f_{\text{ohm}} \frac{\ell_B^2}{\lambda} \frac{q_c}{H_T}, \qquad (7.16)$$

where f_{ohm} is the fraction of the energy dissipated by Ohmic dissipation rather than by viscous dissipation. It is thought to be close to one in planetary dynamos.

The magnetic length scale ℓ_B depends on the magnetic Reynolds number. At high R_{m} the folding of field lines in the flow can continue to smaller scales before reconnection occurs. The flow velocity and hence R_{m} depend on the available energy flux as well. Here, we consider the scaling law based on this concept in the form proposed by Christensen *et al.* (2009), without going into the intricate scaling arguments that can lead to it:

$$\frac{B^2}{2\mu_o} = c \, f_{\text{ohm}} \, \bar{\rho}^{1/3} (F q_o)^{2/3}, \qquad (7.17)$$

where $\bar{\rho}$ is the mean density, c is a constant prefactor, q_o a reference value for the heat flux (for example the surface flux) and F is a dimensionless efficiency factor of order one. The necessary averaging of radially varying quantities, such as ρ, q_c or H_T, is condensed into the efficiency factor, which can be calculated for a given planetary or astrophysical object from a structural model and assumptions on the radial distribution of the convected flux. A remarkable point about Eq. (7.17) is that it predicts the surface-averaged magnetic field strength (or flux density) to be independent of rotation rate and of electrical conductivity. There are obvious limits to that; for example, for very low conductivity the magnetic Reynolds number would be subcritical and the field strength must be zero.

A fairly large number of geodynamo calculations are currently available that cover a decent range of the numerically accessible control parameter space. This allows scaling laws to be tested. Non-dimensionalizing Eq. (7.17) by dividing it by $\bar{\rho}\Omega\lambda$ leads to a non-dimensional energy flux q^* and a scaled magnetic energy density that is identical to the Elsasser number $E_{\text{m}}^* = \Lambda$. The dependence on density drops out in the non-dimensional form of Eq. (7.17). Figure 7.10 compares E_{m}^* to the right-hand side of the non-dimensionalized Eq. (7.17) for dynamo simulations with a dipolar magnetic field. The efficiency factor F, which results from the model setup, has been calculated analytically. The fraction of Ohmic dissipation has been

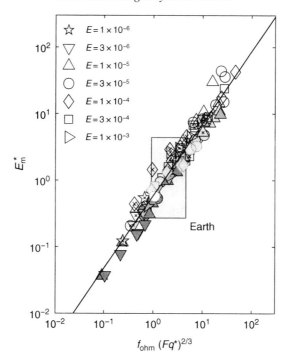

Fig. 7.10. Non-dimensional magnetic energy density in numerical dynamo models versus 2/3-power of the non-dimensional available energy flux. Symbol shape indicates Ekman number (Eq. 7.11), shading of the symbols magnetic Prandtl number (Eq. 7.13); darker means a lower value. Crosses inside the main symbols indicate $P_r > 1$ and circles indicate $P_r < 1$. Symbols with black edges are models driven by an imposed temperature contrast, those with grey edges are for compositional convection. The line represents the fit for a forced slope of one, equivalent to an exponent of 2/3 in Eq. (7.17). The location of the geodynamo in the diagram is shown by the grey rectangle.

recorded for each dynamo run, where it is typically in the range 0.3–0.8. While the exponents in Eq. (7.17) come from scaling theory, the prefactor $c = 0.63$ is obtained by fitting the numerical model results.

Figure 7.10 shows a number of important points. (1) The Elsasser number is not always close to unity in the different dynamo models, but varies over three orders of magnitude. (2) The model results fall on a single line reasonably well, irrespective of the value of Ekman number (which differs by three orders of magnitude) and of the two Prandtl numbers (which differ by two orders of magnitude each between different models). The Ekman and Prandtl control parameters describe the influence of viscosity, magnetic, and thermal diffusivity, and of the rotation rate. Therefore, these quantities do not play an important role in the field strength. The fear that the dynamical regime in the models differs fundamentally from that in the Earth's core (and could be dominated by viscosity) is probably unfounded. (3) The model results agree reasonably well with the 2/3-power scaling law. A best fit results in a slightly larger exponent of 0.71. (4) The magnetic field inside Earth's core (estimated to be in the range 1–4 mT) agrees well with the prediction for current estimates of the heat flow at the core–mantle boundary and the associated compositional driving due to inner core growth (grey rectangle in Fig. 7.10).

To estimate the magnetic field strength inside the dynamos of other planets, the observed (low-order) magnetic field must be downward continued to the outer boundary of their dynamo region and an assumption must be made on the factor by which the internal field is stronger, which is guided by the ratios found in the numerical dynamo models. Jupiter's field strength agrees well with the prediction (Fig. 7.11). Other solar system bodies are more problematic. For Mercury

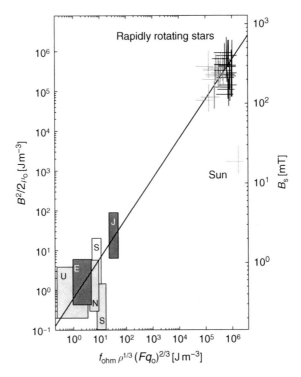

Fig. 7.11. Magnetic energy density in the dynamos of planets and certain stars versus the predicted dependence on a function of available energy flux and density. E: Earth; J: Jupiter; S: Saturn; U: Uranus; N: Neptune. The grey rectangle for Saturn is for a dynamo boundary at $0.62R_p$ and the white rectangle for $0.4R_p$. Black crosses for rapidly rotating main-sequence stars of low mass and grey crosses for classical T Tauri stars. The scale on the right is the average field strength (or flux density) at the surface of the dynamo. The black line is taken from Fig. 7.10 and converted here to physical units.

and Ganymede the available energy flux is very uncertain. The strengths of the multipolar fields of Uranus and Neptune may not fall on the line that has been calibrated with dipolar dynamo models, although within the uncertainties it seems to be compatible. Saturn's field is too weak for a dynamo that extends to the top of the metallic hydrogen layer at approximately 60% of the planetary radius. If the top of the active dynamo region, below a stably stratified region, is put at $0.4R_\mathrm{p}$, the result can be brought into line with the prediction (compare Sections 7.4.3 and 7.7.3).

Is the scaling law applicable to other convection-driven spherical dynamos in cosmic bodies? The solar dynamo seems fundamentally distinct from the dynamos of Earth or Jupiter, because the Sun rotates too slowly and/or because of the special role played by the tachocline in the field generation process. Main-sequence stars with less than 0.35 solar masses (M-type dwarfs) are fully convective, and so are very young contracting T Tauri stars (see Chapters 2 and 3). Hence, they lack a tachocline. Furthermore, these stars generally rotate much more rapidly than the Sun does. The rapid rotators have strong surface magnetic fields in the range of several tenths of a tesla (several kilogauss). Recent observational evidence shows that the magnetic flux of M-type dwarfs increases with decreasing rotation period (decreasing Rossby number) up to some threshold, but becomes independent of rotation rate for the more rapidly rotating stars (Reiners *et al.*, 2009; compare Section 2.3.2). This is akin to the independence of the field strength on rotation rate found in the numerical geodynamo models. Furthermore, the mapping of the magnetic field topology of some mid-M dwarfs by a technique called Zeeman–Doppler tomography shows strong large-scale magnetic field components at the surfaces of these objects, which are often dominated by the axial dipole (see Section 2.6 and Fig. 2.16; also, e.g. Morin *et al.*, 2008).

Christensen *et al.* (2009) found that the observed magnetic flux density of rapidly rotating M-type dwarf stars and of T Tauri stars agrees with the prediction of Eq. (7.17), as shown in Fig. 7.11. The slowly rotating Sun does not fall on the same line. Note that the solid line in the figure is *not* a fit to the various objects, but both its slope and the prefactor c have been taken from scaling theory and from the fit to the results of numerical geodynamo models shown in Fig. 7.10. The fact that rapidly rotating stars fall on the same line as the planets strongly supports the validity of the scaling law. Furthermore, it suggests that dynamos in rapidly rotating stars are not fundamentally different from planetary dynamos, despite the much higher energy flux and far greater magnetic Reynolds number.

A more detailed account on scaling laws for planetary dynamos, including the scaling of velocity and ohmic dissipation is found in Christensen (2009).

7.7 Dynamo models for Mercury and the gas planets

Existing dynamo models for planets other than Earth are derivatives of geodynamo models, where some parameter or other has been adapted to the assumed conditions in the specific planet. This does not concern the fundamental control parameter – they cannot be made to match planetary values anyway. Several of these models assume the existence of stably stratified layers in the fluid core of the planet.

7.7.1 Mercury

The main task of a dynamo model for Mercury is to come up with an explanation for the weakness of the observed magnetic field. Stevenson (1987) suggested that the flow in Mercury's core would be too slow for the magnetic Reynolds number to exceed the threshold of a standard hydromagnetic dynamo, but that a "thermoelectric" dynamo may generate the observed weak field. Topography on Mercury's core–mantle boundary would imply slight variations of its temperature. These are accompanied with spatial differences of the thermoelectric EMF arising from the contact of two different materials. The associated currents would set up a toroidal magnetic field (that is invisible from outside). The α-effect of helical flow in Mercury's core acting on this toroidal field would generate poloidal field even at modest values of R_m. Conditions for this model to work are that the lower mantle of Mercury is a fairly good conductor, so that the thermocurrents can close, and that the core topography is of large scale, in order to explain the large-scale external field. It is highly uncertain whether these conditions are met.

Mercury probably has a solid inner core. Its size is very uncertain, but may be quite large, with only a thin fluid shell remaining. Some thin-shell dynamo models (Stanley *et al.*, 2005, Takahashi and Matsushima, 2006) succeeded in producing relatively weak magnetic fields outside the dynamo region. However, these fields are either still too strong by a factor of ten or more, or they are rather multipolar, in contrast to the observed dipole dominance.

The outer part of Mercury's fluid core is probably stably stratified, because thermal conduction along a distinctly subadiabatic temperature gradient is sufficient to carry the expected modest core heat flow. A deep sub-shell of the fluid core might then convect, driven by the latent heat of freezing of the inner core and the associated light element flux. Christensen (2006) and Christensen and Wicht (2008) presented numerical dynamo simulations for such a scenario and found that in the dynamo layer a strong magnetic field is generated. Mercury's slow rotation with a period of 59 days implies that the local Rossby number is larger than one. Consequently, the internal dynamo field is small scaled, not dipolar. The small-scale field varies rapidly with time. Therefore, it is strongly attenuated by a skin effect in the conducting stable layer above the dynamo and is virtually unobservable outside of the core. The dipole component makes only a small contribution inside the dynamo, but it varies more slowly with time. Hence, it can penetrate through the stable layer to some degree and dominates the structure of the very weak field at the planetary surface. Some of these models match the observed field strength and geometry.

Mercury's magnetosphere is very small (see Vol. I, Chapter 13). At the top of Mercury's core the external field created by the magnetospheric current systems is much stronger relative to the internal field than it is in case of the Earth. Glassmeier *et al.* (2007) suggest that the feedback of the external field may play an important role for the dynamo. Including it in a kinematic $\alpha\Omega$ dynamo model, they find two branches of solution, with a strong and a weak magnetic field, respectively, of which the latter might represent the situation at Mercury. The hypothesis needs to be tested in MHD dynamo models.

7.7.2 *Jupiter*

The semblance in the geometry of Jupiter's field to that of the Earth's field and the finding that the field strength of both planets is well explained by the energy flux rule (Section 7.6.5) suggest that Jupiter's dynamo and the geodynamo are generically very similar. Some potentially important differences may exist, however. Unlike in the case of the Earth, where the electrical conductivity changes abruptly at the core–mantle boundary, in Jupiter it varies more gradually in the transition region between molecular and metallic hydrogen. An open question is whether the strong zonal wind circulation that is seen at Jupiter's surface is deep-rooted in the molecular hydrogen envelope, and if so, how it interfaces with the dynamics of the dynamo region. Other differences are that the radial density variation in Jupiter's metallic hydrogen shell is more pronounced than it is in the Earth's core and that the flow in Jupiter's dynamo is not strongly driven from below. Dedicated MHD simulations for Jupiter's dynamo are underway. NASA's foreseen Juno mission will characterize Jupiter's magnetic field with a much better resolution than was possible up to now. Having a closer look might reveal some significant differences from the geomagnetic field.

7.7.3 *Saturn*

The challenge in the case of Saturn's dynamo is to explain the high degree of axisymmetry of the magnetic field. The only conceptual model so far is that by Stevenson (1980, 1982), who suggested that the hypothetical stable layer, caused by the helium immiscibility in the upper part of the metallic hydrogen shell, plays the essential role. The density stratification suppresses convection, but is still compatible with toroidal flow, such as differential rotation. Let us assume for simplicity that the whole stable layer rotates like a uniform shell with respect to the underlying dynamo region and that the dynamo field is stationary. Seen from a reference frame that is fixed to the rotating shell, the non-axisymmetric field components is time dependent, whereas the axisymmetric part is stationary. If the magnetic Reynolds number characterizing the shell motion is large enough, a skin effect eliminates the non-axisymmetric parts of the field, but leaves the axisymmetric components unaffected.

In this concept the role of the stable conducting layer, which shields the dynamo, is somewhat akin to that in the Mercury model by Christensen (2006). Differences are that in the Saturn case the primary dynamo field is dipolar because Saturn is a rapid rotator and that motion in the stable layer is important. Christensen and Wicht (2008) find in their dynamo models with a dipole-dominated field that latitudinal differences in the heat flux from the dynamo region into the overlying stable shell drive strong differential rotation as a thermal wind circulation. The magnetic field has significant non-axisymmetric components inside the dynamo region, but the field outside the core is very axisymmetric. The axisymmetry becomes much less when in a control experiment differential rotation in the stable layer is suppressed. These full MHD dynamo simulations support Stevenson's concept.

7.7.4 Uranus and Neptune

In the case of Uranus and Neptune it must be explained why their dynamos generate a multipolar field. Both planets rotate rapidly and the estimate by Olson and Christensen (2006) for the local Rossby number would put them into the dipolar regime. Either the local Rossby number rule for the selection of the field morphology fails, or the dynamos in these two planets are rather distinct from the geodynamo. Stanley and Bloxham (2004, 2006) present a dynamo model with a thin convecting shell that surrounds a conducting, but convectively stable, fluid core region. Some of their dynamo models generate magnetic fields that agree well with the spectral power distribution in the lower-order harmonic field components. It is also not clear if the much lower electrical conductivity in the interior of Uranus and Neptune, compared to that of other planetary dynamos, plays a role. Gómez-Pérez and Heimpel (2007) suggest that a less dipolar field may result in this case.

7.8 Outlook

Recent decades have been an exciting time for planetary magnetism. In addition to better characterizing the geomagnetic field in space and time, the exploration of the magnetic fields of other planets has brought some surprises. There is more diversity than previously thought. At the same time, dynamo theory has matured, and modeling by direct numerical simulation is now feasible.

Geodynamo models are remarkably successful in reproducing many observed properties of the geomagnetic field. In this respect, planetary dynamo modeling seems to be more advanced than that of the solar dynamo. To some extent the task is easier for the geodynamo – our ignorance of the small-scale structure of the geomagnetic field implies that a model can be declared successful when it captures the crude properties. Our conceptual understanding of how exactly the geodynamo works has not quite kept pace with the modeling attempts. The reasons for the success of geodynamo models are a matter of speculation, but the following points may be essential: (1) It is possible to fully resolve the magnetic field structure and hence the details of the magnetic induction process. Put differently, direct numerical simulations at the correct value of the magnetic Reynolds number are feasible. (2) Although the model viscosity and thermal diffusivity are far larger than realistic microscopic values, they seem low enough to not alter the dynamical regime. (3) The flow at large and intermediate scales, which is responsible for magnetic induction, may be realistic in the model. This is made possible because rotation has a stronger influence than it has on the flow in the solar convection zone and imposes some order on the circulation. Also, the strong radial differences of density in the solar convection zone, which lead to large variations in length scales and velocity scales, are not a problem in planetary dynamos.

Future work on the geodynamo must improve our understanding of what the essential conditions are for an Earth-like model. The validity of the proposed scaling laws in a parameter range closer to Earth must be tested in simulations at lower values of the Ekman number and magnetic Prandtl number and in laboratory dynamo experiments. Their theoretical foundation must be put on firmer ground.

Dynamo models for explaining the magnetic fields of planets other than Earth have had some successes, too, but progress here is hampered by our very rudimentary knowledge of the relevant conditions inside these bodies. Cosmic objects such as rapidly rotating low-mass stars may provide a bridge between planetary dynamos and the solar dynamo. Modeling studies of convection-driven dynamos in a wide range of objects, together with improvements of our knowledge of their magnetic field properties from observation, may ultimately lead to a unifying dynamo theory explaining the commonalities and differences found in all these various objects.

8

The structure and evolution of the three-dimensional solar wind

John T. Gosling

8.1 Introduction

Parker (1958) showed that the natural state of a hot and extended stellar corona is one of supersonic, super-Alfvénic expansion. In the case of the Sun, the evolution of the strong magnetic field that permeates the corona modulates this expansion (e.g. Pneuman and Kopp, 1971). Indeed, it is the interplay between the coronal magnetic field and the expansion that produces both a highly structured solar corona and a spatially variable solar wind (see Vol. I, Chapter 9). For example, the combed-out appearance of the outer solar corona is a product of the coronal expansion. Because the solar wind plasma is an excellent electrical conductor, the coronal magnetic field is "frozen" into the solar wind flow (Vol. I, Chapter 3) as it expands away from the Sun, forming what is now commonly called the heliospheric magnetic field, HMF.[†] A simple model of the HMF (Parker, 1958) predicts that solar rotation causes the HMF in the solar equatorial plane to be bent into Archimedean spirals (Vol. I, Section 9.2) whose inclinations relative to the radial direction depend on heliocentric distance and the speed of the wind.

The Sun's magnetic field evolves continually, the most pronounced changes being those associated with the advance of the $\sim$11-year solar activity (sunspot) cycle and the $\sim$22-year magnetic cycle. As the Sun's magnetic field evolves on a variety of temporal scales, so too does the structure of the corona and the related coronal expansion. The most dramatic changes in the expansion occur during coronal mass ejections, CMEs (e.g. Vol. II, Chapters 5 and 6; Crooker *et al.*, 1997; Kunow *et al.*, 2006) in which solar material is propelled outward into

---

[†] The term heliospheric magnetic field, HMF, is used throughout this chapter to denote what is often referred to as the interplanetary magnetic field, IMF. The use of HMF rather than IMF avoids the potential ambiguity that IMF might refer to the field within the ecliptic, which is historically what it often was as most spacecraft sampling the solar wind properties were confined to the disk containing the planetary orbits.

Heliophysics: Evolving Solar Activity and the Climates of Space and Earth, eds. Carolus J. Schrijver and George L. Siscoe. Published by Cambridge University Press. © Cambridge University Press 2010.

the heliosphere from closed magnetic field regions in the solar atmosphere not previously participating in the solar wind expansion.

This chapter is concerned with the global structure and evolution of the solar wind that arise from a combination of spatial and temporal variability in the solar magnetic field (and hence also in the coronal expansion) and solar rotation. We will not specifically address the physics of the coronal expansion itself (see Vol. I, Chapter 9) or the cause of CMEs (see Vol. II, Chapter 6). Rather, we will take it as a given that fast solar wind originates from open-field regions associated with X-ray dark regions called coronal holes, that the slow wind originates near or within the streamer belt that commonly wraps around the Sun (at least during the decline and near the minimum of the solar activity cycle; compare Fig. 8.1 in Vol. I for an approximation of this structure with a potential-field source-surface model), and that CMEs with a variety of masses and speeds occur at a rate that varies roughly with the phase of the solar activity cycle.

8.2 The heliospheric current sheet

To a first approximation the magnetic field in the corona well above the photosphere is roughly that of a dipole tilted with respect to the rotation axis of the Sun, particularly on the declining phase of the solar activity cycle. The tilt of the dipole varies as the solar activity cycle progresses. As illustrated in Fig. 8.1, the dipole moment **M** tends to be inclined substantially relative to the solar rotation axis Ω on the declining phase of the solar cycle, but tends to be nearly aligned with the rotation axis near solar activity minimum. Near the solar magnetic equator and immediately above the solar photosphere the dipole field is transverse to the radial

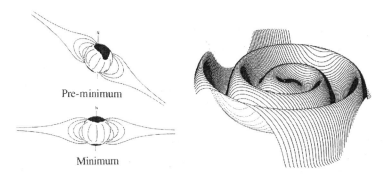

Fig. 8.1. (Left) Schematic illustrating the changing tilt of the solar magnetic dipole and related coronal structure relative to the rotation axis of the Sun. (From Hundhausen, 1977.) (Right) Idealized configuration of the heliospheric current sheet in the solar wind when the tilt of the solar magnetic dipole is substantial. (From Jokipii and Thomas, 1981. Reproduced by permission of the AAS.)

direction and is sufficiently strong to constrain the plasma from expanding outward. Thus closed magnetic arcades commonly straddle the magnetic equator. At greater heights the field weakens sufficiently that the pressure of the coronal plasma opens up the arcades, and the plasma is able to expand outward. The resulting outflow produces helmet-like streamers in the corona and a relatively slow and dense solar wind flow (e.g. Gosling *et al.*, 1981). Embedded within the low-speed flow is a magnetic field polarity reversal that reflects the magnetic control of the expansion. At higher magnetic latitudes the coronal expansion is relatively unconstrained by the magnetic field and produces coronal holes in the solar atmosphere and unipolar high-speed flows of relatively low density (e.g. Krieger *et al.*, 1973).

The magnetic polarity reversal that is embedded within the low-speed flow is known as the heliospheric current sheet, HCS, and is effectively the extension of the solar magnetic equator into the heliosphere. When the solar magnetic dipole and the solar rotation axis are nearly aligned, the HCS coincides roughly with the solar equatorial plane. On the other hand, at times of substantial dipole tilt solar rotation causes the HCS to become highly warped, as illustrated in the right portion of Fig. 8.1. Moving outward from the Sun, successive ridges in the HCS correspond to successive solar rotations and are separated radially by about 4.7 AU when the flow speed of the solar wind at the HCS is 300 km/s. The maximum solar latitude attained by the HCS in this simple picture is directly related to the tilt of the solar dipole. Of course this picture is over-simple because the solar magnetic field is almost never truly a dipole and, at times, the HCS is fragmented into more than one current sheet (e.g. Crooker *et al.*, 1993). Moreover, dynamic processes in the solar wind eventually severely distort the shape of the HCS far from the Sun when the dipole tilt is substantial (Pizzo and Gosling, 1994). Finally, although the HCS is arguably the largest and most important current sheet in the solar wind, it is only one of many current sheets present in the solar wind at any given time. In practice the HCS is best identified (e.g. Kahler and Lin, 1994) in solar wind data by reversals of the solar wind suprathermal electron strahl flow polarity (parallel or anti-parallel to the heliospheric magnetic field). The strahl is a nearly collisionless electron beam that is focused along the HMF and that carries the solar wind electron heat flux (e.g. Rosenbauer *et al.*, 1977); thus, it is always directed away from the Sun along the HMF regardless of the field polarity.

8.3 Latitudinal and solar-cycle variations of the solar wind

Observations reveal that solar wind properties, in particular the solar wind speed, vary strongly as a function of distance from the heliospheric current sheet (e.g. Zhao and Hundhausen, 1981; Bruno *et al.*, 1986). Thus, because the heliospheric current sheet is usually both warped and tilted relative to the solar equator, both

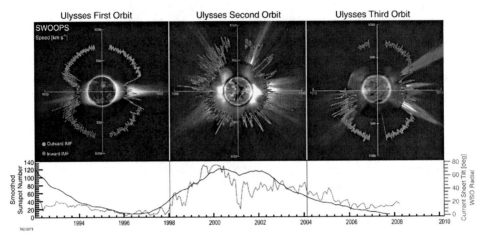

Fig. 8.2. (Top) Polar plots of solar wind speed as a function of latitude for almost three complete Ulysses polar orbits about the Sun plotted over solar images characteristic of solar minimum (8/17/1996), solar maximum (12/07/2000), and again solar minimum (03/28/2006). Color-coding of the speed plots indicates the polarity of the HMF, blue inward directed and red outward directed. In each plot the earliest times are near aphelion on the left (nine o'clock position) and time progresses counterclockwise. The apparent differences in the scale of structures at low latitudes on opposite sides of each panel are artifacts associated with the fact that Ulysses changed latitude very rapidly near perihelion (1.34 AU) but very slowly near aphelion (5.41 AU). The brief low-speed interval near latitude 80 °S in the third orbit plot was associated with Ulysses' encounter with the ion tail of comet McNaught (Neugebauer *et al.*, 2007). The solar images from Sun center outward are from the Solar and Heliospheric Observatory (SOHO) Extreme Ultraviolet Imaging Telescope (Fe XII at 1950 nm), the Mauna Loa K-coronameter (700–950 nm), and the SOHO C2 white light coronagraph. (Bottom) Contemporaneous values for the smoothed sunspot number (black) and the tilt of the heliospheric current sheet (red). (From McComas *et al.*, 2008.) See Color Plate 5.

low- and high-speed wind are commonly observed at low heliographic latitudes as the Sun rotates. As illustrated in the left-most and right-most panels in Fig. 8.2, Ulysses observations reveal that on the declining phase of the solar cycle and near solar activity minimum, substantial variations in solar wind speed at any solar longitude are confined to a relatively narrow latitude band centered on the heliographic equator, whereas a wind of nearly constant high speed (~750−800 km/s) prevails at higher latitudes (e.g. Coles and Maagoe, 1972; Schwenn, 1990; Gosling *et al.*, 1995a). The width of the band of variable wind changes as both the warping of the magnetic equator and the tilt of the dipole change and with the waxing and waning of solar activity. During Ulysses' first polar orbit about the Sun, which occurred on the declining phase of solar cycle 22, the width of the band varied from about ±20° to about ±35°. During Ulysses' third polar orbit during the declining phase of cycle 23 the width of the band appeared to be about $\pm 37°$ (McComas *et al.*, 2008). Numerous crossings of the HCS occurred within the low-latitude band during those orbits, whereas the high-speed wind at high latitudes was magnetically unipolar, with opposite polarities prevailing in the opposite solar hemispheres. The major difference obvious between the first and third orbit plots was the reversal of the magnetic polarities of the high-speed wind at high latitudes in the opposite solar hemispheres, reflecting the reversal of the Sun's magnetic field between the first and third orbits.

In contrast to the relatively simple structure observed during Ulysses' first and third orbits, during the second orbit, which occurred over the rise to and through solar activity maximum in cycle 23, the global structure of the solar wind was considerably more complex, reflecting a much more complex coronal structure (McComas *et al.*, 2001, 2002), consistent with earlier interplanetary scintillation measurements (Coles *et al.*, 1980). Mixtures of low-speed and high-speed wind were observed at all heliographic latitudes owing to a complicated mixture of flows from multiple solar sources, including streamers, CMEs, small coronal holes, and active regions. Throughout this Ulysses orbit, numerous magnetic field polarity reversals were observed at all latitudes; close to solar maximum the apparent dipole axis was oriented nearly perpendicular to the Sun's rotation axis as the magnetic polarity of the Sun's polar regions reversed sign.

8.4 Solar wind stream structure

As noted above, the solar wind at low heliographic latitudes tends to be organized into alternating streams of high- and low-speed flows (e.g. Snyder *et al.*, 1963), particularly on the declining phase of the solar activity cycle and near solar activity minimum. This characteristic pattern of alternating low- and high-speed flows is commonly known as solar wind stream structure. Figure 8.3, which shows 1-hour averages of selected solar wind plasma and magnetic field data obtained during a 42-day interval in 2005 on the declining phase of solar cycle 23, illustrates characteristic aspects of stream structure as observed near Earth's orbit. Five high-speed streams with speeds exceeding 550 km/s are evident in Fig. 8.3. The fourth and fifth streams were actually re-encounters with the first and second streams observed approximately one solar rotation ($\sim$27 d as observed from Earth) earlier. Each high-speed stream was magnetically unipolar with the changes in magnetic polarity associated with crossings of the heliospheric current sheet generally occurring when the wind speed was less than 350 km/s and the particle density was higher than average. Note, however, that the HCS crossing on March 31, 2005 occurred when the wind speed was $\sim$480 km/s, that the nearly 2-day interval of positive polarity of the heliospheric magnetic field on April 9–11, 2005 did not include any

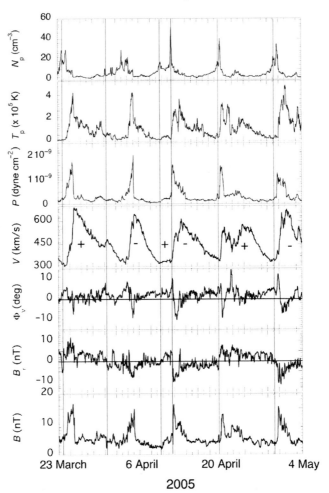

Fig. 8.3. Solar wind stream structure at 1 AU for a 42-day interval in 2005. From top to bottom the quantities plotted are 1-hour averages of the solar wind proton number density, proton temperature, total (field plus plasma) pressure, bulk flow speed, flow azimuth angle (positive in the sense of Earth's motion about the Sun), radial component of the HMF, and the HMF magnitude. Vertical lines mark crossings of the HCS, which were identified by changes in the suprathermal electron strahl flow polarity. Plus and minus signs in the fourth panel mark the magnetic polarities.

high-speed flow, and that the second and third high-speed streams both had the same negative magnetic polarity.

Each high-speed stream in Fig. 8.3 was asymmetric with the wind speed quickly rising to a peak value and then declining more slowly. The magnetic field strength, proton number density, proton temperature, and total pressure (plasma plus field) all peaked on the rising speed portions of the streams, with the density and field peaking prior to the proton temperature. As the speed rose, the flow was characteristically deflected first westward (positive Φ_V) and then eastward. The locations where the flow deflection switched direction occurred close to the positions where the total pressure peaked, were associated with sharp drops in proton density and sharp rises in proton temperature, and are known as stream interfaces (e.g. Belcher and Davis, 1971; Burlaga, 1974). A stream interface is often a discontinuity in the flow, commonly follows the crossing of the heliospheric current sheet by several hours or more and separates what was originally (close to the Sun) slow, dense plasma from what was originally fast, thin plasma (e.g. Gosling *et al.*, 1978; Wimmer-Schweingruber *et al.*, 1997). The overall pattern of variability illustrated in Fig. 8.3 is the inevitable consequence of the evolution of the streams as they progress outward from the Sun.

8.5 Evolution of stream structure with heliocentric distance

8.5.1 A one-dimensional model of high-speed stream evolution

Because of spatial variations in the coronal expansion, solar wind flows of different speeds become radially aligned at low heliographic latitudes as the Sun rotates. It is this rotation-induced radial alignment of flows from different regions on the Sun that produces the characteristic patterns associated with solar wind stream structure. Because radially aligned parcels of plasma within a stream originate from different locations on the Sun, they are threaded by different magnetic field lines and thus cannot interpenetrate one another. Figure 8.4, which shows the result of a simple one-dimensional (1D) gas-dynamic simulation, illustrates the basic reasons why high-speed streams evolve with increasing heliocentric distance. The rising portion of the high-speed stream steepens kinematically with increasing heliocentric distance because gas (plasma) at the peak of the stream is traveling faster than the slower plasma ahead. As the speed profile steepens, material within the stream is rearranged; parcels of plasma on the rising-speed portion of the stream are compressed, causing an increase in pressure there, while parcels of plasma on the falling-speed portion of the stream are increasingly separated, producing a rarefaction.

It is common to refer to the compression on the leading edge of a high-speed stream as an interaction region. Being a region of high pressure, the interaction region expands into the plasma both ahead and behind at the fast mode speed (actually at the sound speed in the calculation shown in Fig. 8.4). The leading edge of the interaction region is called a forward wave because it propagates in the direction of the solar wind flow; the trailing edge is called a reverse wave because it propagates sunward in the solar wind rest frame but is carried away from the Sun

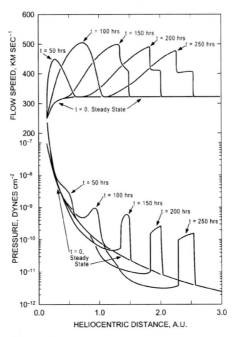

Fig. 8.4. Snapshots of solar wind flow speed and pressure as functions of heliocentric distance at different times during the outward evolution of a high-speed stream as calculated using a simple 1D gas-dynamic code. After obtaining a steady-state solar wind expansion that produced a flow speed of 325 km/s far from the Sun, a high-speed stream was introduced into the calculation by linearly increasing and then decreasing the temperature (and thus also the pressure) by a factor of four at the inner boundary at 0.14 AU over an interval of 100 h. (From Hundhausen, 1973.)

by the highly supersonic flow of the wind. Pressure gradients associated with these waves produce an acceleration of the slow wind ahead and a deceleration of the high-speed wind within the stream. The net result of the interaction is to limit the steepening of the stream and to transfer momentum and energy from the fast wind to the slow wind. The temporal variations of solar wind speed and pressure evident in Fig. 8.3 are in reasonable agreement with the results of this simple simulation; the agreement between simulations and observations improves when the additional two dimensions and the effects of the magnetic field are added to the calculation (e.g. Pizzo, 1980, 1982).

As long as the amplitude of a high-speed stream is sufficiently small, it gradually dampens with increasing heliocentric distance in the manner just described. However, when the difference in speed between the slow wind ahead and the peak of the stream is more than about twice the fast-mode (sound) speed the stream initially steepens faster than the forward and reverse pressure waves can expand into the surrounding plasma; thus, in such cases the interaction region at first narrows with increasing heliocentric distance. The non-linear rise in pressure associated with this squeezing eventually causes the forward and reverse waves bounding the interaction region to steepen into shocks. Because shocks (see, Vol. II, Chapter 7) propagate faster than the fast mode (sound) speed, the interaction region can expand once shock formation occurs. Observations reveal that relatively few stream interaction regions are bounded by shocks at 1 AU (e.g. Gosling *et al.*, 1972; Ogilvie, 1972), but that most are near the equatorial plane at heliocentric distances beyond about 3 AU (e.g. Gosling *et al.*, 1976; Hundhausen and Gosling, 1976; Smith and Wolfe, 1976) because the fast-mode (sound) speed generally decreases with increasing distance from the Sun. At heliocentric distances beyond about 5–10 AU a large fraction of the mass and magnetic field flux in the solar wind at low heliographic latitudes is found within expanding compression regions bounded by shock waves on the rising portions of strongly damped high-speed streams. The basic structure of the solar wind near the solar equatorial plane in the distant heliosphere thus differs considerably from that observed near Earth. Stream amplitudes are severely reduced, and short wavelength structure is damped out. The dominant structures at low latitudes (i.e. within the band of variable wind) in the outer heliosphere are expanding compression regions that interact and merge with one another to form what are commonly called global merged interaction regions, GMIRs (e.g. Burlaga, 1983, 1984).

8.5.2 *Stream evolution in two dimensions*

Should the coronal expansion be time independent but inhomogeneous in heliocentric latitude and longitude, stream evolution proceeds similarly at all longitudes, but the state of a stream's evolution varies with longitude. Because of solar rotation, the interaction region on the leading edge of a high-speed stream is wound into a spiral that at any particular heliocentric distance is inclined to the radial direction at an angle intermediate to that of the magnetic field threading the slow and fast wind flows respectively, as illustrated in Fig. 8.5. The entire pattern of interaction corotates with the Sun and the compression region is known as a corotating interaction region, CIR (e.g. Balogh *et al.*, 1999). It is important to note, however, that it is only the pattern that corotates with the Sun because each parcel of solar wind plasma moves radially outward in this simple picture, except within the interaction region itself where both radial and transverse deflections of the flow occur. Because a CIR is inclined relative to the radial direction, the pressure gradients associated with the interaction region have both radial and azimuthal components. With increasing heliocentric distance, the forward wave propagates both anti-sunward and westward (in the direction of planetary motion about the Sun), whereas the reverse wave propagates both sunward (in the rest frame of the average solar wind) and eastward.

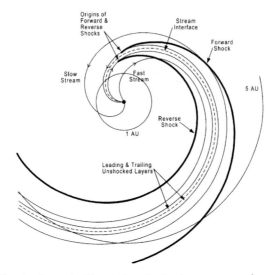

Fig. 8.5. Idealized schematic illustrating the basic structure of a corotating interaction region in the solar equatorial plane. The dashed line threading the middle
of the corotating interacting region (CIR) denotes the stream interface and the
solid heavy lines indicate the forward and reverse shocks. Plasma immediately
surrounding the stream interface is compressed, but not shocked. (From Crooker
et al., 1999. With kind permission of Springer Science and Business Media.)

As a result, the slow wind is accelerated outward and deflected westward within
the interaction region and the fast wind is decelerated and deflected eastward there,
thus accounting for the characteristic westward and then eastward flow deflections
commonly associated with interaction regions on the leading edges of high-speed
streams (see Fig. 8.3 and related discussion). One consequence of the transverse
deflections is that they partially relieve the pressure buildup induced by stream
steepening by allowing the plasma to slip aside. Thus, solar wind streams steepen
less rapidly than is predicted by the simple 1D simulation shown in Fig. 8.4 (e.g.
Pizzo, 1980).

8.5.3 Stream evolution in three dimensions

There is, of course, a three-dimensional (3D) aspect to stream evolution that
becomes most apparent at heliocentric distances beyond about 3–4 AU and at latitudes away from the solar equatorial plane. Ulysses observations have revealed
(1) that the reverse shocks on the trailing edges of CIRs are observed both within
the low-latitude band of solar wind variability and at latitudes 10°–20° above that
band, whereas the forward shocks on the leading edges of corotating interaction
regions are generally confined to the low-latitude band itself; and (2) that in addition to the flow deflections already discussed, the slow wind is usually deflected

in both solar hemispheres toward the opposite hemisphere at the forward shocks, whereas the fast wind is usually deflected poleward at the reverse shocks (Gosling *et al.*, 1993, 1995b). Because a shock propagating through a plasma always deflects the plasma in the direction in which the shock is itself propagating, these observations demonstrate that forward shocks associated with CIRs preferentially propagate anti-sunward, westward (in the direction of solar rotation), and equatorward, while the reverse shocks preferentially propagate sunward (in the solar wind rest frame), eastward, and poleward. Thus CIRs tend to have characteristic north–south tilts that are opposed in the opposite solar hemispheres (e.g. Pizzo, 1991).

The opposed north–south tilts of CIRs arise because (1) the solar wind expansion near the Sun is controlled by the solar magnetic field, (2) the solar magnetic equator commonly is tilted relative to the heliographic equator or contains substantial warps, or both, and (3) solar rotation drives CIRs. The sketch shown in Fig. 8.6 illustrates qualitatively how these factors combine to produce opposed north–south CIR tilts in the opposite solar hemispheres. A band of slow wind emanates from the Sun at low latitudes. This band is associated with coronal streamers that lie above

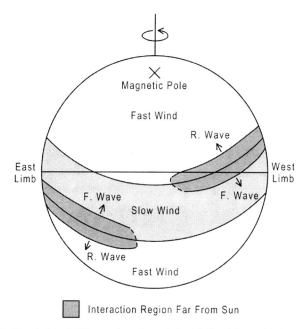

Fig. 8.6. Idealized sketch illustrating the origin of tilted CIRs in the solar wind in terms of a magnetic dipole tilted forward from the plane of the page in this sketch. The band of slow wind (light shaded region) girding the Sun at low latitudes is tilted relative to the heliographic equator in the same sense as is the heliospheric current sheet. The pattern of flow is established close to the Sun and the interaction regions (heavy shade) form well away from the Sun. (From Gosling *et al.*, 1993.)

the solar magnetic equator, which maps outward to form the heliospheric current sheet. Because the magnetic equator and HCS are commonly tilted relative to the heliographic equator, so too is the band of slow wind. As the Sun rotates, the fast wind at low heliographic latitudes overtakes the slow wind well away from the Sun along interfaces (the stream interfaces) that have essentially the same tilts as that of the band of slow wind. A compressive interaction occurs only in regions (heavy shaded region in Fig. 8.6) where the interfaces are inclined away from the equator going from left to right in the figure (in the sense of solar rotation). Where the interfaces are inclined toward the equator the interaction between fast and slow flow produces a rarefaction because in such regions the slow wind trails the fast wind. The interfaces and the CIRs in which they are embedded have opposite north–south tilts in the northern and southern solar hemispheres. The forward and reverse waves bounding the CIRs always propagate roughly perpendicular to the interfaces; thus the forward waves in both hemispheres propagate anti-sunward, westward, and toward the opposite hemisphere, whereas the reverse waves propagate sunward (in the plasma rest frame), eastward, and poleward. Because the HCS is nominally in the center of the band of low-speed wind, it is offset from the interfaces and thus is not initially embedded within the CIRs. However, the forward waves of the CIRs eventually overtake the HCS, which then becomes embedded within CIRs. At 1 AU the HCS is often, but not always, embedded within a CIR (e.g. Borrini *et al.*, 1981) and always precedes the stream interface (e.g. Gosling *et al.*, 1978). At distances beyond about 3 AU the HCS is virtually always embedded within a CIR.

Fully 3D MHD codes have been developed to model the evolution of high-speed streams in the heliosphere (e.g. Pizzo, 1980, 1982). These usually assume quasi-steady flow on the large scale and often restrict the computational domain to regions beyond the point where the solar wind flow is both super-sonic and super-Alfvénic as in the 1D gas-dynamic simulation shown in Fig. 8.4. Pizzo (1991) first applied the geometry shown in Fig. 8.6 to specify the inner boundary conditions for a simulation using such a 3D MHD code. In a later paper (Pizzo and Gosling, 1994) the input flow configuration was that associated with a magnetic dipole tilted 30° to the solar rotation axis (similar to that prevailing during Ulysses' initial transit to high solar latitudes), with fast, hot, low-density flow emanating from regions centered on the dipole axes and slow, cold, dense flow emanating from a belt girding the dipole equator and extending ±20° about it. Parameter values were chosen at the inner boundary of the calculation at 0.15 AU to produce a 310–750 km/s speed range at 1 AU.

Figure 8.7 shows latitude and longitude maps of the resulting outflow velocity at 1 and 5 AU. A spacecraft at rest in the solar wind would pass from right to left through these maps. At 1 AU (upper panel) the pattern of fast and slow wind has not changed very much from the input pattern near the Sun, although CIRs have

SPEED

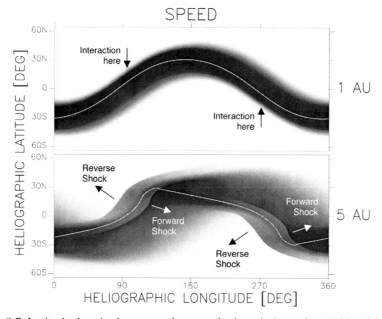

Fig. 8.7. Latitude–longitude grey-scale map of solar wind speed at 1 AU and 5 AU resulting from the 3D MHD simulation described in the text. Light (dark) shading indicates high (low) speed flow. The white line threading the speed plots is the heliospheric current sheet (HCS). In the 5 AU panel, the flow pattern has been shifted 90° in longitude to compensate for corotational drift between 1 and 5 AU. Corotating interacting regions (CIRs) at 5 AU appear as the two medium-tone diagonal features near 90° and 270° longitude. The abrupt speed changes bordering the CIRs are forward/reverse shock pairs. Note that the reverse shocks extend to higher latitudes than do the forward shocks. (Adapted from Pizzo and Gosling, 1994.)

begun to develop in both hemispheres as can be discerned in the displacement of the heliospheric current sheet from the center of the slow-wind band at longitudes where the interaction is occurring. By 5 AU (lower panel) significant structure has developed. A forward–reverse shock pair has formed at the leading edge of the northern stream and, by virtue of the input symmetry, a mirror image CIR has formed at the front of the southern stream. The stream interfaces are just barely visible in the lower panel as changes from slightly darker to slightly lighter grey shading to the left of the HCS in both hemispheres. The gradual darkening in going from right to left at (say) 30° S after passing through the reverse shock near 270° longitude indicates the gradual reduction in flow speed characteristic of the extended rarefactions on the trailing edges of high-speed streams (see, for example, Fig. 8.3). It is clear from the figure that the forward shocks propagate westward (to greater heliographic longitudes as displayed) and toward the opposite hemisphere in the slow wind, while the reverse shocks propagate eastward and poleward into the fast wind and eventually above the latitudes where solar wind variability is most pronounced, consistent with observations. Note also that the CIRs have opposed north–south tilts in the northern and southern hemispheres and that by 5 AU stream evolution has distorted the heliospheric current sheet considerably relative to its shape at 1 AU (e.g. Pizzo, 1994).

A meridional (north–south) cut through the simulated 3D flow structure between 1 and 10 AU, such as that shown in Fig. 8.8, helps to clarify the overall geometry of corotating interaction regions, because at large heliocentric distances such a cut roughly represents a cross section through the 3D spiral CIRs. The figure shows the northern stream pushing into the slow preceding flow inside ∼5 AU, bulging the embedded heliospheric current sheet outward away from the Sun at low latitudes; the bulging effect is even more noticeable within the more evolved southern CIR, which is close to 9 AU at this particular longitude. Note that while fast material continues to pile into the back of the CIR at ∼7 AU, little of the original high-speed flow is left equatorward of ∼30° S; it has been severely eroded in the ongoing interaction with both the CIR and the trailing rarefaction. The pressure enhancement associated with the southern stream can be traced north and sunward along the HCS all the way back to its intersection with the northern CIR near ∼5 AU

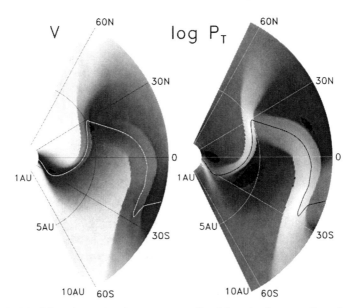

Fig. 8.8. Meridional grey-scale maps at a fixed longitude extending from 1 to 10 AU of the pattern of solar wind speed (left) and log of the total pressure (right) originating from a tilted dipole at the Sun. High (low) values of speed and pressure are indicated by light (dark) shading. The heliospheric current sheet HCS is the light (dark) line weaving back and forth across the equator in the speed (pressure) map. (From Pizzo and Gosling, 1994.)

and $\sim 30°$ N. This extension is a relic of the input conditions, in that the slow flow encasing the HCS is intrinsically dense and the magnetic field wraps relatively tightly within it; hence it maintains a slight pressure excess over the surroundings, even in the absence of any interaction with high-speed wind (for example near the central meridian in Fig. 8.6). The sharp bend in the HCS at the intersection of the northern and southern CIRs develops because near $\sim 30°$ N the interaction between the fast and slow winds is highly oblique and the fast material is deflected poleward off the great curving mass of slow material in its path.

8.6 Transient disturbances in the solar wind

The nearly continuous monitoring of the outer solar corona provided by space-based coronagraphs since the early 1970s reveals that transient ejections of material from the solar atmosphere into the solar wind in events now called coronal mass ejections, CMEs, commonly occur. Observed CME occurrence rates range from ~ 0.1/day near solar activity minimum to more than 3.5 events/day near solar activity maximum (e.g. Webb and Howard, 1994). CMEs originate largely, if not entirely, in closed magnetic field regions in the corona not previously participating in the solar wind expansion. Often these closed field regions are found in the coronal streamer belt that straddles the solar magnetic equator and that underlies the heliospheric current sheet. During a typical CME, somewhere between 10^{15} and 10^{16} g of material is ejected into the solar wind on a time scale of several or more hours (e.g. Hundhausen, 1988). Ejection speeds at ~ 5 solar radii range from less than 50 km/s to greater than 2500 km/s (e.g. Howard *et al.*, 1985; Hundhausen *et al.*, 1994), but by the time they reach 30 solar radii (a heliocentric distance of ~ 0.14 AU) virtually all of the slower CMEs have leading edge speeds comparable to or greater than the minimum solar wind speed (~ 270 km/s) observed near 1 AU (Sheeley, 1999). However, a large fraction of CMEs at 0.14 AU have speeds less than the average solar wind speed of ~ 450 km/s in the ecliptic plane. CME speeds in the corona do not show any obvious latitude trends; however, CMEs occur much more frequently at low than at high heliographic latitudes (e.g. Hundhausen *et al.*, 1994). When observed in the solar wind far from the Sun, CMEs are now commonly called interplanetary coronal mass ejections (ICMEs).

Many ICMEs have sufficiently low speeds relative to the ambient wind ahead that they do not produce large disturbances in the solar wind near or beyond 1 AU in the ecliptic plane. However, the leading edges of the faster ICMEs have outward speeds considerably greater than that associated with the normal solar wind at low heliographic latitudes, and thus commonly drive transient shock wave disturbances in the solar wind at low heliographic latitudes.

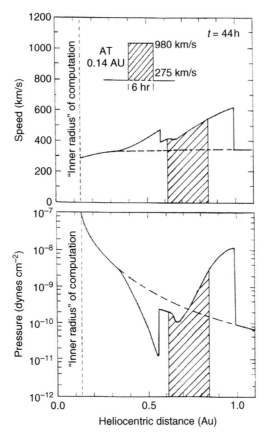

Fig. 8.9. A snapshot of solar wind speed and pressure as functions of heliocentric distance for a 1D gas-dynamic simulation of a solar wind disturbance driven by a fast CME/ICME at the time the leading edge of the disturbance reached 1 AU. The solid-dashed-solid curves indicate the steady state prior to introduction of the temporal variation in flow speed imposed at the inner boundary (0.14 AU) of the simulation shown at the top of the figure. Cross-hatching identifies material that was introduced with a speed of 980 km/s at the inner boundary, and therefore identifies the CME/ICME in the simulation. Temperature and pressure were held constant at the inner boundary during the introduction of the CME/ICME into the simulation. (Adapted from Hundhausen, 1985.)

Figure 8.9 provides a snapshot of a 1D gas-dynamic simulation of a solar wind disturbance driven by a fast ICME at the time the leading edge of the disturbance reaches 1 AU. As for the case of CIRs, this 1D, gas-dynamic simulation predicts too strong an interaction between the newly ejected solar material and the ambient wind because it neglects azimuthal and meridional motions of the plasma that help relieve pressure stresses. Moreover, magnetic forces are not explicitly included, although such forces are expected to play a secondary role in overall ICME evolution. Despite these limitations, the simulation provides a useful starting point for understanding the global evolution of solar wind disturbances driven by fast ICMEs.

A region of high pressure develops on the leading edge of the disturbance as the ICME runs into slower wind ahead. This region of high pressure is bounded by a forward shock on its leading edge that propagates into the slower wind ahead and by a reverse shock on its trailing edge that propagates back into and eventually through the ICME in the case illustrated. Both shocks are, however, carried away from the Sun by the high overall convective flow of the solar wind. A rarefaction develops on the trailing edge of the disturbance in the simulation as the ICME pulls away from slower trailing solar wind. This rarefaction propagates rapidly both forward into the ICME and backward into the trailing wind, producing a deceleration of the rear portion of the ICME and an acceleration of the trailing wind. Thus the ICME slows considerably by sharing its momentum and energy with both the leading and trailing wind. The simulation explains why ICMEs with speeds considerably faster than the high-speed solar wind (∼800 km/s) are only occasionally observed far from the Sun. Only those ICMEs with exceptional momentum contents will not be slowed substantially as they interact with a slower ambient solar wind. To zeroth order the results of this simple simulation agree reasonably well with many observed solar wind disturbances driven by fast ICMEs in the ecliptic plane, although reverse shocks are only rarely detected in practice. Finally, we note that in the case illustrated the interaction produces an expanding ICME (leading edge moving faster than the trailing edge) whose radial width (0.23 AU) at ∼0.8 AU is greater than its width (0.14 AU) at the inner edge of the simulation even though the ICME was not expanding when introduced into the simulation.

Many interplanetary coronal mass ejections observed in the solar wind near the ecliptic plane are expanding when they pass 1 AU. As noted in the previous paragraph, such expansion can be a dynamic response to a rarefaction wave produced by motion relative to slower wind behind. However, an ICME may also expand simply because it initially has an internal pressure (plasma plus field) that substantially exceeds that of the surrounding solar wind plasma. We have used the term "over-expansion" to describe ICMEs in which the expansion apparently is driven by an initially high ICME pressure (Gosling *et al.*, 1994a, b). Figure 8.10 shows an example of an over-expanding ICME that produced a forward–reverse shock pair in the high-latitude solar wind during Ulysses' first polar orbit of the Sun. The shocks were approximately equally offset from the edges of the ICME with the pressure maximizing immediately downstream from the shocks and reaching a minimum value roughly in the center of the ICME. In addition, the speed generally declined from the forward to the reverse shock. Although the ICME had a high speed, it was not traveling faster than the ambient wind ahead of the forward shock. Thus the shock pair could not have been produced by relative motion between the

234234234234234 234 *The structure and evolution of the 3D solar wind*

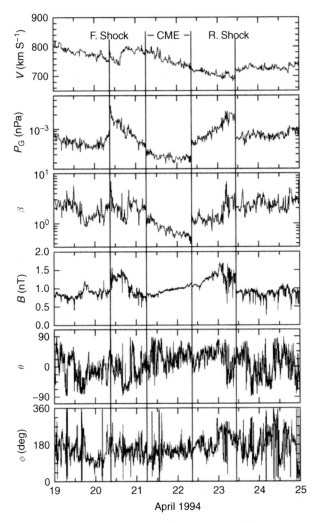

Fig. 8.10. Selected solar wind plasma and magnetic field parameters for a disturbance driven by an interplanetary coronal mass ejection (ICME) observed by Ulysses at 3.2 AU and 61° S. Parameters plotted from top to bottom are the flow speed, gas pressure, plasma beta (ratio of gas to magnetic field pressure), magnetic field strength, and the polar and azimuthal angles of the field. Vertical lines bracket the CME/ICME and also indicate forward and reverse shocks bounding the disturbance. (From Gosling *et al.*, 1994b.)

ICME and the ambient wind ahead. Rather, the observations suggest that the shock pair in this event was produced by over-expansion of the ICME. To the best of our knowledge, shock pairs associated with over-expansion of ICMEs have never been identified at low heliographic latitudes at any distance from the Sun, for reasons yet uncertain. On the other hand, a large fraction of ICME disturbances observed

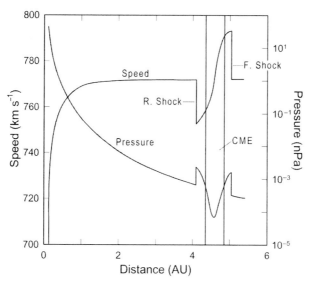

Fig. 8.11. A snapshot of solar wind speed and pressure as functions of heliocentric distance for a simulated (1D, gas-dynamic) ICME-driven disturbance that has just arrived at 5 AU. After establishing a steady-state flow of 760 km/s at large heliocentric distances, the disturbance was initiated at 0.14 AU by increasing the density by a factor of four in a bell-shaped pulse of 10 hours duration while holding the speed and temperature constant there. Vertical lines bracket the plasma introduced into the simulation with higher density and thus identify the ICME in the simulation. (From Gosling *et al.*, 1994b.)

in the high-speed wind at high latitudes appear to be of this nature (e.g. Reisenfeld *et al.*, 2003).

Figure 8.11 shows a snapshot of solar wind speed and pressure for a disturbance initiated in a simple 1D gas-dynamic simulation by increasing the density, and thus also the pressure, in a 10-hour wide bell-shaped pulse while holding the speed and temperature constant at the inner boundary. The radial width of the simulated ICME was 0.17 AU at the inner boundary. Owing to its high internal pressure, the simulated ICME expands as it travels out from the Sun so that at 4.6 AU it has a radial width of 0.5 AU. That expansion produces a forward wave that propagates ahead of the ICME and a reverse wave that propagates back into the trailing wind; thus the width of the overall disturbance at 5 AU is ~0.95 AU. These relatively modest pressure waves steepen into shocks before the disturbance arrives at 3 AU. The expansion causes the density and temperature (and hence also the pressure) within the ICME to be lower than that in the plasma immediately surrounding the disturbance at large heliocentric distances. The temporal signature produced at a fixed point in the outer heliosphere by the simulated disturbance bears a remarkable resemblance to the overall appearance of the event shown in Fig. 8.10. Quantitative differences between the observed and simulated disturbances can largely be attributed to differences between the assumed and the actual initial conditions and to limitations inherent in a 1D gas-dynamic simulation.

Multi-spacecraft observations of ICME-driven disturbances reveal that a given ICME can produce quite different disturbances at different latitudes and/or longitudes owing to spatial variability in the associated CME as well as structure in the ambient solar wind into which the ICME propagates (e.g. Burlaga *et al.*, 1990; Gosling *et al.*, 1995c). Figure 8.12 shows meridional plane snapshots of disturbances initiated 10 days earlier at the inner boundary of a 3D gas-dynamic simulation whose pre-disturbance steady-state solution was nearly identical to that

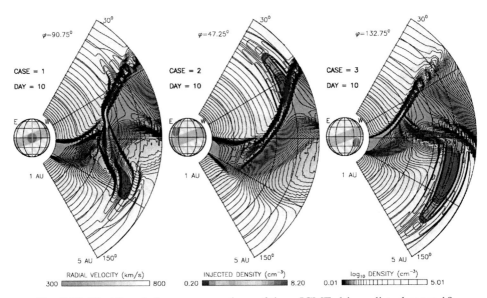

Fig. 8.12. Meridional plane cross sections of three ICME-driven disturbances 10 days after launch at the inner boundary of a 3D gas-dynamic simulation. In the original color version of this figure radial velocity is indicated by a grey-scale, density within the ICME is indicated in color, and log of the density in the ambient wind is indicated by the contour lines. The three cases shown are (left) an ICME injected at the equator within the band of slow wind, (center) an ICME injected to the east of the slow wind band, and (right) an ICME injected to the west of the slow wind band. In each case the ICME was injected as an over-pressured (factor of eight) plasma cloud having a speed identical to that of the high-speed wind. In the first case the pressure enhancement was achieved with density and temperature enhancements two and four times, respectively, that of the slow wind; in the latter two cases the over-pressure was achieved with a density enhancement eight times that of the background fast flow. The clouds had identical 30° circular shapes and 12-hour durations at the inner boundary of the simulations. The ambient slow wind band was tilted 20° relative to the heliographic equator. (From Odstrčil and Pizzo, 1999.) See Color Plate 6.

of Figs. 8.6 through 8.8. In all three cases the ICME undergoes a significant lateral expansion, most of which occurs relatively close to the Sun. In case 1, where the ICME is injected at the center of the band of slow wind, the low-latitude portion of the ICME gets slowed and compressed by its interaction with the slow wind ahead and becomes caught up within the interacting ambient flows there. The portion of the ICME that expands to high latitudes also expands radially and eventually forms forward–reverse shock pairs in the opposite solar hemispheres, similar to those observed at high latitudes by Ulysses. The ICME is strongly distorted by its interaction with the ambient wind structure and the overall disturbance structure is highly complex. In case 2, where the ICME is injected to the east of the band of slow wind, the interaction with the ambient wind is initially dominated by the over-expansion of the ICME. At later times that portion of the ICME that expands to high latitudes continues to expand radially, eventually producing a forward–reverse shock pair characteristic of over-expansion; however, the greater portion of the ICME at lower latitudes becomes entrained, slowed, and compressed within the pre-existing CIR. Finally, in case 3, where the ICME is injected into the fast wind just to the west of the band of slow wind, most of the ICME is convected outward within the fast wind at high latitudes and produces a forward–reverse shock pair there; only the portion of the ICME at the lowest latitudes gets distorted by its interaction with the underlying stream structure.

The results of the 3D simulations shown in Fig. 8.12 are instructive and provide guidance for interpreting observations of ICME-driven disturbances at various heliocentric distances and latitudes. They illustrate that even relatively simple initial conditions can lead to complex disturbance structures in the solar wind far from the Sun. The real solar wind and real CMEs are, of course, typically not as simple as is assumed in the simulations, so it is reasonable to expect that real ICME-driven disturbances are even more complex than illustrated here. Nevertheless, the simulations provide basic understanding of the physical processes that determine how ICME-driven disturbances evolve in the heliosphere.

8.7 The evolving global heliospheric magnetic field

Parker's simple model of the heliospheric magnetic field assumes that field line footpoints are fixed on the solar surface and that the solar wind flow is laminar, time stationary, and spherically symmetric with flow speed being independent of heliocentric distance. Those assumptions, when coupled with uniform solar rotation, lead to a heliospheric magnetic field in the solar equatorial plane that is bent into Archimedean spirals whose inclinations to the radial direction depend only on the speed of the wind and heliocentric distance. Out of the equatorial plane HMF lines in this model take the form of helices wrapped on conical surfaces of constant heliographic latitude. These helical field lines are ever more elongated at higher solar latitudes and eventually become radial lines over the solar poles.

Various averages of the HMF orientation obtained over a wide range of solar wind speeds, heliocentric distances, and heliographic latitudes appear to be in reasonable agreement with Parker's highly idealized model (e.g. Smith, 1979; Thomas and Smith, 1980; Burlaga *et al.*, 1982; Forsyth *et al.*, 1996; Smith and Phillips, 1997). However, there is a considerable variability relative to the model values at all wind speeds, radial distances and latitudes. Much of this variability is associated with violations in the assumptions used to derive the simple model and in many cases were anticipated by Parker (e.g. Parker, 1963). For example, footpoints of field lines in the heliosphere typically are not fixed on the solar surface, but rather move about in response to convection in the solar photosphere (e.g. Jokipii and Parker, 1968), differential solar rotation (e.g. Fisk, 1996), and probably also because of magnetic reconnection between open and closed field lines (now commonly called interchange reconnection) in the solar atmosphere (e.g. Gosling *et al.*, 1995d; Fisk *et al.*, 1998b; Crooker *et al.*, 2002). Further, as illustrated in Fig. 8.5, the interaction of low- and high-speed flows at low latitudes produces radial and transverse accelerations of the wind and corresponding changes in the HMF orientation. The solar wind outflow is also often non-steady. As illustrated schematically in Fig. 8.13, temporal decreases in flow speed at the coronal base can produce significantly underwound magnetic fields on the trailing edges of high-speed streams (e.g. Gosling and Skoug, 2002; Schwadron, 2002; Schwadron and McComas, 2005; Riley and Gosling, 2007). CMEs are prominent examples of non-stationary flow. As illustrated in Fig. 8.14, CMEs not only carry new, initially closed, magnetic flux into the heliosphere, often in the form of twisted flux ropes, but also produce changes in flow speed and field orientation as they interact with the ambient HMF. Changes in field topology associated with reconnection in the magnetic legs of CMEs (e.g. Gosling *et al.*, 1995d; Crooker *et al.*, 2002) and in the solar wind itself (e.g. Gosling *et al.*, 2005) can also seriously modify the structure of the HMF. Finally, large-amplitude Alfvénic fluctuations and turbulence fill much of the solar wind at all heliocentric distances and latitudes and contribute substantially to local variations in the HMF orientation (e.g. Belcher and Davis, 1971; Roberts *et al.*, 1987; Jokipii and Kota, 1989; Marsch, 1991).

As an alternative to Parker's model of the heliospheric magnetic field, Fisk (1996) proposed a model that at high magnetic latitudes incorporates a rigidly rotating magnetic dipole tilted relative to the solar rotation axis (as is assumed in our 3D model of CIRs (Figs. 8.6 through 8.8), differential rotation of the footpoints of HMF lines in the photosphere, and the subsequent non-radial expansion of these same field lines close to the Sun as shown in the left panel of Fig. 8.15. The non-radial expansion close to the Sun is consistent with the observation that the dipolar

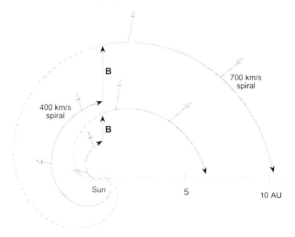

Fig. 8.13. Snapshots from above the north solar pole of a highly idealized magnetic field line in the solar equatorial plane shown at two different times on which
a sudden and semi-permanent decrease in solar wind speed from 700 to 400 km/s
occurred close to the Sun when the field line footpoint was at 0° clock angle on the
trailing edge of a high-speed stream. The snapshots are obtained at successively
later times when the field-line footpoint has rotated to clock angles of −130°
and −270°, respectively. Dashed lines indicate the Parker spirals that would have
resulted had the wind speed on the field line remained constant at 700 km/s. Open
arrows indicate flow vectors. An ever-growing radially oriented kink in the field
line develops at 0° clock angle connecting the pre-decrease and post-decrease 400
and 700 km/s spirals in this kinematic sketch. The sketch ignores dynamic effects
associated with the rarefaction produced by a sudden drop in speed that serve to
bend the radial field segment eastward, yet still produce a field line segment that
is less tightly wound than is a Parker spiral field line. (From Gosling and Skoug,
2002.)

pattern of field strength close to the Sun is not observed in the solar wind far from
the Sun because averages of $r^2 B_r$, where r is heliocentric distance and B_r is the
radial component of the HMF, tend to be nearly constant at latitudes above the band
of solar wind variability (e.g. Smith *et al.*, 1995).

In order to understand the essence of Fisk's model, consider the field line in the
left panel of Fig. 8.15 whose photospheric footpoint is originally slightly (∼8°) to
the left of the dipole axis, **M**, which in the example shown is tilted 20° relative
to the direction opposite to the solar rotation axis (i.e. −Ω; the figure was made
for southern heliographic latitudes). Such a field line initially extends outward at
a heliographic latitude of approximately −85°. Differential rotation of the field-
line footpoint about −Ω within the essentially rigidly rotating perimeter of the
open-field region (see Section 2.2.4 or Wang and Sheeley (1993) on the rotation
profiles of coronal-hole boundaries) causes that same field line at a later time to be
displaced 12° on the opposite side of the rotation axis from the dipole axis. There it

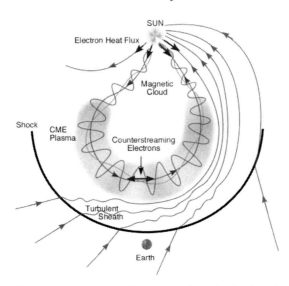

Fig. 8.14. An idealized 2D sketch of a solar wind shock disturbance produced by a fast ICME, depicted as a magnetic flux rope or magnetic cloud, running into slower wind ahead and directed toward Earth. The heavy curved line running in front of the ICME indicates the forward shock, the thin lines with small arrows indicate magnetic field lines, and the heavy arrows indicate the direction of the suprathermal electron strahl directed outward from the Sun. The ambient magnetic field is compressed by its interaction with the ICME and is forced to drape about it. (From Zurbuchen and Richardson, 2006. With kind permission of Springer Science and Business Media.)

expands to a much lower southern heliographic latitude of $-30°$ before extending out into the heliosphere. The rate of differential rotation is such that some field lines emanating from an open-field region in this model can wander 40° or more in heliographic latitude on heliocentric distance scales of the order of 15 AU.

The right panel of Fig. 8.15 compares predictions of Fisk's model with Parker's model for field lines originating at heliographic latitudes of 70 °S. Support for field lines wandering in latitude as in Fisk's model can be found in observations of corotating energetic particle events well above the latitudes where CIR shocks are observed (e.g. Sanderson *et al.*, 1995; Simnett *et al.*, 1995), perhaps suggesting that field lines in the high-speed wind at high latitudes often extend down to latitudes where CIR shocks are directly observed. On the other hand, observations of corotating energetic particle events at high latitudes might equally well be explained by a combination of field-line braiding that occurs close to the Sun (e.g. Kóta and Jokipii, 1995) and cross field diffusion (e.g. Giacalone and Jokipii, 1999). For reasons outlined earlier in this section, it is really quite difficult to distinguish between the various HMF models in actual measurements of the HMF.

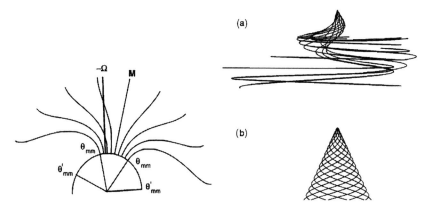

Fig. 8.15. (Left) A schematic illustration of the expansion of magnetic field lines from a south-polar coronal hole centered on a dipole axis (**M**) tilted relative to the solar rotation axis (parallel to $-\Omega$). θ_{mm} is the magnetic co-latitude of the last field lines that extend from the coronal hole out into the heliosphere and θ'_{mm} is the magnetic co-latitude of those same field lines where they extend nearly radially outward into the heliosphere. All field lines drawn thread the high-speed wind. (Right) HMF field lines that originate from 70 °S latitude, projected onto a meridional plane. (a) The field lines resulting from a combination of differential rotation and a rigidly rotating dipole in Fisk's model. (b) The field lines in Parker's model. (Adapted from Fisk, 1996.)

8.8 Long-term changes in the heliospheric magnetic field

Figure 8.3 reveals systematic variations of the order of a factor of three in the magnitude of the heliospheric magnetic field, B, associated with solar wind stream structure in the ecliptic plane at 1 AU. A large fraction of that variation is produced by compressions and rarefactions as discussed in Section 8.5 and thus does not accurately reflect variations in the HMF close to (say at 0.1 AU) the Sun. Even larger variations in B often occur during passage of disturbances driven by interplanetary coronal mass ejections. As illustrated in Fig. 8.16, solar rotation averages of (say) 1-hour averages of B in the ecliptic plane at 1 AU have revealed variations roughly of order two and roughly in phase with solar activity during solar cycles 21–23 (e.g. Svalgaard and Cliver, 2007), although such variations were not apparent in solar cycle 20 (e.g. King, 1979). There appears to be a minimum value or "floor" of ~ 4.6 nT in the solar rotation averages that occurs each solar minimum. However, when extended through 2008, the floor during the recent solar minimum, an interval marked by a relative absence of CMEs, appears to be closer to ~ 4.0 nT (Owens *et al.*, 2008).

A number of authors have suggested that these solar cycle variations in solar rotation averages of B in the ecliptic plane are largely caused by ICMEs, the overall global HMF at any given moment being a mixture of a nearly constant open

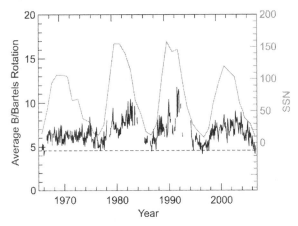

Fig. 8.16. Bartels (27-day) rotation averages of the magnitude of the HMF in the ecliptic plane (black line) from 1965 to 2006 covering most of sunspot cycles 20–23 along with the smoothed sunspot number (grey line). (From Svalgaard and Cliver, 2007. Reproduced by permission of the AAS.)

flux component (e.g. Fisk and Schwadron, 2001) and a time-varying closed component associated largely with CMEs/ICMEs (e.g. McComas *et al.*, 1992; Webb and Howard, 1994; Owens and Crooker, 2006; Owens *et al.*, 2008). As illustrated in Fig. 8.14, the closed field component is identified in solar wind data by counterstreaming suprathermal electron strahls (e.g. Gosling *et al.*, 1987). Each CME injects closed magnetic flux into the heliosphere; the long-term effect of these injections is determined by the rate at which the closed flux is opened up by magnetic reconnection in the magnetic legs of ICMEs and by reconnection elsewhere in the solar atmosphere inside the point where the solar wind flow exceeds the local Alfvén speed. Owens and Crooker (2006) found they could roughly reproduce the solar cycle variations evident in Fig. 8.16, as well as the rate at which closed magnetic flux appears to open up during ICME journeys out to 5 AU, if interchange reconnection in the magnetic legs of ICMEs is the dominant balancing process and if it occurs on a characteristic time scale of ~50 days. Whether or not that time scale for opening up closed flux in the solar wind is consistent with solar observations remains to be determined.

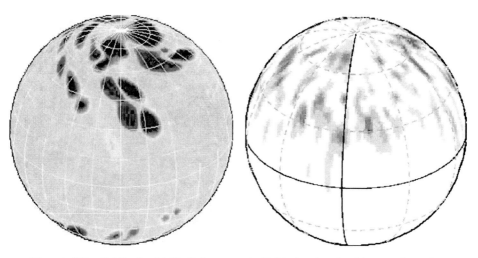

Plate 1 (Fig. 2.16). (Left) Radial magnetic field simulated with a surface flux transport model assuming solar-like transport parameters, but for a flux emergence rate that is 30× solar, and a larger latitudinal range for flux emergence, combined with a meridional flow peaking at 100 m/s (∼5× solar). (Right) Observed radial magnetic field distribution for the rapidly rotating star AB Doradus (P_{rot} = 0.51 d). (From Holzwarth *et al.*, 2007. Reproduced with permission © Wiley-VCH Verlag GmbH & Co. KGaA.)

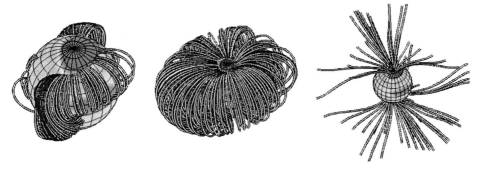

Plate 2 (Fig. 3.9). Inferred magnetic field structure of the classical T Tauri star BP Tau. Surface shading shows photospheric magnetic field strength; the three figures from left to right show estimated near-field closed, far-field closed, and open magnetic field lines. Red and blue tones indicate oppositely directed radial magnetic field strengths. (From Gregory *et al.*, 2008.)

Plates 1-23 are available for download from www.cambridge.org/9780521130202

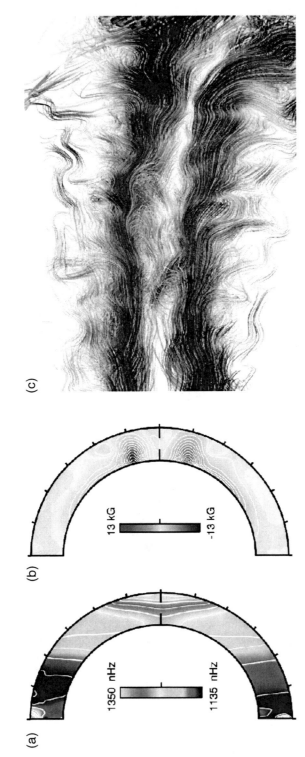

Plate 3 (Fig. 5.4). Rotational shear and toroidal fields in a global 3D dynamo simulation of a solar-type star rotating at three times the solar rate. (a) Angular velocity Ω and (b) toroidal field $\langle B_\phi \rangle$, averaged over longitude and time. (c) A 3D rendering of magnetic field lines in a portion of the convection zone, spanning about $50°$ in longitude. Red and blue lines denote positive and negative B_ϕ respectively. The view is radially outward from a vantage point under the convection zone and slightly above (north of) the equator. (From Brown *et al.*, 2009.)

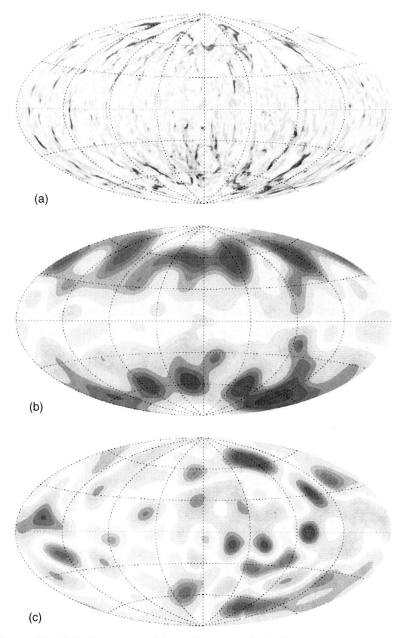

Plate 4 (Fig. 7.7). Snapshots of the radial magnetic field component on the outer boundary from numerical dynamo models. Red areas are used for outward flux and blue for inward flux (arbitrary contour steps in each panel). Color-scale indicates absolute amplitude. (a) Model parameters $E = 10^{-5}$, $R_a^* = 0.12$, $P_m = 0.8$, $P_r = 1$. (b) Same field low-pass filtered to harmonic degrees $n < 14$. (c) Model parameters $E = 10^{-5}$, $R_a^* = 0.17$, $P_m = 0.5$, $P_r = 1$, low-pass filtered. R_m is approximately 900 in both cases; R_{ol} is 0.125 in (a) and (b), and 0.19 in (c).

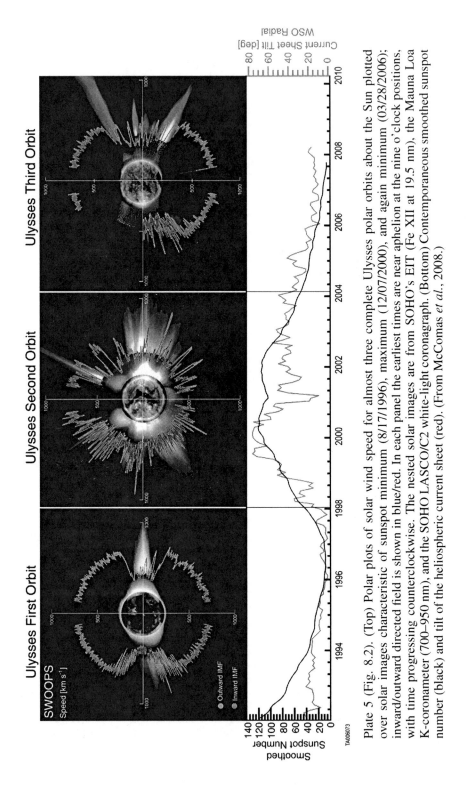

Plate 5 (Fig. 8.2). (Top) Polar plots of solar wind speed for almost three complete Ulysses polar orbits about the Sun plotted over solar images characteristic of sunspot minimum (8/17/1996), maximum (12/07/2000), and again minimum (03/28/2006); inward/outward directed field is shown in blue/red. In each panel the earliest times are near aphelion at the nine o'clock positions, with time progressing counterclockwise. The nested solar images are from SOHO's EIT (Fe XII at 19.5 nm), the Mauna Loa K-coronameter (700–950 nm), and the SOHO LASCO/C2 white-light coronagraph. (Bottom) Contemporaneous smoothed sunspot number (black) and tilt of the heliospheric current sheet (red). (From McComas *et al.*, 2008.)

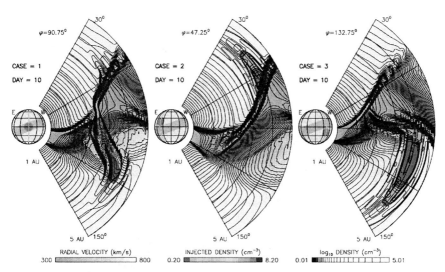

Plate 6 (Fig. 8.12). Meridional plane cross sections of 3D gas-dynamic ICME simulations viewed 10 days after launch at the inner boundary, initially moving at the speed of the fast wind. Radial velocity is shown as grey scale, ICME density in color, and the ambient wind density by contours. Left: ICME injected at the equator within the band of slow wind with density and temperature 2x and 4x, respectively, that of the slow wind. Center and right: ICMEs injected to the east or west of the slow wind band, respectively, with density 8x that of the background fast flow. (From Odstrčil and Pizzo, 1999.)

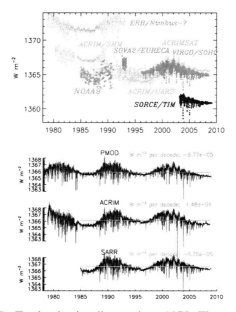

Plate 7 (Fig. 10.6). Total solar irradiance since 1978. The top panel compares these on their "native" calibration scales, above three different composite records constructed with different calibration assumptions (PMOD: C. Fröhlich; ACRIM: R. C. Willson; SARR: S. Dewitte); the slopes of these time series are shown.

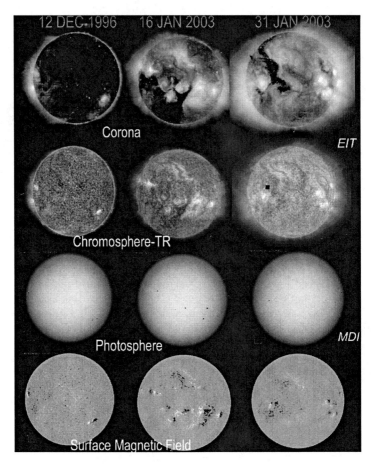

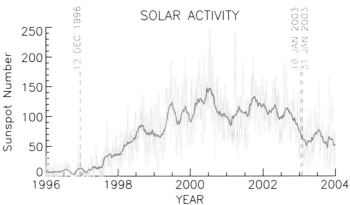

Plate 8 (Fig. 10.2). Images of the Sun's surface: magnetic map (bottom row), visible light (second row up), chromosphere/transition region (third row up), and corona (top row). The left column (December 1996) is typical of quiet cycle minima, and the two righthand columns of higher-activity states (January 2003) at different rotation phases. The bottom plot shows the sunspot number.

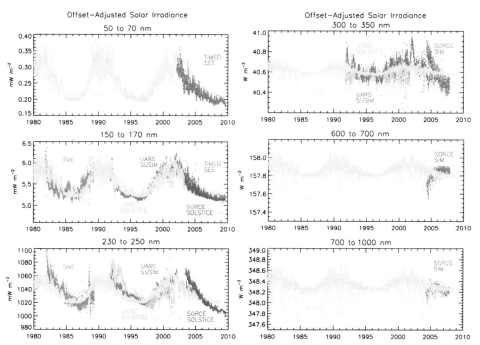

Plate 9 (Fig. 10.8). Assorted space-based observations made of the solar spectral irradiance during the past three solar cycles are compared in selected wavelength bands. From top left to bottom right: EUV (50–70 nm), FUV (150–170 nm), MUV (230–250 nm), NUV (300–350 nm), visible (600–700 nm), and IR (700–1000 nm); the curves shown have offset-adjustments to account for their different absolute calibration scales. Also shown, as the grey time series, are models of the irradiance variations in the same wavelength bands, derived by scaling the observed rotational modulation variations to the Mg (and $F_{10.7}$ for the EUV band) proxy indicators. Note the lack of daily measurements shortward of 110 nm until TIMED SEE observations commenced in 2002, and longward of 400 nm until SORCE SIM observations commenced in 2003.

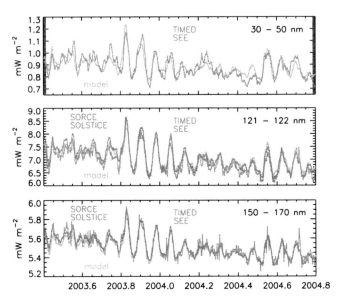

Plate 10 (Fig. 10.12). Comparison of SORCE/SOLSTICE and TIMED/SEE observations and empirical variability model estimates of irradiance in selected wavelength bands including EUV (30 to 50 nm), the Lyman α line (121–122 nm), and the far-UV (150–179 nm) wavelength bands. The comparisons indicate the good agreement among the irradiance measurements and models on the short time scales of the 27-day rotational modulation.

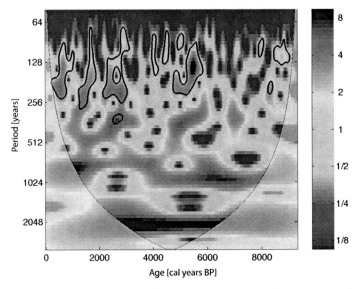

Plate 11 (Fig. 11.17). Wavelet analysis (Grinsted, 2002–4) of the solar modulation function Φ from Fig. 11.16. The color scale is a measure of the spectral power relative to the spectral power of white noise, thus measuring signal significance.

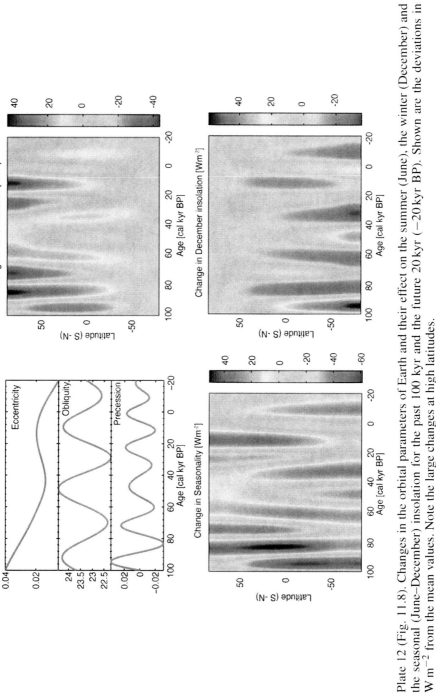

Plate 12 (Fig. 11.8). Changes in the orbital parameters of Earth and their effect on the summer (June), the winter (December) and the seasonal (June–December) insolation for the past 100 kyr and the future 20 kyr (−20 kyr BP). Shown are the deviations in W m^{-2} from the mean values. Note the large changes at high latitudes.

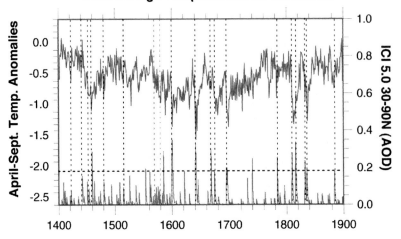

Little Ice Age Temperatures vs Volcanism

Plate 13 (Fig. 12.5). Tree-ring-based reconstruction of Northern Hemisphere (land) summer half-year temperatures (blue curve, left-hand axis) with an index of volcanism (red) for 30°–90° N in units of AOD (aerosol optical depth). The green line shows the AOD for the 1883 CE Krakatau eruption. Seventeen volcanic eruptions coincident with cooling events are marked by dashed black lines; a cluster of small eruptions is contained between the magenta lines in the late 1500s.

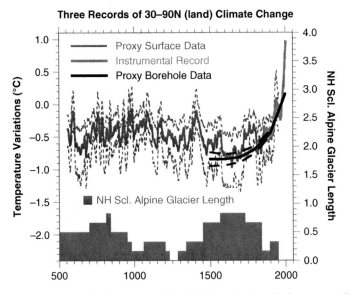

Three Records of 30–90N (land) Climate Change

Plate 14 (Fig. 12.6). Scaled Northern-Hemisphere alpine glacier extent (Section 12.4) compared with two distinct estimates of mean annual temperature changes over land for 30°–90° N: one based on surface proxy data (tree rings, ice cores, etc., from Hegerl *et al.*, 2007), and a "borehole estimate" (based on geothermal measurements of heat flux in boreholes, from Porter and Smerdon, 2004). Figure adapted from results in Hegerl *et al.* (2007), updated through 2008, shown up to 2003 with a 10-year smoothing.

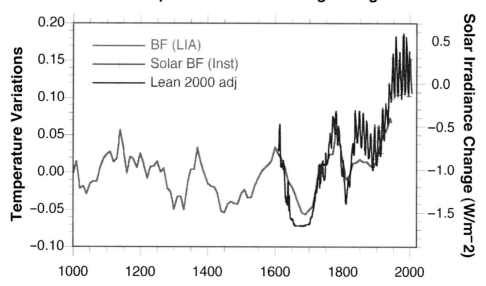

Best Fit Solar Scaling,
with Implied Radiative Forcing Change

Plate 15 (Fig. 12.12). Comparison of independent estimates of solar component for the instrumental interval and Little Ice Age. The right-hand scale indicates the estimated change in total solar irradiance for the background component of the re-scaled Lean (2000) record. See Section 12.5 for a discussion of the method.

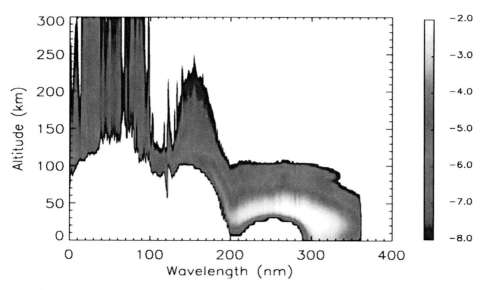

Plate 16 (Fig. 13.3). Altitude of penetration of the solar radiation as a function of wavelength. The color range shows the amount of energy deposited in the different layers of the atmosphere for the different parts of the solar spectrum (on a logarithmic scale, in units of mW m^{-3} nm^{-1}).

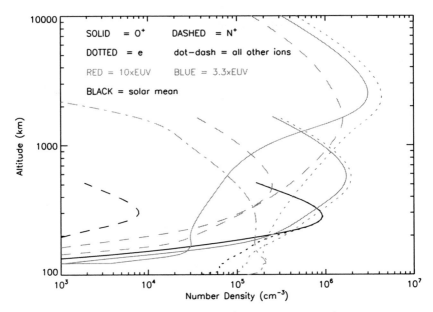

Plate 17 (Fig. 14.9). Density profiles of O^+ (solid curves), N^+ (dashed curves), and electrons (dotted curves) under different solar EUV conditions. The total density curves of all ions other than O^+ and N^+ in the $10\times$ present EUV case is presented with the dot-dashed curve. (From Tian *et al.*, 2008b.)

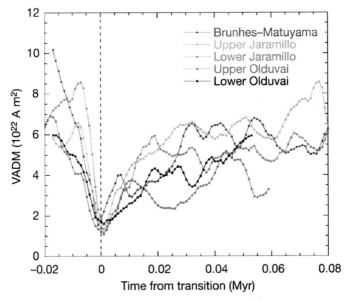

Plate 18 (Fig. 14.11). Variations in the virtual axial dipole moment across the five reversals occurring during the past 2 Myr. These are superimposed about their respective reversal epoch (with time running from right to left). A 60–80 kyr long decrease precedes each reversal. (From Valet *et al.*, 2005.)

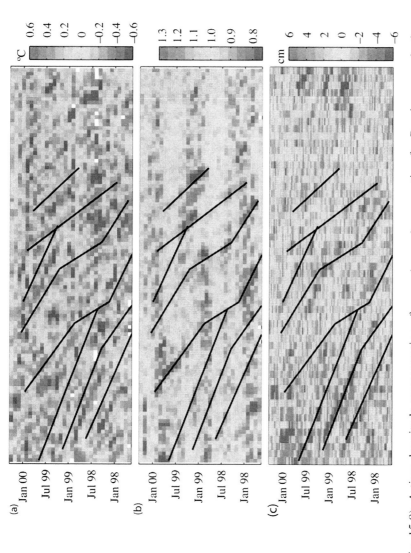

Plate 19 (Fig. 15.8). A time–longitude representation of group (energy) propagation for Rossby waves in the ocean (Hovmöller diagram), showing the variation of (a) sea surface temperature anomalies, (b) chlorophyll concentration anomalies, and (c) sea surface height anomalies. The tilt of features in this diagram reflects the westward propagation of wave groups. (From Quartly *et al.*, 2003.)

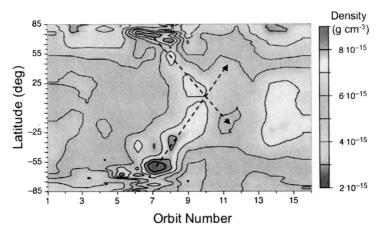

Plate 20 (Fig. 15.21). An example of a traveling atmospheric disturbance seen in density near 400 km measured by the accelerometer on the CHAMP satellite in connection with a geomagnetic disturbance. The disturbances appear to penetrate into opposite hemispheres from their origins in the Northern and Southern Hemisphere auroral zones. The simultaneous appearance in both hemispheres is due to conjugate activity. (From Forbes, 2007.)

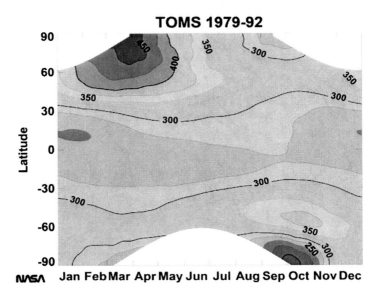

Plate 21 (Fig. 16.4). Vertically integrated ozone concentration (expressed in Dobson units or DU; at 1 DU the column depth of ozone only would equal 10 μm at sea-level pressure and average temperature) represented as a function of latitude and month of the year. The distribution is established on the basis of observations made by the spaceborne TOMS instrument between 1972 and 1992. High values at the end of the winter are visible in the Arctic (March and April) and around 60°S in September and October. The presence of the Antarctic ozone hole is visible in the Antarctic in September–November. The value of 300 DU corresponds to an ozone layer of 3 mm under STP conditions. (From NASA.)

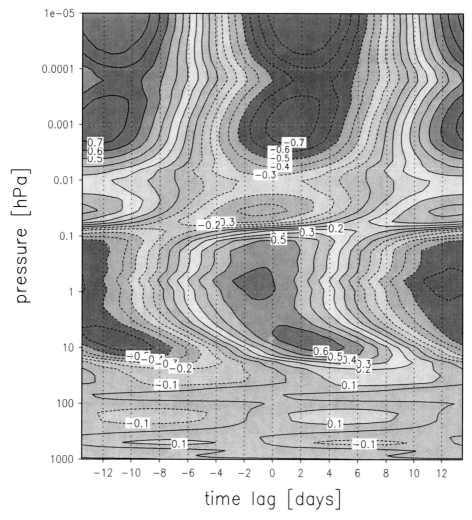

Plate 22 (Fig. 16.10). Correlation coefficient between the ozone concentration and the 27-day solar variation as a function of atmospheric pressure (hPa) for different time lags (days) as calculated by the HAMMONIA model. One notes that, at 1 hPa (about 50 km altitude) for example, the highest correlation is found when ozone and solar radiation are in phase. At 10 hPa (about 30 km), the ozone signal with a 4-day phase lag is best correlated with the 27-day solar signal. Above 0.1 hP (~65 km), the ozone signal appears to be out of phase with the solar periodic variation. (From Gruzdev *et al.*, 2009.)

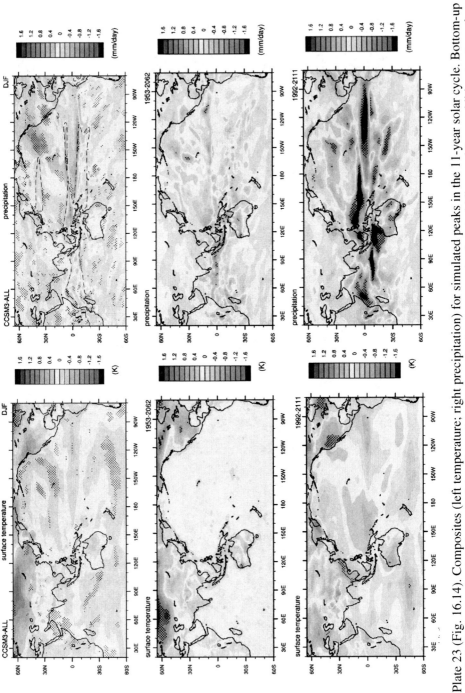

Plate 23 (Fig. 16.14). Composites (left temperature; right precipitation) for simulated peaks in the 11-year solar cycle. Bottom-up coupled air-sea mechanism (top panels) and top-down stratospheric ozone mechanism (middle panels) are additive to strengthen convection in the tropical Pacific and produce a stronger La Niña-like response to peaks in solar forcing (bottom panels). (From Meehl *et al.*, 2009.)

9

The heliosphere and cosmic rays

J. R. Jokipii

9.1 Introduction

The heliosphere is a vast spheroidal cavity in the local interstellar plasma, some 150–200 AU in size, created by a supersonic flow of plasma called the solar wind, which flows radially outward from the Sun in all directions (Fig. 9.1). The scale of the heliosphere is determined by both the solar atmosphere and the pressure of the surrounding interstellar plasma and magnetic field. Far enough from the Sun the solar wind is spread over such a large volume that it can no longer push back the interstellar plasma. Because the wind is flowing out much faster than waves can propagate inward, the solar wind flow ends at a spheroidal shock wave, which is called the termination shock, where the supersonic flow changes suddenly to a subsonic outward flow.

The interstellar plasma is moving at about 26 km/s relative to the heliosphere, pushing it in on one side, as shown in Fig. 9.1. Beyond the termination shock, the solar plasma continues to flow outward, but it is deflected and eventually turns to flow in the same direction as the interstellar plasma, forming a large, trailing, heliospheric tail (see Vol. II, Fig. 7.1b). The interstellar medium also contains neutral atoms, and while these play a role in the interaction, they may be neglected in the lowest order.

Energetic particles or cosmic rays pervade the heliosphere, as they do all regions of low-enough density in the universe. The energetic particles are of four basic types: (i) galactic cosmic rays, (ii) anomalous cosmic rays (ACR), (iii) interplanetary energetic particles, and (iv) solar energetic particles (SEP; see Vol. II, Section 9.1). The ambient thermal fluid contains most of the mass and momentum and, because of its low density, does not interact collisionally with the energetic particles, but rather through the magnetic field carried with the fluid. The energetic charged particles have energies above the thermal energy of the background fluid, and those observed in the heliosphere originate either

Heliophysics: Evolving Solar Activity and the Climates of Space and Earth, eds. Carolus J. Schrijver and George L. Siscoe. Published by Cambridge University Press. © Cambridge University Press 2010.

Fig. 9.1. Artist's rendition of the heliosphere, including the positions of the Voyager 1 and 2 spacecraft in 2009. (Copyright P. C. Frisch, University of Chicago and A. Hanson, Indiana University.)

at the Sun (solar cosmic rays), the heliosphere, or in interstellar space (galactic cosmic rays).

Here, I concentrate primarily on galactic cosmic rays (see Vol. II, Chapters 5–9, for descriptions of the other three populations). They come from the galaxy, where they are thought to be accelerated at supernova blast waves. They are confined to the galaxy by a random walk in the interstellar magnetic field for ~20 million years before leaking out. They envelop the heliosphere with a very nearly constant, isotropic bath. These particles are partially excluded from the inner parts of the heliosphere (see Fig. 9.3 for an illustration of a model calculation of their radial variation and Fig. 9.7 for a cartoon illustration of the particle motions). Therefore, their intensity reflects the varying properties of the heliosphere. Galactic cosmic rays have a typical energy of 1 GeV and are present continuously, but fluctuate on a variety of time scales.

Solar cosmic rays are produced sporadically in solar flares, and are intense mainly at considerably lower energies than the galactic particles (see Vol. II, Fig. 9.1). Their spectrum is also a much more rapidly decreasing function of energy. After a solar flare produces solar cosmic rays, they are present in the solar system only for a relatively short time and decay away with a time scale of days (at low energies) to hours (for GeV particles). The time-averaged intensity of these two types of cosmic rays, as a function of energy, is illustrated in Fig. 9.2, where the solar particles are a solar-cycle average. Also shown is an approximation

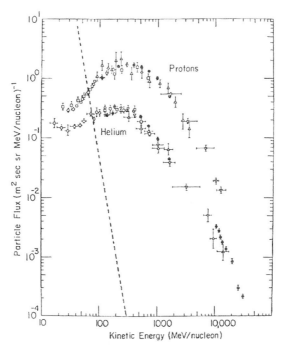

Fig. 9.2. Observed spectra of cosmic rays at Earth. The data points show the quiet-time spectrum of galactic cosmic-ray protons and helium nuclei, and the steeper, dotted line is a characteristic spectrum of solar energetic particles for an average event in the solar cycle. Different symbols reflect results from different instruments; see Jokipii (1986) for details. See Figs. 3.1 and 9.1 in Vol. I for the continuation of the particle spectra from different sources at lower energies. (From Jokipii and Marti, 1986.)

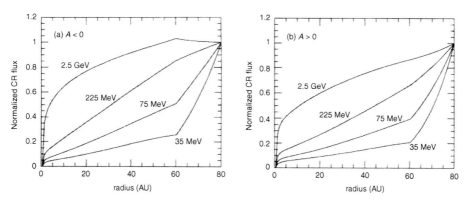

Fig. 9.3. Model calculation for the radial variation of cosmic rays for field configurations in which the magnetic field is either inward (a: $A < 0$) or outward (b: $A > 0$) in the northern hemisphere of the heliosphere for an early model with the radius of the termination shock of 60 AU. Observations from spacecraft generally confirm this picture. The intensity decreases inward away from the particle source (see Vol. II, Section 9.5.2) and goes to zero near the Sun because of absorption by the solar atmosphere. (From Jokipii *et al.*, 1993.)

to transient solar-energetic-particle spectra, which have fewer high-energy particles. The average spectrum over time is dominated by the galactic cosmic rays,
although for short periods (hours to days) the solar particles can be quite intense.
Figure 9.3 illustrates the basic radial variation of cosmic rays for the two signs of
the interplanetary magnetic field.

The intensity of galactic cosmic rays in the inner solar system varies over a wide
variety of time scales. The time variations of galactic particles are in part due to
variations in the solar wind and its entrained magnetic field, which are accessible –
albeit sparsely – to direct measurement. There exists a generally accepted physical
model that quantitatively accounts for these modulations.

It is demonstrated below in Section 9.3.1 that cosmic-ray variations on time
scales less than about one hundred thousand years must be caused by changes in
the heliosphere, caused by changes in the Sun.

Interstellar variations over longer time periods can be caused by either the
motion of the solar system through the interstellar medium or by transient changes
in the interstellar cosmic-ray intensity caused by supernova blast waves, etc. In
addition, the heliospheric structure, and hence its effects on cosmic rays, can be
affected by changes in the interstellar medium caused by, for example, interstellar
clouds.

9.2 Observed cosmic-ray time variations

In this section the cosmic-ray variations observed in the inner solar system, using a
variety of techniques and over a variety of time scales, is discussed. Their relation
to the basic physics of cosmic-ray transport is discussed in Section 9.4.

9.2.1 Direct measurements

Direct measurements of the time variations of cosmic rays are available for the time
period spanning the last four sunspot cycles. The largest known periodic variation
is the variation of galactic cosmic rays in anti-phase with the 11-year sunspot cycle.
The variation of the galactic cosmic-ray intensity over the past five sunspot cycles
is illustrated in Fig. 9.4, and there is no doubt about the variation and its association
with the sunspot index. The interpretation in terms of transport theory is discussed
in detail in Section 9.3.

Figure 9.5 shows a power spectrum of the variation in the intensity of galactic cosmic rays observed at Earth with the Climax neutron monitor. The main
conclusion one can draw from this plot is that there is a continuous spectrum of
fluctuations. If one were to obtain such a curve covering all temporal frequencies,
it would be similar – consisting of a smooth background of variations, with stronger

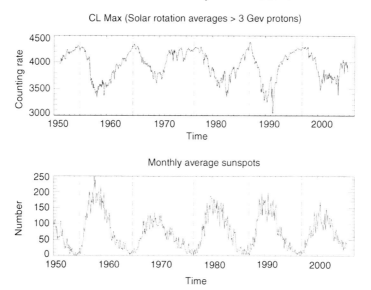

Fig. 9.4. The modulation of galactic cosmic rays during five sunspot cycles. Note the alternation in the cosmic-ray maxima between sharply peaked and more-rounded shapes. (Data courtesy of the University of Chicago.)

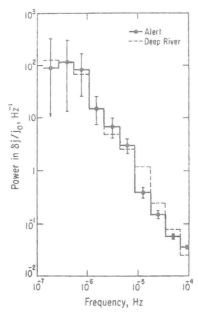

Fig. 9.5. The temporal power spectrum of the observed variations of $\sim 1\,\mathrm{GeV}$ galactic cosmic rays at Earth. Note that 1 Hz or one cycle per second is $\sim 3 \times 10^7$ cycles per year. (From Jokipii and Owens, 1973.)

or weaker peaks at certain frequencies (not shown in Fig. 9.5), corresponding, for example, to the 11-year sunspot cycle. The broad spectrum of variations illustrated in the figure are real and can be shown to be the result of the continuous bubbling and variation of the solar wind with its entrained magnetic field.

At this time it is not completely understood how solar activity changes the interplanetary medium to produce the observed temporal variations. There are five principal elements. The physical picture underlying these is discussed in Sections 9.3 through 9.5. Some notable transient or periodic variations in the cosmic-ray intensity are:

  (i) Co-rotating high-speed wind streams (Chapter 8) produce nearly periodic variations at the solar rotation period.
 (ii) Coronal mass ejections and high-speed solar wind streams combine to produce large-scale, long-lived structures (known as global merged interaction regions or GMIRs; see Section 8.5.1) of enhanced magnetic field that propagate out to the termination shock and into the heliosheath. There is a strong correlation between the rate of CMEs and sunspot numbers that have been observed over periods of high and low solar activity. From this, one may conclude that the almost 400 years of sunspot observations provide a useful tool to study the levels of solar activity over that time.
(iii) Changes in magnitude of the heliospheric magnetic field (HMF) over many scales (see Chapter 8). The gyroradii of cosmic rays are inversely dependent on the strength of the HMF, and this, together with the observed turbulence of interplanetary plasma (Vol. I, Chapter 7) produce changes in the cosmic-ray diffusion coefficients that are approximately inversely proportional to the magnetic field magnitude B (see Vol. II, Section 9.4).
 (iv) The changing inclination of the heliospheric current sheet that changes from a nearly flat configuration in the equatorial plane at solar minimum to a 90° inclination at solar maximum and then with decreasing solar activity returns to its near equatorial position at the next solar minimum (cf. Vol. I, Figs. 8.1 or 9.3, or Fig. 8.1 in this volume). This is associated with a change in magnetic polarity, leading to a 22-year solar magnetic cycle.

9.2.2 Terrestrial carbon and beryllium

The isotopes ^{14}C and ^{10}Be produced in the Earth's atmosphere by cosmic rays are a major source of information about variations in the past several thousand years (see Chapter 11 and e.g. McCracken *et al.*, 2004). Observations of these cosmogenic species show that heliospheric modulation processes that change the incident flux have varied during the past. There is clear evidence from this that the level of solar activity has changed substantially over the past 400 years; the sunspot numbers have exhibited a rising trend since the seventeenth century, and the peak annual sunspot numbers in the late twentieth century are among the highest ever recorded (see Fig. 2.2). Our current understanding of these processes therefore applies to the recent past, and the dominant processes may be different at times of sustained lower solar activity when key parameters, such as the strength of the heliospheric magnetic field, may have been different. To fully understand the modulation processes and the factors driving them, we need to test and extend them using the behavior of the galactic cosmic rays (GCR) during periods of lower solar activity. The cosmogenic isotopes stored in polar ice and in biological materials provide the data required to do this (Chapter 11).

9.2.3 The Maunder minimum

The Maunder minimum is the name given to the period spanning roughly the years 1645 to 1715, when sunspots were exceedingly rare, as noted by solar observers of the time (Section 2.8). The cosmogenic ^{10}Be and ^{14}C data provide useful tools to study the associated effects on the cosmic-ray intensity in the GeV energy range, and hence the effects of the minimum on the heliosphere. ^{14}C data for this period were reported by Stuiver and Quay (1980) and are shown in Fig. 9.6. Also, Miyahara *et al.* (2008) argued that the fluctuations in the period of the Maunder

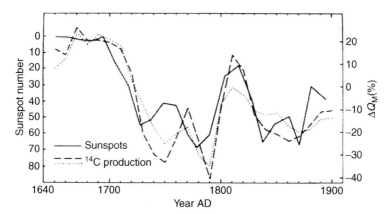

Fig. 9.6. Carbon-14 concentrations around the time of the Maunder minimum. Shown are the sunspot number (inverse scale) and the production rate of ^{14}C inferred from tree-ring studies. During the Maunder minimum period in the 1600s, when the sunspot number was lowest, the ^{14}C production rate was highest. The 11-year running mean sunspot number is given by the solid line (left axis) and the ^{14}C measurements are given with the dotted line (assumed residence time 60 yr) and the dashed line (assumed residence time 20 yr), both with the right axis. (From Stuiver and Quay, 1980.)

minimum arguably have different structure (in terms of both period and amplitude) than those before or after.

The Spörer minimum was a 90-year span from about 1460 until about 1550, which occurred before sunspots were routinely observed. The existence of this extended minimum state was inferred instead by analysis of the proportion of ^{14}C in tree rings.

It is likely that the Maunder and Spörer minima each provide valuable insights into the modulation processes of the heliomagnetic field and cosmic rays during extended periods of low solar activity. A possible mechanism to produce the observed changes is discussed in Section 9.7.

9.2.4 Measurements using extraterrestrial matter

This section summarizes measurements of cosmic-ray variations on a range of time scales, following the discussion in Jokipii and Marti (1986), to which the reader is referred for more detail.

Galactic cosmic-ray particles have sufficient energy ($\sim$ GeV) to induce nuclear reactions in solid solar system matter. On the other hand, as was discussed above, the solar cosmic-ray flux at these energies has much lower intensity and effects of the solar cosmic rays can only be seen in surface layers of extraterrestrial matter. The unstable isotopes produced can be studied to determine the cosmic-ray intensity, and hence provide information on temporal variations on time scales of the order of the decay life of the unstable species.

In addition, the record regarding the cosmic-ray flux can only be obtained if the exposure geometry is known and has remained constant; limitations are set, for example, by erosional processes. Information on the longer time scales of 10^8–10^9 yr can be obtained from iron meteorites, because the times of exposure of their material to cosmic rays range up to 2×10^9 yr. The energy of an interacting particle and the chemical composition of the target determine the cascade of nuclear reactions taking place and the distribution of reaction products. An important question which is addressed later is how the galactic cosmic-ray flux is modulated inside the heliosphere and how one can separate interstellar flux variations and solar modulation effects.

9.2.4.1 Variations on the 10^1–10^4-year time scale

Correlations are known among solar activity indices and ^{14}C activity: ^{14}C increases during solar minima and decreases during solar maxima, reflecting solar modulation effects on the galactic cosmic rays. On a 10^3–10^4-year time scale, the dominating effect appears to be a 10^4-year period. The largest amplitude, called the "Suess wiggle", is observed at about 200 years. The observed radio-activities, when corrected for target chemistry and for shielding differences, indicate variations of the spallation production rates by the cosmic rays of a factor of two. These variations are correlated with the sunspot cycle. Therefore, the production rates of cosmic-ray-produced nuclides also vary with the solar 11-year cycle, suggesting that they are related to solar modulation of galactic cosmic rays.

9.2.4.2 Variations on the 10^5–10^7-year time scale

The record of the galactic cosmic-ray flux on a million-year time scale can be inferred from induced nuclear reactions in extraterrestrial matter of known exposure geometry, such as lunar rocks or meteorites. Nuclear reactions produce a variety of radioactive and stable nuclei that can be measured and related to the incident cosmic ray flux. The radionuclides ^{81}Kr (2.1×10^2 yr half life), ^{36}Cl (3.0×10^5 yr), ^{26}Al (7.2×10^5 yr), ^{10}Be (1.6×10^6 yr) and ^{53}Mn (3.7×10^6 yr) represent a good set of monitors for cosmic-ray flux variations on this time scale. Among the chondritic meteorites, which were studied extensively, the production rates of the above radionuclides can vary because of differences in size and shielding conditions. These, when analyzed, reflect a constant (± 10–15%) galactic flux over the 10^5–10^7-year time scale, which matches the average present-day flux.

9.2.4.3 Variations on the 10^7–10^9-year time scale

There are few radioisotopes with appropriate half lives that can be used for this time scale and only ^{129}I (1.6×10^7 yr) and ^{40}K (1.3×10^9 yr) have been studied so far.

Chondritic meteorites cannot be used to study variations in the cosmic ray flux on longer time scales, because their exposure ages (time between being formed and striking the Earth) are typically less than a few tens of million years. Fortunately, there are numerous recovered iron meteorites that were exposed in space as small bodies for up to two billion years since being formed, and which are well suited for this purpose. The measurement of all three isotopes of potassium permits the detection of the cosmic-ray-produced component which is superimposed on potassium initially present in the meteorite. For the time period of 0.2–1.0 Gyr ago, essentially constant ^{38}Ar production rates are observed, and agreement between ages determined from ^{38}Ar and from ^{40}K and ^{41}K.

9.3 The physics of heliospheric cosmic-ray temporal variations

We next consider how to link observed temporal variations in cosmic rays to the physics of cosmic-ray transport in either the heliosphere or in the local interstellar medium.

The physical processes underlying the transport of cosmic rays in the interstellar medium and in the heliosphere are basically the same. These fluids are always turbulent, and collisions with particles are very rare, so the motion of cosmic rays must be described statistically. The most basic approximation is that the cosmic rays *diffuse* through the turbulent magnetic field with a diffusion coefficient κ determined by the magnetic fluctuations. Thus, the time to move a distance L is $\sim L^2/\kappa$. Because the turbulent magnetic field is advected with the ambient average velocity, the transport must include this as well (see Vol. II, Chapter 9). This, then, leads to the Parker transport equation, which is discussed further in Section 9.4.

9.3.1 Interstellar causes of cosmic-ray variations

We first consider the possibility that the observed variations are the result of variations in the *interstellar* flux enveloping the heliosphere, rather than variations related to the Sun. Interstellar variations can be of two kinds: the Earth could pass through effectively static cosmic-ray variations in its motion relative to the nearby interstellar medium, or dynamical cosmic-ray variations associated, for example, with shock waves in the interstellar gas could cross the solar system. It is not likely that cosmic-ray intensity variations observed at Earth on time scales of less than of the order of 10^5 yr could be be produced by the solar system passing through quasi-static variations. For such variations to exist long enough for the motion of the solar system to bring the Earth through them, the transport of galactic cosmic rays would have to be much less rapid than is currently thought to be possible.

The near isotropy of the cosmic rays implies *diffusive* transport (see Section 9.4). Consider a fluctuation in the cosmic-ray density of characteristic scale L, which would have a diffusive lifetime τ of order L^2/κ, where κ is the cosmic-ray diffusion coefficient. If the solar system is moving at a speed v_E, it takes a time L/v_E to cross this fluctuation. Therefore, we require

$$\frac{L^2}{\kappa} \gg \frac{L}{v_E}. \tag{9.1}$$

Setting $v_E \approx 20$ km/s and $\kappa = (1/3)\lambda w$, where w is the particle speed (essentially the speed of light) and λ is the diffusion mean free path (λ must at least exceed several cosmic-ray gyroradii in the interstellar magnetic field of microgauss), we find that $L \gg 3 \times 10^{17}$ cm, which would be crossed in a time $\tau \approx 10^5$ yr or more.

In view of this, the only way that short time-scale variations (i.e. of less than a few times 10^5 yr) could be observed in the interstellar flux would be for intense, probably localized, *dynamical* processes to propagate past the Earth. This would include, for example, strong shock waves from a nearby supernova. But, in general, the recurring fluctuations on time scales $< 10^5$ yr which one sees in many cosmic-ray records probably are not the result of plausible interstellar processes.

However, the converse is not necessarily true, in that it is reasonable that the heliosphere affects the charged-particle environment in the local interstellar medium, to several times the scale of the heliosphere, or several hundreds of AU. Heliospheric effects caused by the solar wind and its embedded magnetic field may distort and modify the intensities of the cosmic rays well outside the heliosphere, in the local interstellar medium (LISM). The nature of the effects depends in large part on the poorly determined transport parameters in the LISM. As far as we know, these effects have not been seriously addressed before in the published literature. I present here a brief summary of the general nature of the expected effects.

If the interstellar magnetic-field spectrum is such that the *local* interstellar cosmic-ray diffusion mean free path is of the order or less than the scale of the heliosphere, or a few hundred AU, the effects of the heliosphere on the cosmic-ray flux extend several hundreds of AU out into the LISM. On the other hand, if the local interstellar diffusion mean free path for cosmic rays is much larger than the scale of the heliosphere, the effects are more complex. The cosmic rays will propagate quite rapidly away from the heliosphere and in most cases the interstellar intensity will be unchanged up to quite close to the heliosphere. However, because of the guiding of the particles by the magnetic field, there will likely be localized interstellar regions where the effects of the heliosphere are seen far from the heliosphere.

9.3.2 Cosmic-ray variations of heliospheric origin

We are left, then, with attempting to interpret the observed cosmic-ray variations on time scales less than some 10^5 yr in terms of physical processes associated with changes in the heliosphere. Figure 9.7 is a schematic illustration of the heliospheric structure generated by the flow of the solar wind and its interaction with the interstellar gas. The heliospheric cavity is the result of a radial outflow of solar wind from the Sun. The wind flows supersonically out to a spheroidal "termination" shock, now known to be at some 80–100 AU from the Sun, where the supersonic flow becomes subsonic. Beyond the shock the flow undergoes a gradual deflection to a direction downwind from the solar system. Downwind here refers to the relative direction of motion of the interstellar plasma which confines the solar wind.

The separatrix between the solar plasma and the interstellar plasma, the heliopause, is ideally a thin contact surface, but probably this will have some as yet unknown thickness because of non-ideal processes such as instabilities, interdiffusion, etc. This issue is not yet well studied, as spacecraft have not explored beyond some 120 AU. However, the uncertainties should not affect the general nature of

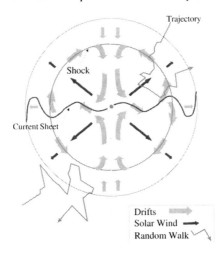

Fig. 9.7. Schematic illustration of the motions of galactic and anomalous cosmic rays in the heliosphere. The directions of the arrows are for a sunspot cycle when $A > 0$ (northern hemisphere magnetic field out). The direction of the drift arrow would be reversed for the opposite sign of the magnetic field. The galactic cosmic rays enter from the outer boundary and the anomalous cosmic rays are accelerated at the termination shock. Both are affected by the shock and the current sheet.

the cosmic-ray modulation models discussed in Section 9.4. The crossing of the termination shock by the two Voyager space probes suggests that, although details in the outer heliosphere are puzzling, the general picture developed over the past few decades is a good approximation to observations in the inner heliosphere, far from the termination shock. Because the inner heliosphere is the principal focus of this chapter, we ignore the uncertainties near the termination shock. I note that a thorough review of cosmic rays in the heliosphere is given in the volume *Cosmic Rays in the Heliosphere* (Fisk *et al.*, 1998a).

Figure 9.4 illustrates one major effect of the solar wind on galactic cosmic rays. Shown is the counting rate of the neutron monitor at Climax, which responds essentially to the flux of ~ 1.6 GeV protons just outside of the Earth's magnetosphere, since 1953. The 11-year, sunspot-cycle-related variation is clearly seen. Deep minima in the cosmic-ray intensity occur at the maxima in the sunspot count. Also apparent is a possible 22-year variation, in that the shapes of successive maxima are alternately flat topped or sharply peaked. As we shall see in Section 9.5, such a 22-year variation is a natural result of current understanding (see also Vol. II, Section 9.5.4). From such observations, independent of specific models, we may readily infer that longer-scale variations observed in various proxy indicators of the cosmic-ray flux could be produced by physical processes similar to those which cause the sunspot-cycle-related variations.

9.4 The transport of cosmic-ray particles in the heliosphere

The first problem before us then is to obtain a transport equation that governs the motion of fast charged particles in a magnetic field, which is advected outward with the solar wind. This magnetic field has an average given by the Archimedean spiral originally suggested by Parker (Vol. I, Section 7.3). Superposed on this average field are magnetic irregularities which scatter the cosmic-ray particles and cause a random walk. It is important to note that collisions of cosmic rays with the ambient plasma are completely negligible, so that the magnetic field determines the motion. This transport problem has been studied intensively over the past few decades, with the equation that is now used being first written down by E. N. Parker (Parker, 1965). In this approximation, particle distribution is taken to be nearly isotropic in pitch angle due to scattering by magnetic irregularities. The resulting transport is basically the combination of four major physical effects (see also Vol. II, Section 9.3). These are:

(i) *Advection*: The solar wind advects the magnetic field and cosmic rays outward at the solar wind velocity $\mathbf{v}$. (The solar wind is hydromagnetic and the magnetic field is frozen in – see Vol. I, Section 3.2.3.1.)

(ii) *Diffusion*: A random walk or diffusion caused by the scattering of the cosmic rays by the irregularities in the magnetic field. This scattering produces the observed directional near-isotropy of the cosmic-ray flux. The associated diffusion tensor is denoted κ_{ij}, which is anisotropic, being significantly larger along the magnetic field than perpendicular to it; κ_{ij} can, in principle, be calculated in terms of the irregularity spatial power spectrum. In practice, it is an assumed parameter, adjusted to fit observations.

(iii) *Drifts*: The large-scale spatial variations of the average magnetic field cause coherent guiding-center drifts of the energetic cosmic-ray particles in addition to the random walk. These are the usual guiding-center drifts, which one studies in basic plasma physics problems and are well understood (see Vol. II, Section 9.2). The drift velocity (averaged over the isotropic pitch angle distribution), is given in terms of the average magnetic field $\mathbf{B_0}$ and the particle charge q by $\mathbf{v}_{\mathrm{d}} = (pcw/3q) \nabla \times \left(\mathbf{B_0}/B_0^2\right)$. The drifts, for a Parker spiral magnetic field, are illustrated in Fig. 9.8.

(iv) *Adiabatic energy change*: In addition to the wind's outward advection, its radial divergence or expansion $\nabla \cdot \mathbf{v}$ causes the cosmic rays to continuously lose energy to the wind as they propagate in the solar system. This energy loss is a significant factor in the modulation process. In contrast, where the fluid is compressed, such as at the termination shock, the particles gain energy from the flow.

The Parker transport equation, for the quasi-isotropic distribution function (or phase space density) $f(x_i, p, t)$ of cosmic rays of momentum p at position x_i and time t, combines the above physical effects and may be written

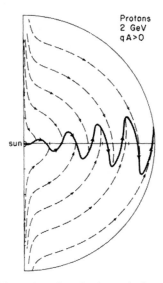

Fig. 9.8. Cosmic-ray drift motions in a Parker spiral magnetic field with a current sheet. The arrows shown correspond to the time when the northern-hemisphere heliospheric magnetic field is outward from the Sun (1975, 1996) for positively charged particles. The arrows reverse for the alternate sign of the magnetic field (1986, 2007) and for the opposite sign of the particle's electric charge. (From Jokipii and Thomas, 1981.)

$$\frac{\partial f}{\partial t} = \frac{\partial}{\partial x_i}\left[\kappa_{ij}\frac{\partial f}{\partial x_j}\right] \qquad (\textit{diffusion})$$

$$- v_i \frac{\partial f}{\partial x_i} \qquad (\textit{advection})$$

$$- V_{di}\frac{\partial f}{\partial x_i} \qquad (\textit{guiding-center drift})$$

$$+ \frac{1}{3}\frac{\partial v_i}{\partial x_i}\left[\frac{\partial f}{\partial \ell n p}\right] \qquad (\textit{energy change})$$

$$+ Q. \qquad (\textit{source})$$

(9.2)

Here, the diffusion tensor κ_{ij} may be written in terms of the magnetic field **B**, the parallel diffusion coefficient $\kappa_\parallel$, and the perpendicular diffusion coefficient $\kappa_\perp$ as

$$\kappa_{ij} = \kappa_\perp \delta_{ij} + (\kappa_\parallel - \kappa_\perp)\frac{B_i B_j}{B^2}. \qquad (9.3)$$

The associated streaming flux of the particles may be written

$$S_i = -\kappa_{ij}\frac{\partial f}{\partial x_j} - \frac{U_i}{3}\frac{\partial f}{\partial \ell n(p)} \qquad (9.4)$$

with the associated anisotropy

$$\delta_i = \frac{3S_i}{wf}, \qquad (9.5)$$

where w is the particle speed.

The labels next to each of the various terms in Eq. (9.2) indicate the associated physical process. This transport equation is remarkably general, and has been used in most discussions of cosmic-ray transport and acceleration over the past two decades. It appears that it is a good approximation if there is enough scattering by the magnetic irregularities to keep the distribution function nearly isotropic. It requires that the particles have random speeds substantially larger than the background fluid advection speed. The fluid velocity need not be a continuous function of position, so shocks can be discussed within the framework of this equation. In fact, all of the standard theory of diffusive shock acceleration is contained in this equation.

Parker's transport equation, applied to the heliosphere, provides much of the basis for our current understanding of the solar modulation of galactic cosmic rays and the acceleration and transport of anomalous cosmic rays. A simplified model, often used in simulations is illustrated in Fig. 9.7.

9.4.1 The boundary conditions

When Eq. (9.2) is applied to the heliosphere, as described above, it is solved subject to a boundary condition where the distribution function f approaches the (assumed) interstellar value at some outer boundary often taken to be a sphere of radius R, which is typically taken to be some 30–50% larger than the radius of the termination shock.

Computational models of cosmic-ray transport in the heliosphere require knowledge of the boundary conditions at this interface with the interstellar medium and its cosmic-ray flux, as well as at the Sun. The specification of the boundary conditions is a complex physical problem that has not been much discussed in the literature. This has traditionally been handled by assuming that at the outer boundary the value of f is some specified interstellar value (steady over the time scales under study). Although the form of this spectrum is not well determined, a reasonable guess can be made. In early models, this boundary was taken to be a sphere where the supersonic solar wind ended – the termination shock. In the 1980s it was realized that the termination shock should be part of the heliosphere, because it accelerates the anomalous cosmic rays and because it significantly affects the galactic cosmic rays. In most applications the termination shock was simply set to be a spherical surface at some suitable radius, where the radial wind speed was reduced by the shock ratio, r_{sh}. Florinski and Jokipii (1999) included the influence of cosmic rays on the shock. In these models, the boundary then moved outward to some sphere beyond the termination shock, where again the physics was not clearly determined. Fortunately, many of the consequences for the inner heliosphere are not very sensitive to the details of the transport beyond the shock. However, as we now have observations at the termination shock and beyond, it has become important to treat this boundary more correctly.

Just what this boundary corresponds to must be clarified if it is to be treated accurately. There are four relevant spatial regions: the solar wind, the termination shock, the heliosheath (a region of subsonic solar plasma beyond the termination shock), and the local interstellar plasma, each with its own associated physical properties. Within each of the regions the values of the flow velocity and diffusion coefficient vary with position. More importantly, the diffusion coefficients may vary considerably as a function of energy, usually increasing with increasing energy.

Important parameters are κ_{ISM}, the diffusion coefficient in the local ISM, and v_{ISM}, the velocity of the local ISM relative to the Sun. There are two important dimensionless parameters which determine the nature of the boundary condition, given a typical dimension ℓ of the system. They are

$$\eta(E_{kin}) = \frac{\ell v_{ISM}}{\kappa_{ISM}} \tag{9.6}$$

and the contrast

$$r_\kappa(E_{kin}) = \frac{\kappa_{HS}}{\kappa_{ISM}} \tag{9.7}$$

between the interstellar (ISM) and heliospheric (HS) diffusion coefficients, where the possible dependence on the particle kinetic energy (E_{kin}) of the numbers must be considered.

It is readily seen that if κ_{ISM} is large enough that both $\eta(T)$ and $r_\kappa(T)$ are much less than unity, the diffusion in the local ISM is very rapid, and the boundary may be taken to be the standard "free escape" boundary at the heliopause. Beyond this point the particles move so rapidly that the cosmic-ray intensity may be taken to be the full interstellar value. This is essentially what has been done in simulations carried out up to now. Putting numbers into Eqs. (9.6) and (9.7), we find that this would require that $\kappa_{ISM} \gg 10^{23}$ cm^2/s.

If κ_{ISM} were to be of the order of 10^{23} or less, then the boundary condition would be much more complicated, and depend on the local structure of the interstellar medium, its velocity and its magnetic field. In practice, for the energies of interest here, κ_{ISM} is large enough that this problem does not arise.

In the following, we ignore subtleties in the model calculations, and simply require that the intensity at the outer boundary is equal to the assumed interstellar spectrum. Many of our conclusions are not very sensitive to this assumption.

9.5 Solar modulation of galactic cosmic rays

The essential problem in the solar modulation of galactic cosmic rays is to solve the differential equation (9.2) by analytical or numerical methods, subject to the boundary condition that the intensity at the boundary of the heliosphere is the galactic cosmic-ray spectrum, which is assumed to be constant in time. It may be shown that the characteristic time scale for these various transport processes to bring the cosmic-ray distribution into equilibrium throughout the heliosphere is of the order of, or less than, a year. Hence, the study of cosmic-ray variations on scales of the 11-year sunspot cycle or longer requires only that we solve the differential equation in the stationary limit (setting the term in $\partial f/\partial t = 0$), and then obtain the intensity at various times by looking at the solution for the different magnetic field configurations or different values of the parameters expected at different times. Simulations of solar-cycle or longer-period variations which retain the full time dependence do not differ significantly from these. The study of some observed time-dependent effects (such as a possible differential time dependence for particles having different energies) may require use of the full time-dependent equation, but modulation studies have not utilized time dependence in any detail.

In order to proceed quantitatively we first specify the solar and heliospheric plasma and magnetic field configuration. The large-scale structure of the heliospheric magnetic field has been clarified considerably by observations carried out on space probes, as has the inferred relationship to observed coronal structure (see Vol. I, Chapters 8 and 9). It is found that during the years around each sunspot minimum, the field is generally organized into two hemispheres separated by a thin current sheet across which the field reverses direction. In each hemisphere the field is a classical Parker Archimedean spiral, with the sense of the field being outward in one hemisphere and inward in the other. Because the current sheet is not a plane at the Sun, and on average is inclined to the rotational equator, it is twisted by solar rotation into a spiral wave which oscillates above and below the solar equatorial plane (cf. Fig. 8.1). Observations indicate that the waviness, or inclination, of the current sheet is a minimum at sunspot minimum and *increases* as a function of time away from sunspot minimum. Around the time of sunspot maximum the sense of the field in the northern and southern hemispheres changes sign. The overall magnetic field direction therefore alternates with each 11-year sunspot cycle.

In the rest of the present discussion, it is generally assumed that the overall magnetic structure is given by this model, recognizing that it may not be a particularly good representation in the few years around maximum sunspot activity. Jokipii and Kota (1989) pointed out that the magnetic field near the solar rotation axis, at large heliospheric radii, may differ significantly from this, but it is sufficient for present purposes to ignore this additional complication.

With the heliospheric structure established, we next examine Eq. (9.2) to determine whether any of the transport effects are small enough to be neglected. We can estimate the drift velocity magnitude, the gradient of the cosmic rays, and the diffusion coefficient. All four terms – advection, diffusion, energy change, and drifts – are comparable in magnitude. They are all complicated and one really must do a full numerical solution to gain even qualitative insight into the solutions.

A major effect in the transport comes from the cosmic-ray drift motions. Figure 9.8 shows drift streamlines for $\sim$GeV protons in the 1996 heliospheric magnetic field configuration. It is readily apparent that the particles drift in over the poles and are ejected very rapidly along the current sheet which separates the regions of opposite sign of the magnetic field.

The heliospheric magnetic field structure alternates with each subsequent sunspot minimum, and the directions of the arrows reverse with that. In current models, this particle drift turns out to be probably the most significant effect on the particle transport in the standard model. There is no simple way to avoid the fact that because the drifts are coherent over large distances in the heliosphere, in some cases they can be more important than the random walk and advection terms. For the standard spiral we find that nearly all of the particles we see in the inner solar system during the 1975 and 1996 sunspot minima came from a very small region near the poles of the boundary. Conversely, the ones in 2006 have come in more or less along the current sheet.

Application of this differential equation to the modulation of galactic cosmic rays has shown that the modulation around 2009 seems to reflect similar contributions of diffusion and drift.

The random walk or diffusion of the cosmic rays through the magnetic field as they are advected and cooled in the expanding solar wind is effectively all that was contained in the initial view of modulation. It is useful to discuss this simplified model first, neglecting the drift motions. In this picture, the solar wind flows outward to a boundary at some distance D from the Sun. At this point the solar wind ceases, the advection ceases, and the intensity of cosmic rays is equal to its time-invariant galactic value. The cosmic rays tend to diffuse, or random walk, into the inner solar system from this point, but their progress is impeded by the outward advection of the magnetic field and plasma of the solar wind. In addition, they are cooled in the expanding flow. This competition between the inward diffusion and the outward advection, the so called *diffusion/advection/adiabatic-cooling* model, gives rise to a depressed intensity in the inner heliosphere. Clearly, as one increases the radial wind velocity v, as might happen during periods of higher solar activity, one lowers the intensity at a given point in the solar system. Similarly, if one decreases the diffusion coefficient κ, corresponding to more scattering by magnetic irregularities, we would also decrease the cosmic-ray intensity in the inner solar system. Hence, if one is willing to vary v or κ in an appropriate way by using this procedure one can easily get the 11-year solar cycle variation. However, there is no clear observational evidence for such a systematic variation in v or κ which is correlated with the sunspot cycle. In addition, because the random walk does not depend on the sign of the interplanetary magnetic field, any 22-year variation obtained from this model would be ad hoc.

A further approximation to the above simplified model is often used, even today, because of its exceedingly simple form. If one assumes spherical symmetry, and then considers only high-energy cosmic rays, for which the dimensionless modulation quantity rv/κ, which measures the strength of the modulation, is small, a very simple analytic solution can be obtained. The form of this solution corresponds exactly to that obtained for charged particles influenced by a force field with a potential energy given as a function of heliocentric radius r by

$$\Phi(r) \propto \int_{r}^{D} \left[V_w/\kappa \right] dr. \tag{9.8}$$

Note that this cannot be a real electrostatic potential because it affects positively and negatively charged particles in the same way. Attempts to fit the data yield values of $\Phi \approx 300\,\text{MeV}$ near 1 AU (Caballero-Lopez *et al.*, 2007). Because of the use of an effective potential energy, this approximation is called the "force-field" solution. Although such fits are made, it is not clear what meaning to attach to Φ, since the model is so restrictive. For example, observations show a strong dependence of the cosmic-ray intensity on heliographic latitude, which is assumed to be absent in the force-field model. Nonetheless, observers often parameterize their observed time variations with the parameter Φ (e.g. Section 11.4.3; see Fig. 11.16 and Table 11.5).

9.6 Sample model simulations using the full transport equation

The full differential equation (9.2), containing all of the effects, can be solved numerically with two- or three-dimensional, time-independent heliospheric models, although proper inclusion of the termination shock appears to require a full time-dependent code. The effects of helicity and cosmic-ray viscosity have not yet been shown to be significant and are not discussed further here. A number of computer codes to accomplish this have been developed over recent years. It is quite clear that these can quite naturally reproduce many of the phenomena observed in the cosmic-ray flux, including many 22-year cyclic effects.

The computed intensity as a function of time for two succeeding sunspot cycles is illustrated in Fig. 9.9. This should be compared with the observations in Fig. 9.4. Clearly, the model is capturing the essential 22-year cycle. The alternation of sharply peaked with more-rounded cosmic-ray maxima is a robust prediction of the theory if the drift motions and the interplanetary current sheet play a significant role.

A basic prediction of the theory including the coherent drift motions was that, because the sense of the particle drift changes from one sunspot cycle to the next, one expects changes in the cosmic-ray intensity and its spatial distribution from one sunspot cycle to the next. In particular, in a sunspot minimum such as around 2008, where the magnetic field is inward in the northern hemisphere of the heliosphere (denoted $A < 0$), protons drift inward along the current sheet and then are scattered off of it to drift outward towards the pole. As they drift away from the current sheet they lose energy (cool) because of the divergence in the flow velocity of the wind. Another, equivalent way of saying this is that in both cases they are drifting against the interplanetary $\mathbf{v} \times \mathbf{B}$ electric field. Because of the cooling, the intensity *decreases* for some distance away from the current sheet. Conversely, in the preceding minimum around 1996 (with $A > 0$) the particles drifted in from

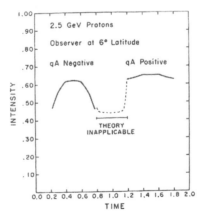

Fig. 9.9. Illustration of the results of a simple model calculation for the galactic cosmic ray flux where only the current sheet tilt and the sign of the magnetic field varied. The model is inapplicable near sunspot maximum. The plots are for cosmic-ray ions (with charge $q > 0$) which comprise the majority of observed cosmic rays. (From Jokipii and Thomas, 1981.)

the *polar* regions, striking the current sheet, and then were ejected outward along the sheet. As they drift in from the polar regions they are again drifting against the electric field, and the intensity decreases as they get closer to the current sheet. This produces a *positive* latitudinal gradient away from the vicinity of the current sheet. For a flat current sheet one can show that there is a general relationship between the radial and latitudinal gradient (Jokipii *et al.*, 1977).

One expects that this relationship also holds in general order of magnitude for a warped current sheet, provided it is not warped too much. The basic prediction, then, by Jokipii *et al.* (1977) was that the *sign* of the gradient with respect to heliographic latitude (near the equatorial regions) would change from being positive in the years around the 1996 sunspot minimum to being negative during the 2008 minimum. The effect alternates, with A changing sign during each sunspot maximum.

9.6.1 Comparison of the GCR transport model with observations

For a general review of the comparison of the model predictions with observations see Jokipii (1989).

Measurements generally confirm theoretical predictions. In the 1970s, the negative gradient toward the poles was observed to be consistent with theory. But a gradient of this sign would be expected even if drifts were neglected, because of the rapid diffusion in from the poles. When the current sheet flattened out for the 1986 sunspot minimum, a number of experiments on the Voyager and Pioneer spacecraft reported a negative latitudinal gradient. Moreover, the values of the radial and latitudinal gradients were consistent with expectations.

Another interesting point is consistent with theory (Jokipii, 1989). As pointed out above, during the 2008 minimum, the diffusion and the drifts are competing whereas during the preceding minimum they reinforced each other. One might expect the results for the 1986 sunspot minimum to be very fragile, in the sense that a small change in parameters could change the negative gradient. There is evidence from Voyager 1 in the mid 1980s that the negative gradient is observed for the $A < 0$ period only for the relatively short period during which the current-sheet maximum latitude was below that of the spacecraft, and changed back to a small positive gradient when the current sheet latitude was larger than that of the spacecraft during part of the solar rotation.

We may conclude from these analyses that there is a very plausible comprehensive model of the effects of the solar wind on galactic cosmic rays, which is based on sound physical principles. Detailed numerical simulations of the fundamental equations plus consideration of the basic physics involved has led to a picture in which both drifts and diffusion play important roles. The model accounts well for a variety of phenomena observed over the past two sunspot cycles.

In a sense then we may say that the net or total modulation of galactic cosmic rays at the present time can be regarded effectively as a combination of drift effects and diffusion–advection effects. The diffusion–advection effects can naturally produce only an 11-year periodicity, which is the dominant variation observed today. The drifts also play a significant role in this variation, but in addition they predict significant 22-year variations. These 22-year solar magnetic cycle periodicities are observed, indicating that the effects of the change in sign of the drifts every 11 years are also present.

As we go back further in time where we have no direct interplanetary measurements, and we consider variations on time scales larger than the 22-year solar magnetic cycle period, we cannot look at details. We only have measurements which effectively give us the cosmic-ray flux at the orbit of the Earth, so we will have to interpret any variations observed in terms of the physical effects discussed above. Isolating the various physical effects in the historical record is difficult.

9.7 The Maunder minimum

As one example of the application of these ideas to pre-historical observations of the cosmic-ray flux, I discuss the Maunder minimum. Prior to 1640, a period of presumably normal solar activity, an approximately 11-year periodic variation can be discerned in the isotope records (e.g. Fig. 11.10). We also know from other data that since the Maunder minimum there is a clearly observable 11-year cyclic dependence. However, these data suggest that, during the Maunder minimum (taken here to be from 1660 to 1720), the variation actually had a larger period. A power spectral analysis of these data lends some support to this in that there is an indication of two periods in the data, one of which is close to the 11-year sunspot cycle. The other period is consistent with a substantially longer period, perhaps twice as long.

In the light of the discussion in Section 9.6, it is instructive to try to interpret this variation of the intensity in terms of the drift component of the modulation during this period of the Maunder minimum, which would naturally have a the 22-year period suggested by the data.

This possibility has been investigated in a series of model calculations (Jokipii, 1986) and compared with observations by Miyahara *et al.* (2008). The model is summarized in Fig. 9.10 from Miyahara *et al.* (2008). Shown in the upper panel is the sunspot cycle and (the alternating vertical areas) the simultaneous magnetic field cycle. The lower panel shows two separate corresponding time histories of 1 GeV protons in the inner solar system. The dashed line shows the intensity for

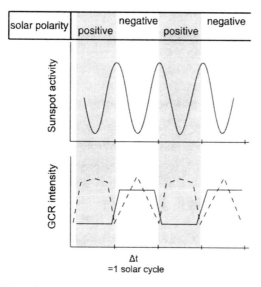

Fig. 9.10. Model results presented by Miyahara *et al.* (2008), showing the two effects of the heliosphere on galactic cosmic rays. Sunspot activity is represented by the upper curve. The solid line in the lower curve represents the effect of the changing sign of the heliospheric magnetic field (positive and negative for $A > 0$ and $A < 0$, respectively, as defined in the caption to Fig. 9.3; plots are shown for positively charged ions). The dashed line in the lower panel is the normal situation which combines both the sunspot activity and the change in sign of the heliospheric magnetic field.

a normal model, in which all of the variation comes from both varying the tilt of the current sheet and changing the sign of the magnetic field. Clearly apparent is the 11-year variation with the alternating flat and sharp profiles discussed above. This behavior is consistent with the neutron monitor data for the past few sunspot cycles.

The solid line in the lower panel, on the other hand, shows a 44-year time span in the model, for the model in which the current sheet was assumed to remain flat at all times, but there was a change in sign of the field every 11 years as at present. Also, it is thought that the heliosphere was quiet during the Maunder minimum, so the diffusion coefficients were changed to be those appropriate to a quiet wind ($\kappa_\parallel$ increased by a factor of two and $\kappa_\perp$ decreased by a factor of three). It is clear that the 11-year variation has been suppressed and the 22-year variation is all that remains.

Miyahara *et al.* (2008) present observations which support this view, in general. They conclude that the Maunder minimum shows a suppressed 11-year variation, consistent with the picture in which solar activity was less. However, the 22-year period suggests that, during this period, the heliospheric magnetic field still changed sign. This suggests that a scenario in which the major heliospheric effect on cosmic rays during the Maunder minimum was the 22-year solar magnetic cycle.

9.8 On the heliospheric modulation of galactic cosmic rays

Observations of galactic cosmic rays together with theory and modeling have served for decades as unique and valuable remote probes of the heliosphere in regions which are difficult or impossible to observe *in situ*. Much of what we know about the Sun and heliosphere has been contributed by cosmic-ray studies. A generally accepted view of energetic particles in the heliosphere has evolved in which the fluctuating and turbulent solar magnetic field was drawn out radially from the Sun by the supersonic solar wind and twisted, on average, into the well-established Parker spiral configuration by solar rotation. This fluctuating magnetic field acts to impede cosmic-ray access to the inner heliosphere, resulting in the "modulation" of galactic cosmic rays in anti-phase with solar activity. The supersonic solar wind ends in a termination shock, and the plasma flows subsonically into the heliosheath.

This paradigm has had a number of significant successes in explaining a number of diverse observations over the past few decades, including the 22-year modulation cycle, charge-sign dependence, energy spectra, and radial and time-varying latitudinal gradients.

9.8.1 Physics of longer-term variations

Relating the longer-term cosmic-ray variations considered in Section 9.2 to detailed physical processes is more difficult than on the time scale of a few decades. It is clear that variations in the solar wind and its fluctuations are a likely cause. Beyond some 50 000 yr, interstellar causes remain likely. Interstellar clouds and supernova blast waves undoubtedly contribute to variations seen at these time scales.

For plausible values of the various parameters, the guiding-center drifts due to the large-scale structure of the interplanetary magnetic field are the dominant effect in determining the motion of the particles. Hence, the magnitude and structure of the large-scale magnetic field, which determines the drifts, is the most important parameter. In Fig. 9.11 only the magnetic field magnitude was varied, with the wind velocity, diffusion coefficient, etc., held constant. From this one may conclude that it is possible that the magnetic field strength is a major factor in determining the level of cosmic-ray intensity in the inner heliosphere, and that this may change significantly over long time scales.

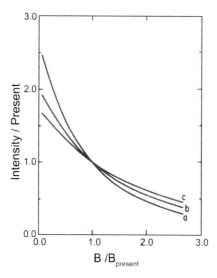

Fig. 9.11. Illustration of the computed value of the cosmic-ray intensity for various particle energies (a: 0.5 GeV, b: 1 GeV, and c: 1.5 GeV) as a function of the magnetic field strength. (From Jokipii and Marti, 1986.)

9.9 The sunspot minimum between cycles 23 and 24

The 2008 solar minimum is anomalous, in that it is has lasted longer than any previous minima over the past 100 years. Moreover, the observed solar wind dynamic pressure (ρv^2) is the lowest that has been observed with *in-situ* sensors. This pressure determines the extent of the heliosphere. Also, the magnitude of the interplanetary magnetic field was observed to be very low. The picture of modulation developed in this chapter suggests that, because of these observed changes, the cosmic-ray intensity should be quite high. This is borne out by the observations shown in Fig. 9.12. Because the time scale for effects to propagate through the heliosphere is months to years (Jokipii, 1989), a continuing increase in the intensity of cosmic rays should be expected.

The longer-term effects of the prolonged 2008 minimum will be interesting to see. They will help to determine the effects of solar variations on cosmic rays and perhaps their potential effects on climate.

9.10 In conclusion

The study of the modulation of galactic cosmic rays by the solar wind has progressed to the point where the basic physical ideas seem to be in place. Most of the observations obtained in the past few decades, both from Earth and from

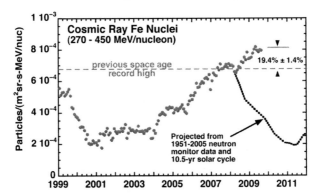

Fig. 9.12. Intensity of cosmic-ray iron from the NASA ACE spacecraft. Note that the rate increases beyond the previous space high. (Courtesy R. A. Mewaldt.)

spacecraft, seem to be reasonably well explained by a plausible model in which diffusion–advection effects and drift effects play significant roles.

Application of these ideas to ^{14}C data obtained during the Maunder minimum suggests that the expected smoother, quieter solar wind at that time would result in cosmic-ray variations dominated by a 22-year variation, with the 11-year solar activity cycle playing a smaller role than at present.

10

Solar spectral irradiance: measurements and models

Judith L. Lean and Thomas N. Woods

10.1 Introduction

Solar photons are Earth's primary energy source: Earth is habitable only because the Sun shines, radiating energy throughout the entire heliosphere. The Sun shines because its surface, warmed by energy produced in its nuclear burning core (Bahcall, 2000), is hotter (5770 K) than the surrounding cosmos (4 K). Electromagnetic energy traveling radially outward from the Sun illuminates the heliosphere with a flux of photons that diminishes inversely with the square of increasing distance. Earth, which has only an insignificant internal energy source (see Section 7.4.1), intercepts solar radiant energy, collecting photons emitted from all locations of the solar disk. Unimpeded in their journey to Earth, solar photons take eight minutes to establish the fastest and most direct of all Sun–Earth connections.

The photon energy incident on the Earth at its average distance from the Sun of one astronomical unit (AU), and prior to absorption in the Earth's atmosphere, is called the solar irradiance (e.g. Fröhlich and Lean, 2004). When the energy of the photons is integrated over all wavelengths across the electromagnetic spectrum, the average total solar irradiance is $1361 \pm 4\,\mathrm{W\,m^{-2}}$. Solar photons at visible wavelengths have the largest flux (Fig. 10.1) because the emission spectrum of a blackbody near 5770 K peaks in this region. Although really a high temperature plasma, the Sun's "surface" is defined as that layer of the solar atmosphere for which optical depth is unity for photons near 500 nm (cf. Vol. I, chapter 8).

Wavelength-dependent variations in photon fluxes are driven by the solar dynamo, which generates differing levels of magnetic flux in the Sun's atmosphere. The combined energy of all photons within a specified wavelength band per unit time at 1 AU is the spectral irradiance at that wavelength. Near-ultraviolet (UV), visible and near-infrared (IR) radiations emanate from the vicinity of the Sun's surface and lower atmosphere (the photosphere). Ions, atoms, and select simple molecules in the overlying solar atmosphere absorb this radiation in certain spectral

Heliophysics: Evolving Solar Activity and the Climates of Space and Earth, eds. Carolus J. Schrijver and George L. Siscoe. Published by Cambridge University Press. © Cambridge University Press 2010.

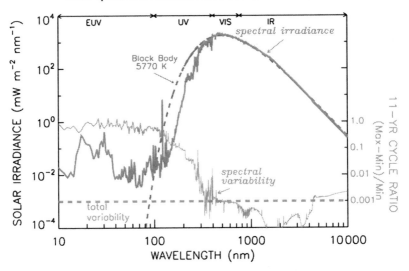

Fig. 10.1. Comparison of the solar spectrum and the blackbody spectrum for radiation at 5770 K (the approximate temperature of the Sun's visible surface). Also shown is an estimate of the variability of the solar spectrum during the 11-year solar cycle, inferred from measurements (at wavelength below 400 nm) and models (at longer wavelengths) and, for reference (dashed line), the solar cycle 0.1% change in the total solar irradiance.

regions, imposing a complex spectral character to the spectrum of radiation that reaches the Earth. Thus, Fraunhofer and other absorption lines, and a blanket of myriad absorptions in the near-UV spectrum, suppress the continuum blackbody radiation (e.g. Foukal, 1990).

Emissions at shorter wavelengths, in the far and extreme ultraviolet (EUV) spectrum below about 160 nm, and at longer radio wavelengths, emerge from increasingly hotter and more tenuous upper layers of the Sun's atmosphere (the chromosphere and corona). As a result, the Sun's radiant flux of EUV photons exceeds by many orders of magnitude that of a blackbody at 5770 K (Fig. 10.1). These emissions are produced *in situ* by radiation from gases in the hot outer solar atmosphere. The radiation is a mix of lines and continua, becoming increasingly line-dominated at shorter wavelengths, with a few weak EUV continua. The Sun's EUV radiative output is thus determined by emission processes of hot (10^5 to a few times 10^6 K) plasma, and depends on the plasma's electron and ion densities, on temperature, and on the atomic properties of the composite gases (such as H, He, O, Fe, Mg, Ca, etc.; cf. Vol. I, Section 8.4.1).

Until about 30 years ago the variability of the Sun's total radiative output – the total irradiance – was unknown, hence the term "solar constant" for its magnitude. Nevertheless, features such as sunspots had been observed to emerge and disappear from the Sun's disk for centuries, so it was accepted that the Sun is a variable star (see, for example, Abbot, 1958, and references therein). In reality, the Sun's irradiance at all wavelengths changes continuously in response to magnetic activity which alters the temperature (and density) in local regions of the solar surface and atmosphere, from where the photons emerge. The distribution of magnetic "active regions", where the local brightness differs from the average background of the Sun's surface and atmosphere, evolves with the progression of the Sun's nominal 11-year activity cycle (Fig. 10.2; cf. Chapter 2). The term "active region" generically refers to a collection of distinct solar spatial features whose properties depend on the wavelength of light with which they are observed.

Photons at different wavelengths emerge from different regions of the Sun's atmosphere, and are modulated by solar activity in different ways. When viewed in the visible light that dominates the total solar irradiance, the solar disk displays two prominent types of active features, called sunspots and faculae, which are respectively darker and brighter than the surrounding solar surface (Fig. 10.2). In ultraviolet light, magnetically active regions are composed of large bright complexes that tend to overlie their smaller, more compact photospheric counterparts. These regions – called plage – often congregate in complexes that persist for longer than any single individual region. Some coronal features, called coronal holes, are darker than the surrounding solar atmosphere because of the low density of the emitting plasma, which streams away from the Sun along the open magnetic field.

Solar irradiance is the integral of emission from all features on the entire hemisphere of the Sun visible at the Earth. It therefore changes perpetually, in response to solar activity. And because the contrast of the different features relative to the background Sun depends on wavelength, the spectral irradiance varies by different amounts at different wavelengths. Larger spectral irradiance variations occur in those emissions that originate in hotter (generally higher) regimes of the solar atmosphere. Note that luminosity, for comparison, is the integral over the entire surface, including the far side not seen at Earth, a quantity not measured.

Solar irradiance measurements must be made from space, and require instruments that are both accurate and precise. Because current measurement uncertainties are significantly larger than true solar irradiance variations, a reliable database requires overlapping observations for cross-calibration of both the total and spectral irradiances. To provide the inputs needed for understanding, simulating and forecasting Earth's response to solar irradiance variations on multiple time scales, irradiance variability models of various complexity have been developed that link the extant irradiance databases to solar activity, using images and proxies.

When solar photons arrive at Earth, some are absorbed by gases in the Earth's atmosphere (primarily O_2, N_2, O, and O_3), at different altitudes depending on their wavelengths (Fig. 10.3). The visible spectrum reaches the surface and troposphere because the Earth's overlying atmosphere is largely transparent to these

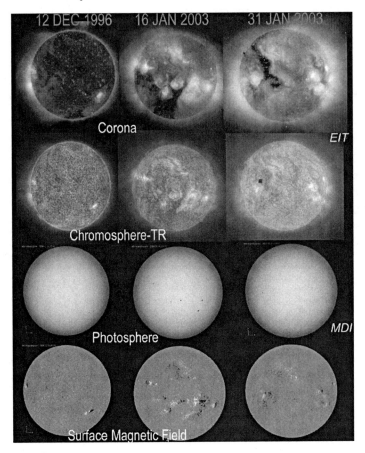

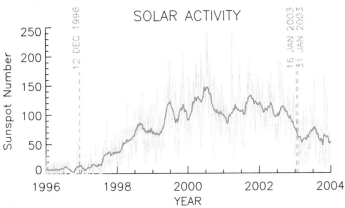

Fig. 10.2. Images of the Sun's surface seen in magnetic flux (bottom row) and visible light (second row up) are compared with images of the solar chromosphere/transition region (third row up) and corona (top row). The images, made using different wavelengths of the solar spectrum, are compared at different levels of solar activity. The images in the left column (December 1996) are typical of quiet conditions during solar activity minima, and those on the right two columns illustrate conditions near high solar activity (January 2003) at two different phases of the Sun's 27-day synodic rotational cycle. The bottom plot shows the sunspot number time series. See Color Plate 8.

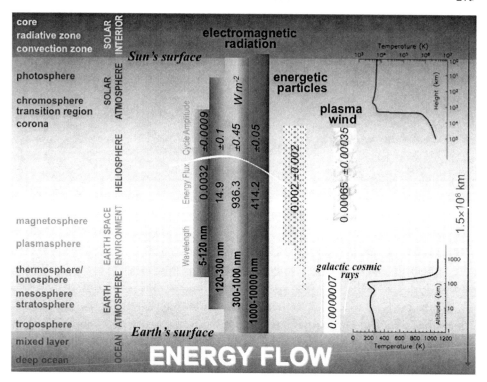

Fig. 10.3. The flow of energy from the Sun to the Earth is compared for photons in four different wavelength bands, energetic particles, and the plasma wind. The numbers are approximate energies with their variations (following the ± symbol) during an 11-year solar activity cycle, in $W\,m^{-2}$. Visible radiation connects the surfaces of the Sun and Earth while ultraviolet radiation connects their atmospheres. Particles and plasma connect the outer solar atmosphere primarily with Earth's magnetosphere and high-latitude upper atmosphere. Shown on the right are approximate temperatures of the various Sun and Earth regimes. (Adapted from Lean, 2005.)

photons. But the Earth's atmosphere absorbs all photons at wavelengths less than about 300 nm, so that solar UV, EUV, and X-ray emissions never reach the surface. Rather, these photons deposit their energy in increasingly higher regions of the Earth's atmosphere – the middle UV (MUV: 200–300 nm) spectrum in the stratosphere, the far-UV (FUV: 120–200 nm) and soft X-ray (XUV: 1–10 nm) spectra in the upper stratosphere and lower thermosphere, and the extreme UV (EUV:10–120 nm) spectrum in the thermosphere.

There is thus a broad-brush, systematic mapping of the solar atmosphere with the terrestrial atmosphere that establishes the particular wavelength bands of solar irradiance most relevant to terrestrial phenomena (Fig. 10.3). Lower solar atmosphere emissions map to lower terrestrial atmosphere absorption, and outer solar atmosphere emissions to outer terrestrial atmosphere absorption. Specifically the visible, near-UV and near-IR regions (which dominate the total irradiance) are relevant for climate, the UV bands for the stratosphere and ozone, and the FUV and EUV spectra for the upper atmosphere, ionosphere, and space weather (see Chapters 13 and 16; also Lean, 2005).

10.2 Historical perspective

The Sun, it seems, is always changing, its disk an evolving kaleidoscope of different features such as those evident in Fig. 10.2. The amount and patterns of dark sunspots, observed since the early 1600s, provide solar astronomers with the longest of all records of the Sun's activity, displaying clearly the predominant 11-year cycle, first detected as a 10-year cycle by Heinrich Schwabe (1844). As defined by the varying strength and length of the 11-year Schwabe sunspot cycle (Fig. 10.4), the Sun has evolved over the past four centuries from a state of anomalously low activity in the Maunder minimum (Eddy, 1976) through the more moderate Dalton minimum to the present-day modern maximum. While the average period for the sunspot cycle is 11 years, its period actually ranges from 9 to 13 years. Ground-based telescopic solar observations during the early twentieth century posited "the probable existence of a magnetic field in sunspots" (Hale, 1908), thereby establishing magnetism as the fundamental element of solar activity.

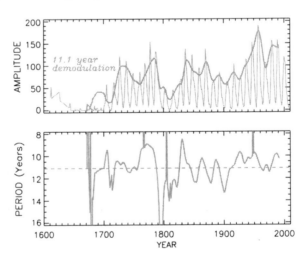

Fig. 10.4. Annual values of the historical sunspot record during the past four centuries (top panel), indicating the evolution of the 11-year activity cycle. The amplitude of the cycle obtained by demodulation of the annual time series is also shown. In the bottom panel, the instantaneous period of the cycle, similarly obtained from demodulation, indicates the inconstancy of the nominal "11-year" sunspot cycle period.

This magnetic activity is most clearly seen as strong magnetic fields of 1000 gauss or more in large sunspots and as a reversal of polar magnetic fields roughly every 11 years, which is also known as the 22-year Hale magnetic cycle.

With the advent of the space era in the 1960s, instruments were able to observe the Sun beyond the Earth's atmosphere. Myriad different solar phenomena and emissions were recorded, and found to similarly exhibit pronounced 11-year (Schwabe) cycles. Like sunspot numbers, most (if not all) solar phenomena also exhibit pronounced shorter-term variations generated by the Sun's ~27-day synodic rotation on its axis (which alters the projection of magnetic active regions toward Earth), and by the appearance and disappearance of features generated by magnetic flux emergence, eruption, and cancellation. The shorter-term variations are superimposed on background levels that rise and fall as solar activity evolves throughout the 11-year sunspot cycle (e.g. Cox *et al.*, 1991; Sonett *et al.*, 1991).

That the Sun's irradiance at certain wavelengths varies significantly in concert with solar activity cycles and solar rotation was actually known before routine space-based measurements commenced. The longest irradiance record is that of the 10.7 cm radio flux, $F_{10.7}$, for which ground-based measurements commenced in 1947 (Fig. 10.5). Seeking an explanation for observed geophysical phenomena that appeared to be related to solar activity, mainly in the ionosphere, the first space-based irradiance measurements of X-rays were made in 1964 onboard SOLRAD. Operational monitoring of the X-ray fluxes commenced in 1974 to provide knowledge of eruptive solar activity indicated by strong enhancements in flares, responsible for sporadic ionospheric disturbances (Donnelly *et al.*, 1977).

Intermittent EUV and UV irradiance measurements made by rockets and the AE-E satellite demonstrated that, like the X-rays and the 10.7 cm radio flux, these short-wavelength solar emissions also vary significantly (Hinteregger *et al.*, 1981). However, the lack of continuous measurements and inadequate instrument calibration and stability precluded reliable knowledge of solar irradiance cycle amplitudes. Even in the strongest emission line in the entire spectrum below 200 nm, the H I (i.e. neutral hydrogen) Lyman α line at 121.6 nm, the AE-E measurements were unable to establish true irradiance variations because of calibration offsets and instrumental drifts (Fig. 10.5). Furthermore, attempts to link the various EUV measurements with solar activity indicated by sunspots or the 10.7 cm radio flux were inconclusive as to the suitability of assumed linear relationships (e.g. Hedin, 1984). Nevertheless, solar activity proxies such as $F_{10.7}$ (whose measurements were more reliable) were input to the developing Jacchia and MSIS geospace models (Jacchia, 1970; Hedin *et al.*, 1977) to represent the geophysically relevant EUV irradiance variations.

Even though solar emissions corresponding to the short X-ray and EUV emissions and very long radio waves – at the wavelength extremities of the blackbody

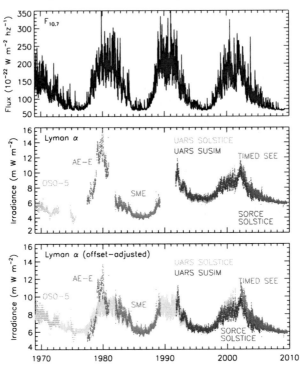

Fig. 10.5. The longest and most widely used index of solar extreme ultraviolet (EUV) irradiance variability is the 10.7 cm radio flux, $F_{10.7}$, whose daily values are shown in the top panel for more than three solar cycles. Observations of the Lyman α irradiance, the strongest emission line in the solar UV spectrum, are shown in the middle panel on their "native" calibration scales. Adjusting individual Lyman α observations by constant offsets to account for differences in absolute calibrations, in the bottom panel, reveals more clearly the actual Lyman α solar cycle variability. Short-term rotational modulation variations are superimposed on both the 10.7 cm and Lyman α time series.

curve – were recognized as being highly variable, with flux variations related to the solar activity cycle, detecting fluctuations in Earth's dominant energy input, the total solar irradiance, proved a much greater challenge. A century of ground-based observations failed to reveal unequivocal variations associated with solar activity because of noise in the observations induced by the Earth's overlying atmosphere, which scatters, reflects and absorbs solar photons (Foukal and Vernazza, 1979). Only when space access afforded continuous, atmosphere-free solar observing was the magnitude and nature of solar irradiance variability finally established (Willson *et al.*, 1981). Since the late 1970s, space-based radiometers have monitored the Sun's total irradiance continually, producing a record over multiple solar sunspot cycles (Fig. 10.6).

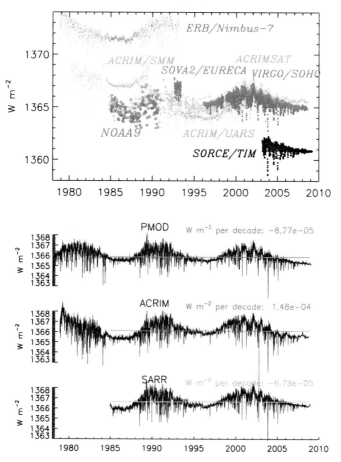

Fig. 10.6. Individual measurements of the total solar irradiance, made since 1978 from a variety of space platforms, are compared on their "native" calibration scales in the top panel. Shown in the lower three panels are different composite records of total solar irradiance constructed from the observational database, using varying assumptions about calibration offsets and in-flight sensitivity changes (PMOD: C. Fröhlich; ACRIM: R. C. Willson; SARR: S. Dewitte). For quantitative comparison, the slopes of the time series are also indicated. See Color Plate 7.

Growing awareness that Sun-as-a-star (i.e. irradiance) variability pervades extensive regions of the solar spectrum coalesced in the 1970s (White, 1977) to the recognition that the entire electromagnetic spectrum – at all wavelengths – likely varies. Systematic ground-based observations of selected solar flux emissions and indices commenced at the Kitt Peak National Observatory (KPNO) during the solar minimum of 1975 in a concerted effort to quantify and understand the entire solar radiative output and the responsible solar processes in a comprehensive way (White and Livingston, 1981). This task, for which accurate and precise long-term measurements is essential, are still ongoing, from the ground (e.g. SOLIS, Synoptic Optical Long-term Investigations of the Sun[†]) and space, to achieve coverage and exploration of ever-longer time scale domains.

Concurrent with the various quests to determine the Sun's irradiance variations, stellar astronomers sought clarification of cycles in other stars, similarly recognizing the need for ongoing long-term observations. Following a broad initial survey of chromospheric variability of stars (Wilson, 1978), subsequent studies attempted to quantitatively link the Sun with Sun-like stars (Radick *et al.*, 1998), to advance understanding of long-term changes in solar activity and irradiance.

With the recognition that the Sun's irradiance varies at all wavelengths and on multiple time scales, and that this variability is apparently common behavior in the fluxes of Sun-like stars, attempts were made to reconstruct past solar irradiance changes for use in climate change studies (Hoyt and Schatten, 1993; Lean *et al.*, 1995; Lockwood and Stamper, 1999). Proxies of solar activity such as the ^{14}C and ^{10}Be cosmogenic isotopes in tree-rings and ice-cores, respectively, were meanwhile emerging as reliable indicators of past solar activity (Chapter 11; also Beer *et al.*, 1990; Berggren *et al.*, 2009). Despite known different processes of terrestrial deposition, the different isotope records nevertheless exhibit considerable common variance, thereby implicating a common source, i.e. the solar-driven variations in the heliosphere which modulate the flux of incoming galactic cosmic rays that produce the isotopes. The paleoclimate community makes extensive use of cosmogenic isotopes to detect linkages of climate and solar activity. With their "calibration" against actual solar irradiance, new capabilities ensued to enable model simulations of past changes with actual (albeit speculative) solar irradiance inputs (Chapter 12; Bard *et al.*, 2000; Bard and Frank, 2006).

10.3 Measuring solar irradiance and its variations

Measuring solar irradiance and its variations is a difficult task from a metrology perspective. Required are instruments that are both accurate and precise at levels sufficient to detect real solar variations without contamination by instrumental effects. Because the Earth's atmosphere absorbs, reflects, and scatters solar photons, the instruments must operate on space platforms where they typically encounter varying thermal, radiation, pointing, and contamination environments that can alter the instrument's accuracy and precision through wavelength-dependent processes that are still poorly known. Furthermore, pre-flight calibrations are difficult to validate because post-flight calibrations are (generally) precluded.

[†] URL: http://solis.nso.edu/solis_data.html

A measurement of irradiance has high accuracy (and low uncertainty) when its difference from the "truth" (defined by an absolute radiometric standard at the National Institute of Standards and Technology, NIST) is small. A measurement is precise when repeated observations of the same irradiance differ minimally, i.e. the spread (or deviation) of multiple, repeated observations is small (Taylor and Kuyatt, 1994). Precision applies to a range of time scales, and long-term precision (on annual to decadal time scales) is sometimes referred to as relative accuracy.

The uncertainty and repeatability required to measure solar irradiance depends on the amplitude and time scales of the irradiance changes, which are largest at the very shortest and longest wavelengths of the electromagnetic spectrum, and smallest in the visible and near-IR spectra. To be capable of characterizing the Sun's 11-year solar cycle changes, the instruments that measure solar irradiance must have high precision over time scales of many years. The expectation for modern instruments is a relative accuracy of at least one-tenth of the solar cycle variability and repeatability/stability (relative precision per year) at least one-tenth of the accuracy (i.e. one-hundredth of the solar cycle variability). These desired values depend strongly on wavelength. For example, the requirement objectives for future operational monitoring of total solar irradiance, which varies by about 0.1% over the solar cycle, by the National Polar-orbiting Operational Environment Monitoring System (NPOESS) are 0.01% accuracy and 0.0005% per year precision. As another example, the requirement objectives for the solar EUV irradiance, which varies by about a factor of three over the solar cycle, by NASA's Solar Dynamics Observatory (SDO) are 20% accuracy and 2% per year precision.

10.3.1 Total irradiance

Self-calibrating "active" electrical substitution radiometers measure total solar irradiance by comparing with equivalent electrical power the radiant power (W) illuminating an aperture of known area (m^{-2}) absorbed by a cavity of known optical–electrical equivalence. The radiometers collect solar photons at all wavelengths and record the total energy without wavelength discrimination (Willson, 1979; Kopp and Lawrence, 2005). A significant source of uncertainty arises from inaccuracies in the measurement of the entrance aperture area (Butler *et al.*, 2008). Other sources of uncertainty include the (lack of) equivalence of optical and electrical power and the need for corrections to account for aperture diffraction and Sun–Earth distance. Long-term precision is affected by, for example, changes in the optical properties of the cavity surface (and therefore in the amount of radiant power that the cavity captures), drifts in electrical components that define power (resistance and volts) and changing thermal and contamination environments. The most advanced solar radiometric instruments include multiple cavities, only one of which is used routinely; the others are exposed to the Sun much less frequently, and on differing duty cycles to provide in-flight calibration tracking of time-dependent exposure-related degradation (Kopp *et al.*, 2005a,b).

With the launch in 1978 of the Hickey–Friedan radiometer on the Nimbus 7 spacecraft (Hickey *et al.*, 1988), followed soon after (in 1980) by the Active Cavity Radiometer Irradiance Monitor (ACRIM[†]) on the Solar Maximum Mission (SMM; Willson *et al.*, 1981), regular measurements of total solar irradiance commenced a long-term database that continues to the present (Fig. 10.6). Evidence for true variations in total solar irradiance (TSI[‡]) emerged rapidly as the two instruments simultaneously recorded significant decreases in irradiance when large sunspots traversed the disk of the Sun projected to Earth (Hudson *et al.*, 1982). As the database lengthened, such decreases were found to be the dominant mode of irradiance variability on time scales of days to months and they are evident in Fig. 10.6 throughout the entire record. In October 2003 the Total Irradiance Monitor (TIM) on the Solar Radiation and Climate Experiment (SORCE) recorded the largest such decrease (of 5 W m^{-2}, or 0.4%) in the space era (Fig. 10.7; see Woods *et al.*, 2004).

Absolute values of the total solar irradiance measured by different space-based radiometers can differ by up to 10 W m^{-2} (0.7%, Fig. 10.6). By the mid 1990s it was generally considered that the measurements were converging to a "true" TSI absolute value of 1366 ± 1 W m^{-2} (Crommelynck *et al.*, 1995; Fröhlich and Lean, 1998), but according to TIM on SORCE, the TSI is closer to 1361 ± 0.4 W m^{-2}. Subsequent calibration comparisons and tests at NIST (Butler *et al.*, 2008) have identified previously undetected uncertainties in the prior measurements (for example, inadequate corrections for aperture diffraction) and generally favor the lower irradiance value, but some uncertainty remains in knowledge of the Sun's true brightness.

There is now no doubt that the total solar irradiance varies continuously with a pronounced 11-year cycle. In addition to recording large irradiance dips, all the radiometers capture an overall increase in total solar irradiance during solar activity maxima, relative to minima (Fig. 10.6). The radiometers are able to reliably detect true solar fluctuations on time scales of days to years (Fig. 10.7), despite the spread in their absolute scale, because their precisions are typically an order of magnitude higher than their accuracies. Nevertheless, once in orbit, solar radiometers invariably undergo a range of instrument sensitivity changes that produce additional uncertainties in their long-term precisions. Exposure to solar photons, which alters the absorptance of the interior cavity surfaces, is a primary source of long-term instrument degradation. Myriad other technical issues further

[†] URL: http://acrim.com/
[‡] URL: http://www.pmodwrc.ch/pmod.php?topic=tsi/composite/SolarConstant

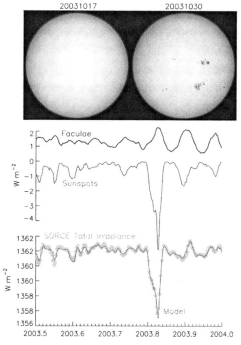

Fig. 10.7. Continuum light solar images made by the Michelson Doppler Imager (MDI) instrument on SOHO are compared for October 17 and 30, 2003, when total solar irradiance underwent one of the largest decreases observed in the space era. Shown in the lower panel are the SORCE/TIM daily mean observations (symbols), compared with an empirical variability model developed from sunspot and facular influences. The middle panel compares the respective bolometric facular brightening (upper curve) and sunspot darkening (lower curve) that together produce the changes in total solar irradiance observed by TIM, using linear scaling to determine their relative contributions.

mitigate long-term repeatability, including changing thermal environments (which alter electrical circuitry that measures power), spacecraft pointing (which introduces sensitivity changes as the rays enter the instruments in different ways and traverse different optical paths, and parts of the components), and inadequate correction of distance to 1 AU (because the radiometers actually measure the solar power at the spacecraft location).

Extracting the "true" long-term (multi-annual to decadal time scale) behavior of solar irradiance during the past three solar cycles has proven difficult, and is still not conclusive. To investigate such changes, three different composite records of total solar irradiance have been constructed (Willson and Mordvinov, 2003; Fröhlich and Lean, 2004; Dewitte *et al.*, 2004b) by cross-calibrating and cross-validating the multiple independent observations to adjust the different absolute scales and correct in-flight trends arising from different long-term precisions (repeatability).

Although each composite (Fig. 10.6) captures the higher irradiance values near solar maxima, there are distinct differences in their respective levels during solar minima. Taking into account the standard deviation of the three composites, it is not possible to confirm directly from the database whether the different solar minimum levels are the result of real solar irradiance changes or remnant instrumental effects in the composite compilations.

With the launch of SORCE in 2003, the total solar irradiance observations have entered a new phase that presages future operational solar monitoring by NPOESS, commencing in 2013. The TIM total irradiance monitor on SORCE is a state-of-the art instrument with accuracy (0.035%) and long-term stability (10 ppm per year) a factor of three better than prior measurements (Kopp *et al.*, 2005a,b). Technological advances include phase-sensitive detection, the replacement of painted interior surfaces with etched nickel phosphorous (NiP) black, and forward placement of the defining aperture to limit diffraction and scattered light from baffles. Even higher accuracy and precision are expected to be achieved with extended NIST-traceable radiometric calibration and characterization for the second TIM to be launched on the Glory spacecraft, to link the existing 30-year TSI database with NPOESS.

10.3.2 Spectral irradiance

To measure solar spectral irradiance (SSI[†]) an instrument must be capable of discriminating the wavelength of incident photons and recording this signal, significantly weaker than the total solar irradiance, with the needed accuracy and precision. Dispersion is typically achieved with an optical element such as a grating (Brueckner *et al.*, 1993; Rottman *et al.*, 1993) or prism (Harder *et al.*, 2005), although (metal) transmission gratings are increasingly used to disperse EUV radiation (Judge *et al.*, 1998). The dispersing element directs photons at different wavelengths along different optical paths in the instrument, typically via additional optical elements, to where detectors are positioned. Practical application usually necessitates some combination of selective filters, reflective elements, and baffles to block or capture "stray" light (at wavelengths outside the range of interest). The detectors must be sufficiently sensitive to measure radiant energy three or more orders of magnitude smaller than the total solar irradiance. The various elements introduce additional uncertainties to the spectral irradiance measurements, which are uniformly an order of magnitude or more less accurate and less precise than those of total solar irradiance. Ultimately, the NPOESS operational monitoring requires solar spectral irradiance measurements at wavelengths longward of 200 nm with 1% accuracy and 0.02% per year precision.

[†] URL: http://lasp.colorado.edu/lisird/

The longest space-based spectral irradiance record is that of the H I Lyman α emission line at 121.6 nm (Fig. 10.5), which commenced in the 1970s and continues to the present, although with continuous (overlapping) data only since 1991 (Woods *et al.*, 2000). At the time of this writing, two instruments measure the Lyman α irradiance; the Solar EUV Experiment (SEE) on the Thermosphere Ionosphere Mesosphere Energetics and Dynamics (TIMED) spacecraft (Woods *et al.*, 2005) and the Solar Stellar Intercomparison Experiment (SOLSTICE) on SORCE (McClintock *et al.*, 2005). Although the absolute values in the database can range by a factor of three, the Lyman α irradiance is most likely near 6 mW m^{-2} at solar minimum, increasing by about 50–60% to near 10 mW m^{-2} at solar maximum (Fig. 10.5).

The SEE measures not just the Lyman α irradiance, but the entire EUV and FUV spectral irradiance from 0 to 195 nm in 1 nm bins (Fig. 10.8) using NIST-calibrated spectrometers with in-flight sensitivity tracking by redundant optical elements and rocket under-flight calibrations. Prior to SEE, the AE-E spacecraft last measured the solar EUV spectrum continuously from 1976 to 1980; there is thus a 22-year gap in the solar EUV irradiance database. Current understanding therefore relies primarily on the SEE observations, which cover only the descending phase of solar cycle 23. The irradiances exhibit pronounced modulation by both solar rotation and the solar cycle (Fig. 10.1), with variations in 1 nm bins in the range 50 to 100%, i.e. comparable to or greater than that of Lyman α. The variability depends strongly on wavelength, with emissions from the hotter corona varying more than those from the chromosphere and transition regions. Most variable of all are X-rays below about 2 nm, which exhibit order-of-magnitude cycle changes (Woods *et al.*, 2004).

Spectroradiometric measurements of the solar UV spectral irradiance at wavelengths from 120 to 300 nm with spectral resolution 1 nm (or better) commenced with the Solar Mesosphere Explorer (SME) in 1980 (Rottman, 1999) and are currently being made by the SOLSTICE on SORCE (McClintock *et al.*, 2005) (Fig. 10.8). Both the Solar Ultraviolet Spectral Irradiance Monitor (SUSIM) (Brueckner *et al.*, 1993) and SOLSTICE (Rottman *et al.*, 1993) on the Upper Atmosphere Research Satellite (UARS) measured solar spectral irradiance at wavelengths in the region 120 to 400 nm from 1991 to 2005. These UARS solar UV irradiance observations were validated to an accuracy of 2–10% (wavelength dependent) with the NOAA SBUV (DeLand and Cebula, 1998) and space shuttle ATLAS observations (Woods *et al.*, 1996). Both SUSIM and SOLSTICE had sufficient precision for the reliable detection of true irradiance variations for the first time in the far and middle UV spectrum, but not in the near-UV solar spectrum from 300 to 400 nm (Fig. 10.8; Lean *et al.*, 1997).

The capability for measuring solar spectral irradiance variations in the near-UV, visible, and near-IR regions has been enacted only recently, with the development,

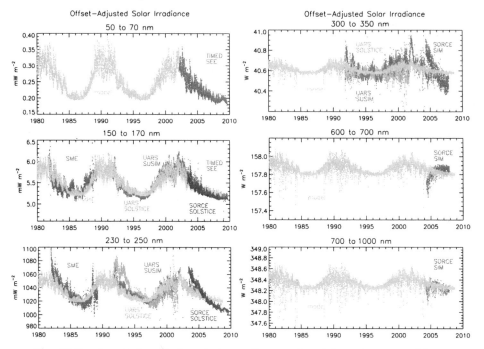

Fig. 10.8. Assorted space-based observations made of the solar spectral irradiance during the past three solar cycles are compared in selected wavelength bands. From top left to bottom right: EUV (50–70 nm), FUV (150–170 nm), MUV (230–250 nm), NUV (300–350 nm), visible (600–700 nm), and IR (700–1000 nm); the curves shown have offset-adjustments to account for their different absolute calibration scales. Also shown, as the grey time series, are models of the irradiance variations in the same wavelength bands, derived by scaling the observed rotational modulation variations to the Mg (and $F_{10.7}$ for the EUV band) proxy indicators. Note the lack of daily measurements shortward of 110 nm until TIMED SEE observations commenced in 2002, and longward of 400 nm until SORCE SIM observations commenced in 2003. See Color Plate 9.

calibration, and launch of the Spectral Irradiance Monitor (SIM) (Harder *et al.*, 2005). Accompanying the TIM on SORCE, SIM is making the first measurements of solar spectral irradiance with sufficient precision to detect real solar changes (Fig. 10.8), and sufficient spectral coverage to account for 95% of the spectral energy that composes the total. NPOESS will continue the database of these measurements that SORCE commenced in 2003.

As with total solar irradiance, spectral irradiance variations have been determined most reliably on short time scales (e.g. during the 27-day solar rotation) because these short time-scale variations are least susceptible to contamination by instrumental sensitivity drifts. Solar spectral irradiance variations have now been characterized on time scales of days to months at wavelengths from 0 to 2000 nm, confirming and characterizing the variability of the entire electromagnetic spectrum. At wavelengths below about 300 nm (i.e. in the X-ray, EUV, FUV, and MUV spectral regions) solar spectral irradiance increases when there are large active regions present on the Sun's disk. At the same time, the spectral irradiance at longer wavelengths, like total irradiance, decreases.

And, as with the total irradiance radiometers, solar spectroradiometers measure additional, longer-term irradiance changes during the solar cycle but are impaired by instrumental changes in their ability to quantify variability on longer time scales. Determining the true wavelength-dependent magnitude of the spectral irradiance cycle is confounded by gaps in the data records, absolute calibration offsets, and sensitivity drifts over time scales of many years. Even though databases of spectral irradiance at some wavelengths extend over more than two solar cycles, insufficient long-term stability from absent or inadequate in-flight calibration tracking precludes the detection of possible longer term changes underlying the solar cycle. Because of the relatively short (5 years) database, knowledge of the longer-term (solar cycle) variability of the solar spectrum longer than 400 nm is least certain of all.

10.4 Understanding and modeling solar irradiance variations

Solar irradiance variability is a fundamental manifestation of the Sun's magnetic activity. As with other variable solar phenomena, the driver of irradiance variations is the subsurface dynamo (Chapters 2, 5, and 6; see also, e.g. Ossendrijver, 2003) that alters the amount, distribution, and strength of magnetic flux that erupts into the solar atmosphere from the convection zone below (Weiss and Tobias, 2000). An ultimate goal of understanding and modeling solar irradiance variations is therefore to establish physical descriptions of the eruption and transport of magnetic flux on the Sun in the present, past, and future (Wang *et al.*, 2005; Dikpati, 2005), and of the processes through which this flux alters the photon emission from the Sun's surface and atmosphere (Warren, 2005; Schrijver *et al.*, 2006; Fontenla *et al.*, 2006, 2007). In reality, work on this daunting task, which in essence requires a complete understanding of solar behavior, is only just beginning.

There is also a practical need to specify solar irradiance continuously over many decades for purposes of augmenting, interpreting, and extending the observational database, and specifying solar irradiance for geophysical applications. This has motivated construction of relatively simple empirical and semi-empirical models that relate observed irradiance variations to judiciously selected proxy indicators of the magnetic sources of irradiance variability, accounting for wavelength dependence by incorporating different source regions for emissions in different spectral regions that emerge from different solar regimes (Hoyt and Schatten, 1993; Lean *et al.*, 1997, 1998, 2005; Warren *et al.*, 2001; Krivova *et al.*, 2007).

In the near-UV, visible, and IR spectrum two primary surface features – sunspots and faculae – modulate solar irradiance (Fig. 10.7). Sunspots, which are cooler (by about 1800 K in their darkest regions, the umbrae) than the surrounding photosphere, are regions where radiation from the Sun's surface is diminished locally. In spectrally integrated (bolometric) radiation, sunspots are about 35% darker than the "quiet" (i.e. background) solar surface. Faculae are regions on the Sun's surface where the magnetic flux produces local heating, and they are bolometrically a small percentage brighter than the surrounding solar surface (Schatten *et al.*, 1986; Spruit, 2000).

By specifying the bolometric sunspot darkening and facular brightening it is possible to account for a large proportion of the variations observed in total solar irradiance. The sunspot darkening (also called the photometric sunspot index) time series is calculated using information about sunspot areas and locations extracted from white-light images of the solar surface. The facular brightening (or photometric facular index) can be calculated similarly (Lean *et al.*, 1998). However, a "flux" (i.e. disk-integrated) proxy of facular brightening provides higher fidelity for the model in tracking the temporal changes in total solar irradiance (e.g. Lean *et al.*, 2005). This is because facular characteristics are poorly observed in the images and inadequately specified compared with the more compact, darker and relatively well-defined sunspot regions. Furthermore, whereas sunspot regions are typically discrete and therefore relatively easily quantified, faculae occur with a continuous distribution of sizes and contrasts, so that statistical definitions (which can be ambiguous) are needed for practical quantification.

A useful proxy for global facular brightening is the Mg index, which is the ratio of core emission in the Mg Fraunhofer line to that in the nearby continuum (Viereck *et al.*, 2004). Because the core emission is enhanced in magnetically active bright regions, this index is a sensitive indicator of the total (net) emission from all bright regions on the solar disk. As the ratio of absolute fluxes, the Mg index is less susceptible to instrumental sensitivity changes that potentially contaminate the temporal fidelity of time series and, because it has been measured routinely from space since 1978, it covers the epoch of available irradiance observations.

A model of total irradiance variability that specifies the sunspot and facular influences explains 93% of the variance in the high precision SORCE/TIM observations since 2003 and permits the calculation of daily total solar irradiance since 1978 (Fig. 10.9), for comparison with the different observational composite records (Fig. 10.6). When the sunspot and facular components are specified in absolute terms these components are combined independently of the measured irradiance variations, aside from specifying the value of the "quiet" (solar minimum, or background, level; Lean *et al.*, 1998). In practice, a proxy model derived using multiple regression apportions the relative sunspot and facular contributions (Lean, 2000;

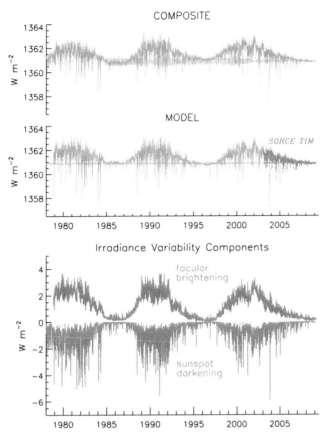

Fig. 10.9. Shown in the upper panel is a record of total solar irradiance obtained as an average of three different observational composites (shown separately in Fig. 10.6). In the second panel are irradiance variations estimated from an empirical model that combines the two primary influences of facular brightening and sunspot darkening. The symbols indicate direct observations made by the TIM instrument of the SORCE mission, used to determine the relative sunspot and facular components in the model, shown separately in the lower panel.

Fröhlich and Lean, 2004). Another approach utilizes the features in magnetograms to specify the irradiance components (Krivova *et al.*, 2007).

Irradiance variability models provide useful practical tools for investigating and exposing instrumental differences in the various measurements (Fig. 10.10). For example, modeled total solar irradiance variations closely track those measured by SORCE/TIM throughout the entire period of the SORCE mission thus far (2003 to 2008). The ratios of the measurements and model have a trend of 8 ppm per year, which is less than the SORCE/TIM long-term precision of 10 ppm per year. This excellent agreement provides confidence that the variations measured by SORCE/TIM are indeed real solar variations, and are not the result of instrumental

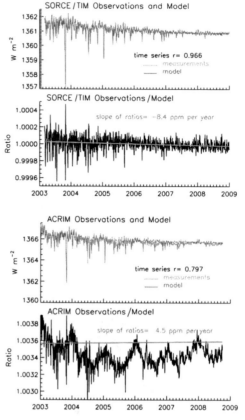

Fig. 10.10. A model of total solar irradiance variations, obtained from multiple regression of sunspot and facular indices with the SORCE/TIM observations, compared in the top panel with the direct TIM measurements, explains a large fraction (93%) of the observed variance. The second panel shows the ratios of the observations and model, indicating that the slope of the ratios is within the TIM long-term precision. In the two lower panels, the same total irradiance variability model is compared with the ACRIM observations, the ratios indicating significant, fluctuating deviations not present in either the model or the TIM measurements, and therefore likely of instrumental origin.

differences. A contrary situation is evident when the same irradiance variability model is compared with the total solar irradiance measured by the ACRIMSAT. In this case the ratios exhibit a large and semi-regular oscillation (with possibly an annual period) that is clearly not of solar origin, being absent in both the model and the SORCE/TIM (and also the PMOD) observations.

The solar spectral irradiance variations (Fig. 10.11) that compose the total are modeled analogously to total irradiance variability by specifying the sunspot and facular components, but combined in different proportions according to the wavelength-dependent contrasts of these sources, relative to the bolometric

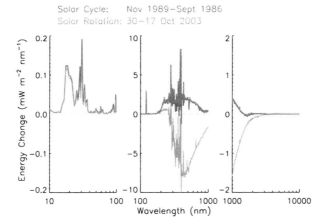

Fig. 10.11. Estimated changes in solar spectral irradiance are shown across the electromagnetic spectrum, from the EUV to the IR, for both the 11-year solar activity cycle (maximum in November 1989 and minimum in September 1986) and the 27-day solar rotational cycle (peak on October 30 and valley on October 17). These results indicate how the solar cycle and rotation cycle variations are in-phase at short wavelengths (below about 300 nm) and out of phase at longer wavelengths.

quantities (Lean, 2000; Lean *et al.*, 2005; Krivova *et al.*, 2006). In the UV spectrum below about 300 nm, the variations are adequately represented by a proxy for just the bright active regions (Fig. 10.12). Sunspots play a negligible role in variability at these shorter wavelengths because the faculae are much brighter than sunspots are dark. Historically, empirical EUV irradiance models have used $F_{10.7}$ for this purpose (Richards *et al.*, 1994, 2006), although more recent evidence from modeling the SEE EUV spectral irradiance observations suggests that the Mg index is preferable for wavelengths longer than about 30 nm. In the spectrum longward of 300 nm, both sunspot and facular proxies are needed to reproduce the observed changes (Fig. 10.13). The other solar features, and their associated proxies, that play an important role in solar UV variability are the less bright active network and coronal holes (e.g. Woods *et al.*, 2000; Warren *et al.*, 2001).

Semi-empirical approaches that employ additional physical insight about the source regions of the irradiance variations are also used to model solar irradiance variations. In the EUV spectrum, the widely used "emission measure" determines the (changing) electron density of the emitting plasma region, which fundamental (unchanging) atomic physics of the plasma composition then converts to spectral output. This approach has been utilized to model the solar EUV spectral irradiance variations by combining emission measures of representative quiet and magnetically active features with information from solar imagery about the fraction of these

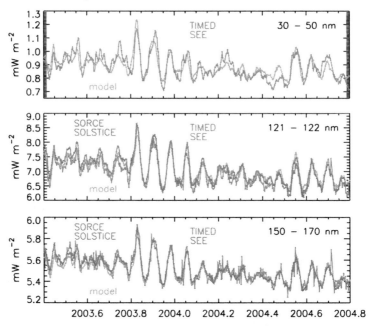

Fig. 10.12. Comparison of SORCE/SOLSTICE and TIMED/SEE observations and empirical variability model estimates of irradiance in selected wavelength bands including EUV (30 to 50 nm), the Lyman α line (121–122 nm), and the far-UV (150–179 nm) wavelength bands. The comparisons indicate the good agreement among the irradiance measurements and models on the short time scales of the 27-day rotational modulation. See Color Plate 10.

features on the solar disk (Warren *et al.*, 2001; Lean *et al.*, 2003). Solar irradiance is then calculated as the net radiative output from all locally emitting regions along the line of sight from the solar atmosphere to the Earth. Also under development are more complex models of the UV, visible, and IR spectrum that allow for multiple magnetic features in the solar atmosphere and directly utilize atomic properties of the gases and plasma temperatures to calculate how altered emission in individual features modulates the (disk-integrated) signal, i.e. irradiance (Fontenla *et al.*, 2006, 2007).

10.5 Reconstructing historical irradiance changes

No two solar activity cycles are alike, whether in the rate of their increase to maximum and decay to minimum, or in the shape and duration of the maximum and minimum epochs. Although quasi-regular, the solar activity cycle is by no means deterministic (stationary). The varying amplitude and length of the sunspot cycle in the past 400 years (Fig. 10.4) exemplify this; cycle amplitudes vary from 0 to 180 and cycle lengths from 9 to 13 years.

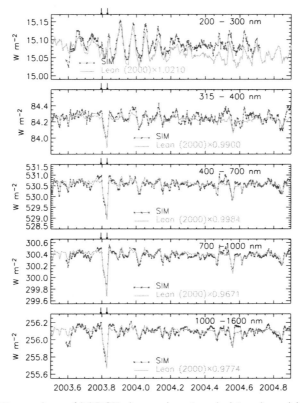

Fig. 10.13. Comparison of SORCE observations (symbols) and empirical variability model values (solid line) of irradiance in selected wavelength bands including the middle UV (200–300 nm), near-UV (315–400 nm), visible (400–700 nm), and near-IR (700–1000 nm and 1000–1600 nm) wavelength bands. SORCE observations are made by SIM, except at wavelengths between 200 and 300 nm, which are made by SOLSTICE. The SORCE time series have been detrended by subtracting a 30-day running mean, to remove known instrumental drifts not yet incorporated in the data reduction algorithms. The model time series are scaled by the values shown in each panel to agree with the SIM absolute scale. The arrows indicate the times of the two images in Fig. 10.7.

The sunspot number is a generic indicator, only, of solar activity; a numerical construct – rather than a true geophysical quantity – defined as the number of sunspots plus ten times the number of sunspot groups. In reality, the activity cycle evolves because fluctuations in the subsurface dynamo alter the amount of magnetic flux in the solar atmosphere, which is then transported by meridional flow, differential rotation, and diffusion (Chapter 2). But as the only directly observed historical solar quantity, sunspot numbers are relied upon for broad indication of the Sun's magnetic flux evolution on centennial time scales. Thus, on the basis of sunspot numbers, the extended period of few or infrequent sunspots from 1645 to

1715 is recognized as one of anomalously low solar activity, called the Maunder minimum (cf. Chapter 2, in particular Section 2.8; Eddy, 1976). A subsequent more modest minimum, the Dalton minimum, occurred during the nineteenth century. For comparison, the database of irradiance observations made in the space era since the 1960s coincides with relatively high solar activity, termed the modern maximum.

What do the fluctuations in the sunspot record in recent centuries imply for past solar irradiance variations? From the association of increased total solar irradiance with high solar activity cycle in the past three 11-year cycles, it is surmised with some confidence that solar irradiance has undergone similar cycles of activity during the past 400 years. Hence, one approach for reconstructing historical solar irradiance is to simply scale the irradiance according to sunspot numbers (Schatten and Orosz, 1990). This approach produces a record in which total irradiance returns to essentially the same value during every cycle minimum (Fig. 10.14, upper envelope). Another approach is to utilize different historical indices of solar activity such as cosmogenic isotopes or geomagnetic activity (Chapter 11; Beer *et al.*, 1990, Lockwood and Stamper, 1999) producing, instead, irradiance variations in which local cycle minima are superimposed on a slowly varying background of increasing irradiance since the seventeenth century Maunder minimum (Fig. 10.14, lower envelope). The relationship of sunspot numbers (or derived irradiance) with cosmogenic isotopes further suggests that over the past 10 000 years (of Earth's current interglacial climate epoch) sporadic episodes of solar "grand" minima have punctuated activity levels comparable to that of the present day (Fig. 10.14).

Faculae, the dominant determinant of solar cycle irradiance variations, are implicated as a primary source of longer-term irradiance trends. Specifying the past evolution of the facular signal is therefore crucial for reconstructing historical irradiance variations. But unlike the sunspot signal, which is suggested by direct observations of sunspot numbers, the facular component is highly uncertain and dependent on assumptions inferred from circumstantial evidence. For example, based on current observations of facular contrast and disk coverage, the disappearance of all faculae from the Sun's surface is estimated to decrease total solar irradiance about 0.1% (Lean *et al.*, 1992). Attempts have also been made to translate variations in the chromospheric activity in Sun-like stars to a plausible range of the facular influence on solar irradiance (Lean *et al.*, 1992, 1995), with results broadly consistent with inferences from the cosmogenic and geomagnetic indices (Fig. 10.14, lower envelope in the top panel). Changes in solar structure are also considered as possible sources of long-term irradiance variations in addition to, or instead of, facular variations (Hoyt and Schatten, 1993; Tapping *et al.*, 2007) producing levels as much as 0.3% below contemporary solar minima values (e.g. review of Maunder minimum levels in Lean *et al.*, 2005).

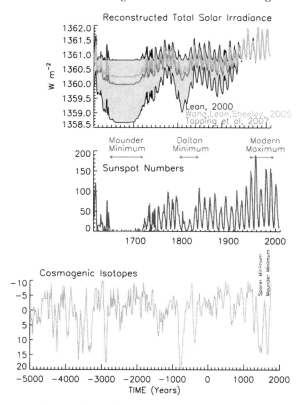

Fig. 10.14. (Top) Comparison of estimates of total solar irradiance since the Maunder minimum, reconstructed with different assumptions about the strength of the background component that underlies the activity cycle. The upper boundary of the grey shaded region indicates the solar irradiance reconstructed subject to the assumption that the 11-year activity cycle is the only source of variability. The lower boundary of the grey shaded region indicates the solar irradiance reconstructed (by Lean, 2000) subject to the assumption that there is an additional long-term component whose magnitude is derived from the chromospheric Ca II K variations that Baliunas and Jastrow (1990) reported for Sun-like stars. (Middle) The annual sunspot numbers (Fig. 10.4) are shown for comparison, indicating epochs of anomalously low activity during the Maunder and Dalton minima. (Bottom) Episodes of low activity are seen to punctuate the 7000 year record of the ^{14}C cosmogenic isotopes (a proxy for long-term solar activity archived in tree rings), shown (on an inverted scale) in the bottom panel.

With solar irradiance reconstructed over the past four centuries from sunspot numbers, estimates of irradiance variations during the past 10 000 years can then be made using cosmogenic isotopes (Bard *et al.*, 2000; Steinhilber *et al.*, 2009), which are the only type of solar activity proxy available prior to the 1600s. Levels of high solar activity (and irradiance) correspond to low fluxes of cosmogenic isotopes in tree-rings and ice-cores, but the relationship is not simply linear. In both the ^{14}C and

[10]Be isotope records, solar activity cycles are superimposed on a slowly varying background component that is absent in sunspot numbers. Because solar activity modulates cosmogenic isotopes by altering the complexity and structure of the heliosphere, the varying background in cosmogenic isotopes is physically related to variations in "open" flux – that component (10 to 20%) of the Sun's total magnetic flux which pervades the heliosphere (Wang *et al.*, 2005). In comparison, sunspots and faculae, the primary sources of irradiance variations, are features defined by "closed" flux, anchored below the Sun's visible surface.

A flux transport model with variable meridional flow that simulates the eruption, transport, and accumulation of magnetic flux has been used to estimate variations in both open and total magnetic flux arising from the deposition of bipolar magnetic regions (active regions) and smaller-scale ephemeral regions on the Sun's surface, in strengths and numbers proportional to the sunspot number (Wang *et al.*, 2005). The open flux compares reasonably well with the geomagnetic and cosmogenic isotopes, which gives confidence that the approach is plausible. A small accumulation of total flux (and possibly ephemeral regions) produces a net increase in facular brightness that, in combination with sunspot blocking, permits the reconstruction of total solar irradiance (Fig. 10.14). The increase from the Maunder minimum to the present-day quiet Sun is about 0.04%.

10.6 Forecasting irradiance variations

Forecasts of solar irradiance are sought on a wide range of time scales for multiple geophysical applications ranging from space weather (National Space Weather Strategic Plan, 1995[†]; Lean *et al.*, 2009) to recovery of the ozone layer and near-term climate change (Lean and Rind, 2009). Satisfying the needed range of forecasts therefore requires multiple approaches, typically some combination of statistical tools, numerical models, and physical understanding (Hathaway, 2009). The statistical tools harness the recurrence of solar activity patterns and solar rotation (Hathaway *et al.*, 1999), the numerical models describe this behavior, and the physical understanding utilizes knowledge of the dynamo, active region evolution, and transport by rotation, diffusion, and meridonal flow (Dikpati *et al.*, 2006).

Currently, solar activity is forecast daily, monthly, and for future solar cycles in terms of sunspot numbers and $F_{10.7}$. Associations between these indices and the observed irradiance variations, such as encapsulated in irradiance variability models, then translate the solar activity forecasts into accompanying irradiances. Recognizing that the Mg index is superior to $F_{10.7}$ as an indicator of the facular source of irradiation variations, 1 to 10 day forecasts have been recently developed

---

[†] URL: http://www.ofcm.gov/homepage/text/pubs.htm

for this index, permitting an improved practical capability for forecasting the solar EUV and UV spectrum (Lean *et al.*, 2009).

Forecasting total solar irradiance and the solar spectrum at wavelengths longer than 300 nm is considerably more difficult than forecasting the facular-dominated EUV and UV emissions. Irradiance variations at these longer wavelengths are the net effect of sunspot and facular influences, each of which is related to magnetic activity, making the relationship of irradiance and sunspot numbers more complicated than simply linear. That total solar irradiance has increased approximately the same amount during each of the past three cycles, even though the sunspot numbers were less in cycle 23 than in cycles 21 and 22, demonstrates this nonlinearity. Forecasting total (and near-UV, visible, and IR) solar irradiance therefore encompasses the significant challenge of forecasting, in addition to the overall strength of the magnetic activity, how this flux is apportioned into sunspot and facular regions.

10.7 Summary

Direct, overlapping observations of total irradiance now exist for about 30 years. Large irradiance decreases occur when sunspots are present on the solar disk, but from the minimum to maximum of the 11-year solar cycle the overall decrease due to sunspots is more than compensated (by a factor of two) by the increase in emission from bright faculae. As a result, there has been a net increase of about 0.1% in total solar irradiance at cycle maximum during the past three solar cycles. Whether or not additional, longer-term irradiance changes are occurring cannot be determined from the extant 30-year database because of instrument sensitivity irregularities among the various measurements (e.g. Mekaoui and Dewitte, 2008). It is likely that during the past three solar minima, the level of the total irradiance has remained approximately constant. Circumstantial evidence suggests that solar irradiance may have been somewhat lower during the Maunder minimum than during current solar minima. One recent estimate from simulations using a solar flux transport model is a decrease of about 0.04%, but alternative scenarios have been presented to support both negligible and larger decreases.

Spectral irradiance databases are less comprehensive than that of total irradiance, composed of more sporadic measurements and extended epochs with no observations, especially at wavelengths longer than 400 nm, for which adequate SORCE observations commenced only in 2003, and at wavelengths shorter than 110 nm, for which TIMED observations began in 2002. The general lack of continuous, overlapping spectral irradiance observations during the past 30 years makes the construction of composite spectral irradiance time series difficult, and requires assumptions about the irradiance variations in intervening periods without observations, such as estimated by irradiance variability models. In the EUV spectrum between about 30 and 100 nm, solar spectral irradiance increases (on average) by almost a factor of two during the solar cycle. In the X-ray region the increase is larger, by as much as an order of magnitude. Lyman-α irradiance (at 121 nm), the strongest emission line in the solar electromagnetic spectrum, has a solar cycle increase of about 50%. The spectral irradiance cycle becomes increasingly smaller at increasingly longer wavelengths, of order 10% near 200 nm, a few percent near 250 nm and tenths percent in the visible and IR regions. The actual magnitude (and even the phase) of the near-UV and visible regions is under debate since, relative to irradiance variability models, the recent SORCE spectral irradiance observations indicate near-UV and visible spectral variations larger and out-of-phase during the solar cycle (Harder *et al.*, 2009). On the shorter time scale of solar rotational modulation the observations and models are in good agreement (Lean *et al.*, 2005; Unruh *et al.*, 2008), and closely track the evolution of the competing wavelength-dependent sunspot and facular sources.

Many remaining questions about solar irradiance and its variations challenge future research. What are the actual spectral irradiance changes that compose the total? Why do magnetic fields that emerge from the underlying convection zone coalesce into spots or faculae? What determines the current proportions of these sources of irradiance variations? How and why does this change, and on what time scales? How much dimmer or brighter than current observations might the Sun be? Do the high activity cycles of the modern maximum represent maximum irradiance levels, or might even higher levels be possible? Does solar irradiance decrease during grand minima, and if so, by how much; for example, when entering, during, and exiting the Maunder minimum? Are there long-term – 80 and 200 year – irradiance cycles concurrent with those present in cosmogenic isotopes? How and why does the surface distribution of magnetic flux produce regions of open (coronal holes) versus closed (extended bright active regions) flux in the outer solar layers? How do closed and open flux relate on long time scales, and how do they each evolve under conditions of long-term solar activity? Thus, what is the actual relationship between solar irradiance variations and the cosmogenic isotope archives of solar activity? What can we learn about irradiance variability from the variations observed in Sun-like stars?

While estimates and ideas are being developed to address many of these questions, for others conflicting information or concepts currently preclude reliable resolution. Ultimately, the understanding and specification of solar irradiance variations that is crucial for geophysical applications requires more complete and more accurate future observational databases, along with continued improvements in modeling capabilities and techniques. Now, with the first ever set of solar irradiance measurements covering the full spectral range from 0.1 nm to 2400 nm since 2003, multiple total solar irradiance instruments making concurrent observations, and a plethora of solar imagers, we potentially are at the dawn of a new era from which will emerge answers for the many remaining questions about solar irradiance variability.

11

Astrophysical influences on planetary climate systems

Jürg Beer

11.1 Introduction

The planets and the Sun together form a coupled system, the so-called solar system, which is located in a spiral arm of the Milky Way galaxy. The solar system has existed for 4.6 billion years. Its formation took only between 50 and 100 million years (Chapter 3). According to the nebular hypothesis, a large cloud of gas started to contract under self-gravity. Conservation of angular momentum led to a rotating disk. In the center of this disk mass concentrated into a so-called proto-Sun which grew larger and larger. After reaching a temperature of about 15 million K in the core, nuclear fusion processes started turning hydrogen into helium.

In the inner part of the disk, small planetesimals were formed, which by aggregating more mass became the terrestrial planets (Mercury, Venus, Earth, and Mars). The release of potential energy and the impact of particles produced molten spheres causing a chemical differentiation with denser material sinking to the center and with a loss of volatile components. In the outer disk, lower temperatures prevailed allowing the aggregation of volatile matter such as ices and gases. The result was several larger planets with lower densities (Jupiter, Saturn, Uranus, and Neptune). For a more detailed discussion of the formation and evolution of stars and their planets we refer to Chapter 3.

In summary, the formation of the solar system out of a nebular gas was governed mainly by gravitation and the conservation of mass, and the redistribution and expulsion of angular momentum. While 99.8% of the mass is concentrated in the Sun, 98% of the angular momentum is distributed among the planets, with Jupiter and Saturn taking the largest share (see Table 11.2). Therefore, the solar system has to be considered as a coupled system. The coupling takes place through gravitation, conservation of angular momentum, electromagnetic radiation, particle fluxes, and magnetic fields. In the following we will discuss some basic external influences on planets. Most of them are very small and therefore often negligible. So why

Heliophysics: Evolving Solar Activity and the Climates of Space and Earth, eds. Carolus J. Schrijver and George L. Siscoe. Published by Cambridge University Press. © Cambridge University Press 2010.

do we discuss them anyway? Well, first of all we have to show quantitatively how small they really are. Second, in coupled non-linear systems sometimes even small causes can have large effects. Third, based on pure correlation analysis, weak influences are often claimed to be responsible for observed effects. An example is the cyclic variability of solar activity, which has sometimes been attributed to planetary effects. Last but not least, it is always a good approach to consider first all potential effects and then to eliminate the ones that are negligible.

11.2 External influences

11.2.1 Gravitational influence

If a planet of mass m_2 is orbiting around the Sun with mass m_1 at a distance r, the gravitational force that the Sun exerts on the planet ($\mathbf{F}_{12}$) is given by the universal law of gravitation:

$$\mathbf{F}_{12} = -G\frac{m_1 m_2}{r^2}\mathbf{e}_r, \tag{11.1}$$

where G is the universal gravitational constant and $\mathbf{e}_r \equiv \frac{\mathbf{r}}{r}$ is the unit vector pointing away from the Sun.

Let us consider, in general, two astrophysical bodies as depicted in Fig. 11.1. According to Newton's second law ($\mathbf{F} = m\ddot{\mathbf{r}}$) and third law ($\mathbf{F}_{12} = -\mathbf{F}_{21}$), the equation of motion for each body is given by

$$m_2\ddot{\mathbf{r}}_2(t) = -G\frac{m_1 m_2}{r^2}\mathbf{e}_{r_2} = -G\frac{m_1 m_2}{r^3}\mathbf{r}, \tag{11.2}$$

$$m_1\ddot{\mathbf{r}}_1(t) = -G\frac{m_1 m_2}{r^2}\mathbf{e}_{r_1} = G\frac{m_1 m_2}{r^3}\mathbf{r}. \tag{11.3}$$

To determine the trajectory of body 2 in a reference system centered on body 1, Eqs. (11.2) and (11.3) can be combined to read

$$\ddot{\mathbf{r}}_2(t) - \ddot{\mathbf{r}}_1(t) \equiv \ddot{\mathbf{r}}(t) = -\mu\frac{\mathbf{r}}{r^3}, \tag{11.4}$$

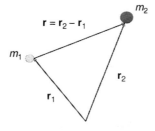

Fig. 11.1. Movement of two bodies m_1 and m_2 under their mutual gravitational interaction.

where $\mu \equiv G(m_1 + m_2)$. Equation (11.4) describes the motion of body 2 relative to the new reference frame.

Now we consider the angular momentum of body 2 regarding the origin:

$$\mathbf{L} = m_2 \mathbf{r} \times \dot{\mathbf{r}} \Rightarrow \mathbf{L}/m_2 \equiv \mathbf{l} = \mathbf{r} \times \dot{\mathbf{r}}. \tag{11.5}$$

The time derivative of this equation combined with Eq. (11.4) can be used to show that the motion takes place in a plane:

$$\dot{\mathbf{l}} = \dot{\mathbf{r}} \times \dot{\mathbf{r}} + \mathbf{r} \times \ddot{\mathbf{r}} = \mathbf{r} \times \ddot{\mathbf{r}} = \mathbf{r} \times \left(-\mu \frac{\mathbf{r}}{r^3} \right) = 0 \Rightarrow \mathbf{l} = \text{const}. \tag{11.6}$$

We now proceed with the calculation of the geometrical form of the orbit. We need to compute the two quantities: $\mathbf{l} \times \ddot{\mathbf{r}}$ and $\dot{\mathbf{r}} \cdot \ddot{\mathbf{r}}$:

$$\mathbf{l} \times \ddot{\mathbf{r}} = (\mathbf{r} \times \dot{\mathbf{r}}) \times \ddot{\mathbf{r}} = (\mathbf{r} \cdot \ddot{\mathbf{r}})\dot{\mathbf{r}} - (\mathbf{r} \cdot \dot{\mathbf{r}})\ddot{\mathbf{r}}$$

$$= -\frac{\mu}{r^3}(\mathbf{r} \cdot \mathbf{r})\dot{\mathbf{r}} + \left(\frac{\mu}{r^3}\mathbf{r} \right) r\dot{r} = \frac{d}{dt}\left(-\mu \frac{\mathbf{r}}{r} \right). \tag{11.7}$$

Also

$$\mathbf{l} \times \ddot{\mathbf{r}} = \frac{d}{dt}(\mathbf{l} \times \dot{\mathbf{r}}), \tag{11.8}$$

where the vectorial identity $(\mathbf{a} \times \mathbf{b}) \times \mathbf{c} = (\mathbf{a} \cdot \mathbf{c})\,\mathbf{b} - (\mathbf{a} \cdot \mathbf{b})\,\mathbf{c}$ was employed. Then, from Eqs. (11.8) and (11.7) we get

$$\mathbf{l} \times \dot{\mathbf{r}} + \mu \frac{\mathbf{r}}{r} = \text{const} \equiv -\mu\mathbf{e}. \tag{11.9}$$

We write the constant as $-\mu\mathbf{e}$, where $\mathbf{e}$ is known as the Laplace–Runge–Lenz vector. The meaning of $\mathbf{e}$ and the convenience of this choice will become clear later.

Further,

$$\dot{\mathbf{r}} \cdot \ddot{\mathbf{r}} = -\dot{\mathbf{r}} \cdot \left(\mu \frac{\mathbf{r}}{r^3} \right) = \frac{d}{dt}\left(\mu \frac{1}{r} \right),$$

$$= \frac{1}{2}\frac{d}{dt}(\dot{\mathbf{r}} \cdot \dot{\mathbf{r}}) = \frac{1}{2}\frac{d}{dt}(\dot{r}^2). \tag{11.10}$$

From Eq. (11.10) we obtain the following identity:

$$\frac{1}{2}\dot{r}^2 - \mu\frac{1}{r} = \text{const} \equiv h, \tag{11.11}$$

where h is another constant of integration and it represents the total energy, i.e. potential plus kinetic energy. Therefore, Eq. (11.11) expresses the conservation of energy. Next we combine Eq. (11.10) with Eq. (11.11) to derive the relation between $\mathbf{e}$ and h:

$$(\mu\mathbf{e}) \cdot (\mu\mathbf{e}) = \left(\mathbf{l} \times \dot{\mathbf{r}} + \mu \frac{\mathbf{r}}{r}\right) \cdot \left(\mathbf{l} \times \dot{\mathbf{r}} + \mu \frac{\mathbf{r}}{r}\right) \Rightarrow$$

$$\mu^2 e^2 = l^2 \dot{r}^2 + \mu^2 \frac{r^2}{r^2} + 2\frac{\mu}{r}(\mathbf{l} \times \dot{\mathbf{r}}) \cdot \mathbf{r} \Rightarrow$$

$$\mu^2 (e^2 - 1) = l^2 \dot{r}^2 + 2\frac{\mu}{r}(\mathbf{l} \times \dot{\mathbf{r}}) \cdot \mathbf{r} \Rightarrow$$

$$\mu^2 (e^2 - 1) = l^2 \dot{r}^2 - 2\frac{\mu}{r}l^2 = 2l^2 h \Rightarrow$$

$$\mu^2 (e^2 - 1) = 2l^2 h, \tag{11.12}$$

where we have used the vector identity $(\mathbf{a} \times \mathbf{b}) \cdot \mathbf{c} = -\mathbf{a} \cdot (\mathbf{c} \times \mathbf{b})$.

Using Eq. (11.9) we now compute the form of the orbit of a planet around the Sun:

$$\mathbf{r} \cdot \mathbf{e} = r\, e\, \cos\theta, \tag{11.13}$$

$$\mathbf{r} \cdot \mathbf{e} = \mathbf{r} \cdot \left(-\frac{1}{\mu}\mathbf{l} \times \dot{\mathbf{r}} - \frac{\mathbf{r}}{r}\right) = \frac{l^2}{\mu} - r. \tag{11.14}$$

Combining Eqs. (11.13) and (11.14) we finally obtain

$$r(\theta) = \frac{l^2/\mu}{1 + e\, \cos\theta}. \tag{11.15}$$

This is the equation of a conic section, schematically shown in Fig. 11.2 for the elliptic case, where $P \equiv l^2/\mu$ is the parameter of the curve, and e is the eccentricity (we now see that the Laplace–Runge–Lenz vector is related to the eccentricity). The form of the curve depends on the eccentricity so that if $e = 0$ the orbit is circular; $e < 1$ the orbit is elliptic; $e = 1$ the orbit is parabolic; $e > 1$ the orbit is hyperbolic. Because eccentricity e and total energy h are related by Eq. (11.12), the total energy determines the form of the orbit. Typical eccentricities of the

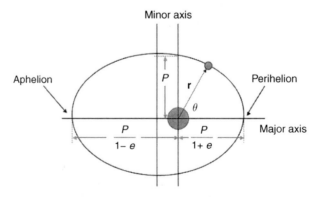

Fig. 11.2. Elliptical orbit of a planet around the Sun which is located in one of the two focal points. The form of the ellipse is determined by the eccentricity e.

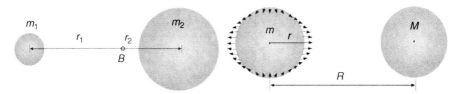

Fig. 11.3. (Left) Two bodies orbiting around the barycenter B. (Right) Tidal acceleration induced by the body with mass M on the body with mass m with the distance R between their centers.

planets are 0.0068 (Venus), 0.0167 (Earth), 0.0483 (Jupiter), and 0.2056 (Mercury). Therefore, in most cases a circular orbit is a good first approximation.

As a result of the gravitational coupling between the bodies in the solar system, the assumption that the Sun represents an inertial system around which the planets orbit is not exactly correct: in spite of the fact that 99.8 % of the total mass of the solar system is concentrated in the Sun, the barycenter of the solar system is not identical with the center of the Sun. Depending on the relative positions of the bodies in the solar system, the center of the Sun moves around. To illustrate this let us assume a simple example of two bodies in space as displayed in Fig. 11.3:

$$m_1 r_1 = m_2 r_2, \qquad (11.16)$$

with $a = r_1 + r_2$ being the distance between body 1 and body 2, so that

$$r_2 = \frac{a}{1 + \frac{m_2}{m_1}}. \qquad (11.17)$$

Taking Jupiter as body 1 and the Sun as body 2, r_2 turns out to be 740×10^6 m. Because the radius of the Sun is 696×10^6 m this means that the barycenter is outside the Sun by about 40 000 km.

Tidal effects A well-known gravitational influence is the tidal force of the Moon and Sun on Earth. To calculate the tidal acceleration, let us consider two masses M and m with the distance R between their centers as shown in Fig. 11.3. According to Newton's law of gravitation, the mass m feels the gravitational acceleration a_g:

$$a_g = -G \frac{M}{R^2}. \qquad (11.18)$$

However each point of a body with mass m feels a different gravitational acceleration depending on the effective distance to mass M, which ranges from $R - r$ to $R + r$. For the two extreme cases we find:

$$a_g = -G \frac{M}{(R \pm r)^2}, \qquad (11.19)$$

Table 11.1. *Tidal acceleration*
on the Sun induced by the planets

Planet	Tidal acceleration $(10^{-10} \, \mathrm{m\,s^{-2}})$
Mercury	1.6
Venus	3.6
Earth	1.6
Mars	0.05
Jupiter	3.7
Saturn	0.2
Uranus	0.003
Neptune	0.001

$$a_g = -G\,\frac{M}{R^2(1 \pm r/R)^2}. \tag{11.20}$$

Because r is much smaller than R this equation can be expanded into a Taylor series:

$$\frac{1}{(1+x)^2} = 1 - 2x + 3x^2 - \cdots, \tag{11.21}$$

$$a_g = -G\,\frac{M}{R^2} \pm G\,\frac{2\,M}{R^2}\frac{r}{R} \mp \cdots. \tag{11.22}$$

The tidal acceleration a_t is the difference between the effective and the gravitational acceleration:

$$a_t \approx \pm G\,\frac{2\,M}{R^2}\frac{r}{R}. \tag{11.23}$$

Note that a_t decreases with the third power of R. As a result of this the tidal accelerations are relatively small. On Earth the tidal acceleration is about $1.1 \times 10^{-6} \, \mathrm{m\,s^{-2}}$ due to the Moon and $0.5 \times 10^{-6} \, \mathrm{m\,s^{-2}}$ due to the Sun, compared to the gravitational acceleration of about $10 \, \mathrm{m\,s^{-2}}$. This corresponds to an expected lunar tidal effect of about 70 cm. In reality, the average tide is about 30 cm because of a slight deformation of the Earth. In the case of the Sun, the tidal effects caused by the planets are very small. Table 11.1 shows the tidal acceleration by the planets on the Sun, which has to be compared to the gravitational acceleration of $270 \, \mathrm{m\,s^{-2}}$. The largest effects are due to Venus and Jupiter with a theoretical tide in the order of 1 mm.

As a result of the friction between the tide and the planet, the rotation slows down. In the case of Earth this is about one second per year. Some 2.5 billion years ago the length of a day was only about 6 hours. Because the angular momentum must be conserved this leads to a corresponding increase in the distance between

Moon and Earth (4 cm per year) as measured by laser technique (Dickey *et al.*, 1994). The tidal friction generates a power of 3×10^{12} W which is mostly dissipated in the ocean. There are indications that this tidal power affects the global ocean circulation which plays a crucial role in the climate system by transporting energy from low to high latitudes (Egbert and Ray, 2000; Keeling and Whorf, 2000; Wunsch, 2000). The tides also act in the atmosphere causing changes in pressure, temperature, and wave propagation (Geller, 1970; Camuffo, 2001).

There are climatic effects on Earth related to the lunar tides. The plane in which the Moon moves is inclined to the ecliptic by about 5°. The points where the lunar orbit crosses the ecliptic are called nodes. As a result of the gravitational force of the Sun on the Moon the orbital spin axis of the Moon precesses, which leads to a continuous slight shift of the nodes. After 18.6 years the nodes are back to their original position. The inclination of the Moon's rotation axis has an effect on the amplitude of the tides. The amplitude of the lunar nodal tide is only about 5% of the daily diurnal tide, but integrated in space and time it becomes significant. The 18.6-year cycle and sometimes also its second subharmonic of 74 years have been found in the Arctic Ocean temperature and sea-ice extent (Yndestad, 1999) and in drought records (Cook *et al.*, 1997; Currie and Fairbridge, 1985).

The dynamics of a multi-body system such as the solar system is largely determined by gravitation. The bodies orbit around the barycenter. In the case of a two-body system with a large body (Sun) and a small body (planet) the orbit is an ellipse with the large body in one of the focal points. In a multibody system (solar system) the gravitational interaction between the bodies slightly disturbs their orbital parameters. For example, the planets (mainly Jupiter and Saturn) change the eccentricity of the Earth's orbit with periodicities of about 100 000 and 400 000 years which has an effect on the amount of solar radiation received from the Sun. These small changes are argued to be the main cause for the observed sequence of glacial and interglacial periods during the past million years (see Section 11.3.3).

The tidal effect depends on the mass (M) and strongly on the distance (R^3). Therefore, tidal effects on planets are quite common and mainly caused by the large mass of the Sun and even more so by nearby satellites. On the other hand, tidal effects of the planets on the Sun seem negligible. Nevertheless, solar cycles found in the sunspot record and in the cosmogenic radionuclide data from ice cores (^{10}Be) and tree rings (^{14}C) are sometimes attributed to planetary effects on the Sun (Jose, 1965; Charvátová, 2000; De Jager and Versteegh, 2005).

11.2.2 Angular momentum influences

The angular momentum of a planet is mainly due to its orbit around the center of mass, which in the case of the solar system is to a good approximation the Sun.

Table 11.2. *Orbital and spin angular momentum of the main bodies in the solar system*

Body	Mass (kg)	Orbital radius (10^9 m)	Orbital period (days)	Spin radius (10^6 m)	Spin period (days)	L_{orbital} (10^{40} J s)	L_{Spin} (10^{40} J s)
Sun	$1.99 \cdot 10^{30}$			696	27		93.08
Mercury	$3.30 \cdot 10^{23}$	58	88	24.4	58.6	0.09	$8.74 \cdot 10^{-9}$
Venus	$4.87 \cdot 10^{24}$	108	225	6.05	243	1.84	$1.91 \cdot 10^{-9}$
Earth	$5.97 \cdot 10^{24}$	150	365	6.38	1	2.68	$6.33 \cdot 10^{-7}$
Mars	$6.42 \cdot 10^{23}$	228	687	3.4	1	0.35	$1.93 \cdot 10^{-8}$
Jupiter	$1.90 \cdot 10^{27}$	778	4333	71.5	0.41	1930.70	0.061
Saturn	$5.68 \cdot 10^{26}$	1429	10760	60.3	0.44	784.14	0.012
Uranus	$8.68 \cdot 10^{25}$	2871	30685	25.6	0.72	169.61	0.0002
Neptune	$1.02 \cdot 10^{26}$	4504	60190	24.8	0.67	250.07	0.0002
Pluto	$1.27 \cdot 10^{22}$	5914	90800	1.17	6.39	0.04	$7.09 \cdot 10^{-12}$

Neglecting the fact that most orbits deviate slightly from a circular form, the orbital angular momentum is given by

$$L_{\text{orb}} = m\, r^2\, \omega, \tag{11.24}$$

with m being the mass of the planet, r the orbital radius and ω the angular velocity which is related to the orbital period P_{orb} by

$$\omega = 2\pi / P_{\text{orb}}. \tag{11.25}$$

A smaller contribution to angular momentum is due to the fact that all bodies in the solar system spin around their rotation axis. Assuming that all bodies are spherical and homogeneous the corresponding angular momentum L_{spin} is

$$L_{\text{spin}} = \frac{4\pi\, m\, R^2}{5\, P_{\text{spin}}}. \tag{11.26}$$

Table 11.2 shows both the orbital and the spin angular momentum for the main bodies in the solar system. Because the angular momentum depends on the square of the orbital radius, it is mostly the outer planets that contribute to the total angular momentum of about 3.2×10^{43} J s.

If we assumed that there were no planets and all the angular momentum resided in the Sun, this would lead to an increase in the angular velocity by about a factor of 35. As a result the Sun would spin around its axis in about 18 hours instead of 27 days. Because the assumption of a homogeneous sphere underestimates the effective momentum of inertia, the Sun would complete a rotation within about 12 hours, which corresponds to a velocity of the photosphere in the order of $100 \, \text{km s}^{-1}$. By emitting a magnetized solar wind, the Sun continuously loses angular momentum with a present-day rate of approximately 5×10^{12} J. The transport of angular momentum is described in Section 3.2 and in Vol. I, Chapter 9.

11.2.3 Electromagnetic influences

The Sun continuously emits electromagnetic radiation from the photosphere, which has an effective temperature T_{eff} of about $5770 \, \text{K}$. Assuming isotropy, the total amount of emitted radiation, called luminosity $L_\odot$, is

$$L_\odot = 4 \, \pi \, R_\odot^2 \, \sigma \, T_{\text{eff}}^4 = 3.8 \times 10^{26} \, \text{W}, \tag{11.27}$$

with $R_\odot$ the solar radius, and σ the Stefan–Boltzmann constant,

$$\sigma = \frac{2\pi^5 \, k_{\text{B}}^4}{15 \, h^3 \, c^2} = 5.67 \times 10^{-8} \, \frac{\text{W}}{\text{m}^2 \, \text{K}^4}, \tag{11.28}$$

with k_{B} the Boltzmann constant, h the Planck constant, and c the speed of light in vacuum.

Potential energy: In 1854, Hermann von Helmholtz proposed that the origin of the solar luminosity is the contraction of the solar mass. In 1862, Lord Kelvin concluded that the energy must come from the coalescence of comparatively small bodies, which means that the age of the Sun is less than one hundred million years. His final statement was quite visionary: "for the future, we may say, with equal certainty, that inhabitants of the earth can not continue to enjoy the light and heat essential to their life for many million years longer unless sources now unknown to us are prepared in the great storehouse of creation" (Thomson, 1862). As we will see below, the unknown energy source was nuclear energy. But first we want to estimate the contribution of the potential energy. During the formation of the Sun from a proto-stellar cloud, a large amount of gravitational potential energy was turned into kinetic energy. If we just consider the Sun and calculate the energy released by concentrating the corresponding amount of gas ($M_\odot = 1.98 \times 10^{30} \, \text{kg}$) in a sphere with the volume of the Sun ($R_\odot = 6.9 \times 10^8 \, \text{m}$) and assume for simplicity a constant density, the potential energy is given by

$$E_\odot = G \frac{3 \, M_\odot^2}{5 \, R_\odot} = 2.3 \times 10^{41} \, \text{J}, \tag{11.29}$$

with the gravitational constant $G = 6.67 \times 10^{-11} \, \text{m}^3 \, \text{kg}^{-1} \, \text{s}^{-2}$.

A quick calculation of the expected lifetime of the Sun shows that we need a much stronger energy source to keep the Sun shining for billions of years:

$$\tau = \frac{E_\odot}{L_\odot} = \frac{2.3 \times 10^{41} \, \text{J}}{3.82 \times 10^{26} \, \text{J s}^{-1}} = 5.9 \times 10^{14} \, \text{s}. \qquad (11.30)$$

This corresponds to only about 20 million years, as estimated by Lord Kelvin.

Nuclear energy   The only energy source efficient enough to produce such a huge amount of power over billions of years is nuclear fusion. As a result of thermonuclear reactions in the core of the Sun, hydrogen is fused into helium. Even with nuclear energy the consumption of fuel is impressive. Every second the Sun loses 4.2 million tons of its mass according to $E = m \, c^2$, all of which are radiated into space.

The energy produced in the core is transported outwards using different mechanisms. The interior of the Sun can be divided into three main zones (see Table 5.1 for some characteristic physical properties in these zones). In the core, thermonuclear reactions take place. Temperature and density are very high. In the radiative zone between about $0.25 \, R_\odot$ and $0.7 \, R_\odot$ ($R_\odot$: solar radius) the energy is transported radiatively. Temperature and density are dropping considerably. The energy transport is very slow because the photons are absorbed and re-emitted about once each centimeter, causing a long random walk taking several $10\,000$ years to cross the radiative zone. During this random walk, the wavelength increases which causes a reduction of the mean free path and makes the radiative energy transport increasingly inefficient. At about $0.7 \, R_\odot$ the temperature gradient becomes large enough for convection to set in and to transport the energy to the photosphere from where it is radiated into space.

The nuclear reactor in the core of the Sun is able to maintain this huge power generation for some 10 billion years. The standard solar model shows that the luminosity is steadily increasing from about 80% 4 Gyr ago to about 130% of its present values in 4 Gyr from now. According to this model the change occurs very smoothly and slowly ($10^{-8}\%$ per year) as shown in Fig. 11.4. The implications of such a low luminosity increase are discussed in Section 11.3.1.

Without the electromagnetic energy from the Sun the temperatures of the planets would be significantly lower. The only remaining sources of energy on planets are cosmic rays, geothermal energy as a result of radioactive decay and gravitational energy from the time of formation of the solar system, tidal energy from moons and other bodies, and in some cases gravitational energy released by compression (Jupiter). In the case of the Earth, these contributions amount to 10^{10} W, 10^{13} W, 10^{11} W, and 0 W respectively, compared to the 10^{17} W obtained from the Sun.

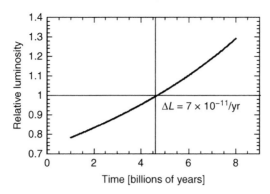

Fig. 11.4. Luminosity of the Sun according to the standard solar model in units relative to the present. (Newkirk, 1983.)

In the following, we investigate the effect of the electromagnetic coupling between the Sun and the planets by estimating the planetary temperatures. To simplify this problem we consider all the planets and the Sun as blackbodies and calculate their mean temperature.

The fraction of the solar luminosity $L_\odot$ that is absorbed by a planet is given by the ratio of the planet's cross section πR_p^2 to the area $4\pi d_\odot^2$ of a sphere containing the planet at distance $d_\odot$ from the Sun, corrected for the albedo a (total reflected power):

$$P_{abs} = \frac{L_\odot\, \pi\, R_p^2\, (1-a)}{4\,\pi\, d_\odot^2}. \tag{11.31}$$

If we assume as a first approximation that a planet is an atmosphere-free blackbody and that the climate machine distributes the incoming solar radiation uniformly, the emitted power is given by the Stefan–Boltzmann law:

$$P_{emi} = 4\,\pi\, R_p^2\, \sigma\, T_e^4. \tag{11.32}$$

Under steady-state conditions absorption and emission are equal and the temperature T can be calculated:

$$T_e = \left(\frac{L_\odot(1-a)}{16\,\pi\,\sigma\, d_\odot^2} \right)^{1/4}. \tag{11.33}$$

Note that the temperature of a planet does not depend on its size. Under the given assumptions, it is only determined by the solar luminosity $L_\odot$, the planetary albedo a, and the distance $d_\odot$ from the Sun.

Figure 11.5 shows the dependence of a planet's temperature on the distance for different albedos (upper panel) and luminosities (lower panel). The distance is given in astronomical units covering the range of the planets from 0.38 AU

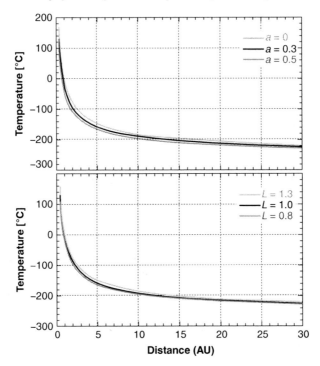

Fig. 11.5. Dependence of planetary temperatures in the absence of an atmosphere on the distance from the Sun (in astronomical units) for different values of the albedo a (upper panel) and the stellar luminosity L_*, which is assumed to be radiated isotropically (lower panel).

(Mercury) to 30 AU (Neptune). The luminosity is given in units relative to the present. In the upper panel the luminosity is set to 1 (present value) and in the lower panel an albedo of 0.3 is assumed. Note that the key parameter is the distance from the Sun. Both albedo and luminosity play only a small role.

In Table 11.3 the calculated equilibrium temperatures for the eight planets in the absence of atmospheres are compared to the measured ones. For each planet a lower value with an albedo of 0.5 and a luminosity of 0.8, an average value with $a = 0.3$ and $L = L_\odot$, and an upper limit with $a = 0.1$ and $L = 1.3L_\odot$ are given.

Overall there is a reasonable agreement between the estimated and the observed temperatures. The largest discrepancy is observed for Venus. The reason is that Venus has a very dense atmosphere which consists of 96% carbon dioxide with clouds of sulfur dioxide generating the strongest greenhouse effect in the solar system. In the case of Earth, the difference between calculated (using the present values $a = 0.3$ and $L = 1$) and measured mean global temperature is $33\,°C$. This difference is also due to the natural greenhouse effect (see Section 16.2). It is important to note that the Earth needs the natural greenhouse effect to be

Table 11.3. *Comparison of the calculated temperatures of the planets for different combinations of planetary albedo a and stellar luminosity L in the absence of atmospheres, compared with the observed temperatures*

Planet	Distance (AU)	Temperature (°C)			
		$a = 0.5$ $L = 0.8$	$a = 0.3$ $L = L_\odot$	$a = 0.1$ $L = 1.3L_\odot$	Observed
Mercury	0.38	77	130	175	180 to 420
Venus	0.72	−10	30	66	460
Earth	1	−50	−18	11	15
Mars	1.52	−95	−65	−40	−87 to 5
Jupiter	5.2	−175	−160	−150	−110
Saturn	9.54	−200	−190	−180	−180
Uranus	19.18	−220	−215	−210	−210
Neptune	30.06	−230	−225	−220	−210

habitable, but not necessarily an additional anthropogenic increase. The range of observed temperatures on Mars is very large because Mars has only a very thin atmosphere (0.3 hPa compared to 1000 hPa of Earth) and no liquid water to transport and distribute energy. Jupiter is considerably warmer than calculated (−110 °C instead of −160 °C). Most likely, this difference is due to gravitational compression which provides an additional power at least as large as the solar insolation.

11.3 Variability of influences

So far, we have assumed that all the conditions determining the climate of a planet are constant. Obviously this is not the case in reality. There are many different sources of variability which ultimately cause deviations from the steady-state conditions or, in other words, climate variability. If we stick to our simplified approach by Eq. (11.33), we already know that the temperature of a planet depends strongly on the luminosity, the distance from the Sun, and the albedo.

11.3.1 Luminosity

The luminosity is the total power emitted over the entire electromagnetic spectrum by the Sun. Figure 11.4 shows that the luminosity increases smoothly and very slowly with solar age. This is due to the fusion process in the core, which is very stable on time scales of millennia. The low luminosity after the formation of the solar system about 4 Gyr ago poses an interesting question called the "faint young

Sun paradox". According to Eq. (11.33) a reduction of the luminosity by 25% leads to a decrease of the mean global temperature on Earth by 18 °C. Under the present conditions such a temperature drop would turn the Earth into a "snowball" with a much larger albedo. This would make a transition to current conditions rather difficult. The generally accepted main reason why snowball conditions did not happen when the Sun was young is a higher content of greenhouse gases in the atmosphere at that time. For a detailed discussion of the topic of "habitability" of planets we refer to Chapter 4.

From the derivative of Eq. (11.27) we obtain

$$\frac{dT_e}{T_e} = \frac{1}{4} \frac{dL_\odot}{L_\odot}. \tag{11.34}$$

The relative change in temperature is 1/4 of the relative change in L. In other words, a change of $L_\odot$ of 0.1% as typically measured between solar minimum and maximum during an 11-year Schwabe sunspot cycle corresponds to a temperature change of the photosphere of about 1.5 K.

It is believed that the radiative energy transport is very stable on time scales shorter than 100 000 years. It is not known to what degree this is also true for the convective transport. However, it cannot be excluded that the magnetic fields generated by the dynamo at the tachocline below the convective zone have some influence on the convection (Kuhn, 1988; Kuhn and Libbrecht, 1991). The observed changes in the annual mean emission from the photosphere account for only about 0.1 % during an 11-year Schwabe cycle (see Fig. 10.6), and there-fore even very small fluctuations of the convective energy transport could have comparable effects.

By far the largest part of the solar power is emitted by the photosphere in the form of electromagnetic radiation. The spectrum resembles that of a blackbody with a temperature of about 5770 K. Only in the UV region of the spectrum are there larger contributions from very high temperatures in the corona, probably induced by reconnections of strong magnetic field lines. The total electromagnetic radiation arriving at the top of the Earth's atmosphere perpendicular to an area of 1 m^2 at the distance of 1 AU is called total solar irradiance (TSI). Its spectral distri-bution is called the solar spectral irradiance (SSI). Direct satellite-based monitoring of TSI over the past 30 years reveals clear variations in phase with magnetic activ-ity of the 11-year Schwabe cycle (see Fig. 10.6; Fröhlich and Lean, 2004; Fröhlich, 2006). The TSI curve is a composite of corrected data from different instruments as indicated by different colors. There are three different composites based on dif-ferent data and corrections. Although different in the long-term trend depending on the applied corrections and the instruments used, all composites show consistently lower values for the present solar minimum than for the previous ones. A detailed discussion of the measurements and models of the TSI is given in Chapter 10.

Simple models describing the TSI as the sum of a constant quiet-Sun component, a positive component due to bright faculae and the magnetic network, and a negative component composed of the dark sunspots and their penumbra are very successful in explaining all the observed short-term fluctuations on time scales from days to solar cycles (Unruh *et al.*, 1999; Solanki and Fligge, 2002; Krivova *et al.*, 2003; Wenzler *et al.*, 2006). However, it is not yet clear whether these models are also applicable to periods of much lower solar activity such as the Maunder minimum, when almost no sunspots were observed for about seven decades. The most recent decline in TSI since 2006 raises some serious doubts. Other potential sources of variability in the solar emission are changes in the solar radius and anisotropic emission. The solar radius is a crucial parameter (Sofia and Li, 2006). However, observations do not provide clear evidence for changes in the radius so far (Thuillier *et al.*, 2005). Clarification is expected from the Picard mission (Thuillier *et al.*, 2006). Even without changes in the luminosity, anisotropic emission of the total power can lead to changes in the TSI. The fact that sunspots and faculae are more prevalent at lower latitudes clearly points to an anisotropic emission. Whether this is also true for the solar disk free of visible magnetic activity remains to be verified.

11.3.2 Distance

As we have already mentioned, the distance from the Sun D is a prime parameter for the temperature of a planet. The first reason is that the solar power decreases with the square of the distance or in other words that the relative change of the temperature is 1/2 of the relative change of the distance:

$$\frac{\mathrm{d}T}{T} = -\frac{1}{2}\frac{\mathrm{d}D}{D}. \tag{11.35}$$

The second reason is that the distance of the planets ranges over almost two orders of magnitude from 0.38 AU (Mercury) to 30 AU (Neptune). Because all the planets have elliptical orbits the distance is continuously changing. The eccentricity (see Fig. 11.1) ranges from 0.0068 for Venus to 0.2056 for Mercury. The eccentricity of the Earth's orbit is 0.017. That means the distance between Earth and Sun is 1.017 AU at aphelion compared to 0.983 AU at perihelion. This difference results in a change of insolation by about 100 W m^{-2}.

11.3.3 Orbital or Milankovic forcing

As we have discussed, all the bodies in the solar system are gravitationally coupled. This has been known since Newton's time. Adhémar (1842) and Croll (1875) were among the early pioneers in this field. However, it was Milutin Milankovic (1930) who, for the first time, worked out the mathematical details of these disturbances, followed more recently by Berger (1978) and Laskar *et al.* (2004). It took a long time until the scientific community accepted this theory. The breakthrough was achieved by the success of the orbital theory in explaining the ice ages of the past several hundred thousand years. There are three orbital parameters of the Earth that are affected by the other planets, the Sun, and the Moon.

Orbital eccentricity   The elliptic nature of the Earth's orbit around the Sun means that the distance between Sun and Earth continuously changes between 0.983 AU at perihelion and 1.017 AU at aphelion. Integration over a full annual cycle reveals the following relationship between the relative change in the total amount of solar radiation S received by Earth and the relative change in the eccentricity e:

$$\frac{\mathrm{d}S}{S} = e^2/(1 - e^2)\,\frac{\mathrm{d}e}{e}. \tag{11.36}$$

The largest change in e (0.06) that the Earth experienced over the past million years (Fig. 11.6) therefore leads to a very small change of 0.36% in the annual mean

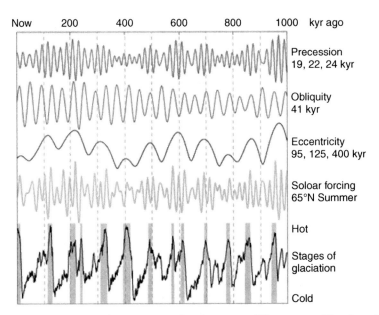

Fig. 11.6. Earth's orbital parameters for the past million years. The first three panels show the three orbital parameters influenced by the other planets (mainly Jupiter and Saturn) and the Moon (precession). The fourth panel exhibits the calculated solar forcing at 65° N. The lowermost panel shows sea-level changes derived from stable isotope measurements on benthic foraminifera indicating glacial (cold) and interglacial (warm, grey bands) periods.

insolation, which corresponds to a mean global forcing of less than $1 \, \text{W} \, \text{m}^{-2}$. The changes in the eccentricity occur on time scales of 100 000 and 400 000 years. It is interesting to note that it is exactly this small change in the eccentricity that seems responsible for the 100 000-year cycle in the sequence of glacial and interglacial periods during the past 1 000 000 years (Fig. 11.6). This is a nice example that climate is a non-linear system and that even a small forcing can cause a large effect if feedback mechanisms are involved. Such a feedback mechanism could be that, although a larger eccentricity does not change the mean annual insolation much, with it the seasonality changes: colder summers in the Northern Hemisphere may result in a reduced melting of the winter snow, enlarging the ice sheets and the albedo which further reduces the insolation (Kawamura *et al.*, 2007).

The obliquity   The tilt angle of the Earth's spin axis relative to the ecliptic plane varies between 22.1° and 24.5° with a periodicity of about 41 000 years. Contrary to the eccentricity changes the obliquity does not change the total amount of received solar radiation but only its latitudinal distribution. The larger the obliquity the stronger is the seasonality. A smaller obliquity reduces both the mean insolation and the summer insolation at high latitudes, thereby providing favorable conditions for ice ages.

Precession   The precession is a wobbling of the Earth's axis of rotation caused by the tidal forces associated with the Moon and the Sun. Because the Earth is spinning, its shape deviates slightly from a sphere leading to an equatorial bulge. Tidal forces act on the bulge and force the axis to precess. The periods of precession range from 19 000 to 24 000 years.

The calculated values of the three orbital parameters are plotted in Fig. 11.6, together with the corresponding summer insolation at 65° N, a latitude that is considered as critical for the formation of ice sheets as a result of cold summers. The bottom panel shows a compilation of $\delta^{18}\text{O}$ records from deep-sea sediments (Lisiecki and Raymo, 2005). Benthic foraminifera live in the deep sea and form calcium carbonate shells. After death, the shells are buried in the sediment layer by layer for millions of years. Measuring the $^{18}\text{O}/^{16}\text{O}$ isotope ratio with a mass spectrometer relative to a standard, expressed as $\delta^{18}\text{O}$, reflects the sea level. Water evaporating from the sea preferentially contains the lighter molecules $H_2^{16}\text{O}$. If the evaporated water stays on the continents forming glacial ice sheets, the ocean becomes depleted in ^{16}O. Warm interglacial periods are indicated by grey bands. They normally last 10 000 to 20 000 years and occur with a typical periodicity of 100 000 years when the eccentricity is large.

Figure 11.7 shows the Fourier power spectrum for the calculated summer insolation record for the latitude of 65° N (fourth panel in Fig. 11.6). The main

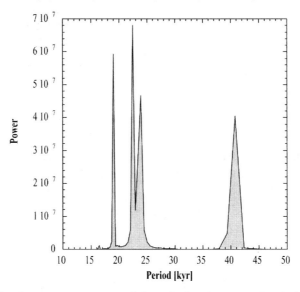

Fig. 11.7. Fourier power spectrum of the calculated summer insolation at 65° N in Fig. 11.6. The precessional periodicities between 19 and 25 kyr and the periodicity of 41 kyr due to the obliquity are clearly visible. The 100 kyr periodicity (eccentricity) is too weak to be seen.

periodicities due to precession (19–25 kyr) and obliquity (41 kyr) are clearly visible. However, the periodicities of the eccentricity (100 and 400 kyr) are not visible. This is in contrast with the observed climate record (lowermost panel in Fig. 11.6) characterized by the 100 kyr cycle of glacial and interglacial periods. As mentioned earlier this is a good example of the non-linearity of the climate response to forcing.

It is important to note that the changes in the distribution of the incoming solar radiation are complex and large. This is shown in Fig. 11.8, where the changes in insolation are shown as functions of latitude and time for the past 100 kyr and the future 20 kyr. The first panel shows the three orbital parameters while the other panel depicts the changes relative to the mean values for the insolation in June, in December, and the June–December difference as a measure of the seasonality. The temperature is given by the color code. There are many interesting details to be seen. For example, the insolation in June in the Northern Hemisphere was very low at 20 kyr BP (before present). This coincides with the last glacial maximum that was followed by a strong warming. A unique property of the orbital forcing is that it allows us to make reliable predictions for some million years. The coming 20 kyr are characterized by relatively small changes in orbital forcing. This relates to the decreasing trend in eccentricity which affects the seasonality.

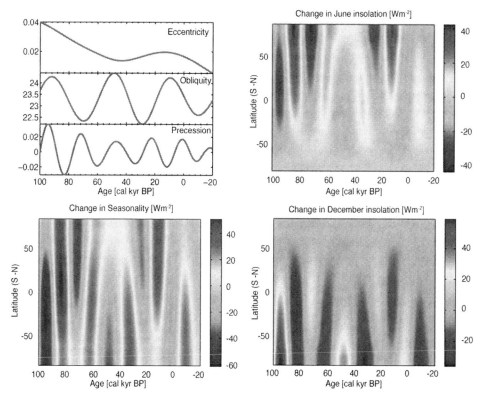

Fig. 11.8. Changes in the orbital parameters of Earth and their effect on the summer (June), the winter (December) and the seasonal (June–December) insolation for the past 100 kyr and the future 20 kyr (−20 kyr BP). Shown are the deviations in W m^{-2} from the mean values. Note the large changes at high latitudes. See Color Plate 12.

To see how well the orbital forcing of the past 100 kyr is reflected in paleoclimatic records we compare in Fig. 11.9 the δ^{18}O record from the GRIP ice core drilled in central Greenland at 72 °N (Dansgaard *et al.*, 1993) with the summer insolation at this latitude. In this example δ^{18}O is a measure of the atmospheric temperature when water vapor condenses and forms snow flakes. The overall agreement between the two curves is good as far as the long-term trend is concerned. The glacial period is characterized by strong, very rapid changes (so-called Dansgaard–Oeschger events) that are most probably related to abrupt changes in the thermohaline circulation of the ocean. During the Holocene (the past 11 kyr) the climate was comparatively stable and has not clearly followed the insolation curve at this particular site. Other paleoclimate records do reflect the Holocene trends (Wanner *et al.*, 2008). This is another implication of the non-linearity of the climate system: the response to an external forcing shows spatial and temporal variability

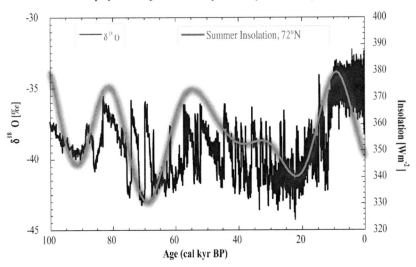

Fig. 11.9. Comparison of the $\delta^{18}O$ record from the GRIP ice core in central Greenland (72° N) with the corresponding summer insolation for the past 100 kyr. (Dansgaard *et al.*, 1993.)

that can only be understood by using coupled ocean–atmosphere global-circulation models.

11.3.4 Albedo

The albedo is defined as the ratio of diffusely reflected to incident electromagnetic radiation and, therefore, lies in the interval 0–1. It is difficult to determine the total albedo of a planet because it is highly variable, ranging from less than 0.1 for water and forests to more than 0.8 for fresh snow. On Earth, the largest contribution comes from the clouds that cover about 50% of its surface. For the Earth an average albedo of 0.3 is usually assumed. Interestingly, the albedo gets much less attention than the TSI although both are equally important as far as solar forcing is concerned. The albedo of clouds plays a central role in the cosmic-ray cloud hypothesis put forward by Danish scientists (Svensmark, 1998). They claim that the Earth's cloud cover is modulated by the cosmic-ray-induced ion production in the atmosphere. Later they reduced the effect to low-altitude, low-latitude clouds (Marsh and Svensmark, 2003). This issue is still debated in papers supporting (Usoskin, 2008) and contradicting (Kernthaler *et al.*, 1999; Wagner *et al.*, 2001) cosmic-ray-induced climate change. Other climate relevant effects related to strong atmospheric electrical currents have been proposed by Tinsley (2000). Although there is no doubt that many more different effects take place in the atmosphere, so far there is no clear evidence that these processes play a significant role in global climate change.

11.4 Reconstruction of long-term solar variability

11.4.1 Proxies of solar activity

There are several proxies of solar activity that are based on direct measurements or on observations and are therefore called direct proxies. They have in common that they are all related in different ways to magnetic processes on the Sun with temporal resolutions of weeks to years and that they are generally more reliable for more recent times. Depending on their length, all the records show similar features (Fig. 11.10): the 11-year Schwabe sunspot cycle, a long-term trend from

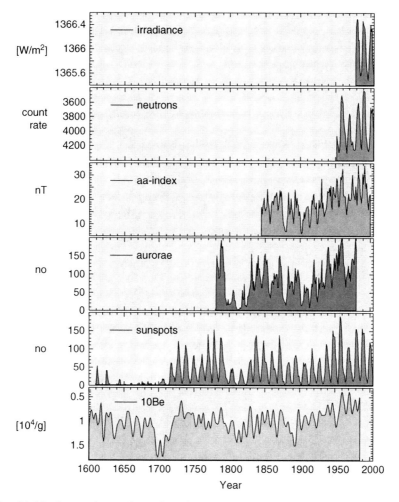

Fig. 11.10. Comparison of proxies of solar activity. Each proxy reflects a specific aspect of solar activity. As a result, all proxies show some common features (11-year cycle, trends, solar minima), but also differences. The length is limited to the periods of instrumental observations and ranges from 30 (TSI) to about 400 years (sunspots). ^{10}Be is the only record that is not based on direct observations and, therefore, has the potential to be extended over at least 10 000 years.

1900 to 1950 with a generally increasing activity with time, and some periods (1–7 decades) with reduced solar activity. The most pronounced of the latter is the Maunder minimum (1645–1715), which is characterized by an almost complete absence of sunspots (cf. Section 2.8; Eddy, 1976). A more detailed comparison shows that, apart from these common features, there are also clear differences, mainly due to the fact that all these proxies are related in various ways to different aspects of magnetic processes taking place on the Sun. For a detailed discussion of the solar magnetic processes we refer to Chapter 2. The geospace climate is the subject of Chapter 14.

The only record in Fig. 11.10 that is not based on direct observations is the ^{10}Be record (Beer *et al.*, 1994). ^{10}Be and other cosmogenic radionuclides, such as ^{14}C and ^{36}Cl, offer the unique opportunity to extend the reconstruction of solar activity back to at least 10 000 years.

11.4.2 Cosmogenic radionuclides

Cosmogenic radionuclides are produced by nuclear interactions of the galactic cosmic rays (GCR) with atoms (N, O, Ar) in the atmosphere; the effect of solar cosmic rays is negligible because of their low energies. To reach the atmosphere, the GCR have to propagate through the heliosphere, which forms a bubble with a radius of about 100 AU around the Sun that is filled with solar plasma-carrying magnetic fields. The propagation of the cosmic rays is described by the transport equation derived by Parker (1965). For a detailed description of the heliosphere and the particle transport we refer to Vol. I, Chapter 9, and to Chapter 9 in this volume. It is difficult to use the transport equation to parameterize the intensity of the GCR; however, the so-called force-field approximation (Gleeson and Axford, 1967) has proven to be a good approximation near Earth. This approximation describes the modulation effect of the Sun on the energy spectrum of the GCR in terms of a parameter Φ called the solar modulation function. The solar modulation function Φ basically corresponds to the average energy lost by a cosmic-ray proton on its way to the Earth (cf., Section 9.5, Eq. (9.8)).

Figure 11.11 shows the differential energy spectrum of the GCR proton flux for different levels of solar activity. A value of $\Phi = 0$ MeV corresponds to the local interstellar spectrum outside the heliosphere. This spectrum is an estimate because no space probe has left the heliosphere yet and actually measured this spectrum; Voyagers 1 and 2 are close, as they have crossed the termination shock and are passing through the heliosheath. Figure 11.11 shows that the shielding effects of the open solar magnetic field and the advecting solar wind are most pronounced at the low-energy end of the spectrum. As a consequence, GCR particles above about 20 GeV are hardly affected by the varying heliospheric magnetic field.

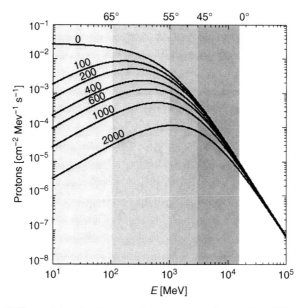

Fig. 11.11. Differential galactic cosmic-ray proton fluxes for different levels of solar activity ranging from a value of the solar modulation function $\Phi = 0\,\mathrm{MeV}$ (Eq. 9.8), corresponding to the local interstellar spectrum arriving at Earth without any solar influence, to $\Phi = 2000\,\mathrm{MeV}$ which corresponds to a very active Sun. There are similar curves for cosmic-ray alpha particles and heavier nuclides. The vertical bands illustrate the effect of the geomagnetic field, which cuts off all protons approaching vertically with an energy below about 100 MeV for a geomagnetic latitude of 65°; below 1 GeV for 55°, and below 3 GeV for 45°. At 0° the cutoff energy is 13.9 GeV for the present geomagnetic field.

Before reaching Earth, the cosmic-ray particles have to overcome a second barrier, the geomagnetic field. This field prevents particles with too low rigidity (momentum per unit charge) from reaching the top of the atmosphere. In a first approximation, the geomagnetic field is considered as a dipole and in this case the cutoff rigidity depends only on the angle of incidence and the geomagnetic latitude. At low latitudes the cutoff rigidity for vertical incidence is presently ∼14.9 GV. This means that a cosmic-ray proton needs a kinetic energy of at least 14 GeV ($14.9\,\mathrm{m_p}c^2$) to reach the top of the atmosphere (see shaded bands in Fig. 11.11). The solar modulation is a monotonically decreasing function of particle energy and consequently the modulation is small near the equator (∼14 GeV) and large at high latitudes which are accessible to the strongly modulated energies near 1 GeV.

If a primary cosmic-ray particle makes its way through the heliosphere and the geomagnetic field and enters the atmosphere it will interact quickly with an atom of oxygen, nitrogen, or argon. Because the energies of incoming particles are generally very high, only part of their kinetic energy is transferred to the first atom

Table 11.4. *Main properties for some cosmogenic radionuclides includ-*
ing nuclear production reactions and mean global production rates
for the present geomagnetic field strength and a solar modulation of
$\Phi = 550 \, MeV$

Isotope	Half life (y)	Decay	Target	Nuclear reaction	Prod. rate $(cm^{-2} s^{-1})$
^{14}C	5730	β^-	N, O	$^{14}N(n, p)^{14}C$ $^{16}O(p, 3p)^{14}C$ $^{16}O(n, 2p1n)^{14}C$	2.02
^{10}Be	1.5×10^6	β^-	N, O	$^{14}N(n, 3p2n)^{10}Be$ $^{14}N(p, 4p1n)^{10}Be$ $^{18}O(n, 4p3n)^{10}Be$ $^{18}O(p, 5p2n)^{10}Be$	0.018
^{36}Cl	0.30×10^6	β^-, EC	Ar	$^{40}Ar(n, 1p4n)^{36}Cl$ $^{40}Ar(p, 2p3n)^{36}Cl$ $^{36}Ar(n, p)^{36}Cl$	0.0019

EC: electron capture.
All nuclear reactions are induced by high-energy secondary particles generated
by the primary cosmic-ray particles (so-called spallation reactions). The only
exception is ^{14}C, which is almost totally produced by thermal neutrons
interacting with nitrogen (Masarik and Beer, 1999, 2009).

they hit. They continue their travel and hit a few more atoms until their energy is
dissipated. Each collision results in the generation of secondary particles covering
the full spectrum of hadrons and leptons, which either decay or interact with other
atoms of the atmosphere. In this way, a cascade of secondary particles develops
which can be simulated using the Monte Carlo technique (Masarik and Beer, 1999,
2009). Table 11.4 shows the different production reactions for the radionuclides
^{14}C, ^{10}Be, and ^{36}Cl, and the resulting mean global production rates for the present
geomagnetic field intensity and a solar modulation function equal to 550 Mev.

The simulations show that the majority of the secondaries are neutrons followed
by protons. Both, in turn, collide with atmospheric atoms initiating spallation reactions (Masarik and Beer, 1999, 2009; Webber and Higbie, 2003), that generate the
cosmogenic nuclides that are archived for us in ice (^{10}Be, ^{36}Cl) or tree rings (^{14}C).
In addition, the cosmic-ray produced neutrons have been monitored continuously
since 1951 by so-called neutron monitors. In panel 2 from the top of Fig. 11.10
the count rate of the Oulu neutron monitor clearly shows the modulation of the
GCR by the 11-year Schwabe sunspot cycle. Whenever the magnetic activity is
high (large sunspot numbers) the shielding is strong and the neutron flux is low.

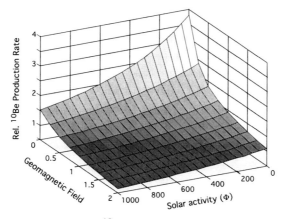

Fig. 11.12. Dependence of the ^{10}Be production rate on the geomagnetic field intensity (in units relative to the today's field) and the solar activity (expressed by the solar modulation function Φ, Eq. (9.8)). The production rate is normalized to the present strength of the geomagnetic field and solar activity corresponding to a solar modulation function of 550 MeV (matching the long-term average).

As we discussed above, the solar modulation of the GCR can be described by the modulation function Φ. Many studies have shown that the 11-year and longer-term variations are faithfully reproduced in the cosmogenic data, and they and the neutron monitor data have been inter-calibrated to yield a continuous cosmic-ray record for the past 10 000 years (McCracken and Beer, 2007).

As an example, the combined effect of solar activity and geomagnetic field on the relative production rate of cosmogenic nuclides is shown for ^{10}Be in Fig. 11.12. The relative dipole component of the geomagnetic field M varies between 0 and 2, 1 being the present field. For $M = 1$ and $\Phi = 550$ MeV (long-term average), the ^{10}Be production rate is normalized to 1. It should be noted that the dependence of the production rate on M and Φ is non-linear. A more detailed overview on the different aspects of cosmogenic nuclide production in the atmosphere is given in Masarik and Beer (1999, 2009).

11.4.2.1 Transport and deposition

The fate of a cosmogenic nuclide after its production in the atmosphere depends strongly on its geochemical properties. Within a short time, ^{10}Be becomes attached to aerosols and follows their pathways. ^{14}C on the other hand oxidises to ^{14}CO$_2$ and exchanges between atmosphere, biosphere, and ocean. After a mean residence time of 1 to 2 years, ^{10}Be is removed from the atmosphere mainly by wet precipitation. On a large scale, the flux F of cosmogenic nuclides from the atmosphere into a polar ice sheet is proportional to the atmospheric production rate P:

$$F = \alpha\, P. \tag{11.37}$$

Locally and temporally, α can vary due to changes in the atmospheric transport and deposition processes. The degree of variability depends very much on how well the atmosphere is mixed. In the case of ^{14}C, the large atmospheric $^{14}CO_2$ reservoir leads to an atmospheric residence time of 6 to 7 years and therefore to a complete mixing. In the case of the aerosol-bound nuclides the residence times are shorter, roughly $1 - 2$ years. Mixing in the troposphere is no longer complete. After deposition, some of the nuclides become incorporated into natural archives such as ice sheets, glaciers, sediments, and tree rings. Important aspects of the physical and chemical processes taking place in the atmosphere are discussed in Chapters 13, 15, and 16.

For our purpose, a useful archive stores the complete flux of nuclides from the atmosphere in a stratigraphically undisturbed way and records the time accurately. Excellent archives in this respect are ice sheets, which directly collect the atmospheric precipitation containing ^{10}Be. Typically, they cover the last several hundred thousand years with a time resolution per sample ranging from one year at the top to decades or centuries near the bottom. However, due to the flow characteristics of ice, dating is difficult, especially in the deeper part of ice cores.

Tree rings represent an ideal archive for the atmospheric $^{14}C/^{12}C$ ratio. So far, by chronologically matching trees of different ages, the atmospheric $^{14}C/^{12}C$ ratio has been reconstructed back to approximately 12 kyr BP (Stuiver *et al.*, 1998). Potentially, the full range covered with today's measuring techniques (40 to 50 kyr) will be traceable in tree rings in the future. What can be learned by measuring cosmogenic nuclides in ice? The answer to this question is summarized in Fig. 11.13. In an archive, changes in the concentration can result from changes either in the production rate or in the Earth system processes (transport and deposition). Changes due to radioactive decay can be corrected for, if a reliable time scale is available. Changes in the production rate can be caused by heliomagnetic and geomagnetic modulation of the cosmic-ray flux. Episodic solar proton events can cause short but intense cosmic radiation, but do not contribute much to the total production rate due to the low proton energies. Changes in the system on the other hand are related to the atmospheric transport and mixing processes as well as to the local precipitation rate.

Now, the question arises how the different causes of concentration changes can be separated. A straightforward answer to this question is to combine several nuclide records from different sites. Comparing ^{10}Be with ^{14}C permits separating production from system effects. Changes in the production rate due to helio- and geomagnetic modulation of the cosmic-ray flux are reflected both in ^{10}Be and in ^{14}C in a very similar way. Changes within the Earth system, however, are expected to affect ^{10}Be and ^{14}C in a completely different way because the geochemical behavior of these nuclides is fundamentally different. One way to test whether a

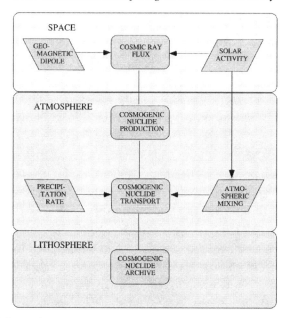

Fig. 11.13. Schematic overview of the production of cosmogenic nuclides and the processes influencing their concentration found in an archive such as a polar ice sheet.

^{10}Be signal reflects a production signal is to use the ^{10}Be data as a proxy for the ^{14}C production rate in a carbon cycle model (Oeschger *et al.*, 1975) and to calculate the corresponding ^{14}C response signal. The degree of similarity between the calculated and the measured ^{14}C signal is a measure of the strength of the production signal. Of course, this method is limited to the range of the radiocarbon dating (last 50 kyr) and requires a high-precision Δ^{14}C record that is not yet available for the period 13–50 kyr BP. The next step is to separate heliomagnetic and geomagnetic signals. In principle, these two signals could be separated by looking at two radionuclide records, one from the equator and one from the regions of the magnetic poles. Without latitudinal atmospheric mixing, the record from the magnetic pole would only reflect solar modulation because geomagnetic shielding disappears at high latitudes, whereas the signal in the equatorial record would be dominated by geomagnetic modulation. However, as a result of atmospheric mixing, this is not the case.

Solar modulation effects have been found in cores from Greenland (Beer *et al.*, 1990, 1994; Wagner *et al.*, 2001) and Antarctica (Beer *et al.*, 1991). The same is true for geomagnetic modulation effects such as the Laschamp event at about 40 kyr BP, when the magnetic dipole field was close to zero (Baumgartner *et al.*, 1998; Wagner *et al.*, 2000a,b). This event is present in the high-latitude ice-cores from the Arctic and from Antarctica (GRIP, Vostok, Byrd, Dome C; Raisbeck *et al.*,

1987; Beer *et al.*, 1992). Radionuclide records from low-latitude ice-cores are still rare (Thompson *et al.*, 1997).

Another approach is to assume that solar modulation effects generally occur on shorter time scales than geomagnetically induced production changes. Applying low-pass filters with cutoff frequencies in the range of 1/2000 and 1/3000 yr^{-1} on cosmogenic nuclide fluxes provides production signals in good agreement with paleomagnetic intensity records based on remanence measurements (Laj *et al.*, 2000).

The task of separating the different causes of variability observed in radionuclide records is complicated by the fact that some of the causes are coupled. For example, changes in solar activity affect atmospheric processes and possibly also, to a smaller extent, climatic changes (Tinsley, 1996). Therefore, additional information from other measured parameters should be included to obtain a complete and consistent picture of what happened during the period of investigation. In the following, we discuss how using the intensity of the geomagnetic dipole field and the solar variability can be derived from cosmogenic nuclides.

11.4.2.2 Geomagnetic field

In Fig. 11.14a a compilation of ^{10}Be data from the GRIP and the GISP ice cores drilled in central Greenland are presented covering the past 60 000 years (Muscheler *et al.*, 2005). To correct for the lower precipitation rate during glacial times (10–60 kyr BP) the ^{10}Be flux has been calculated and smoothed (grey band). The plot shows a significant peak at about 40 kyr BP. To check whether the smoothed curve does reflect the geomagnetic dipole field as expected from Fig. 11.14, the corresponding changes in the dipole field intensity have been calculated based on its relationship with the ^{10}Be production shown in Fig. 11.12. The result is compared in Fig. 11.14b with the completely independent reconstruction NAPIS-75 (Laj *et al.*, 2000), which was derived from remanence measurements in Atlantic sediment cores. Overall the agreement is good and confirms that the ^{10}Be peak at 40 kyr BP corresponds to the Laschamp event when the dipole field intensity was almost zero but did not reverse. An overview of planetary magnetic fields is given in Chapter 7; see also Chapter 14.

11.4.3 Solar variability

We return now to solar variability and discuss to what extent cosmogenic radionuclides can expand our knowledge about long-term solar variability. In a first step we compare annual ^{10}Be data with the sunspot record, which represents the longest observational data of solar variability. A resolution of one year is about the limit because it corresponds to the mean travel time for a ^{10}Be atom produced in the

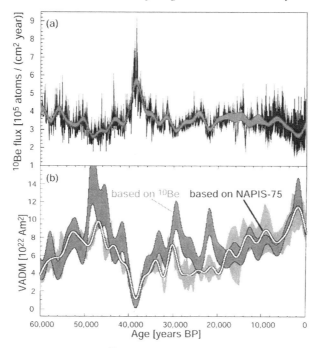

Fig. 11.14. Comparison of (a) ^{10}Be data with (b) the geomagnetic dipole field for the past 60 000 years. Panel (a) shows a compilation of ^{10}Be data from the GRIP and GISP ice cores in Greenland. Panel (b) compares the dipole field derived from ^{10}Be (panel a) to that from remanence data (NAPIS-75) measured in ocean sediment cores (Laj *et al.*, 2000). (Modified after Muscheler *et al.*, 2005.)

atmosphere to reach the Earth surface where it is stored in, for example, an ice sheet. Figure 11.15 shows a comparison of the ^{10}Be concentration from Dye 3, Greenland, with the sunspot number. Both records have been band-pass filtered (8–16 yr). While during the Maunder minimum (shaded area between 1645 and 1715) hardly any sunspots could be observed, the solar dynamo clearly continued to produce open magnetic field modulating the cosmic rays and the ^{10}Be production (see also Fig. 10.4). The present status of the solar dynamo is discussed in Chapters 5 and 6.

The overall good agreement between ^{10}Be and sunspots gives us confidence to extend the time interval over the Holocene, i.e. about the last 10 000 years. During this period the climate was relatively stable compared to glacial times and therefore we can assume that transport and deposition effects did not disturb the production signal in the ^{10}Be record. This assumption is confirmed by GCM model runs which show that the transport effects were relatively stable during the climatic conditions prevailing during the Holocene. So to a first approximation they can be neglected (Heikkila *et al.*, 2008). This is not the case for the geomagnetic field,

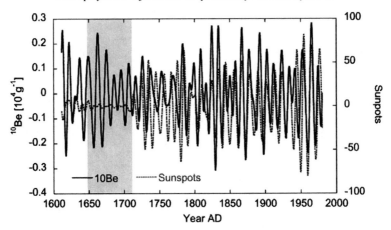

Fig. 11.15. Comparison of the ^{10}Be concentration measured in the Dye 3 ice core from Greenland (Beer *et al.*, 1994) with the sunspot number after applying a band-pass filter (8–16 years). Note that during the Maunder minimum 1645–1715 (shaded area), when almost no sunspots were observed, ^{10}Be shows a clear 11-year Schwabe cycle.

which exhibits significant long-term changes (Muscheler *et al.*, 2005; Vonmoos *et al.*, 2006).

Using our Monte Carlo simulations (Masarik and Beer, 1999, 2009), the effect of the geomagnetic dipole field has been removed and we are left with the solar modulation function Φ (Fig. 11.16). The GRIP ice-core record is limited to the period from 1640 to 9300 BP and has recently been complemented by the most recent 360 years, which are a composite of Φ derived from neutron monitor data and those from a shallow ice-core (Steinhilber *et al.*, 2008). The data of Fig. 11.16 have been low-pass filtered with a 150-year cutoff. The most striking features of the Φ record are the many distinct minima which correspond to grand solar minima such as the Maunder (M), Spörer (S), Wolf (W), and Oort (O). The fact that Φ never reaches zero means that there is always some residual open magnetic flux; in other words the solar dynamo seems to weaken from time to time, but, as a close inspection of the unfiltered data shows, it never stops. The two exceptions in Fig. 11.16 are due to uncertainties in the data.

The maxima are less pronounced. It is interesting to note that the present level of solar activity is comparatively high, although there were earlier periods with similar or possibly even higher activity around 2000, 4000, and 9000 BP. There is also a clear long-term trend indicated by the thick line that is low-pass filtered with a cutoff of 1000 years.

For a more detailed analysis, we calculate the power spectrum using wavelet analysis (Grinsted, 2002–2004). Figure 11.17 shows the wavelet spectrum of Φ. There are several distinct periodicities, some of which are listed in Table 11.5.

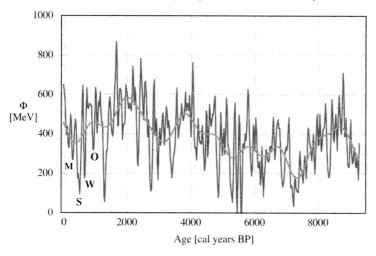

Fig. 11.16. Solar modulation function Φ from the present (0 BP corresponds to 1950) back to 9350 BP (Steinhilber *et al.*, 2008). The black curve shows data that have been low-pass filtered with a cutoff of 150 years; the smooth grey curve with 1000 years. The most recent solar minima are indicated: M: Maunder; S: Spörer; W: Wolf, and O: Oort.

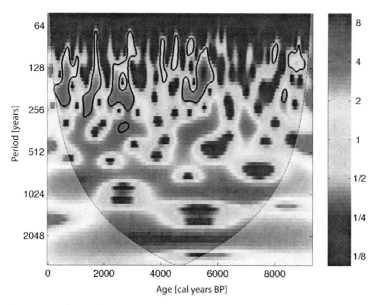

Fig. 11.17. Wavelet analysis (Grinsted, 2002–4) of the solar modulation function Φ from Fig. 11.16. The color scale is a measure of the spectral power relative to the spectral power of white noise, thus measuring signal significance. See Color Plate 11.

Table 11.5. *Prominent common periodicities (in years) found in the Φ record based on ^{10}Be in the GRIP ice core, the measured $\Delta^{14}C$ record from tree rings and the calculated ^{14}C production rate $Q^{14}C$ based on $\Delta^{14}C$ and a carbon cycle model*

Cycle / Period	Φ	$\Delta^{14}C$	$Q^{14}C$
Hallstatt	2194	2275	2424
	982	984	957
De Vries, Suess	207	208	208
	352	350	350
	704	714	713
	497	512	512
	105	105	105
Gleissberg	86	87.9	87.0

Since the time scales for ^{10}Be in ice cores are not as easily established as those for ^{14}C in tree-rings we also give the corresponding periodicities for Δ^{14}C (Reimer *et al.*, 2004) and Q^{14}C calculated for almost the same time interval (1750–9300 BP). Q^{14}C is the ^{14}C production rate, which was calculated using the Intcal04 calibration curve and the Siegenthaler–Oeschger carbon cycle model (Oeschger *et al.*, 1975). An interesting feature of Fig. 11.17 is that the cycles wax and wane during the Holocene: there are periods when most cycles show large amplitudes (between 2000 and 3000, and between 5000 and 6000 BP) and times when the amplitudes are generally low (between 3000 and 4000 BP).

11.4.3.1 Solar variability and climate change

Solar variability and climate change is the topic of Chapter 12. Therefore we only provide a short description of how the record of solar activity derived from cosmogenic radionuclides can be used to estimate the long-term variability of the TSI. Then we use reconstructed advances and retreats of the largest glacier in the Alps as an example to illustrate how well solar forcing and climate response compare.

As we have discussed, the past solar activity expressed by the solar modulation parameter Φ can be reconstructed from ^{10}Be measurements in polar ice cores (see Fig. 11.16). Using basic physics it is possible to derive from Φ the open magnetic field (Steinhilber *et al.*, 2009). The difficult step is to relate the open magnetic field to the TSI. As discussed in Chapter 10, various approaches have been used by different authors. In a very recent one, Steinhilber *et al.* made use of the fact that both TSI and the open magnetic field show a linear correlation during the solar minima from 1978 to present (Fröhlich, 2009) and derived the first TSI record for the past 9300 years (Steinhilber *et al.*, 2010).

To illustrate the relationship between TSI and climate change we use as an example the fluctuations in size of the Aletsch glacier in the Swiss Alps. The size of a glacier is strongly determined by the winter precipitation and the summer temperature. It is rather inert and does not respond immediately to single climatic events, but records the climate changes averaged over the last few decades. By dating trees that were buried during glacier advances it is possible to reconstruct fluctuations in the length (Denton and Karlén, 1973; Holzhauser *et al.*, 2005; Hormes *et al.*, 2006; Joerin *et al.*, 2006).

Figure 11.18 shows the reconstruction of the length fluctuations of the Aletsch glacier (Holzhauser *et al.*, 2005). The length is given relative to current length. The curve shows that the Aletsch glacier was longer during the Little Ice Age (about 1350–1850 CE) by almost 3500 m. However, the present situation is not unique. There were earlier periods when its length was comparable to today's length. These times coincide with warm epochs such as the Roman and the Medieval Warm Period. However, taking into account the delayed response (corrected for

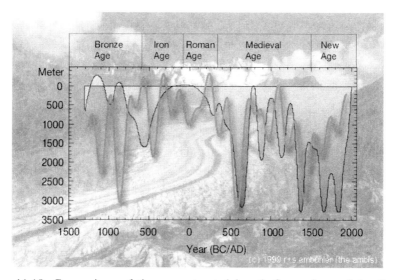

Fig. 11.18. Comparison of the reconstructed length fluctuations of the Great Aletsch glacier (shown in the background) relative to its present length (filled black contour) in the Swiss Alps (from Holzhauser *et al.*, 2005) with the reconstructed total solar irradiance (grey curve with shadow; from Steinhilber *et al.*, 2010). Lower TSI coincides generally with larger extensions of the glacier. Note that the dating of the record of glacier length fluctuations has some uncertainties. The TSI record is shifted by about 80 years to account for the time lag of the Aletsch glacier.

in Fig. 11.18), the Aletsch glacier does not yet reflect the global warming since 1970 and will therefore continue to melt in the future.

This example illustrates the difficulties we face when we try to attribute a climate record to solar forcing: the solar forcing has considerable uncertainties, the responses of the climate system in general and of the glacier length in particular are non-linear, the dating of the length fluctuations and the solar activity have uncertainties, and there are volcanic and other forcings involved as well. Unquestionably, complex Global Circulation Models are needed to account for all the non-linear couplings and feedback mechanisms within the climate system.

12

Assessing the Sun–climate relationship in paleoclimate records

Thomas J. Crowley

12.1 Introduction

One could write an interesting essay on the twists and turns in Sun–Earth climate science, addressing both the instrumental record and the much longer interval of paleoclimate records. The conclusion at the time of this writing with respect to the importance of low-frequency solar variability in the most recent decades, and perhaps up to centuries, might be "Perhaps, but probably small". The main reasons why uncertainties persist regarding this issue include:

(i) The ∼150-year instrumental record is too short to draw definitive statistical conclusions about the connection of any relation existing on the multi-decadal time scale.

(ii) Forcing from anthropogenic greenhouse gases represents a significant overprint on trends since about 1850 CE. Because to first order the trends in proxies for solar activity indices and in greenhouse gas concentrations are similar, there is a statistical degeneracy that leads to ambiguous, and thus potentially misleading, conclusions unless great care is taken.

(iii) A similar problem of statistical degeneracy applies to the Little Ice Age interval of cool conditions during the last millennium (main phase about 1450–1850 CE), when mountain glaciers advanced in many regions and planetary temperatures were about 0.5 °C lower (e.g. Jones and Mann, 2003; Hegerl *et al.*, 2007). During the Little Ice Age, solar activity, as inferred from changes in radiogenic isotopes such as ^{14}C and ^{10}Be, appears to have varied similarly to pulses in volcanism and slightly lower carbon dioxide (CO_2) levels (further discussed below). Ignoring this similarity in patterns of variability in internal and external (in planetary terms) climate drivers can lead to erroneous conclusions.

In this overview I give some examples that highlight what we can say from studies of the last two millennia, and conclude with some results for time scales longer than ∼500 years. I focus on inferred temperature responses in different

Heliophysics: Evolving Solar Activity and the Climates of Space and Earth, eds. Carolus J. Schrijver and George L. Siscoe. Published by Cambridge University Press. © Cambridge University Press 2010.

systems, ignoring the interesting but more disparate evidence for solar influence with respect to, for example, cosmic rays and clouds, western-US drought rhythms, and engineered systems (for more information on those aspects, see e.g. Cook *et al.*, 1997; Camp and Tung, 2007; Thomson *et al.*, 2007). In this chapter the focus is on trends and statistical arguments; Section 16.6 discusses detailed climate ocean–atmospheric modeling.[†]

12.2 The instrumental record of climate change

The main purpose of this chapter is to update conclusions based on the paleo-climate record (e.g. Hegerl *et al.*, 2007). Hence, I only briefly recount the status of the problem as it relates to the relatively short instrumental climate records. Figure 12.1 illustrates an updated assessment of the well-known radiative forcing from CO_2 through 2008 CE, along with a plausible, although not proven esti-mate of solar variability based on Lean (2000) and updated through 2008 CE by splicing in recent sunspot data and scaling to the 1970–2000 CE overlap interval in Lean (2000). Other greenhouse gases are not included, because to a

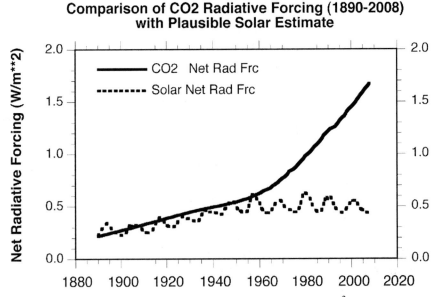

Fig. 12.1. Comparison of well-known radiative forcing $(W\,m^{-2})$ of the anthro-pogenic carbon dioxide increase (solid) to possible lower-frequency changes in solar irradiance (dotted).

[†] Many detailed discussions, and references, regarding the many processes involved in the evolving climate can be found in the reports of IPCC Working Group I, focusing on the physical science basis of climate modeling at http://www.ipcc.ch.

first approximation the radiative effects of these additional species appear to be largely countered by the reflecting effects of tropospheric aerosols (i.e. aerosols from smokestack emissions; compare Fig. 16.1, in particular the top and bottom entries).

An estimate of net radiative forcing changes from the Sun is also illustrated in Fig. 12.1 and based on the observed relationships between changes induced by the 11-year solar cycle over the satellite era (i.e. since 1979). About a 0.1% change in total solar irradiance is observed (see Chapter 10), which translates into 0.34 W/m^2 averaged over the surface of the Earth (i.e. multiplying the 0.1% difference by the area of the Earth facing the Sun divided by area of the whole Earth). After factoring in the additional effects of the Earth's planetary albedo (or wavelength-integrated reflectivity) due to cloud and ice cover (of order 30%, see Fig. 16.2), the net effect of changes in the 11-year sunspot cycle is about 0.24 W/m^2, assuming no significant long-term trend beyond the time scale of the sunspot cycle. For comparison, a doubling of the Earth's carbon dioxide level would cause about a 3.0–3.7 W/m^2 change in radiative forcing due to increased absorption of long-wave emission in the Earth's troposphere.

Also shown in Fig. 12.1 is a speculative lower-frequency trend in solar irradiance (Lean, 2000) that is postulated to occur as a result of longer-wavelength changes in the Sun (on a century scale or longer) based on an extrapolation of variations observed over the past few decades. Because climate models respond similarly to changes in short-wavelength and in infrared radiation, this comparison is as close to a like-for-like comparison as can be obtained using simple approximations as discussed in this chapter.

The figure clearly indicates that known net radiative forcing from CO_2 is more than three times larger than a possible change in solar forcing over the twentieth century (here, I momentarily suspend discussion of the revision of the solar estimate by Lean *et al.*, 2002). This is the first and most economical argument that should be used in weighing the relative importance of the two mechanisms for climate change in the twentieth century.

Next, I consider statistical arguments. Figure 12.2 illustrates the global temperature record from 1890 through 2008 versus known forcing converted to temperature, and a scaling of the solar forcing in Fig. 12.1 also to temperature. If one were to consider just correlations, then it would appear that solar forcing might indeed play a larger role than Fig. 12.1 suggests, because the correlation between the solar-irradiance curve and temperature could explain 49% of the variance in the latter. Such a conclusion would ignore a very big problem, however. Because the variance correlated with CO_2 is even larger at 85.6%, a naive assessment of the joint effect would be to conclude that the two forcings explain 134.6% of the covariance in the temperature record. Obviously, this is impossible; the paradox arises

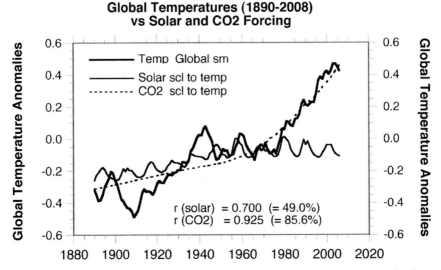

Fig. 12.2. Comparison of plausible temperature changes from radiative forcing of solar variability (thin solid) and carbon dioxide (dashed) against the 5-year smoothed global temperature record (thick solid). The panel shows the correlation coefficients (and, between parentheses, the percentage of the covariance) of temperature records and solar variability (r(solar)) and for CO_2 concentration (r(CO2)).

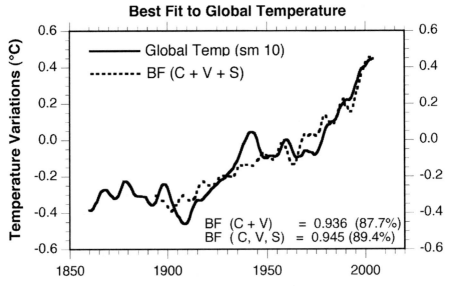

Fig. 12.3. Best-fit (BF) model (dotted) compared with the observed global temperature (solid) for three contributions: C for CO_2, V for volcanism, and S for solar. The correlation coefficients and relative covariances are given for a best fit with CO_2 and volcanism only and for a best fit including all three effects (the difference between the two compilations is so small that only the total forcing is included in the figure).

because one of the curves projects in part onto the other (mainly in the long-term trend).

By first choosing the forcing of the best-known records of CO_2 and volcanism (discussed further below), and then adding the effects of solar changes (Fig. 12.3), the problem of statistical degeneracy is addressed to first order, and the result suggests that the amount of additional covariance explained by solar forcing is likely only 1.7% – a very small, although potentially detectable, effect. Calculations by North and Stevens (1998) support this interpretation. They demonstrated that for the 11-year cycle there is a degeneracy between the sunspot cycle and volcanism in the twentieth century. Once total forcing is considered, the 11-year solar signal is not detected at a significant level in their analysis.

12.3 Results from the climate records for the past 2000 years

In this main section of the chapter, I present three tests that attempt to sort out the relative importance of solar forcing versus other sources of forcing for northern-hemispheric climate change over the last 2000 years. First, however, I provide some background information about paleo-data and principal climate trends over the past two millennia.

12.3.1 Sources of paleo-information

It is necessary, first, to briefly discuss the choice of solar index against which results are to be compared (see further discussion in Chapter 11). Since the pioneering study of Stuiver and Quay (1980), fluctuations in ^{14}C have been known to reflect a variable Sun (in terms of solar magnetic field as manifested in both sunspot number and auroras, coupled to the cosmogenic isotopes through the magnetic modulation of the cosmic ray flux). ^{10}Be is also a cosmogenic isotope and has a different geochemical cycle. After removing the effects of changes in the Earth's magnetic field, there is quite a good residual relationship between the ^{14}C and ^{10}Be solar indices over the last 10 000 years (Fig. 12.4), thereby indicating that solar variability is the prime cause of fluctuations in the two records (remaining offsets could possibly reflect differential responses such as changes in deep-water circulation that would affect ^{14}C more than ^{10}Be). Below, primarily the ^{14}C record is used because of its higher temporal resolution.

A standard approach to testing the significance of different forcing mechanisms is to compare them against some reconstruction that has annual resolution. Tree-ring records are by far the best source of such information, because the methodology has been tested for decades and there are usually many sites that go into a composite record for any one region. Because the number of records (e.g.

C14 vs Be10 Over the Last 9000 Years

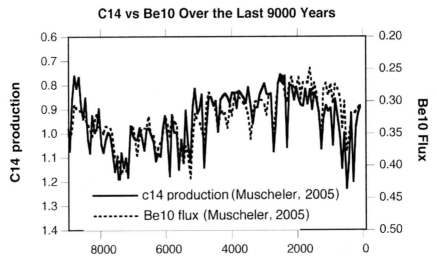

Fig. 12.4. Comparison of two independent estimates of solar variability deduced from relative changes in ^{14}C to ^{12}C ratios and from ^{10}Be concentrations (in units of 10^6 atoms/cm^2 per year).

trees) at a particular site decreases as one goes back in time, the uncertainty and the potential bias in the reconstruction of a particular site increase going back in time. Almost all reconstructions have another source of bias: there are generally many more sites in younger parts of any reconstruction, so that trends and variability can potentially change because of the added information from the greater number of sites in going from distant past to present.

One way to address this problem is to use only the sites represented in the oldest part of the record. This is a called a "frozen-grid" approach. Comparison of frozen-grid reconstructions with variable-grid reconstructions often shows that the frozen grid captures to a large degree the patterns of change in the more heavily sampled younger sections of many reconstructions (e.g. Crowley and Lowery, 2000).

The approach taken here is to use only tree-ring records, because their chronology is excellent (independent repeat analyses indicate that the chronology is reproducible to within a year). Variability in tree-ring width also correlates well with summer half-year temperatures. Fortunately, this is the season that might most readily record changes in volcanism and/or solar variability, because the potential for radiative forcing perturbations is of course greater when total insolation is greatest. Detection of the signal is also easier because there is less climate "noise" in summer than in winter.

Many reconstructions often extend to regions outside the domain of a data set, such as the tropics. Although appropriate statistics can be employed to make such estimates, the approach taken in the present comparison is to restrict the

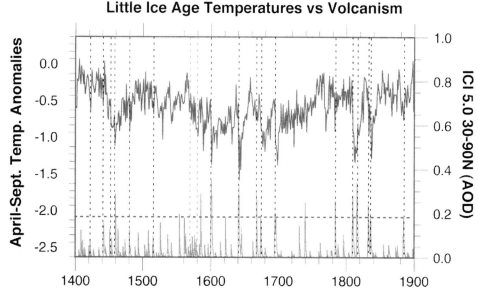

Fig. 12.5. Comparison of tree-ring-based reconstruction of Northern Hemisphere (land) summer half-year temperatures (upper curve, left-hand axis) with an index of volcanism (lower curve, right-hand axis) for 30°–90° N (AOD denotes aerosol optical depth). Seventeen volcanic eruptions coincident with cooling events registered by tree rings are marked by dashed black lines. A cluster of small eruptions in the late 1500s is contained between the two magenta dashed lines. The AOD of the great 1883 CE Krakatau eruption is highlighted by a horizontal green dashed line. See Color Plate 13.

reconstruction to where the tree-ring record is best, i.e. for latitudes of 30°–90° N and for land areas only. Because we are interested in possible detection of individual volcanic eruptions (see below), the new reconstruction is not smoothed in time (see Fig. 12.5).

The tree-ring estimate used in this chapter incorporates data from more sites than used by Jones *et al.* (1998), with the instrumental calibration being for the interval 1885–1960. This and other decadal-to-multidecadal scale reconstructions (e.g. Fig. 12.6) consistently show that, for especially the Northern-Hemisphere mid- and high-latitudes (for which tree-ring data are most abundant), there is a general pattern of cooling in the early Middle Ages (approximately 500–940 CE), followed by an interval of warming from about 950 to 1200 that is often called the Medieval Warm Period (MWP). Then follow some stepwise coolings and oscillations, with the first peak cooling occurring in the middle of the fourteenth century, a stronger cooling in the middle of the fifteenth century, and then a late-sixteenth century cooling that culminates in what is usually termed the main phase of the Little Ice Age – the seventeenth century (with a notable warm interlude around 1650 CE).

Three Records of 30–90N (land) Climate Change

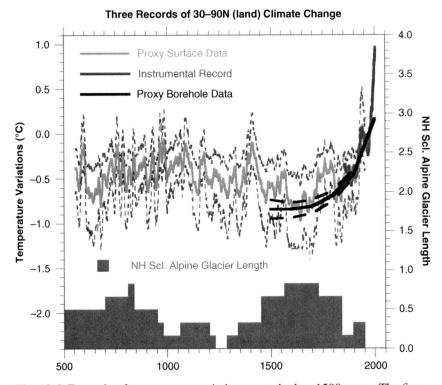

Fig. 12.6. Example of temperature variations over the last 1500 years. The figure illustrates a comparison with scaled Northern-Hemisphere alpine-glacier extent (see Section 12.4) of two estimates of completely different data types for mean annual temperature changes over land for 30°–90° N. The estimate of surface proxy data (tree rings, ice cores, etc.) is from Hegerl *et al.* (2007). The "borehole estimate" is based on geothermal measurements of heat flux in boreholes, in which a deconvolution method has been used to remove the heat flux gradient in the Earth's interior in order to determine an estimate of the downward diffusion of century-scale land surface temperature changes over time (Pollack and Smerdon, 2004). To my knowledge, the approach in the Hegerl *et al.*. (2007) paper is the only time these two techniques have been quantitatively compared over the same time and space domains (where most of the data are located). Figure adapted from results in Hegerl *et al.* (2007), with the unsmoothed instrumental record updated through 2008, and shown up to 2003 after a 10-year smoothing filter was applied. See Color Plate 14.

Warming starts round 1700 CE, but is interrupted by a sharp phase of cooling in the 1810s and 1830s.

12.3.2 Tests for the role of solar forcing in climate change

Next, I discuss three different tests to examine more closely the evidence as to a potential Sun–climate relationship in the paleoclimate record of the last 2000 years.

Test 1 I first compare volcanism with a new estimate of 30°–90° N temperatures on land for the summer half-year (Crowley *et al.*, 2010). Volcanism is chosen as the first test because the information is less conjectural than solar forcing; in some cases there are observations of an important historical eruption (examples include Huaynaputina in the Andes in 1600, Laki in Iceland in 1783, Tambora in Indonesia in 1815, Krakatau in 1883, and a number of twentieth century eruptions). Sulfur injected into the stratosphere is the primary cause of cooling; a fraction of the volcanic sulfur is deposited in polar ice sheets (e.g. Zielinski, 1995). The new volcanism reconstruction (Crowley *et al.*, 2008) has been calibrated against satellite aerosol optical depth estimates (Sato *et al.*, 1993) of the 1991 eruption of Mt. Pinatubo in the Philippines, which was the largest sulfur-producing eruption of the twentieth century.

In our first test (Fig. 12.5) the new compilation of high-latitude Northern Hemisphere volcanism with the new tree-ring reconstruction indicates that, during peak Little Ice Age cooling (~1450–1850 CE), 17 eruptions coincide with cooling events registered by tree rings. Many of these eruptions were larger than the great 1883 CE Krakatau eruption, the magnitude of which is highlighted by a horizontal green dashed line in the figure.

A further cluster of small eruptions in the late 1500s (between the two magenta dashed lines) suggests another influence, wherein the accumulated effect of a number of small eruptions can have a comparable effect to a bigger eruption. The illustrated relation does not take into consideration additional damped effects from changes in ocean heat storage (see Crowley *et al.*, 2003). The main story of the Little Ice Age is that most of the coolest years or decades correlate to either single eruptions or clusters of smaller volcanic eruptions (see Crowley, 2000).

Test 2 Another way to evaluate the Sun–climate relationship is to redo the same test used for the instrumental era: take the forcing data (Fig. 12.7) that are best known – CO_2 and volcanism – and first determine the best-fit combination of these forcings for an interval when the latter is strong (1450–1850 CE). Only then is the solar-activity proxy included to determine whether there is any additional covariance explained as a result of this new forcing term. This test is different than earlier tests (e.g. Crowley, 2000) because a climate model is not used; for the first time, we directly compare the forcing and the response, bypassing uncertainties that might be raised by use of models. Although we cannot estimate the temperature response to the forcing change, the comparison bypasses the usual arguments about quality of climate models, etc., to get at the main issue of forcing and response (feedbacks just amplify or decrease the response to forcing, linearly or non-linearly).

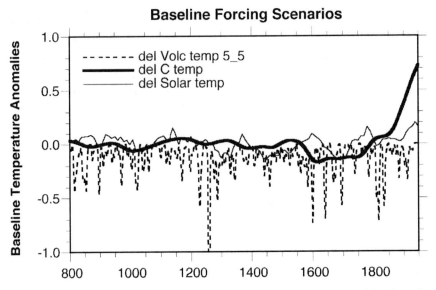

Fig. 12.7. Plausible temperature variations for three different kinds of forcing (C: CO_2; V: volcanism; S: solar) that were used as a baseline for best-fit comparisons to tree-ring reconstruction.

In addition to the volcanic and solar-activity records already described, small variations of CO_2 have occurred in the pre-industrial era. The CO_2 forcing record used here is a splice of the more finely sampled Law Dome (Antarctica) record (Etheridge *et al.*, 1996) back to about 1000 CE with the lower resolution Taylor Dome (Antarctica) record (Indermuhle *et al.*, 1999), with a slight adjustment of baseline in the latter to account for a small offset between the two records in the overlap interval.

The solar record is a linear adjustment of the residual ^{14}C record of Stuiver *et al.* (1998). I assume that the ^{14}C record is a proxy for any type of Sun–climate connection, be it direct irradiance or some more complicated feedback that nevertheless correlates with ^{14}C. Of course it is possible that some component of solar influences on the climate system would not correlate with ^{14}C or ^{10}Be variations, but that would be hard to test for with paleoclimate records at hand, so this is not considered further in this chapter.

Using an iterative approach, a best-fit correlation (Fig. 12.8) of 0.686 (or 47.1% of the covariance) is obtained for the trial interval with just CO_2 and volcanic forcing (the latter with 10-year smoothing). Adding the solar proxy increased the correlation to 0.711 (or 50.6% of the covariance). The increase in covariance is therefore 3.5%, which is broadly comparable to that obtained for the instrumental interval. Replacing ^{14}C with ^{10}Be variations from Bard *et al.* (2000; using data updated in December, 2007, in the National Geophysical Date Center website)

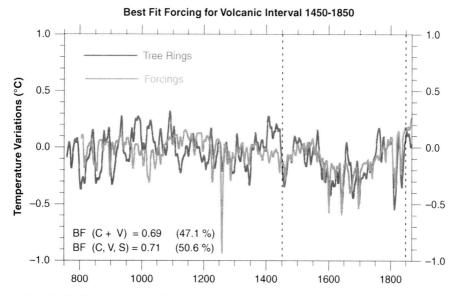

Fig. 12.8. Comparison of best-fit forcing run (grey) over the interval 1450–1850 CE (boundaries of vertical dashed lines) with paleoclimate observations (black) for both the calibration interval and the remaining part of the reconstructed records (back to 800 CE). The correlation coefficients and percentage of total variance are shown for two- and three-component fits (the two curves are almost indistinguishable, thus only one is plotted).

resulted in a slight decrease by about 1% in the amount of solar covariance over the best-fit interval.

These results, and detection and attribution work by Hegerl *et al.* (2003), leave little doubt that volcanism, and a small but significant CO_2 decrease (see below), played the major role in peak Little Ice Age cooling. In the next test, I present a way to determine whether there is any additional effect from inclusion of solar forcing, as characterized by residual changes in the ^{14}C tree-ring records (Stuiver *et al.*, 1998).

How can the above results be reconciled with the oft-cited relationship with ^{14}C during, particularly, the Maunder minimum interval of the late seventeenth century (1640–1710)? A virtually unrecognized problem is that the issue of statistical degeneracy also raises its head in the paleoclimate interval of the Little Ice Age (a problem independently recognized by C. Amman). Consider Fig. 12.9, which compares a smoothed version of the volcanism record against ^{14}C. Careful statistical analyses would have to be done to test for a significant relationship between these two records. Regardless of the statistical significance of the results, it is clear that the coincidence between the two can confuse the proper attribution as to cause of the cooling events.

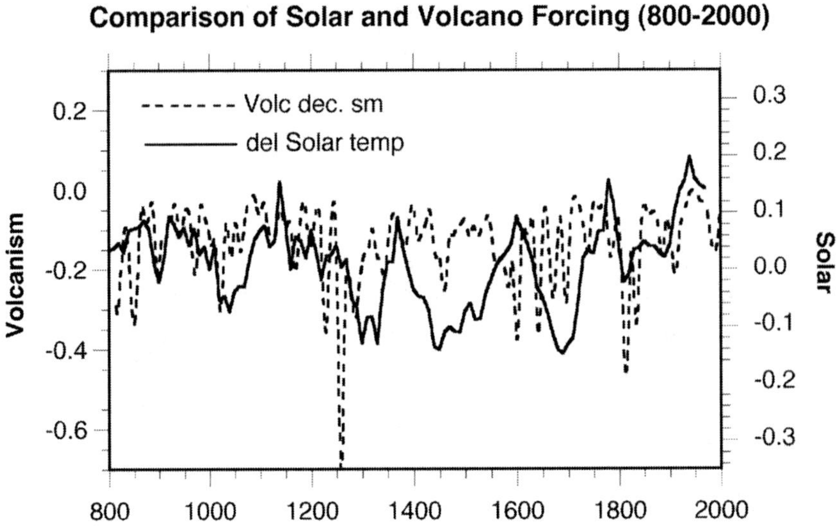

Comparison of Solar and Volcano Forcing (800-2000)

Fig. 12.9. Comparison of solar forcing and smoothed volcanic forcing indicating that there is covariation over a number of time intervals, which can lead to erroneous interpretations of causality.

Test 3   Another approach is to examine the correlation between proxies for solar activity and temperature in an interval not dominated by volcanism. As illustrated above, the sustained period of enhanced volcanism that characterizes the Little Ice Age began in the thirteenth century. Examining moderately long time series before that time enables another test of the potential effect. There are three time series that have been produced spanning different lengths: (1) a new reconstruction of annual temperatures back to 744 CE (Crowley *et al.*, 2010) that has been decadally smoothed to better highlight the low-frequency signal; (2) a previously published, decadally averaged estimate of zonal-averaged (30°–90° N) temperatures back to 550 CE (Hegerl *et al.*, 2007) and (3) an independent estimate from Jones and Mann (2003) back to 200 CE.

   In order to avoid the very large offset between forcing and response for the 1258 CE volcanic eruption (an event that has some nonlinear responses, see Timmreck *et al.*, 2009), the comparisons were restricted to time intervals up to 1250 CE (Fig. 12.10). Whether using original reconstructions or smoothing to enhance low-frequency correlations, there is an insignificant correlation (Table 12.1) between the ^{14}C residual and the climate reconstructions: $r = 0.0 \pm 0.2$. These tests do not support any inference of a linear relationship between solar irradiance, geomagnetic variability and climate. Because the longest of these time series is 1000 years, the conclusion becomes even more compelling and possibly even suggests that the very weak relationship determined for intervals of the last 1000 years could be a simple coincidence of trends and not a weak positive correlation.

Table 12.1. *Correlations between* ^{14}C *index of solar variability and different climate reconstructions for different levels of smoothing and different intervals preceding the Little Ice Age period of intense volcanism*

Reconstruction	Smoothing (yr)	Interval (CE)	Corr. coeff.
Crowley *et al.* (2010)	10	759–1250	−0.177
Crowley *et al.* (2010)	30	767–1270	−0.228
Hegerl *et al.* (2007)	10	558–1250	+0.004
Hegerl *et al.* (2007)	30	566–1250	−0.006
Jones and Mann (2003)	30	214–1250	−0.136
Jones and Mann (2003)	30	214–589	+0.202
Jones and Mann (2003)	30	590–1250	+0.062

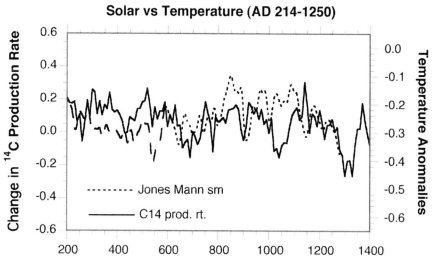

Fig. 12.10. Comparison of solar forcing (solid line; left axis) with a climate reconstruction (dotted and dashed line; right axis) for the interval prior to the time of intense "Little Ice Age" volcanism (214–1250 CE). The dashed-dotted line highlights the first section of the Jones–Mann reconstruction to determine whether there were any differences-of-fit for different intervals (there were not – see Table 12.1).

12.4 Sun and climate beyond the most recent two millennia

What about correlations on millennial scales? Here the situation might be different. Although Bond *et al.* (2001) have discussed a possible correlation at a period of about 1500 years for records of the last 10 000 years, there is no significant spectral peak in the ^{14}C or ^{10}Be records at this period. However, Chapman and Shackleton (2001) have demonstrated a significant covariance at a period of ~550 years between an index of North Atlantic Deep Water (NADW[†]) formation and atmospheric ^{14}C. There is also some evidence for ^{14}C and ^{10}Be fluctuations, at a period of ~ 1000 years, that may be of cosmogenic origin (Debret *et al.*, 2007) and can be found in some paleoclimate records (Obrochta *et al.*, 2010).

Is there evidence for any other parts of the climate system changing at this time scale? At least two studies suggest a relationship between longer-period ^{14}C variations and alpine glacial fluctuations (Denton and Karlen, 1973; Wigley and Kelly, 1990; compare also Fig. 11.18). Here, I conduct an updated assessment. I first developed a composite of alpine glacial fluctuations over the last 8000 years, that is, after the major time of possible influence by residual fluctuations in the terminal Laurentide Ice (the largest ice sheet in North America (and, indeed, in the world) during the last glacial maximum about 20 000 years ago; it had mainly melted away by 8000 years ago, except for some residual components in north-eastern Canada).

The composite was normalized and then averaged. Time–distance plots of component alpine glacier variations (Grove, 1988) are from five regions – the eastern and western Alps, Alaska, Baffin Island, northern Scandinavia, and the Colorado Front Range. Even though there is considerable noise when compositing such records, the composite agrees fairly well with an illustration of alpine glacier fluctuations (Schaefer *et al.*, 2009) and with evidence for glaciation in South America and Antarctica (Clapperton and Sugden, 1988). The ^{14}C dates were converted to calendar years using the standard INTCAL04 chronology (Reimer *et al.*, 2004).

Results (Fig. 12.11) indicate some correspondence between the composite alpine record and atmospheric ^{14}C, especially if an offset is permitted in the records before 6000 BP (before present, which is defined as 1950 CE). Such an adjustment is justified because of higher Northern-Hemisphere summer insolation values in the early Holocene that would likely result in different threshold values for glacial growth (e.g. Weber *et al.*, 2004). The adjustment is better if alpine glaciers are allowed to lag forcing by about 100 years because of the time delays in glacial growth and decline (e.g. Weber *et al.*, 2004).

Because of the different nature of the time series, careful statistical analyses are required to determine the level of statistical significance. Wigley and Kelly (1990) established a significant correspondence between the two. As this conclusion would not be affected by time scale adjustments (both being based on ^{14}C), it should also hold for the new composite. A longer period (~2300 years) in the

---

[†] NADW forms in winter in the high latitudes of the North Atlantic from cooling of salty water of Gulf Stream origin; it then descends to a depth of ~ 2500–4500 m and flows southward to the Circumpolar Southern Ocean as part of the Great Global Conveyor. Changes in water chemistry can be used to detect changes in NADW over time.

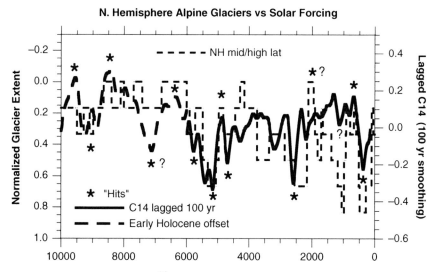

Fig. 12.11. Comparison of ^{14}C estimate of solar variability (solid and dashed curve) with composite record of alpine glaciation (dashed histogram).

^{14}C record discussed by Denton and Karlen (cf. Sonett and Finney, 1990) is also consistent with the conclusion of a 2400-year periodicity in both ^{14}C and ^{10}Be (Debret *et al.*, 2007), if one allows for the more accurate absolute chronology now available.

A critical gap remains in adequate quantification of the millennial-scale variations in temperature, if such fluctuations are to be related to solar forcing and climate sensitivity. Until this is done we cannot fully grasp the enticing information from longer period climate fluctuations and relate them to the types of analysis conducted in the first part of this chapter.

12.5 Discussion and conclusions

An updated analysis of solar and climate records over the last two millennia continues to support the conclusions of Hegerl *et al.* (2003, 2007) that the Sun–climate connection, at least in terms of temperature, is weak on the time scales of decades to centuries. However, there is more support for a linkage at longer periods (of order 550, 1000, and 2400 years) that would benefit from quantification of the magnitude of temperature variations inferred for the climate system over this band.

This conclusion makes some physical sense because there is generally more variance at lower frequencies in geophysical time series. The absence of a significant relationship inferred for higher frequencies may simply reflect that a small forcing signal is combined with a large amount of climate noise. At longer frequencies, a signal might have a greater chance of being detected. Analyses of additional

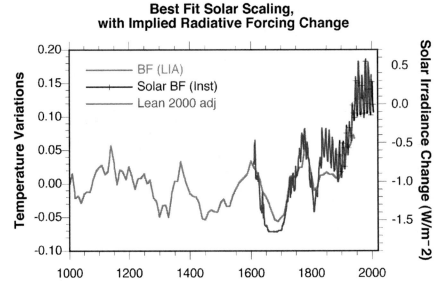

Fig. 12.12. Comparison of independent estimates of solar component for the instrumental interval and Little Ice Age. The right-hand scale indicates the estimated change in total solar irradiance for the background component of the re-scaled Lean (2000) record. See Section 12.5 for a discussion of the method. See Color Plate 15.

long temperature records may provide critical information for further testing this hypothesis, which in turn may untangle some remaining conflicting claims about the reality of the Sun–climate connection on longer time scales.

A final, intriguing consideration involves comparison (Fig. 12.12) of the estimated solar contributions to the instrumental period and the last millennium (based on 1450–1850 CE calibration). Because there are differences in the amplitude of response to the same forcing of global temperatures and of 30°–90° N land temperatures (summer half-year), the solar signal for 30°–90° N land temperatures was scaled by the ratio of the present trend in those temperatures and in the global temperatures. A comparison indicates a rather close agreement in amplitude of the independent estimates. This could of course be coincidental, but the consistency may also support validity of the estimates. The inferred near-present-day peak in solar activity is also consistent with a long record of [10]Be fluctuations from an ice core (Solanki *et al.*, 2004).

A further aspect of the agreement involves comparison with the Lean (2000) reconstruction of the trend (but not amplitude) of plausible solar irradiance changes since 1600 CE. To determine the amplitude of the revised Lean (2000) pattern, the background signal and 11-year signals in these records were treated separately, because it is known that the system response to high-frequency forcing is smaller than for lower-frequency forcing (North *et al.*, 1984; cf. Crowley and Kim, 1993). A standard temperature sensitivity to radiative forcing variations of $0.67\,°C/W\,m^2$ forcing was used, equivalent to a best-guess sensitivity of about $2.5\,°C$ to a doubling of CO_2 (Hegerl *et al.*, 2006), with the 11-year signal being damped by a further factor of about 0.6. The two signals were then iteratively adjusted until they agreed with the temperature estimate from the instrumental record (see Fig. 12.12). The small adjustments are consistent with the best-guess sensitivity. The Lean (2000) reconstruction was then fit to the estimated radiative forcing for the best fit in the instrumental record, with the latter then extended back to 1600 CE.

The revised amplitude of the Lean forcing is about 70% of the original background forcing, yielding about $1.5–1.7\,W/m^2$ range for the low-frequency component of solar variability. Perhaps by chance again, this estimate is strikingly consistent with the work of Foster (2004), as illustrated in Lockwood (2006), of a plausible $1.7\,W/m^2$ range of low-frequency solar variability. The predicted scale of the early twentieth century solar component ($\sim0.1\,°C$) is, however, less than the $2\,°C$ range estimated with the Hadley Centre model (Stott *et al.*, 2003). Because the latter model has a higher temperature sensitivity to radiative forcing perturbations than the one used in the present study, it is not clear that the discrepancy is truly significant.

From this analysis one almost has the perception of some level of convergence between different types of evidence for variations in past solar activity. The troubling fly in the ointment of the present line of argument is the still unconvincing level for any Sun–climate connection in the interval from about 200 to 1250 CE, when pulses of volcanism were much less frequent. Until this puzzle is clarified, some hesitance must be maintained with regards to a completely convincing case for the reality of the Sun–climate connection in paleoclimate records.

13

Terrestrial ionospheres

Stanley C. Solomon

13.1 Introduction

Early investigation of the terrestrial ionosphere through its effect on radio waves resulted in description by means of layers, principally the D, E, and F layers, the latter subdivided into F_1 and F_2 (see Vol. I, Fig. 12.1). This terminology continues to influence our current concept of the nature of energy deposition in atmospheres, although the misleading term "layer" has given way to "region". The term "layer" arose from the observation of systematic variation in the height at which the critical frequency of reflection occurs in ionospheric radio sounding; this method cannot detect ionization above the peak of a region, which explains the appearance of layers. Radar and spacecraft measurements now give a more complete picture of peaks and valleys and reveal the complex morphology of the ionosphere. Chamberlain and Hunten (1987) provide a referenced discussion of the historical literature.

An overview of the altitude dependence and variability of Earth's ionosphere is given in Fig. 13.1, showing the diurnal and solar-cycle changes and the locations of the named regions. Space and planetary exploration has also found that the Earth's ionosphere is unique, just as its atmosphere is unique, and for some of the same reasons as we shall touch upon in this chapter.

An additional historical artifact in terminology is the word ionosphere itself. Because the atmospheric ionization was discovered before the neutral thermosphere in which it is contained, anything above the stratosphere is often referred to as the ionosphere, resulting in a common misconception that this region of the atmosphere is mostly ionized. In fact, it is mostly neutral, ranging from less than a part in a million ionized during the day at 100 km altitude to about 1% ionized at the exobase ($\sim$600 km, depending on solar activity; compare Vol. I, Fig. 12.13). Even at 1000 km, there is only of the order of 10% ionization. At several thousand km, where ions (mostly protons) finally become dominant, the region is defined as the plasmasphere.

Heliophysics: Evolving Solar Activity and the Climates of Space and Earth, eds. Carolus J. Schrijver and George L. Siscoe. Published by Cambridge University Press. © Cambridge University Press 2010.

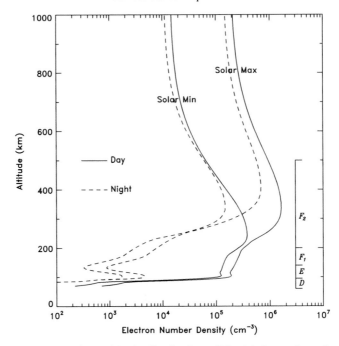

Fig. 13.1. Overview of the altitude distribution of Earth's ionosphere for daytime and nighttime conditions, at high and low solar activity, from the International Reference Ionosphere (IRI) empirical model. (Bilitza, 1990).

Figure 13.2 shows the results of two standard semi-empirical models of the thermosphere and ionosphere, the Mass Spectrometer Incoherent Scatter (MSIS) model (Hedin, 1987) and the International Reference Ionosphere (IRI; Bilitza, 1990), for typical daytime conditions at moderate solar activity. Charge neutrality is assumed, so the electron density profile is the sum of all the ion density profiles. Only the major ions are shown. It is clear from this plot that O^+ is the most important ion, particularly in the extensive F_2 region above $\sim$200 km. The F_1 region from $\sim$150 to $\sim$200 km appears as a mere plateau in the profile, but is distinguished by a transition to molecular ions, particularly NO^+. The low level of N_2^+, given the dominance of N_2 at these altitudes, is noteworthy. The E region from $\sim$100 to $\sim$150 km exhibits a small peak, dominated by O_2^+ and NO^+. The weak $\sim$70 to $\sim$90 km D region is not shown on this plot.

At night, the ionosphere below 200 km almost disappears. Small residuals remain, and sporadic layers of metallic ions in the E region sometimes appear, but basically the night ionosphere is confined to the F_2 region. The peak electron density erodes and its altitude varies, but the F_2 ionosphere persists through the night. At auroral latitudes, a new E region can occur, co-located with auroral emission features. Sometimes the F_2 region is enhanced in the presence of aurora; at other times it is greatly attenuated.

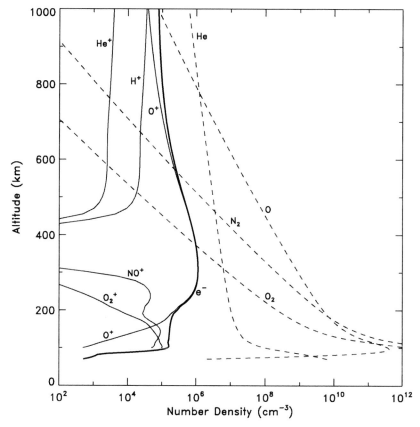

Fig. 13.2. Semi-empirical models of the Earth's thermosphere/ionosphere: the Mass Spectrometer Incoherent Scatter (MSIS) model (Hedin, 1987) and International Reference Ionosphere (IRI; Bilitza, 1990).

13.2 Ionization

The ionosphere is created by ionizing radiation, including extreme ultraviolet (EUV) and X-ray photons from the Sun, and corpuscular radiation that is mostly energetic electrons and mostly occurs at high magnetic latitude as auroral "precipitation". The solar photon output at these wavelengths, from ~ 1 nm to the H I Lyman-α line at 121.6 nm, varies by factors ranging from ~ 2 to > 100 over the 11-year solar activity cycle (see Fig. 10.1), and is additionally variable on shorter time scales, including especially the 27-day solar rotation period. This causes dramatic variations in the temperature and density of the thermosphere and ionosphere. Changes in the solar wind and interplanetary magnetic field also affect the thermosphere/ionosphere through geomagnetic perturbations that result in transfer of energy from the magnetosphere, both in the form of auroral particle ionization and in the form of heat from the resulting currents imposed in the polar regions (see Vol. I, Chapter 10, as well as in Vol. II). An additional form of energy transfer is the generation of energetic electrons released in the ionization process. These electrons, referred to as photoelectrons in the case of photo-ionization and secondary electrons in the case of particle impact ionization, have enough energy to excite, dissociate, and further ionize the neutral atmosphere as well as heat the ambient plasma. Solar ionization and its byproducts provide most of the ionization and heating of the thermosphere, and account for most of its 11-year cyclic and 27-day rotational variation, but geomagnetic activity accounts for much of the shorter term variation on time scales from hours to days.

The details of ionospheric formation can be explained through examination of the photo-ionization and photoabsorption cross sections of thermospheric constituents. The ionization continua of N_2, O, and O_2 all peak in the vicinity of 60 nm at tens of megabarns (10^{-18} cm^2). This causes their energy to be deposited largely in the F_1 region (compare Fig. 13.3). This can be calculated for any particular wavelength by applying Beer's law to an exponential atmosphere, which was done by Chapman (1931; see Vol. II, Section 12.3.2). For a single-constituent, planar, isothermal atmosphere with density n_0 at some reference altitude z_0 and scale height H, a simplified form of the Chapman production function is

$$q_z = I_\infty \sigma_i n_0 \exp\left(-\frac{z - z_0}{H} - \tau_z\right), \qquad (13.1)$$

where q_z is the ionization rate as a function of altitude, I_∞ is the unattenuated ionization frequency, σ_i is the ionization cross section, and τ_z is the optical depth. This function peaks at the altitude of unit optical depth, and reduces rapidly

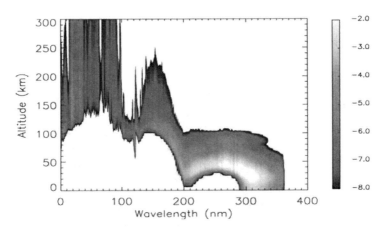

Fig. 13.3. Altitude of penetration of the solar radiation as a function of wavelength. The color range shows the amount of energy deposited in the different layers of the atmosphere for the different parts of the solar spectrum (on a logarithmic scale, in units of mW m^{-3} nm^{-1}). See Color Plate 16.

below that altitude. The smaller the cross section, the larger the column density at unit optical depth, and the lower the peak. Shortward of 60 nm, cross sections decrease and the radiation penetrates to lower altitude. The intense solar He II emission at 30.4 nm deposits most of its energy near 150 km and 1–10 nm soft X-rays can penetrate to 100 km. Most of the E region is produced by longer wavelength radiation, however, particularly the C III line at 97.7 nm and the H I Lyman-β line at 102.7 nm. These do not have enough energy to ionize N_2 and O but penetrate through gaps in the N_2 absorption spectrum to ionize O_2 to O_2^+. Longward of 103 nm, only the important minor species NO has a low enough ionization potential to be ionized by solar radiation. H I Lyman-α happens to fall at a low point in the O_2 absorption spectrum and so penetrates below 90 km, where ionization of NO to NO^+ and subsequent products creates the D region. Thus, while the Chapman production function is approximately correct for any species at each wavelength, ionized regions are created by the superposition of many such functions for different species and wavelengths. In general, the more polychromatic the absorbed irradiance, the broader the resulting region of energy deposition.

13.3 Recombination

Positive ions have generally fast collision rates with electrons, so one would suppose that ionospheric production would be balanced by recombination and that the ions would be short-lived after sunset. However, atomic ions colliding with electrons have the problem common to all two-body reactions that a single molecule is unlikely to result, because there is nothing to carry away surplus kinetic energy. Photon emission following collision of an atomic ion with an electron can stabilize the resulting atom; this radiative recombination is quite slow, with rate coefficients of the order of 10^{-12} cm^3 s^{-1}. Although radiative recombination occurs and is important in the highest reaches of the ionosphere, it is insufficient as a loss mechanism for ions and electrons given their observed F region densities. Because the solar ionization frequency is $\sim 10^{-6}$ s^{-1}, ion densities would be several orders of magnitude larger than observed if radiative recombination were the only loss mechanism. The recognition by Bates and Massey (1947) that atomic ions must yield their charge to molecular ions in order to undergo rapid dissociative recombination is regarded as seminal. Dissociative recombination, schematically $XY^+ + e^- \rightarrow X + Y$, has rate coefficients of the order of 10^{-7} cm^3 s^{-1} and is the fundamental loss mechanism for ions in planetary ionospheres. In order for an atomic ion to become a molecular ion, atom–ion interchange, schematically $X^+ + YZ \rightarrow XY + Z^+$, or charge exchange, schematically $X^+ + YZ \rightarrow X + YZ^+$, must occur. Charge exchange reactions are typically fast if energetically possible, but atom–ion interchange rates depend on the nature of the reacting molecule, because a bond must be broken.

In regions of the atmosphere where molecules dominate, recombination chemistry is simplified because it is essentially a balance between ionization and dissociative recombination. Again following Chapman's work, a common approximation is use of an effective recombination rate coefficient α_{eff}, the density-weighted average of the ion recombination rates. In photochemical equilibrium, the production rate $q = \alpha_{\text{eff}}[M^+][e^-]$, where M^+ is the sum of the ions, and where square brackets denote number densities. Assuming charge neutrality, this yields $[e^-] = (q/\alpha_{\text{eff}})^{1/2}$. Applying the Chapman production function to obtain q results in a Chapman "layer", considering as above the caveats associated with use of that term. Thus, in molecular atmospheres, electron density varies approximately as the square root of the ionization rate profile. This is a particularly useful form for auroral ionization, where electrons (and sometimes protons or heavier ions) penetrate to ~ 100 km or lower.

Although the F_2 region has some of the morphological appearance of this type of layer, it is at the wrong altitude, and in the atom-dominated region. It is not a Chapman layer at all, but a result of diffusive processes. O^+ has an increasingly long lifetime as altitude increases and the molecular fraction of the thermosphere decreases. Above 200 km it becomes subject to diffusion, but is still chemically controlled up to the peak of the F_2 layer near 300 km. Above this altitude, ambipolar diffusion takes over, where "ambipolar" refers to the effect of electrical attraction between the ions and nearly massless electrons, resulting in a scale height for O^+ about twice that of O. The F_2 region varies in response to thermospheric winds and electric fields, so the mid-latitude and equatorial ionosphere can be greatly influenced by auroral processes at high latitudes through their effect on thermospheric dynamics.

Figure 13.4(left) is provided as a guide to understanding the ion–neutral chemical processes described above. It is a greatly simplified schematic, but contains the essential species and reactions necessary to describe ionospheric photochemistry from 100 to 600 km. The vertical direction implies approximate ionization potential. Ionization occurs primarily on the three major species N_2, O, and O_2 by photon, photoelectron, and auroral particle impact. N_2^+ quickly loses its charge through dissociative recombination, atom–ion interchange with O to make NO^+, and, at lower altitude, charge exchange with O_2 to make O_2^+. Thus it is always low in density and negligible in the absence of production. O^+ loses its charge by atom–ion interchange with N_2 and O_2. Reaction with N_2 is slow, $\sim 10^{-12}$ cm^3 s^{-1}, because of the high strength of the triple $N \equiv N$ bond. Reaction with O_2 is faster, $\sim 10^{-11}$ cm^3 s^{-1}, but there is far less O_2 available. This is why O^+ is long-lived in the Earth's ionosphere, and why the F_2 region exists. O_2^+ loses its charge through

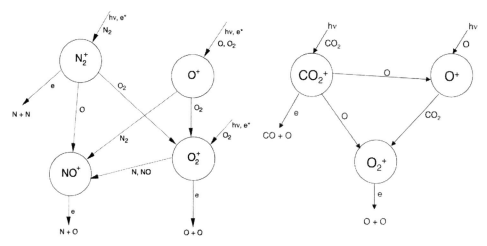

Fig. 13.4. Simplified diagram of ionospheric chemistry in the upper atmosphere of Earth (left) and of Venus and Mars (right). The vertical direction in these diagrams implies approximate ionization potential.

dissociative recombination or through reaction with the odd-nitrogen species NO and N, which control the balance between O_2^+ and NO^+ in the E region. NO^+, daughter of all the above and the "terminal ion", is subject only to dissociative recombination. Figure 13.4(left) thus describes a machine for dissociating molecular gases. Ionization goes in the top, and dissociation comes out the bottom, because dissociative recombination is the only significant way out.

Figure 13.4(left) does not describe the ionosphere at very high altitude because it neglects hydrogen, which reacts by charge exchange with O^+ to make H^+. N^+ is also a significant minor ion, created by photodissociative ionization of N_2, that is neglected here. Doubly ionized species are also ignored in this simplification. Although ground-state $O^+(^4S)$ does not have enough energy to make N_2^+, metastable $O^+(^2D)$ and $O^+(^2P)$ are created by photon and electron impact ionization, and these can charge exchange with N_2 to form N_2^+. It is possible that vibrational excitation of thermospheric N_2 can also accelerate the reaction of O^+ with N_2.

In the E region, there is a complex interplay between O_2^+ and NO^+, due to the involvement of odd-nitrogen species with the ion chemistry, because O_2^+ is converted to NO^+ by reaction with NO and N. NO in particular is highly variable with solar activity, geomagnetic activity, and location, so this is a complicated problem. Older empirical and theoretical models which assumed that O_2^+ is the dominant E region ion, due to its production by solar H I Lyman-β radiation, have been superseded by evidence that NO^+ is generally observed to be the dominant E region ion, and considerable recent observational and modeling advances in understanding the high levels of NO and its importance to radiational cooling as well as ion chemistry have occurred.

In the D region, ion chemistry is entirely different due to the higher neutral density which allows three-body attachment, particularly $2O_2 + e^- \rightarrow O_2 + O_2^-$. This sets in motion a complicated negative-ion chain involving carbon, nitrogen, and hydrogen compounds, including water, that finally results in mutual neutralization of negative and positive ions, schematically $M^+ + M^- \rightarrow M + M$. NO^+ created by H I Lyman-α ionization also initiates an involved positive-ion sequence, again involving hydration processes.

For the F regions, Fig. 13.4(left) is a relatively sound synopsis nonetheless. There have been many recent advances, but most of the fundamental chemistry was worked out during the Atmosphere Explorer program, as reviewed by Torr and Torr (1978).

13.4 Venus and Mars

The terrestrial planets, Venus, Earth, and Mars, are so named because of their fundamental similarity, and are presumed to have had common elemental origins. However, their subsequent evolution differed, due to their differing distance from the Sun, the smaller size of Mars, and the lack of rotation of Venus (see Chapters 4 and 11). Thus, their atmospheres are entirely different, and so are their upper atmospheres and ionospheres. Early exploration of Venus and Mars found that instead of persistent, high-altitude, F_2-type ionospheres, these planets had less-dense, lower-altitude ionospheres (Figs. 13.5 and 13.6) that more resembled Chapman "layers", that were greatly attenuated at night, and consisted mostly of O_2^+ and other molecular ions (e.g. Schunk and Nagy, 1980). The presence of O_2^+ seems especially perplexing, because Earth is the planet we generally associate with the unusual and quite reactive oxygen molecule. At higher altitude, O^+ becomes an important species in the ionospheres of Venus and Mars, as on Earth, but at significantly lower density and without the same degree of persistence throughout the night. CO_2^+ is a minor ion on both planets. There is a basic similarity in their ionospheres, despite the vastly different density of their lower atmospheres. Figures 13.5 and 13.6 provide an overview of the typical altitude structure (to be compared to Earth's in Fig. 13.2).

The reason that the ionospheres of Venus and Mars are different from that of Earth is that the molecular compositions of their atmospheres are different, and therefore the compositions of their thermospheres are different (Figs. 13.7 and 13.8). Table 13.1 gives a simple overview of the abundance of the primary atmospheric gases in the three terrestrial planets.

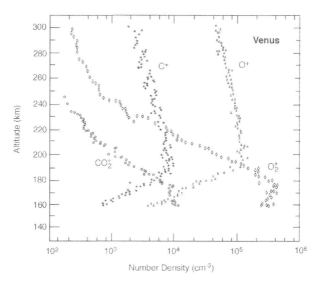

Fig. 13.5. Measurements of the ionosphere of Venus from the ion mass spectrometer on the Pioneer Venus spacecraft. (From Taylor *et al.*, 1980.)

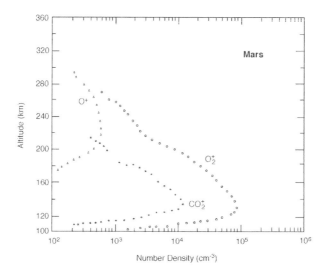

Fig. 13.6. Measurements of the ionosphere of Mars from the ion mass spectrometer on the Viking-1 spacecraft (From Chen *et al.*, 1978.)

Aside from the large differences in surface pressure, the atmospheres of Venus and Mars are similar in composition, and N_2 is an important species on all three planets. N_2 requires more energy to dissociate than the oxygen compounds, however, so at thermospheric altitudes, atomic oxygen becomes important on all three

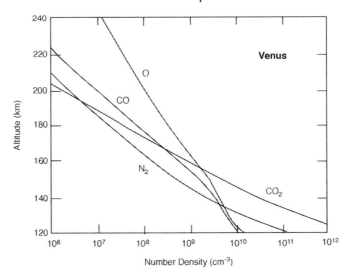

Fig. 13.7. Typical composition of the Venus thermosphere. (From Fox and Bougher, 1991, after Hedin, 1983.)

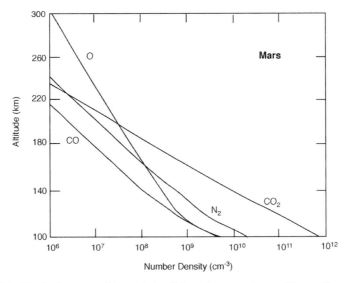

Fig. 13.8. Typical composition of the Mars thermosphere. (From Chen *et al.*, 1978, after Nier and McElroy, 1977.)

planets. The Venus and Mars thermospheres are distinguished by high levels of CO_2 (and also CO) due to the underlying atmospheric composition, as shown in Fig. 13.7 and 13.8.

On Earth, O^+ is a long-lived species in the high ionosphere because the $O^+ + N_2$ reaction is so slow and there are few other molecules to react with to make a short-lived molecular ion. On Venus and Mars, the reaction $O^+ + CO_2$ is quite fast,

Table 13.1. *Atmospheric composition of the terrestrial planets*

Planet	Molecule	Abundance (bars)	Fraction of total
Venus	CO_2	86.4	0.96
	N_2	3.2	0.035
	Ar	0.006	0.00007
	H_2O	0.009	0.00001
Earth	N_2	0.77	0.77
	O_2	0.21	0.21
	H_2O	0.01	0.01
	Ar	0.0093	0.0093
	CO_2	0.0004	0.0004
Mars	CO_2	0.0062	0.95
	N_2	0.00018	0.027
	Ar	0.00010	0.016
	H_2O	4×10^{-7}	0.00006

$\sim 10^{-9}\,\mathrm{cm^3\,s^{-1}}$, because CO_2 is much less strongly bound than N_2. (The triple bond in $N \equiv N$ is the strongest known chemical bond in nature.) The reaction of $O^+ + CO_2$ produces O_2^+, which is also produced by the reaction $CO_2^+ + O$. Ionization of the major thermospheric gases on Venus and Mars, O and CO_2, thus quickly produces O_2^+, which dissociatively recombines, resulting in the observed ionospheric morphology, lacking a significant F_2 region. A simplified schematic of these processes is shown in Fig. 13.4(right). Thus, curiously, although life-supporting Earth is the planet associated with O_2, Venus and Mars are the planets with O_2^+ ionospheres.

The F_2 ionosphere is unique to Earth among the known planets. This is due to its peculiar atmosphere, lacking in CO_2, dominated by N_2, and carrying its oxygen in unusual and reactive states. Venus and Mars have nitrogen as well, but carbon and oxygen dominate their upper atmospheres, so it has little effect. Earth has a significant carbon budget, and once had much higher levels of CO_2 in its atmosphere, but most of its carbon is currently locked up in the crust in the form of carbonate rocks. Thus, the F_2 ionosphere may be a recent event in the history of Earth, an artifact of geology and biology.

13.5 Ionospheres, exoplanets, and signatures of life

Ion density at high altitude is relatively easy to detect and measure in planetary exploration due to its effect on radio signals. Thus, it is intriguing to speculate what an F_2-type ionosphere would signify if observed on some other planet. A full exposition of the processes that removed carbon from the Earth's atmosphere is beyond the scope of this chapter (see Chapter 4), but there are significant inorganic processes that extract dissolved carbon from the ocean and deposit it in rocks, which would continue to dominate even if the well-known biological pathways for absorption of CO_2 from the atmosphere did not exist. Therefore, although it cannot be claimed that the F_2 ionosphere is an indicator of biological processes, it may be related to the presence of an ocean, which is at least a precursor, if not a prerequisite, for the development of replicating or even aerobic chemical compounds. At the current state of Earth's evolution, with a small but ongoing effort underway to restore our planet's carbon balance to its natural state, it may be pertinent to consider just how unusual our atmosphere has become over the past two billion years (cf. Section 4.7), and whether such a configuration of neutral and ionized gases can long endure.

14

Long-term evolution of the geospace climate

Jan J. Sojka

14.1 Introduction

Our understanding of the prehistory of geospace is largely addressed by a series of "what if" scenarios or thought experiments. Hence, this chapter will appear more qualitative than many of the other chapters. As the thought experiments will primarily be based on knowledge of the geospace presented quantitatively in other chapters, these chapters are referenced as the thought experiments are performed. Having freed ourselves from mathematical rigor, we can step backwards into earlier times. As we shall see, it is quite surprising that mankind's earliest records of the cosmos also detail geospace events. Geospace is a very modern term and represents the assimilation of knowledge gained by modern technologies, radio transmitters and receivers, optical and laser systems, and rocket and satellite instruments, whose measurements were made inside geospace. However, prior to this technological momentum of the last century, mankind was aware of geospace phenomena via observations of aurorae, sunspots, solar flares, magnetometer deflections, induced currents in telegraph lines, etc. (see also Vol. II, Chapter 2). This chapter begins by exploring the extent of this human awareness from a geospace climate perspective.

My own introduction to these phenomena occurred while I was a Scottish schoolboy in the Borders town of Gordon. As a 6-year old at a Christmas party on a cold, dark December afternoon, I heard a rendition of the song "The Northern Lights of Old Aberdeen", the lyrics of which were written by Mel and Mary Webb in 1952. The song describes in vivid terms the beauty of the auroral form, a verse of which is:

> When I was a lad, a tiny wee lad,
> My mother said to me,
> "Come see the Northern Lights my boy
> They're bright as they can be."

Heliophysics: Evolving Solar Activity and the Climates of Space and Earth, eds. Carolus J. Schrijver and George L. Siscoe. Published by Cambridge University Press. © Cambridge University Press 2010.

She called them the heavenly dancers,
Merry dancers in the sky,
I'll never forget, that wonderful sight,
They made the heavens bright.

This song is comparatively recent, written at a time when science was well on its way to unraveling the auroral mysteries and to an extent tarnishing the abundant auroral folklore. Robert Eather in his book *Majestic Lights* provides a wealth of information on the aurora in science, history, and the arts, which includes its folklore (Eather, 1980). A century earlier, the connection between the Sun, its dynamics, terrestrial magnetism, and auroral observations was made. Carrington (1860) at his London home observed a "white" light solar flare on September 1, 1859, and at the same time a magnetometer located at the Kew Observatory in London registered significant disturbances (see Vol. II, Chapters 2 and 5). Only 18 hours later, aurorae were reported from many mid- and low-latitude locations such as Puerto Rico, a latitude of only 18 degrees (Kimball, 1960; see Cliver and Svalgaard, 2004, for a review of the 1859 events from Sun to Earth). Indeed, so extensive were these reports that Elias Loomis at the time collected and published eight of these reports, which have recently been reprinted by Shea and Smart (2006).

Mankind's earlier ability to count sunspots is well documented, as is the much reduced number of sunspots during the Maunder minimum (Section 2.8). Somewhere between 1609 and 1611, with the introduction of the telescope, scientists began technically aided observations of sunspots, which can also be viewed as the start of the modern era of geospace science. However, more than a millennium before the introduction of the telescope, the reporting of "naked-eye" sunspots was made, included in literature from China, dating at least as far back as 28 BCE, and observations made by the Greek philosopher Anaxagoras who observed a sunspot as early as 467 BCE. In these countries and probably in other civilized nations two and three millennia ago, astronomers played an essential role since their astrological inferences were considered a critical part of political decision making. These astronomers recorded not only the planetary and other celestial events and extremes in tropospheric weather, but also noted the occurrence of extreme geospace events. Their vocabulary for what we call the aurora or the Northern Lights was not uniform and still proves a challenge in accurately interpreting their records. Stephenson *et al.* (2004) provide an example of these challenges in their analysis of a partial Babylonian clay tablet written in cuneiform. The event occurred on the night of March 12/13, 567 BCE in which an unusual red glow occurred in the night sky. Given that the frequency of major auroral events is and probably always has been small, these records are indeed sparse and mainly confirm that geospace then was qualitatively the same as it is today. For these early astronomers, the observation of these infrequent auroral events was almost always viewed as a negative portent. The human record of geospace, especially its most spectacular manifestation, the aurora, can currently be tracked back 2500 years. It is even argued by some scholars that Cro-Magnon cave paintings, 30 000 BCE, referred to as "macaronis" may comprise the earliest depictions of aurora. Perhaps the most commonly known climate change in geospace is the Maunder minimum period, when there was almost a complete absence of sunspots and mid-latitude auroral observations for about six-and-a-half sunspot cycles. These human records of the geospace climate, spanning only approximately 2000 years, provide a starting point from which to set up more speculative thought experiments.

14.2 Our experience of geospace climate change

The 11-year solar cycle was predicted to have transitioned from its solar minimum end of cycle 23 to the beginning of cycle 24 in the fall of 2008. It did not, however. This effect, or rather the lack of a transition to solar cycle 24, is related to the behavior of the Sun. The solar measure of this is the lack of sunspots of the cycle 24 polarity. Indeed, the solar wind monitor, NASA's ACE satellite, indicates that the solar wind continued to be less dense and its magnetic field weaker than before (Vol. I, Chapters 8 and 9, and Chapter 2) well beyond the anticipated solar minimum in 2008. Satellites have observed the solar wind during the prior three solar minima. Is this difference from expectation only geospace weather or is it a separate climate state of geospace? If it is merely weather, then was 1645 to 1715, the Maunder minimum period, only weather? This question is crucial in addressing the long-term climate of geospace since the Sun is one of the major drivers of the system, i.e. a different solar climate is probably associated with a different geospace climate.

Another present-day change in geospace is that global climate change would be expected to affect the chemical composition of the upper atmosphere. It is proposed that a subset of these changes may lead to enhanced cooling of the upper atmosphere which, in turn, would impact the mesosphere, thermosphere, and ionosphere. The ionosphere has been observed by ionosondes in many cases for more than four complete solar cycles and hence has been identified as a data resource for studying "climate changes". An ionosonde transmits and then receives reflected radio waves from the electrons in the ionosphere whose refractive index makes the medium transparent or not to radio waves. Although the ionosonde was invented early in the twentieth century, it was only during the International Geophysical Year (IGY) beginning in 1957 that large deployments of ionosondes were carried out and the beginning of long-term ionogram records began. Although four sunspot cycles may not be viewed as long-term change, a climate/global change is viewed as having very long-term changes. Hence, this specific response of geospace will also be considered in this chapter.

There are three major elements that define geospace: the Earth's magnetic field, the Earth's atmosphere, and the Sun's electromagnetic and particle emissions. The human record of geospace contains observations of all three. Over the past century, knowledge of these has increased to a point that we reasonably understand how they interact to form geospace. This extends to addressing questions of how a change in one affects the geospace environment. Sections 14.3 and 14.4 discuss how the Earth's upper atmosphere combined with the Sun's photon radiation creates geospace's inner boundary, the ionosphere. It also describes how changes in these two elements affect geospace. Section 14.5 considers how the Sun's solar wind combined with the Earth's magnetic field creates geospace's outer boundary, the magnetopause/magnetosphere. Finally, in Section 14.6 an introductory discussion of the search for how long-term change in the ionosphere has progressed in the past two decades is given.

14.3 Geospace climate response to solar photon irradiation

The solar spectrum provides a very stable irradiance of $\sim 1360\,\mathrm{W/m^2}$ to the Earth's upper atmosphere (see Chapter 10). Geospace responds to about 2–$3\,\mathrm{mW/m^2}$ at solar minimum and up to about 7–$8\,\mathrm{mW/m^2}$ at solar maximum of this solar irradiance. As described in Chapter 13, this fraction of the solar spectrum lies between 3 and 360 nm which extends from the X-ray through the ultraviolet part of the spectrum. The photons in this spectral range may ionize atoms and molecules or may deposit their energy directly into the thermal reservoir of the upper atmosphere. These processes are responsible for the "climate" of the geospace–atmosphere interface whose regions are labeled the ionosphere and the thermosphere (IT). The IT in this sense is only weakly dependent upon the Earth's magnetic field or the solar wind. In this "climate" scenario, the role of the terrestrial dipole field can be viewed as defining the boundary for the plasmasphere and then with the solar wind the magnetosphere. In the case of the Earth's sister planets, Venus and Mars, the absence of a significant intrinsic magnetic field confirms that the IT development has been based upon these three photochemical processes.

Now let us explore the question how geospace climate responds to extremes of the solar photon radiation. Roble *et al.* (1987) and Roble (1995) pioneered the concept of a 1D global mean model of the upper atmosphere. They demonstrated that these 1D models were able to reproduce solar cycle global average trends of the IT. Hence, these models provide the necessary tool to address the question. Smithtro and Sojka (2005a), following the Roble *et al.* (1987) and Roble

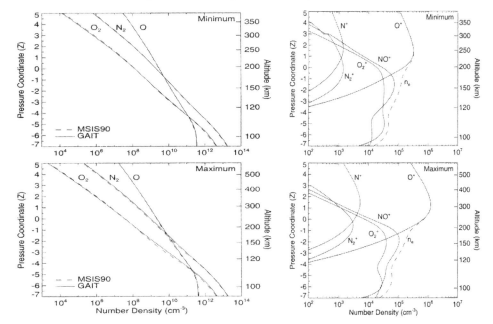

Fig. 14.1. (Left) Global mean number-density profiles under the same conditions for the three major neutral species (N_2, O_2, and O) calculated using the GAIT model (solid lines) and MSIS-90 empirical model (dashed lines). (Right) Global mean number density profiles for five ion species (O^+, NO^+, O_2^+, N^+, and N_2^+) and the total electron density (n_e) calculated using the GAIT model. Solar minimum (top) and solar maximum (bottom; assuming quiet geomagnetic conditions with $Ap = 4$). The discontinuity observed in the NO^+ and N_2^+ profiles at $Z = 3$ corresponds to where the photoelectron calculation stops. (From Smithtro and Sojka, 2005a.)

(1995) work, updated the 1D chemical reaction rates and energy budget calculations and developed the Global Averaged Ionosphere and Thermosphere (GAIT) model for studies of extremes in the solar photon radiation, which were presented in Smithtro (2004) and Smithtro and Sojka (2005b). Figures 14.1 and 14.2 provide a comparison of the GAIT solar cycle climate of the thermosphere's neutral densities, of the ionosphere's plasma densities, and of the neutral and plasma temperatures respectively. Each panel is shown as a function of pressure level defined by

$$Z = \log p_0/P, \qquad (14.1)$$

where p_0 is the reference pressure of $50\,\mu\text{Pa}$. The corresponding altitude scale is also provided. For reference to observations, the dashed lines where present in Figs. 14.1 and 14.2 correspond to profiles obtained from the MSIS-90 empirical model of the thermosphere (Hedin, 1991).

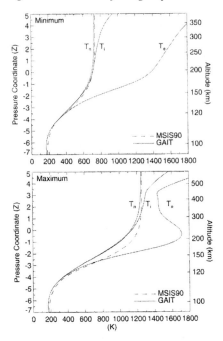

Fig. 14.2. Global mean temperature profiles calculated using the GAIT model (solid lines) and MSIS-90 empirical model (dashed lines). The three profiles correspond to neutral (T_n), ion (T_i), and electron (T_e) gases. Solar minimum (top) and solar maximum (bottom) assuming quiet geomagnetic conditions ($Ap = 4$). (From Smithtro and Sojka, 2005a.)

A simple interpretation of this solar minimum to solar maximum climate change in the IT is that the effective IT energy deposition has almost quadrupled; hence, the neutral atmosphere, which at these heights is in hydrostatic equilibrium, leads to a hotter thermosphere; compare T_n in the two panels of Fig. 14.2. In turn, the hotter thermosphere has redistributed neutrals now with relatively higher densities at higher altitudes; compare neutral densities in the two left-hand panels of Fig. 14.1 using the right side altitude scale. For example, O_2 at 300 km is approximately 10^6 cm^{-3} at solar minimum and quadruples to about 4×10^6 cm^{-3} at solar maximum. A secondary but also important additional effect is that the composition is also being modified because of the different neutral masses. For the ionosphere, the consequences can readily be seen by comparing the two right-hand panels of Fig. 14.1. The dominant density peak in both panels is that associated with the O$^+$ peak, which is called the F_2 layer peak of the ionosphere (the F_1 layer is discussed below). At solar minimum, the density is 3×10^5 cm^{-3} and located at 250 km and at solar maximum the density increases to 1×10^6 cm^{-3} and the peak rises 300 km. A less obvious peak, called the E layer peak lies in the 100–105 km altitude range and is the result of molecular ions and NO$^+$ summing up to form a peak. The long-dashed line in the right-hand panels of Fig. 14.1 corresponds to the electron density, the sum over all ion species. At solar minimum, the E layer peak density is about $4.5 \times 10^4 \, \text{cm}^{-3}$ while at solar maximum it is $6.5 \times 10^4 \, \text{cm}^{-3}$. A comparison of the E and F layer peak density provides another useful scaling law, rule-of-thumb, in that the F layer density scales linearly with the appropriate photon wavelength energy flux while that of the E layer is more like a square-root dependence upon energy flux.

That the different ionospheric layers respond differently to the solar spectrum creates the problem of deciding what the most suitable solar spectral representation is. In fact, even over the limited solar cycle energy flux range of a factor of about four, the spectrum itself is variable and the E and F layers respond to different parts of the spectrum. The thermosphere is a somewhat better integrator as seen in Fig. 14.3 taken from the Smithtro and Sojka (2005b) study in which four distinctly different representations of the solar spectrum were used as drivers for the GAIT model. As each spectral model was driven over the solar cycle range of 2–8 mW/m^2, the GAIT exospheric temperature was determined. The results are that the GAIT-model thermosphere responded linearly to each spectral model, and the same linear dependence was found for each. Note that the exospheric neutral temperature refers to the asymptotic, altitude independent, temperature found at higher altitudes, see Fig. 14.2 for specific solar minimum and maximum examples. The exosphere refers to the ionospheric plasma, whose composition is light ions of hydrogen and helium, that is located in altitude above the F layer.

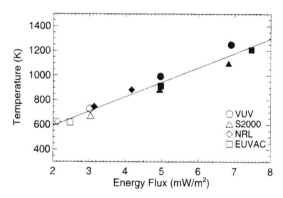

Fig. 14.3. Modeled global mean exospheric temperatures as a function of the EUV energy flux (3–105 nm), calculated using four different irradiance models. The circles, triangles, diamonds, and squares correspond to the VUV (Woods and Rottman, 2002), Solar 2000 version 2.21 (Tobiska *et al.*, 2000), NRLEUV (Warren *et al.*, 2001) and EUVAC (Richards *et al.*, 1994) models respectively. Open symbols represent solar minimum conditions , solid symbols solar maximum conditions, and shaded symbols moderate solar conditions. The solid line indicates a least-squares fit to the results. (From Smithtro and Sojka, 2005a.)

From Fig. 14.3, it is possible to extrapolate how the thermospheric exospheric temperature would trend for lower and higher levels of the solar EUV flux. Indeed, this is the basis of the Smithtro and Sojka (2005b) study aimed at understanding how the Earth's ionosphere would respond to historic conditions such as the Maunder minimum or the Medieval grand maximum between 1100 and 1250 CE (Eddy, 1976; see Fig. 11.18). Their procedure assumed that a linear dependence existed in the relevant EUV energy flux between solar minimum and solar maximum. An index S_{EUV} was defined to be 0 at solar minimum (energy flux = 3 mW/m^2) and $S_{EUV} = 1$ at solar maximum (energy flux = 7 mW/m^2). Then Maunder-minimum type conditions correspond to S_{EUV} values less than zero and grand maximum values correspond to S_{EUV} values greater than 1. Note that the specific response to the solar cycle of each wavelength is different, hence, S_{EUV} is applied to each wavelength separately to generate extreme solar spectra.

14.3.1 Maunder minimum geospace climate

The solar cycle has shown great variability over the human observation time scale, and between 1645 and 1715 it became very much weaker with sunspots largely – but not entirely – absent; this period being the Maunder minimum (Section 2.8; Eddy, 1976). During this time, not only did sunspots largely disappear, but the cosmogenic isotopes and tree-ring data also showed marked differences (see Chapters 11 and 12). Hence, there exists a body of evidence indicating that the solar spectral irradiance was likely quite different (Chapters 2 and 10). Lean *et al.* (2001) used such data to construct a solar chromospheric activity index that could be quantified back to 1610 CE just prior to the Maunder minimum. On this basis, the Maunder minimum S index would be between $S_{EUV} = -0.5$ and -1.0. Figure 14.4 shows the GAIT ionospheric plasma composition for solar minimum ($S_{EUV} = 0$), $S_{EUV} = -0.5$, and $S_{EUV} = -1.0$. The earlier trends concerning the E and F layer are continued as the S_{EUV} index reduces. The F_2 peak density drops from 3×10^5 to 1.2×10^5 to 6×10^4 cm^{-3} as S_{EUV} decreases from 0, to -0.5, to -1.0. The F_2 peak height drops from 250 km to 180 km as S_{EUV} drops from 0 to -1.0. The E layer height remains near 105 km but the density reduces as S decreases. In addition, a number of chemical changes occur, specifically the F_2 peak for $S_{EUV} = -1$ conditions has a significant percentage of molecular ions (NO$^+$). During times of normal solar cycles this condition only happens for extreme auroral precipitation conditions under quite dissimilar physical and chemical conditions. The E region dominant molecular ion is O_2^+ for solar minimum conditions (see Fig. 14.1, upper right panel), but as S_{EUV} decreases towards Maunder minimum conditions NO$^+$ becomes dominant in the E region.

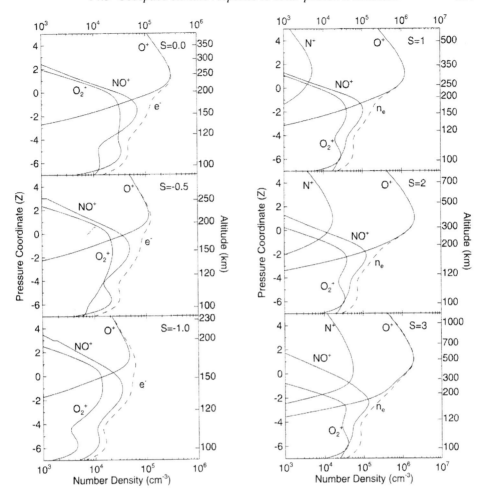

Fig. 14.4. Global mean concentration of the ion and electron (dashed line) gases, calculated using the GAIT model for six different levels of the solar activity increasing from $S_{EUV} = -1.0$ to $S_{EUV} = 3$, clockwise from the bottom left. The profiles are plotted as a function of the pressure coordinate, Z, with the corresponding altitudes provided on the right-hand axis. (From Smithtro and Sojka, 2005b.)

The most significant ionospheric modification during the Maunder minimum period is that the molecular ion NO^+ peak below 200 km becomes significant; compare the top-left panel for $S_{EUV} = 0$ and the bottom-left panel $S_{EUV} = -1$ in Fig. 14.4. This additional structure in the electron density profile is referred to as the F_1 layer. Indeed, in Fig. 14.4 the $S_{EUV} = -1.0$ case almost has this F_1 electron density equal to that of the higher altitude F_2 peak. The F_1 peak height at $S_{EUV} = 0$ (solar minimum) is at 160 km and at $S_{EUV} = -1$ (Maunder conditions) drops to about 135 km. Decreasing S a little below $S_{EUV} = -1$ would lead to F_1 density exceeding the F_2. These Maunder minimum scenarios would result in significantly different environments for modern-day technology. For example:

- using the ionosphere to propagate radio waves over the horizon is restricted to much lower frequencies because the maximum ionospheric density has decreased;
- because the F_1 layer is located significantly lower than the F_2 layer propagation, paths for radio waves are also modified significantly;
- with less ionospheric density in the path of GPS radio waves, the adverse role of the ionosphere in geolocation analysis is reduced.

14.3.2 Grand maximum geospace climate

Smithtro and Sojka (2005b) applied a value of $S_{\mathrm{EUV}} = 3$ as the upper range of enhanced solar EUV flux to simulate grand maximum type conditions. The grand maximum existed between 1100 and 1250 CE (Eddy, 1976). An S_{EUV} value of 3 corresponds to doubling the solar-maximum energy flux from $7\,\mathrm{mW/m^2}$ to just over $14\,\mathrm{mW/m^2}$. The right-hand column in Fig. 14.4 shows the GAIT-model ionospheric plasma distributions at solar maximum ($S_{\mathrm{EUV}} = 1$), $S_{\mathrm{EUV}} = 2$, and $S_{\mathrm{EUV}} = 3$ from top panel to bottom panel. In all cases, the F_2 layer is the dominant layer with O^+ the dominant ion. As predicted from the normal solar cycle trend, this layer will rise, in this case from $300\,\mathrm{km}$ ($S_{\mathrm{EUV}} = 1$) to about $500\,\mathrm{km}$ ($S_{\mathrm{EUV}} = 3$). The F_2 peak density does not increase linearly with S_{EUV}! Between $S_{\mathrm{EUV}} = 2$ and $S_{\mathrm{EUV}} = 3$ the F_2 peak density has remained at $2 \times 10^6\,\mathrm{cm^{-3}}$.

   This maximum in the F_2 peak density is by far the most significant change in the geospace climate in response to solar photon radiation. The processes responsible for this effect are: (i) the production of neutral O and, hence, its concentration non-linearly decreases at altitudes at which the F_2 peak is created as the thermosphere heats up as S_{EUV} increases from 2 to 3; (ii) the O^+ production rate increases linearly as S_{EUV} increases from 2 to 3; and (iii) the competition between these two processes leads to a maximum peak F_2 density at $S_{\mathrm{EUV}} = 2$, and a slight decrease in the peak density as S_{EUV} increases further. The consequences for modern-day technologies under enhanced solar maximum, grand maximum conditions are:

- The changing altitude of the F_2 layer leads to modified radio wave propagation paths.
- That the peak F_2 density saturates only slightly above solar maximum values implies that the "radio" reflection characteristics of the ionosphere are consistent with today's "radio climate".
- The impact on trans-ionospheric radio applications such as GPS geolocation is somewhat adverse since the total electron content (the electron density integrated over a line of sight, i.e. a column density) continues to increase even though the F_2 peak density becomes constant.

- Because the ionosphere is significantly more dense, the absolute magnitude of plasma density irregularities increases, which leads to greater scintillation problems with radio propagation.

14.3.3 Other geospace climate impacts

In modeling the ionosphere and thermosphere as the solar EUV energy flux is changed, there are at least two impacts of significance for the outer reaches of geospace. First, assuming that the magnetosphere is somewhat similar to the state that we are familiar with, then the IT contributes plasma to the magnetosphere/plasmasphere and second, the IT electrical conductivity is a component of the magnetosphere–ionosphere (M-I) electrical coupling. Under the Maunder-minimum type conditions, ionospheric outflow of plasma into the magnetosphere/plasmasphere will decrease because the ionospheric topside is colder and less dense. Under extreme conditions such as $S_{EUV} = -1$, the composition may also begin to change from atomic to molecular. In contrast, under $S_{EUV} = 2$ and upward grand maximum conditions with hotter topside, the outflow would increase and be very much O^+ dominated. Note that in these GAIT-type modeling studies the light ion, H^+, has not been included, and therefore the remarks pertain to O^+ and heavier molecular ions. In contrast, the ionospheric conductivity changes are smaller because the major contribution comes from the E layer whose composition remains molecular. However, the decreasing dayside conductivity during Maunder conditions would raise issues about how this impacts the M-I electric circuit response, i.e. would this modify present-day concepts of voltage versus current generator descriptions of the M-I system? Under the grand maximum with enhanced conductivities and also the assumption of increased solar wind energy, would M-I coupling be characterized by significantly enhanced currents and electric field? Both scenarios would probably impact the morphology of auroral displays! This may lead to the most significant human experience of the geospace climate.

14.4 Geospace climate at earlier terrestrial ages

In Section 14.3, the geospace responses to extremes in solar EUV radiation were considered under present atmospheric conditions. These extremes reflected only the present-day cycle variability of the EUV energy flux, and are thus relatively limited compared to conditions in the distant past or future. Kulikov *et al.* (2007) considered the upper atmosphere's response over 4.6 Gyr for the planets Earth, Venus, and Mars. In earlier times, the solar EUV was more intense and,

Table 14.1. *Historical values of the solar EUV fluxes relative to the present-day value (cf., Fig. 2.14)*

Time	Solar flux multiplier
3.5 Gyr ago	factor ~6
3.8 Gyr ago	factor ~10
4.33 Gyr ago	factor ~50
4.5 Gyr ago	factor ~100

consequently, the thermosphere was much hotter, leading to the dominance of significantly different processes. The issues Kulikov *et al.* (2007) studied pertain to an early period when atomic hydrogen was in a blow-off phase as well as periods when high escape rates for the fastest particles in the energy distribution ("Jeans escape") of heavier species such as H_2, He, C, N, O existed. However, their studies also show that IR-radiating molecules such as CO_2, NO, OH, etc., control the exospheric temperature that, in turn, controls the Jeans escape rates for the neutral constituents. Hence, the results depend not only on a knowledge of solar EUV but also the contribution of molecules such as CO_2 and H_2O in the earlier terrestrial atmosphere. Kulikov *et al.* (2007) developed a diffusive-gravitational equilibrium and thermal balance model to study heating of the earlier thermosphere.

In an initial simulation, this model was used to evaluate the terrestrial exospheric temperature over the past 4.6 Gyr. Significant assumptions were made that the present-day composition as well as that of the lower atmosphere up to 90 km were the same then as they are today. The increased solar flux values at earlier ages (see Chapter 2) were obtained from Ribas *et al.* (2005) and Lundin *et al.* (2007). Estimates of how much higher these earlier EUV fluxes were compared with those of today are summarized in Table 14.1.

Figure 14.5 shows the Kulikov *et al.* (2007) result for the history of the Earth's exospheric temperature. Assuming that the blow-off temperature for atomic hydrogen is about 5000 K, the Earth's first Gyr would exhibit a markedly different upper atmosphere where even the atomic species and molecular hydrogen would be approaching their thermal escape speeds. Kulikov *et al.* (2007) point out that this simple model becomes a rough estimate when the exospheric temperatures exceed 10 000 K.

The major assumption that should be questioned for these earlier Earth ages is the density of the IR radiating molecules such as CO_2. In significantly earlier times, these would be expected to be larger. If this were the case, then their role in "cooling" the thermosphere would increase. Defining the CO_2 mixing ratio relative to present atmospheric level (PAL) as 1, Kulikov *et al.* (2007) studied the

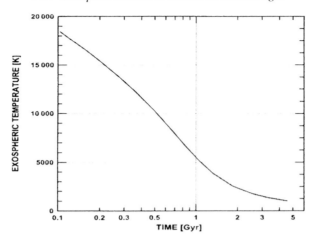

Fig. 14.5. Evolution of the exospheric temperature based on Earth's present atmospheric composition over the planet's history as a function of the solar EUV flux for a strongly limited hydrogen blow-off rate. (From Kulikov *et al.*, 2007.)

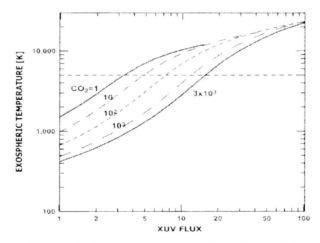

Fig. 14.6. Earth's exospheric temperatures for different levels of CO_2 abundance in units of PAL (Present Atmospheric Level: 1 PAL for $CO_2 = 3.3 \times 10^{-4}$) in the thermosphere as a function of solar XUV flux. The numbers by the curves correspond to CO_2 volume mixing ratios expressed in PAL. The horizontal dashed line shows the blow-off temperature of atomic hydrogen. (From Kulikov *et al.*, 2007.)

exospheric temperature dependence upon increased levels of CO_2. Figure 14.6 shows how, indeed, significant increases in CO_2 would cool the upper atmosphere. In this figure, the "XUV flux" is the scaling ratio of earlier age solar EUV compared to today. The current situation is shown at unit XUV flux. This shows that increasing CO_2 by a factor of 10 (10 PAL) leads to a drop of almost 600 K in the exospheric temperature from 1600 to 1000 K.

As an aside, the present-day discussion concerning global change (climate change) does include the scenario that increased CO_2 would lead to cooling of the thermosphere. Rishbeth (1990) predicted that this cooling would lead to measurable changes in the ionospheric F_2 layer. Specifically, the height of the layer would decrease. Further discussion of the geospace climate response is given in Section 14.6.

Tian *et al.* (2008a, b) developed a multi-component hydrodynamic thermosphere model to study early terrestrial planetary atmospheric conditions when the exosphere temperature could reach and exceed 10 000 K. Their model includes both ions and neutrals, and solves both momentum and energy equations. A total of 154 reactions are included in the model. For the present-day extremes of the solar cycle, this model is in agreement with observations as encapsulated in the MSIS (Hedin, 1991) neutral atmosphere and IRI (Bilitza, 2001) ionospheric empirical models. Furthermore, the Tian *et al.* (2008a) findings are similar to those of Smithtro and Sojka (2005a, b) when the solar EUV flux is increased by a factor of two or three beyond present-day solar maximum conditions. Figure 14.7 shows how the Tian

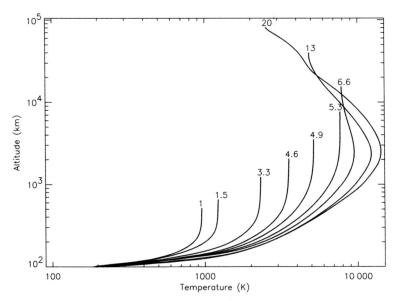

Fig. 14.7. Thermospheric temperature profiles for different solar EUV flux cases (normalized to present-day solar mean energy flux $\sim$1 times present EUV, which represents solar EUV energy flux $\sim$5.2 mW/m^2). It is shown that when solar EUV energy flux exceeds a certain critical value, the upper part of the thermosphere begins to cool as a result of the increasingly significant adiabatic cooling effect. Beyond the critical flux ($\sim$5 times present EUV in this plot), the higher the energy input into the thermosphere, the lower the exobase temperature. (From Tian *et al.*, 2008a.)

et al. (2008a) model of the thermospheric temperature increases as the solar EUV increases. Each curve is labeled with a scaling factor that is referenced to a present-day average solar EUV flux of 5.1 mW/m^2. This baseline value corresponds to an S_{EUV} index value of $S_{EUV} = 0.525$, see Section 14.3. The solar EUV factor of 3.3 corresponds to an S_{EUV} value of about 3.5 and is the upper extent of the Smithtro and Sojka (2005b) study. In Fig. 14.7, apart from the obvious increase in thermospheric temperature, there is increased height in the exobase, the exobase being defined as the altitude at which the exospheric temperature becomes constant. Up to solar EUV factors of about six, a well-defined exobase is found and the height has increased from about 400 km to above 2000 km at solar EUV factors of six. For higher factors of the solar EUV, the temperature has reached levels at which Jeans' escape fluxes are significant, and the "cooling" mechanism is no longer dominated by IR-type cooling. From their earlier work, Kulikov *et al.* (2007) discovered atomic hydrogen had reached the blow-off stage at about 5000 K. Hence, by solar fluxes that are about five times the present average EUV energy flux of 5.1 mW/m^2, the composition of the upper thermosphere will be dramatically different from today as Jeans' escape mechanism becomes effective for hydrogen as well as other atomic species.

Figure 14.8 provides a summary of the ionospheric impact predicted by the Tian *et al.* (2008a) model for solar EUV fluxes that extend up to four times their average solar EUV flux of 5.1 mW/m^2. The top-left panel shows the linear energy flux relationship with the commonly used P index. The P index is an equal weighting of the instantaneous F10.7 and 81-day average F10.7 solar radio flux at 10.7 cm. On this scale, solar minimum is about 70 and solar maximum is about 230, with some solar rotation periods reaching 300. Therefore, the results shown in Fig. 14.8 correspond to energy fluxes that are about three times the present-day solar maximum values. The top-right panel shows that, over this solar flux range, the exospheric temperature is still increasing linearly, reaching 3000 K at a solar EUV energy flux of 20 mW/m^2. In sharp contrast, the ionospheric ion densities are almost constant for solar EUV fluxes above 7 mW/m^2, i.e. the present-day solar maximum values for the dominant ions – O^+, NO^+ and O_2^+ – but N^+ is increasing almost linearly with the energy flux. The bottom-left panel, however, shows that the total electron content (TEC, i.e. the column density for electrons) of the ionosphere is linearly increasing for EUV energy fluxes above 7 mW/m^2. This linear rate of increase predicted by the Tian *et al.* model is significantly faster than observed during the normal solar cycle.

In addition, the layer height of the F_2 peak is increased as the thermosphere heats. Figure 14.9 contrasts two cases of increased solar EUV fluxes to the present-day average conditions of the F_2 layer altitude profiles. These results are from the Tian *et al.* (2008b) study, a specific set of simulations based on the Tian *et al.*

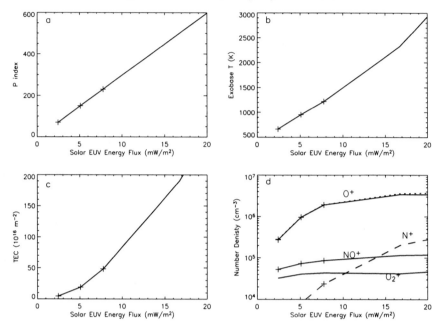

Fig. 14.8. (a) Variations of solar EUV energy flux with P index (an equal weighting of the instantaneous F10.7 and 81-day average F10.7 solar radio flux at 10.7 cm). (b) Variation of exobase temperature with solar EUV flux. (c) Variation of total electron content (TEC) with solar EUV flux. (d) Variation of peak densities of major ion species with solar EUV flux. The dotted curve in panel (d) marks the variation of peak electron densities. In each panel, the crosses correspond to the solar minimum mean and maximum values. (From Tian *et al.*, 2008a.)

(2008a) model. The dotted red line is the electron density profile for a tenfold increase in the EUV flux. The corresponding F_2 peak altitude is about 2500 km, and the composition now has a significant N^+ contribution. Under present-day conditions, N^+ is always a negligible minor ion. This is the case even at an increased solar flux of 3.3, see blue curves in Fig. 14.9. The model studies carried out by Kulikov *et al.* (2007) and by Tian *et al.* (2008a, b) indicate that, at earlier ages of the Earth's upper atmosphere–ionosphere, the response to increased solar EUV flux was a heating of this part of geospace with the following impacts upon the geospace climate:

- For exospheric temperatures of 5000 K and above, the upper atmospheric composition would be dramatically different due to Jeans' escape fluxes of hydrogen and other atomic species.
- The altitude of the F_2 layer would increase to heights above 2000 km.
- The F_2 layer peak density would become constant.
- The total electron content of the ionosphere would increase linearly with increasing solar EUV flux.

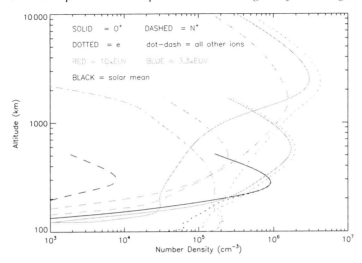

Fig. 14.9. Density profiles of O^+ (solid curves), N^+ (dashed curves), and electrons (dotted curves) under different solar EUV conditions. The total density curves of all ions other than O^+ and N^+ in the $10\times$ present EUV case is presented with the dot-dashed curve. (From Tian *et al.*, 2008b.) See Color Plate 17.

The geospace climate would change from an ionospheric standpoint when the solar energy flux slightly exceeds levels of the present-day solar maximum. From the thermospheric point of view, the geospace climate would change when the solar flux EUV reaches about $20\,\text{mW/m}^2$ (three times the present-day solar maximum values). As a footnote to these trends, if indeed the CO_2 mixing ratio of today's atmosphere is increasing, this will lead to a reduction in the exospheric temperature and an overall lowering of the ionosphere. This would lead to Maunder-type expectations as simulated by Smithtro and Sojka (2005b).

14.5 Geospace climate response to Earth's magnetic field changes

In Sections 14.3 and 14.4, the geospace climate trends were considered from the perspective of the interaction of the solar EUV with the Earth's upper atmosphere that creates the ionosphere, the inner boundary of geospace. Both long-term changes in the solar EUV flux and in the upper atmosphere were considered in discussing the ionospheric long-term changes. Following this interaction methodology, the outer boundary of geospace is defined to be the magnetosphere and, specifically, the magnetopause. It is created by a solar flow, the solar wind, that interacts with an intrinsic property of the Earth, the magnetic field (see Vol. I, Chapters 10 and 13). Consequently, in this section questions concerning how long-term trends of the solar wind and Earth's magnetic field are considered in discussing the long-term geospace climate. Of specific interest are the conditions under which the geospace would be dramatically changed. A human experience perspective would be associated with observations of aurorae and ground-based electromagnetic signals such as magnetometer measurements. The questions to be addressed include:

- How does the Earth's magnetic dipole field reversal affect geospace?
- Was the solar wind's solar cycle variability significantly different leading to a different geospace climate?
- If the Earth's magnetic field were dominated by non-dipolar terms, how would the geospace climate respond?

14.5.1 Geospace climate response to dipole flips

Perhaps the geospace response to flips in the Earth's dominant dipolar field is the most frequently discussed geospace "what if" scenario. Geological evidence obtained in the last century has clearly proven that the Earth's magnetic field, especially its dominant dipole component, has reversed many times during geological times (see Fig. 7.4). The most recent reversal occurred 0.78 Myr ago. Prior to this reversal the six most recent occurred at 0.99, 1.07, 1.19, 1.2, 1.77, and 1.95 Myr ago. Figure 14.10 shows the time-line of these recent reversals indicating that they occur at quite irregular intervals, with the shortest time between reversals being at the Cobb Mountain reversal pair separated by only about 10 000 years. The last reversal was about 780 000 years ago, therefore, is one imminent? The answer to this is "probably not". Valet *et al.* (2005) studied the magnetic field conditions at times of reversals in detail and inferred that reversals occur when the dipole field strength is relatively weak.

   Figure 14.11, from the Valet *et al.* (2005) study, shows the epoch analysis of recent reversals indicating a gradual reduction in the dipole field strength over about 80 000 years, and then a rapid increase in the field following the reversal. It is also significant to note that Valet *et al.* (2005) had to exclude the pair of reversals associated with the Cobb Mountain epoch because they were only separated by 10 000 years. Quantitatively, Knudsen *et al.* (2008) placed a threshold

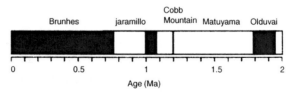

Fig. 14.10. The Earth's virtual axial dipole moment (VADM) orientation over the past 2 million years. The current orientation are the black shaded regions. See Fig. 7.4 for a much longer record. (From Valet *et al.*, 2005.)

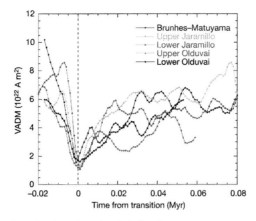

Fig. 14.11. Variations in the virtual axial dipole moment across the five reversals occurring during the past 2 Myr. These are superimposed about their respective reversal epoch (with time running from right to left). A 60–80 kyr long decrease precedes each reversal. (From Valet *et al.*, 2005.) See Color Plate 18.

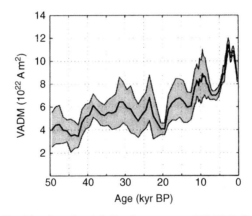

Fig. 14.12. The Earth's virtual axial dipole moment (VADM) for the past 50 kyr (black line) and the associated error estimates (2σ) obtained using a bootstrap approach (grey zone). (From Knudsen *et al.*, 2008.)

value of 4×10^{22} A m^2 on the virtual axial dipole moment (VADM) above which reversals were not eminent. Figure 14.12 shows the Earth's VADM variation over the Holocene period and the past 50 000 years. From this figure it is clear that although the VADM is currently decreasing its value of around 8×10^{22} A m^2, this is double the "reversal" threshold.

With this reversal history, it would seem that for the past 700 000 years geospace dependence upon the Earth's magnetic field has been somewhat similar to the present time dependence as far as reversals are concerned. The specific "N-S" or "S-N" dipole orientation itself would not introduce significant geospace climate changes. Perhaps most obvious is that the solar wind northward versus southward reconnection morphology would be reversed. What is significantly more important is the magnitude of the Earth's field and the orientation of dipole component.

14.5.2 Geospace climate dependence upon dipole strength

Over the past 100 years, the Earth's dipole moment has decreased by about 5% from 8.3×10^{22} to 7.8×10^{22} A m^2, while 3000 years ago, it was at almost 12×10^{22} A m^2, at its highest value during the Holocene era. From earlier chapters on the solar wind and magnetosphere (Vol. I, Chapter 10), it is the balance between the Earth's magnetic field and the solar wind pressure that determines the outer boundaries of the magnetosphere/geospace. Hence, a larger (weaker) dipole moment with otherwise the same solar wind conditions would increase (reduce) the size of geospace. In turn, this would reduce (increase) the size of the polar cap, and auroral regions would move poleward (equatorward). However, a 5% change in the VADM would probably not have a dominant impact on geospace because the solar wind pressure varies by more than this over its normal solar cycle (e.g. Vol. I, Chapter 9). Considering earlier times when the VADM did decrease to values as low as, if not lower than, 2×10^{22} A m^2, the geospace climate may well have been dramatically different, especially during solar maximum type conditions. The magnetosphere would have been severely reduced, and in volume regions such as the plasmasphere it would have almost been reduced to ionospheric altitudes and in the "open" polar regions would extend to mid-latitudes. The effectiveness of plasma sheet energization processes would also have been changed, causing impacts on ring currents and electrojets, as well as the visible aurora. Perhaps the energy transfer to geospace would simply decrease as the magnetosphere's cross section to the solar wind decreased, and consequently, all internal energy processes would be similarly scaled down.

The extreme scenario of the dipole reversal is the idea that the VADM for a time period is extremely small, approximately zero. If the higher order multipole terms are also negligible, then the Earth's atmosphere is unprotected. But this is the Venus and Mars type scenario and extensive analysis has been done on these planetary atmospheres (Nagy *et al.*, 2008). At the present time, the scientific techniques that provide information on the reversals are unable to be specific on this question, but a near-zero magnetic field appears to last no longer than a few thousand years, if that (see Fig. 14.11.)

14.5.3 Geospace climate and the orientation of the Earth's dipole

The scenarios for the geospace climate dependence on tilt angle between the Earth's rotational axis and dipole axis provide vivid geometries of geospace regions

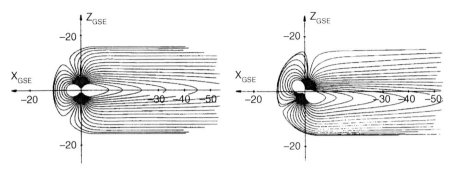

Fig. 14.13. Field lines plotted in the noon–midnight meridian plane for an untilted planetary dipole embedded in the solar wind (left) and the dipole tilted in the noon–midnight meridian at 35° (right). (From Voigt, 1981.)

such as the plasmasphere, plasma sheet, cusps, auroral zones, and open/closed field line regions. For extreme tilt angles, a significant question is how rapidly can these regions evolve and replenish themselves. Many studies of relatively static as well as dynamic magnetospheric configurations have been carried out (e.g. Vol. I, Chapter 11, and Vol. II, Chapter 10). Figure 14.13 provides a pair of noon–midnight cross sections through the Earth's magnetosphere for a zero and 35° tilt (Voigt, 1981). In the left-hand panel, all the conventional magnetosphere regions can be identified and their evolution over a day shown at each time, as seen in this panel for a constant solar wind. Our present-day tilt scenario is somewhat different; in the Northern Hemisphere it is approximately 10° while in the Southern Hemisphere it is almost 15°.

However, even with this tilt, the fundamental magnetospheric regions found in Fig. 14.13(left) are present all day with relatively small wobbles in the geocentric-solar-ecliptic coordinate system (GSE; x, Earth–Sun line, z, ecliptic north pole) of this figure. Both cusps are dayside and wobble in latitude. The plasmaspheric equatorial plane is that of the "average" dipole and it wobbles in 24 hours about the GSE x-axis. Even today, the concept of the plasmasphere's "average" dipole orientation is not fully explored since it is well known that the Earth's equatorial fields are not well represented by a pure dipole component.

Over time scales of decades and more, the tilt angle as well as its geographic longitude wander. Indeed, this has been identified as a major factor in complicating the historic auroral observation database. For example, when an aurora was observed at lower mid-latitudes as described in Section 14.1, was this due to an especially strong or geoeffective solar storm (CME) or did the Earth's dipole tilt have a particularly large value at that time, making this terrestrial location a much higher geomagnetic latitude?

The right-hand panel in Fig. 14.13 shows the magnetospheric geometry for a specific "UT" during Northern Hemisphere winter solstice, when the tilt can reach

$35°$. At other times of the day, as the Earth rotates, this geometry changes significantly. Six hours earlier or later, the x–z GSE cross section might look similar to the symmetric geometry in the left-hand panel. However, the cusps would be displaced in the y GSE direction and the plasma sheet would have a large tilt in y–z GSE cross section. As the tilt angle increases beyond the $35°$, would the normal diurnal independences of the magnetospheric morphologies remain? For example, would auroral zones still be referred to as a north and a south auroral oval? In the extreme case of a tilt approaching $90°$, does the plasma sheet in the x–z GSE cross section have two plasma sheets at certain UTs? Under these conditions with the same VADM and solar wind, dramatically different geospace climate would be observed in the form of auroral sightings as well as terrestrial magnetic field records of the electrojets and ring currents.

14.5.4 Geospace climate dependence upon the solar wind

In Section 14.4 we discussed that the very early Sun's X-EUV emissions were very much stronger (see Table 14.1) and thus led to a different thermospheric–ionospheric geospace climate. Hence, a parallel argument could result: if the Sun's solar wind was different from today's solar cycle variability, would this cause a different geospace climate? Most of today's knowledge of the early Sun's history, normally referred to times when the Sun reached its zero-age main sequence (ZAMS), has been obtained from studies of Sun-like stars, i.e. main-sequence G and K stars. Wood *et al.* (2002) and Newkirk (1980) determined X-ray activity, stellar mass loss, and solar wind velocity relationships. Griemeier *et al.* (2004) obtained time dependences for the solar wind velocity (v_{sw}) and density (n_{sw}) at 1 AU:

$$v_{sw} = v^* \left[1 + \frac{t}{\tau}\right]^{-0.4}, \tag{14.2}$$

$$n_{sw} = n^* \left[1 + \frac{t}{\tau}\right]^{-1.5}, \tag{14.3}$$

where $v^* = 3200 \, \text{km/s}$, $n^* = 2.4 \times 10^{10} \, \text{m}^{-3}$ and $\tau = 2.56 \times 10^7$ yr. Lundin *et al.* (2007) point out that more studies need to be made using a larger pool of Sun-like stars to refine these relationships. Using these relationships with the assumption that the geomagnetic field remains unchanged yields a subsolar magnetopause distance (with Eq. (10.1) in Vol. I) that is as small as $\sim 1/8$ times the present-day value for a young Sun, or $\sim 1.25 R_E$.

Figure 14.14 from Lundin *et al.* (2007) shows the large spread in the range of the n_{sw} dependence since the Sun reached ZAMS about 4.6 Gyr ago. Note

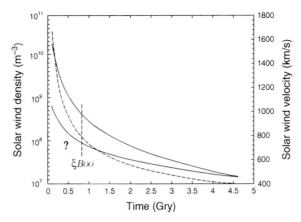

Fig. 14.14. Evolution of the observation-based minimum and maximum stellar wind densities scaled to 1 AU (left scale; solid lines) obtained from several nearby solar-like stars. On the right scale is the solar wind speed for the stellar wind evolution (dashed line). (From Lundin *et al.*, 2007.)

that these expressions and the plots shown in Fig. 14.14 have as a reference a present-day n_{sw} of $20 \, cm^{-3}$ ($2 \times 10^7 \, m^{-3}$) and a $v_{sw} = 400 \, km/s$. Today's solar cycle and solar storms have periods when the density can be almost a factor of 10 higher and the velocity reaches $1000 \, km/s$. These enhanced conditions are associated with storms and superstorms in geospace that can persist for days while the solar wind remains perturbed. If the Earth's intrinsic magnetic field were then as it is today, would geospace at 2 to 3 Gyr ago be in a continuous superstorm state? Figure 14.14 shows that at these times the solar wind's pressure would permanently be at, or exceed, superstorm solar wind conditions. Would the auroral phenomena be permanent displays and exist to very low latitudes, or would perhaps M-I coupling require an unsustainable flow of ionospheric plasma into the magnetosphere? The past geospace climate over the Holocene, human, time period was not significantly affected by the solar wind, while at very early ages it could well have been a very illuminating dynamic M-I coupling environment.

14.5.5 Geospace climate dependence upon non-dipole components

Although the Earth's magnetic field is qualitatively described as a dipole field, it is understood and, in fact, modeled as a much higher order field geometry (Chapter 7). These models are readily available, and on time periods as short as a few years show statistically significant changes in even the dipole components. Collecting historic records of the higher magnetic field moments is still in its early stages. Knudsen *et al.* (2008) provide a glimpse of this over the past 12 000 years. They indicate that improved fits can be made to magnetic field observations if an axial quadrupole moment is included. However, they caution the reader that more work is needed to specify when and how strong these terms were. Higher order moments of the magnetic field are also discussed as a possible scenario for field reversals. At the present time, neither case has compelling evidence for a geospace magnetosphere dominated by higher order magnetic field moments. Such scenarios most certainly would entertain extremely dynamic if not artistic geometries of what we regard as the key features of geospace; plasmasphere, plasma sheet, auroral ovals, cusps, etc. Indeed, one could imagine a cottage industry of MHD modeling of geospace's "aurora" and field dynamics.

14.6 Geospace climate response to anthropogenic change

For the past two decades, the question of global warming (climate change), especially where anthropogenic contributions may be significant, has been extensively studied and debated. In Sections 14.3 and 14.4, the role of the upper atmosphere in the climate of the thermosphere and ionosphere was discussed on long time lines, i.e. Maunder minimum time scales to the very earliest solar system times. The present-day debate on global warming refers to time scales of a century or less. If global warming is sustained, then an additional long-term influence on geospace climate exists.

Roble and Dickinson (1989) studied the consequences for the upper atmosphere of increases in the mixing ratios of ionospheric carbon dioxide and methane that resulted from a global warming–greenhouse scenario. As described in Sections 14.3 and 14.4, the role of carbon dioxide is to cool the thermosphere. Roble and Dickinson (1989) deduced a 50 K cooling for the thermosphere based on current climate trends. This would also lead to a lowering of the ionospheric layer altitudes. Rishbeth (1990) analyzed this ionospheric effect in more detail, making specific predictions for the magnitude of the altitude decrease. For example, using the Roble and Dickinson (1989) cooling of the thermosphere, Rishbeth (1990) predicted that the F_2 ionospheric layer should drop in altitude by 15 to 20 km. Such magnitudes are well within the observational sensitivity of the ionosonde technique. Ionosondes have been operational since the early twentieth century and many more since the IGY in 1957. Therefore, observational data sets exist to address this relatively short time scale, long-term geospace influence on the climate. Bremer (1992) used 33 years of ionosonde observations from Juliusruh (54.6 °N, 13.4 °E) to test the predictions of Rishbeth (1990) for the decrease in the ionospheric layer heights. Indeed, a positive result was obtained for the F_2 layer altitude, which decreased over the 33-year period. Figure 14.15 shows how the annual change in height has progressed over the 33 years at Juliusruh, the

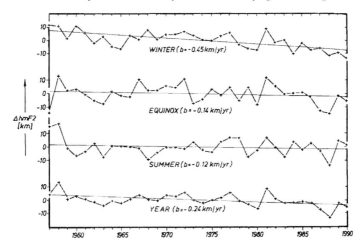

Fig. 14.15. Seasonal and yearly trends of the F_2 layer height (hmF2) at Juliusruh after elimination of solar and geomagnetic influences. (From Bremer, 1992.)

average decrease being about 0.24 km/yr. Over 50 years, this would correspond to 12 km.

However, there are many other ionosondes located around the world in different latitude and longitude regions. These early predictions by Roble and Dickinson (1989) and Rishbeth (1990), followed by observations at Juliusruh, have been followed by almost two decades of controversial ionosonde studies; Ulich and Turunen (1997), Bremer (1998), Jarvis *et al.* (1998), Upadhyay and Mahajan (1998), Danilov and Mikhailov (1999), Marin *et al.* (2001), Mikhailov and Marin (2001), Xu *et al.* (2004), and Yue *et al.* (2006).

This topic in itself deserves a chapter; however, a final conclusion is not forthcoming from these studies. The controversies arise from interpretation of the ionosonde's ionograms and how the true height is inferred. What does appear to hold true is that not all regions have observed a net decrease in the F_2 layer height. The compounding factors in the analysis revolve around the solar cycle, seasonal, and geomagnetic activity influences, all of which produce variability larger than the magnitude of the long-term trend. Other techniques have also been used on this issue. Holt and Zhang (2008) analyzed 29 years of Millstone Hill incoherent scatter radar measurements of the F layer ion temperature. They found a cooling trend over this period of -4.7 K/yr. Keating *et al.* (2000) and Emmert *et al.* (2004) analyzed satellite orbital elements to infer upper thermospheric density decreases. In both studies, the conclusion was that the decay rate was as much as -5% per decade.

In the context of long-term changes in geospace climate, these findings are beginning to assume significance. For example, the standard empirical models of the ionosphere and thermosphere utilized by a wide application community are based on observations spanning three to four sunspot cycles, only 30 to 40 years. However, the models originally did not include long-term trends of the sort being discussed. Efforts are now underway to evaluate the impact of long-term trends on these model specifications, and, most importantly, forecast where such trends become magnified.

15

Waves and transport processes in atmospheres and oceans

Richard L. Walterscheid

Waves in planetary atmospheres are interesting in their own right, but their importance lies mostly in the effects they have on the background atmosphere. Gravity waves may transport momentum, heat, and minor constituents through wave fluxes and may mix the atmosphere through the turbulence they induce when they break down (Lindzen, 1981; Fritts, 1984; Garcia and Solomon, 1985; Walterscheid, 1981, 1995, 2001; Walterscheid and Schubert, 1989). Planetary waves transport heat, momentum, and constituents (e.g. ozone) latitudinally and play a significant role in the heat, momentum, and ozone budgets (Holton and Wehrbein, 1980; O'Sullivan and Salby, 1990; Fusco and Salby, 1999).

15.1 Atmospheric waves

Planetary atmospheres admit a rich variety of waves. These waves involve to varying degrees the rotational, compressional, and buoyant properties of a fluid in motion. Many wave disturbances arise in instabilities, including those that give rise to weather systems. Other motions arise as free waves or waves forced by agents external to the atmosphere (e.g. planetary topography and solar heating). These comprise two broad classes of waves in planetary atmospheres: Rossby waves and gravity waves. Rossby waves are the comparatively low-frequency waves that dominate the ultra-long wave field in the lower atmosphere. Rossby waves are rotational waves where latitudinal displacements are opposed by the latitudinal gradient of planetary rotation. Gravity waves, in contrast, are comparatively high-frequency divergence waves where vertical displacements are opposed by pressure forces induced by gravity. Gravity waves with more-or-less typical phase speeds are vertically propagating. They have the ability to propagate from sources in the lower atmosphere into the thermosphere (see Vol. I, Fig. 12.1 for definitions of atmospheric domains), where they may achieve large amplitudes by virtue of a tendency to maintain constant energy density as they propagate into exponentially

Heliophysics: Evolving Solar Activity and the Climates of Space and Earth, eds. Carolus J. Schrijver and George L. Siscoe. Published by Cambridge University Press. © Cambridge University Press 2010.

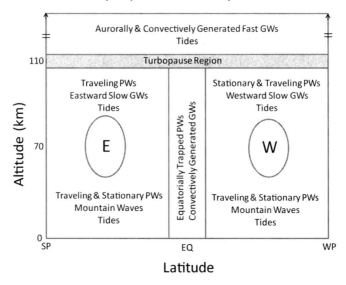

Fig. 15.1. Summary diagram for the dominant wave modes in the Earth's atmosphere as function of altitude and latitude from pole to pole, including gravity waves (GW), stationary and traveling planetary waves (PW), atmospheric tides, and mountain (gravity) waves. The approximate height of the turbopause, below which turbulent mixing dominates, is indicated. Here SP denotes the summer pole and WP the winter pole, E denotes the location of the stratospheric–mesospheric easterly (westward) jet and W denotes the location of the westerly (eastward) jet. The jet maxima are generally found within a few kilometers of 70 km altitude. The mix of waves in various regions indicates the dominant waves. Other waves may be present (for example non-equatorial convectively generated waves below the jets). The directionality associated with slow gravity waves above the jets is caused by the filtering action of the zonal wind system. North–south (meridionally) propagating gravity waves are also present, but the primary source of slow waves is zonal flow over mountains and this generates mostly zonal propagation.

decreasing ambient density. Atmospheric tides (principally those excited by the diurnal solar forcing) may be waves of the Rossby or gravity type. Like gravity waves, they may originate in the lower atmosphere and propagate into the upper atmosphere achieving very large amplitudes. Tides dominate the wave field of the upper atmosphere. Neutral motions in the upper atmosphere associated with waves may couple to the ionosphere and cause disturbances in the plasma. The reverse is also true: the ionosphere can couple to the neutral atmosphere. During geomagnetically active periods transient geomagnetic disturbances excite large amplitude waves mainly originating at high latitudes.

Planetary Rossby and Rossby–Haurwitz waves The adjective "planetary" refers both to the scale of the wave and to dynamics peculiar to rotating bodies.

Planetary waves are waves that depend in some way on the effects of rotation – in particular on the variation of the vertical component of rotation with latitude. These waves are also known as Rossby waves, which broadly speaking depend on the vertical component of rotation with latitude for their existence, or Rossby–Haurwitz waves, when specifically refering to planetary-scale oscillations on a sphere.

Gravity waves   Gravity waves are waves in a stratified medium for which gravity is the restoring force. An initial perturbation in the mass field (caused by, for example, vertical displacements or heating) in conjunction with gravity generates a propagating disturbance via an interplay between pressure forces and the divergence of the horizontal wind.

Atmospheric tides   Atmospheric tides are the response to periodic astronomical forcing. Atmospheric tides are forced primarily by the thermal heating due to the absorption of solar radiation by ozone and water vapor. These tides have periods that are the length of a mean solar day and its harmonics. The dominant tides have diurnal and semidiurnal periods. Tides are global in nature and exhibit the characteristics of Rossby–Haurwitz waves and gravity waves to a greater or lesser extent, depending on the tidal mode in question.

Plasma–neutral interactions   Neutral motions in the upper atmosphere associated with waves may couple to the ionosphere and cause disturbances in the plasma. These disturbances may oscillate (travel) with the wave or they may induce (seed) plasma instabilities. They may also generate electric fields through dynamo action and cause disturbances in the magnetic field and force a redistribution of charge in the ionosphere.

Transient disturbances in the ionosphere–thermosphere system   The Earth's upper atmosphere at high latitudes is the scene of very energetic geomagnetic disturbances with short durations. The perturbing effects of these events may generate disturbances in the neutral atmosphere. Ions driven into rapid motion by electric fields can accelerate neutrals through the drag exerted by the ions. Kinetic heating caused by energetic charged particles and Joule heating by the dissipation of electric currents can strongly heat the neutral atmosphere. Waves generated by impulsive events can travel far from the source region and affect the whole thermosphere. The wave types that are most likely to respond to impulsive events are normal modes of the Rossby type and gravity waves.

15.2 Examples of observed atmospheric waves

Some examples of large-scale atmospheric waves are examined here. For planetary waves one may distinguish two major subclasses: traveling and stationary. Traveling waves are normal modes (free waves) of the atmosphere. Stationary waves are waves forced by topographical features, mainly large mountain ranges, and are fixed to these features.

Traveling planetary waves   Figure 15.2 shows the results of a spherical harmonic analysis of the 24-hour tendency field in the height of the 500 hPa surface ($\sim$5.5 km altitude). The three waves shown have zonal wavenumbers $s = 1$, 2, and 3 (the first index of each index pair). The $s = 1$ wave has a period near 5 days, the $s = 2$ wave around 8 days, and the $s = 3$ wave around 16 days. These periods, along with a period near two days, are the periods of the dominant traveling waves.

Stationary planetary waves   Figure 15.3 shows contours of the height of two different pressure levels: in the troposphere at 500 hPa ($\sim$5.5 km altitude) and in the stratosphere at 10 hPa ($\sim$16 km) for July 15, 1958 and January 15, 1959. These dates were during the International Geophysical Year and were sampled especially well for the time. The large-scale wave structures seen in the winter stratosphere are quasi-stationary waves forced by topography in the lower atmosphere. The smaller-scale features seen in the troposphere are a combination of transient eddies

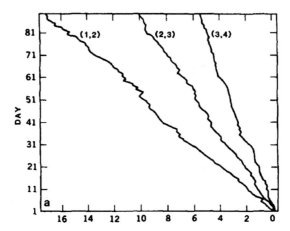

Fig. 15.2. The results of a spherical harmonic analysis of the 24-hour tendency field (i.e. the change over a 24-hour period) in the height of the 500 hPa surface (at $\sim$5.5 km) in the Earth's atmosphere. The abscissa is the number of passages of a fixed phase of a wave. The ordinate is the number of days since the first passage. The three waves shown have zonal wavenumbers $s = 1$, 2, and 3 (the first index of each index pair). (From Lindzen *et al.*, 1984, redrawn from Eliasen and Machenhauer, 1965.)

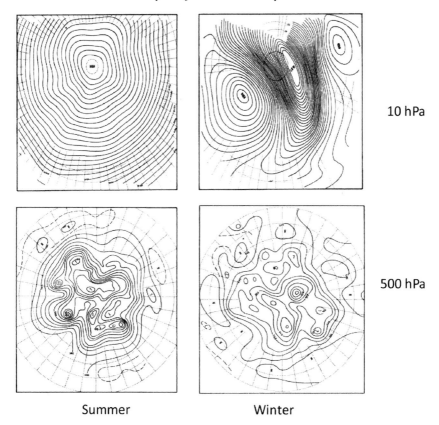

10 hPa

500 hPa

Summer   Winter

Fig. 15.3. Traveling planetary waves: contour maps of the height of two different pressure levels in the Northern Hemisphere: in the troposphere at 500 hPa (~5.5 km altitude) and in the stratosphere at 10 hPa (~16 km) for July 15, 1958 (left) and January 15, 1959 (right). (From Charney, 1973.)

(including waves on the polar front during winter) and the larger-scale stationary waves. Note that the summer-stratosphere flow (which approximately parallels the contours) is almost purely circumpolar with little indication of waves. In the winter, the flow in the stratosphere shows large-scale structure, but the small-scale structure seen in the troposphere has been filtered out.

Tides Figure 15.4 shows a long wave train of gravity waves visualized in clouds over the Indian Ocean. The waves are traveling on an interface between two layers of differing thermal properties.

Figure 15.5 shows the vertical variation of the amplitude of the diurnal component of the meridional wind at a low-latitude location for all four seasons. One notes an overall tendency for tides to increase with height in amplitude superimposed on considerable structure. The growth is common to vertically propagating

Waves and transport processes in atmospheres and oceans

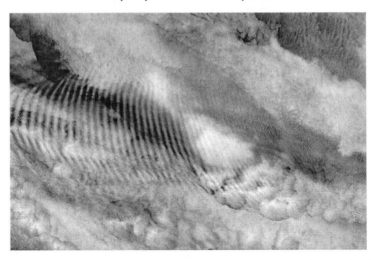

Fig. 15.4. A long wave train of gravity waves visualized in clouds over the Indian Ocean. The waves are traveling on an interface between two layers of differing thermal properties. (Image courtesy NASA/GSFC/LaRC/JPL,MISR team.)

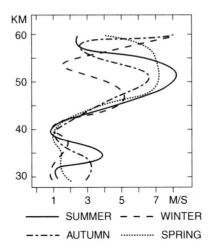

Fig. 15.5. The vertical variation of the amplitude of the diurnal component of the meridional wind at a low latitude location for all four seasons. (From Nastrom and Belmont, 1976.)

planetary waves, gravity waves, and tides. It is a result of waves approximately maintaining constant energy density (energy per unit volume) as they propagate upward through exponentially decreasing ambient density. The structure reflects the fact that tides of a given period comprise a superposition of modes with different wavelengths and these modes may interfere (Chapman and Lindzen, 1970).

15.3 Dynamics of planetary waves

Planetary waves may be regarded as vorticity waves modified by divergence. The prototypical planetary wave (Rossby wave) is non-divergent and is thus a pure vorticity wave.

15.3.1 Vorticity and absolute vorticity

Vorticity is defined as the curl of velocity. It is a three-dimensional vector and is a measure of swirl analogous to angular velocity. The large-scale winds are mainly horizontal, thus one is mainly interested in the vertical component of the vorticity vector, because only motions in a horizontal plane contribute to motion with swirl around a vertical axis. For horizontal motion in solid rotation the vorticity is twice the angular velocity. Henceforward in this chapter, the vertical component of the vorticity is called simply the vorticity.

The Earth's rotation vector may be separated into a component along the local vertical direction and the orthogonal northward direction. The vorticity derived from rotation about the vertical component is here called the planetary vorticity. Planetary and relative vorticity add to give the absolute vorticity. The planetary vorticity equals the Coriolis parameter

$$f = 2\Omega \sin \varphi, \qquad (15.1)$$

where Ω is the Earth's angular speed and φ is latitude. The absolute vorticity is the vertical component of the vorticity that would be measured in an inertial frame. Absolute vorticity in the free atmosphere is approximately conserved on short time scales. We will use this fact to discuss the mechanism that supports planetary waves.

15.3.2 Mechanism for planetary wave motion

Figure 15.6 shows in schematic form the mechanism of planetary wave motion based on conservation of absolute vorticity along a parcel's trajectory. Relative vorticity may be resolved into a component due to curvature along the path and shear normal to the path. Only the interplay between curvature and planetary vorticity is considered in this simplified example shown for the Northern Hemisphere. An

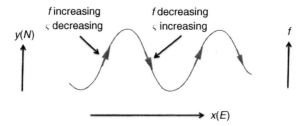

Fig. 15.6. The mechanism of planetary wave motion shown in schematic form based on conservation of absolute vorticity along a parcel's trajectory.

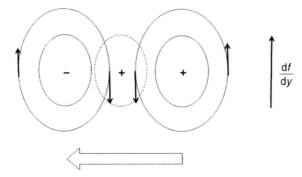

Fig. 15.7. An illustration of Rossby wave propagation. The schematic shows two relative vorticity cells of opposite sign. The contours are stream functions with anticlockwise flow around the positive center and clockwise flow around the negative center (see text for details).

eastward-moving parcel is given a northward displacement at a point where the relative vorticity is nil (an inflection point). As it moves northward, it moves toward larger values of planetary vorticity f. To conserve absolute vorticity the northward moving parcel acquires increasing negative relative vorticity until it reaches a turning point, where it begins to move southward toward decreasing planetary vorticity. As the parcel moves southward the curvature of its trajectory increases and acquires positive curvature after moving through an inflection point. The trajectory again reaches a turning point and begins to move northward. This is repeated to give wave motion.

15.3.3 Mechanism of wave propagation

The discussion in Section 15.3.2 explains how wave motion originates but it does not explain wave propagation. This is addressed in Figure 15.7. The schematic shows two relative vorticity cells of opposite sign. The contours are stream functions with anticlockwise flow around the positive center and clockwise flow around the negative center. The flow is southward between the two centers down the latitudinal gradient df/dy of planetary vorticity. This generates a positive vorticity tendency. The positive cell (the one on the right) will be displaced toward the center of positive tendency. It is easy to see that a negative-tendency center is formed to the left of the negative center, with the effect that cells move to the left, i.e. to the west.

15.3.4 Prototype Rossby waves

15.3.4.1 Non-divergent barotropic motion on a β-plane

As we have seen, Rossby waves depend for their existence on the latitudinal variation of planetary vorticity. The simplest dynamics that supports these waves is non-divergent barotropic motion (i.e. an incompressible fluid in which pressure depends on density only) on a so-called β plane. A β plane is a Cartesian frame where the essential effects of sphericity are retained by expanding the latitudinal variation of planetary vorticity (Coriolis parameter) in a Taylor expansion about fixed latitude φ, thus

$$f = f_0 + \beta y, \tag{15.2}$$

where $\beta = (df/dy)_0$ and $y = a\sin(\varphi_0)(\varphi - \varphi_0)$. The β plane was first used by Rossby (1939).

15.3.4.2 Dispersion relation for Rossby waves

The dispersion relation for Rossby waves is obtained from the linearized non-divergent barotropic vorticity equation (Rossby, 1939). Assuming a waveform solution of the form $\psi = A\exp(i\omega - kx)$ and a uniform zonal background flow $\bar{u}$, one obtains the dispersion relation

$$\omega = \bar{u}k - \beta/k, \tag{15.3}$$

where ω is the wave angular frequency and k is the horizontal wavenumber for the eastward (positive x) direction.

The phase velocity is

$$\frac{\omega}{k} = c = \bar{u} - \beta/k^2; \tag{15.4}$$

hence Rossby waves propagate westward (against the direction of rotation) relative to the mean flow, in agreement with the qualitative discussion above.

The group velocity is

$$\frac{\partial \omega}{\partial k} = u_g = \bar{u} + \beta / k^2, \tag{15.5}$$

whence group velocity is eastward relative to the mean flow. Because wave energy flows with the group velocity it is seen immediately that Rossby waves propagate energy and phase in opposite directions with respect to the zonal current.

15.3.4.3 Rossby-wave group propagation

The propagation of phase and amplitude are closely related; both propagate with the group velocity. A representation of group (energy) propagation for Rossby waves (indeed any waves) is the Hovmöller diagram. In a typical representation, it shows the variation of some field versus time and longitude at fixed latitude. Figure 15.8 shows a Hovmöller diagram of this kind for Rossby waves in the ocean (Quartly *et al.*, 2003). In this diagram, westward tilt of sea features (sea surface temperatures, chlorophyll concentration, and sea surface height) indicates eastward propagation of Rossby-wave groups. This is clearly seen throughout the figure.

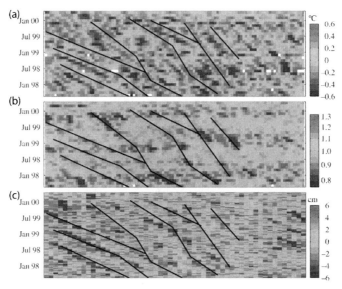

Fig. 15.8. A time–longitude representation of group (energy) propagation for Rossby waves in the ocean (Hovmöller diagram), showing the variation of (a) sea surface temperature anomalies, (b) chlorophyll concentration anomalies, and (c) sea surface height anomalies. The tilt of features in this diagram reflects the eastward propagation of wave groups. (From Quartly *et al.*, 2003.) See Color Plate 19.

15.4 Dynamics of gravity waves

Gravity waves depend for their existence on entropy stratification in a gravitational field (Lighthill, 1978). The stratification can be concentrated at an interface – as in a surface water wave – or can be continuous – as in a mountain lee wave. Whereas Rossby waves propagate vorticity, gravity waves propagate via the interplay of the divergence of the horizontal wind field and horizontal pressure gradients.

15.4.1 Prototype gravity waves

The simplest system that supports gravity waves is an incompressible fluid of constant depth. In the example shown in Fig. 15.9, the fluid has a free surface and an undisturbed average depth of $\bar{h}$. The fluid is initially deformed with a displacement height of η. In a gravitational field, the displacement of the free surface gives a horizontal pressure gradient that acts to the left and right of the position of the maximum displacement and – for wave motion that is not too rapid – is the same at all depths below the undisturbed surface. The pressure gradient drives the fluid to the left and right into still fluid, causing the free surface to fall under the initial displacement and to rise on its flanks, giving traveling disturbances that propagate to the left and right (in contrast to Rossby waves, which propagate only in one direction).

15.4.1.1 Dispersion relation for gravity waves

The dispersion relation for waves that are long compared to the fluid depth is found from the linearized equations of motion to be

$$\omega^2 = g\bar{h}k^2. \tag{15.6}$$

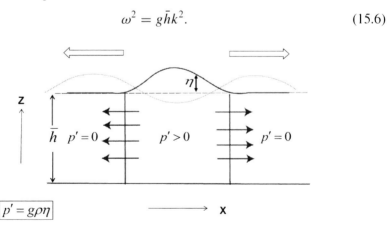

Fig. 15.9. Gravity waves in an incompressible fluid of uniform depth. The fluid has a free surface and an undisturbed average depth of $\bar{h}$. The fluid is initially deformed with a displacement height of η. Wave propagates through interplay between horizontal divergence and pressure gradients in a gravitational field.

The phase velocity is found to be

$$c_x = \pm\sqrt{g\bar{h}}, \qquad (15.7)$$

in agreement with the qualitative discussion above that the wave propagates to the left and right (or east and west, say).

The group velocity equals the phase velocity (these waves are non-dispersive), which implies eastward and westward energy propagation.

15.5 Quantitative theory of oscillations on a rotating sphere

In this section we sketch the quantitative theory for oscillations on an ideal ocean surrounding a rotating planet. This theory will turn out to be better suited to atmospheres than oceans inasmuch as oceans are confined to basins, whereas atmospheres entirely surround planets.

15.5.1 Shallow-water equations

The system we examine is based on the so-called shallow-water equations where the water depth is shallow compared to the horizontal scale of the motion. The motion is also considered to be quasi-static, that is, the motion is not static, but the motions are slow enough that a balance between gravity and the vertical pressure gradient is maintained.

15.5.2 Oscillations of an ideal ocean: Laplace's tidal equation

We study the oscillations of the free surface of an ocean of uniform undisturbed depth $\bar{h}$ (as in Fig. 15.9) covering a rotating planet. This system, though highly idealized, supports a rich variety of motions that are found in both oceans and atmospheres. These include Rossby–Haurwitz waves, gravity waves, Kelvin waves, and tides (Longuet-Higgins, 1968; Flattery, 1967). Laplace's tidal equations are the linearized equations of motion on a background state of rest for an ideal ocean. One eliminates dependent variables in favor of the geopotential oscillation $\phi' = g\eta$, assumes waveform solutions of the form

$$\phi'(\varphi, \lambda, t) = \Phi(\varphi)\exp(i(\omega t - s\lambda)), \qquad (15.8)$$

where s is zonal wavenumber, and λ is longitude, and obtains an equation of the form

$$F\left[\theta^{\omega,s}(\varphi)\right] = \epsilon\theta^{\omega,s}(\varphi). \qquad (15.9)$$

The operator F is defined as

$$F = \frac{d}{d\mu}\left(\frac{1-\mu^2}{f_*^2 - \mu^2}\frac{d}{d\mu}\right) - \frac{1}{f_*^2 - \mu^2}\left[\frac{s}{f_*}\left(\frac{f_*^2 + \mu^2}{f_*^2 - \mu^2}\right) + \frac{s}{1-\mu^2}\right], \quad (15.10)$$

where

$$\epsilon = \frac{g\bar{h}}{(2\Omega R_{\mathrm{p}})^2}. \quad (15.11)$$

The parameter ϵ is known as Lamb's parameter and here and henceforward, except as noted explicitly, f_* is the normalized frequency $f_* = \omega/2\Omega$. The latitudinal coordinate is $\mu = \sin(\varphi)$ and $R_{\mathrm{p}} \equiv a$ is the planetary radius. Equation (15.9) is Laplace's tidal equation (LTE). One solves the LTE subject to boundary conditions as an eigenfunction-eigenvalue problem to obtain the eigenfunction θ and the eigenvalue f_*. The eigenfunctions of the LTE are called Hough functions (Hough, 1898). The solutions depend parametrically on s and ϵ. For every s and ϵ there is a complete set of modes each with a different frequency and latitudinal structure.

15.5.2.1 Wave types for Laplace's tidal equation (limiting cases)

In general, the solutions of the LTE must be obtained numerically and are not simply related to elementary functions. However, the solutions can be classified in broad terms based on some limiting cases. Of particular interest is the small ϵf limiting case that admits close analogs to the simple waves that we considered earlier. There are two subclasses of interest.

High-frequency Class-1 oscillations (irrotational gravity waves on a sphere)
For large f_* the eigenfrequencies satisfy the relation

$$f_n^2 = \epsilon n(n + 1). \quad (15.12)$$

These waves propagate eastward and westward. The eigenfunctions are

$$\theta_n^s(\mu) = P_n^s(\mu), \quad (15.13)$$

where the $P_n^s(\mu)$ are Legendre polynomials. These waves are irrotational gravity waves on a sphere. Waves of this type are referred to as Class-1 waves by Longuet-Higgins (1968). In dimensional terms Eq. (15.12) is

$$\omega_n^s = \frac{g\bar{h}}{R_{\mathrm{p}}^2}n(n + 1), \quad (15.14)$$

which is similar to Eq. (15.6), the dispersion relation for shallow-water gravity waves.

Low-frequency Class-2 oscillations (non-divergent Rossby–Haurwitz waves)
For small f_* the eigenfrequencies satisfy

$$f_n = -\frac{s}{n(n+1)}. \tag{15.15}$$

These waves propagate only westward. As for Class-1 waves, the eigenfunctions are Legendre polynomials. They are Rossby waves on a sphere, also referred to as Rossby–Haurwitz waves. Waves of this type are denoted Class-2 waves (Longuet-Higgins, 1968). In dimensional terms

$$\omega_n = -\frac{2\Omega s}{n(n+1)}, \tag{15.16}$$

which is similar to Eq. (15.3) and analogously implies propagation against rotation (i.e. westward on Earth).

15.5.2.2 Eigensolutions of Laplace's tidal equation

Except for special cases, the first solutions to Laplace's tidal equations were obtained numerically by Longuet-Higgins (1968) and Flattery (1967). Flattery (1967) solved for the Hough functions and the associated equivalent depths for tides. Here, we present the solutions obtained by Longuet-Higgins (1968) which also apply to traveling free waves.

Eigenfrequencies   Figure 15.10 shows the eigenfrequencies for zonal wave number 1 ($s = 1$). Results are shown for eastward and westward traveling waves. The intersections of the plotted curves (with vertical lines denoting fixed ϵ) are the eigenfrequencies of discrete modes of oscillations (normal modes). The curves are labeled with indices that characterize the eigenfunctions in the limit of large and small ϵ. The small-ϵ limit relates to the solutions given above where the solutions are spherical harmonics. The label $n - s$ denotes the number of nodal crossing for Class-1 waves (gravity waves) and $n' - s$ denotes the same thing for Class-2 waves (Rossby–Haurwitz waves). The Rossby–Haurwitz waves (denoted Rossby waves) are the westward-traveling low-frequency waves that become independent of ϵ as ϵ becomes small, in agreement with the limiting solution Eq. (15.12). The gravity waves are the eastward and westward traveling high-frequency waves that slope linearly as $\epsilon^{1/2}$, again in agreement with the limiting solution in Eq. (15.12). The curves denoted Kelvin waves and MRG (mixed Rossby–gravity waves) are special cases that we do not discuss here. Likewise, we do not discuss the limiting cases for large ϵ. For these waves the reader is referred to Longuet-Higgins (1968).

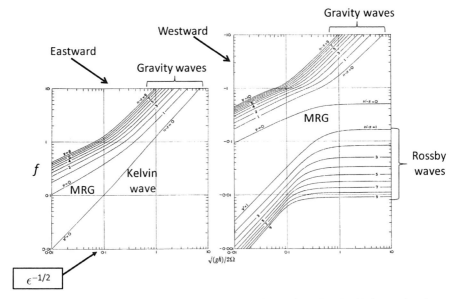

Fig. 15.10. The eigenfrequencies for eigensolutions of Laplace's tidal equation for zonal wave number 1 ($s = 1$). Results are shown for eastward and westward traveling waves. The ordinate is the normalized frequency $f_* = \omega/2\Omega$. The abscissa is the inverse square root of Lamb's parameter ϵ. Mixed Rossby–gravity modes are labeled MRG. (Adapted from Fig. 2 in Longuet-Higgins, 1968.)

Eigenfunctions   The eigenfunctions (Hough functions) for $s = 1$ for the two lowest westward modes ($n - s = 0, 1$) of Class 1 are shown in Figure 15.11. Note that as ϵ increases the oscillations become increasingly equatorially trapped.

15.6 Oscillations of an atmosphere

Oscillations in an atmosphere differ from those in an ocean in the existence of an added dimension. Rather than a layer of fluid of well-defined thickness bounded by a free surface, where the density stratification is concentrated, the vertical stratification of an atmosphere varies continuously and this makes the problem inherently three dimensional.

The tidal equations for an atmosphere are similar to those for an ocean, except for a statement of the first law of thermodynamics (energy conservation) and an equation of state (relating pressure, density, and temperature), while the continuity equation now describes a compressible medium. It is convenient to transform the equations into log-pressure coordinates where the vertical coordinate is $z = -\log(p/p_0)$, where p is pressure and p_0 is a reference value, usually taken to be 1000 hPa for Earth. The vertical velocity is $w = \dot{z}$. The governing equations

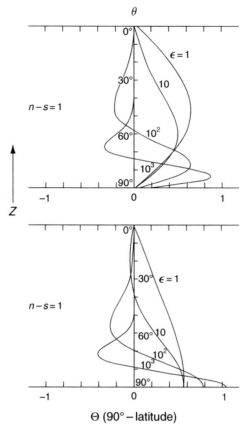

Fig. 15.11. The eigenfunctions (Hough functions) for $s = 1$ for the two lowest westward modes ($n-s = 0, 1$) of Class-1 waves. The curves are for fixed values of Lamb's parameter ϵ. The ordinate is co-latitude ($90° -$ latitude). The abscissa (Z) is a normalized value of the Hough functions. (Adapted from Fig. 8 of Longuet-Higgins, 1968.)

may be linearized and reduced to a single partial differential equation in a single variable, the vertical velocity, say, whence

$$\left[\frac{\partial}{\partial z} - 1\right] \frac{\partial}{\partial z} W(\varphi, z) + \frac{S(z)}{4a^2\Omega^2} F\left[W(\varphi, z)\right] = 0, \qquad (15.17)$$

where solutions of the form

$$w'(\varphi, \lambda, z, t) = W(\varphi, z) \exp\left(\mathrm{i}(\omega t - s\lambda)\right) \qquad (15.18)$$

have been assumed and where w' is the vertical velocity perturbation in log-pressure coordinates.

15.6.1 Separation of variables

One may assume solutions of the form

$$W(\varphi, z) = \hat{w}(z)\Theta(\varphi). \tag{15.19}$$

This separation gives equations that describe separately the vertical (z) and horizontal (latitudinal, φ) structure of the oscillation.

15.6.1.1 Vertical structure equation

It is convenient to weight the dependent variables by density so that the vertical structure equation may be placed in canonical form thus

$$\frac{d^2}{dz^2}\tilde{w}(z) + m^2(z)\tilde{w}(z) = 0, \tag{15.20}$$

where $\tilde{w}(z) = \exp(-z/2)\hat{w}$, and the refractive index (also vertical wavenumber; dimensionless in this coordinate system) is

$$m^2 = \frac{S(z)}{gh} - \frac{1}{4H^2(z)}. \tag{15.21}$$

The quantities S and H are

$$S(z) = \mathcal{R}\bar{T}\frac{d\log(\bar{\theta})}{dz}, \tag{15.22}$$

$$H(z) = \frac{\mathcal{R}T}{g}, \tag{15.23}$$

where $\theta = T\exp(\kappa z)$ is potential temperature, T is temperature, $\mathcal{R}$ is the gas constant, and $\kappa = \mathcal{R}/c_p$, and where c_p is the specific heat at constant pressure. Overbars refer to mean-state quantities. The quantities S and H are, respectively, a measure of the thermal stratification and of the scale height. The quantity h is the separation constant and is called the equivalent depth. This designation is justified by Eq. (15.24) below.

Waves in a continuously stratified fluid (atmospheres) may be internal or external. Internal waves exhibit wave-like behavior in the vertical direction. Conservative, steady-state internal waves maintain nearly constant energy density with altitude. This implies that wave amplitude grows exponentially. Internal waves originating in the lower atmosphere may propagate to great heights and achieve large amplitudes. External waves have constant phase with height and wave amplitude decays in height; these waves are evanescent.

The vertical wavenumber governs whether the wave is internal or external. If $m^2 > 0$ the wave is internal while if $m^2 < 0$ it is evanescent. Equation (15.21) shows that vertical propagation is favored by small h. Evanescence occurs for large h. It may also occur that equivalent depths are negative (see below) in which case the wave is always evanescent.

15.6.1.2 Horizontal structure equation

The separation of variables gives the equation

$$F\left[\Theta(\varphi)\right] = \frac{(2\Omega a)^2}{gh}\Theta(\varphi). \tag{15.24}$$

This is just Eq. (15.9). Thus the horizontal structure of oscillations of an atmosphere is the same as for an ocean of depth h. This is the meaning of the term equivalent depth.

15.6.2 Eigensolutions for atmospheric waves

Whereas for an ideal ocean the depth is a given physical quantity, the equivalent depth for an atmosphere is a separation constant that is obtained as part of the solution. The equivalent depth is found as an eigenvalue of the vertical structure equation. The corresponding eigenfunction gives the vertical structure. For an isothermal atmosphere the eigenfunction is easily obtained and has the value $h = \gamma H$ (see Walterscheid, 1980, and references therein). Here, $\gamma = c_p/c_v$, where c_p and c_v are the specific heats at constant pressure and constant volume, respectively. There is only one eigenvalue and all free waves in log-pressure coordinates have the same vertical structure:

$$\hat{\Psi} \propto \exp\left[\left(\kappa - \frac{1}{2}\right)z\right], \tag{15.25}$$

where ˆ refers to height-dependent amplitude, and where Ψ is any dependent variable. The free modes are Lamb waves for which the vertical velocity in geometric coordinates vanishes.

The solution of the vertical structure equation gives h, whence one may solve the horizontal structure equation for the eigenfrequencies and corresponding Hough functions just as one would for an ideal ocean of depth h. This process is illustrated in Figure 15.12 for an $s = 1$ westward wave. To a good approximation $h = 10$ km. This gives a value of $\epsilon^{-1/2} \approx 0.34$. The intersections of the solution curves with a vertical line at this value of $\epsilon^{-1/2}$ gives the eigenfrequencies for a subset of the infinite solution set. One such intersection is illustrated. It is for the highest-frequency Rossby–Haurwitz mode. The intersection near $f_* = 0.1$ corresponds to a period of 5 days. This is one of the main free waves in the Earth's atmosphere (Eliasen and Machenhauer, 1965; Walterscheid, 1980, and references therein).

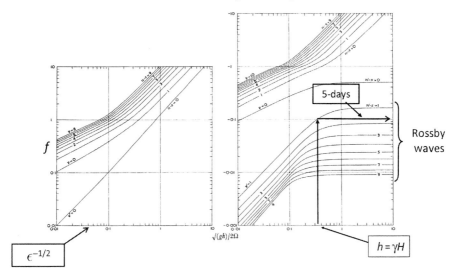

Fig. 15.12. An illustration of the process for determining eigenfrequencies for an atmosphere for an $s = 1$ westward propagating wave. (Adapted from Fig. 2 of Longuet-Higgins, 1968.)

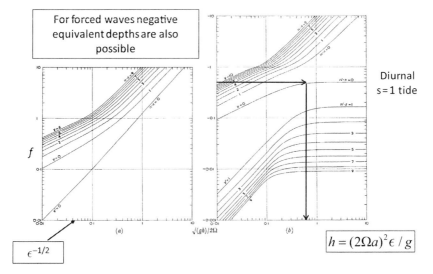

Fig. 15.13. Same as Fig. 15.12 but for a forced wave. Rather than one equivalent depth and an infinite set of associated frequencies there is an infinite set of equivalent depths associated with a given frequency. (Adapted from Fig. 2 of Longuet-Higgins, 1968.)

15.6.3 Main free waves in the terrestrial atmosphere

The main free waves in the terrestrial atmosphere are those found by Eliasen and Machenhauer (1965; the quasi 5, 8, and 16-day waves) plus the quasi 2-day wave. The 5, 8, and 16-day waves are found in the stratosphere and above, while the 2-day

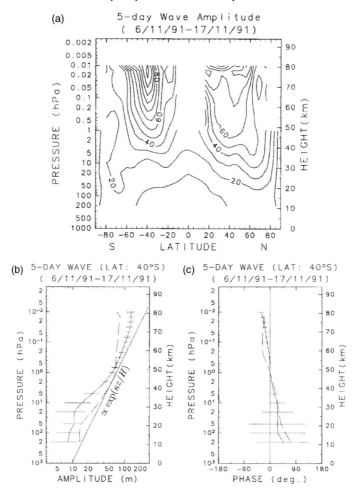

Fig. 15.14. The observed structure of the 5-day free wave in the terrestrial atmosphere November 6–17, 1991. Panel (a) shows the observed structure as a function of height (expressed as pressure) and latitude (from South to North Poles); contour intervals, 10 m. Panels (b) and (c) show the amplitude and phase, respectively, as function of height (pressure) at a latitude of 40 °S. (From Hirooka, 2000. Reproduced by permission of the AAS.)

wave seems to be limited primarily to the mesosphere and lower thermosphere (Walterscheid and Vincent, 1996; Walterscheid, 1980; Madden, 1978).

The observed structure of the 5-day wave is shown in Fig. 15.14. This figure shows the observed height and latitude structure. The wave is symmetric with a mid-latitude maximum in both hemispheres. The oscillation grows with altitude. The altitude growth rate is found to agree well with the theoretical value. The phase is nearly constant, also in agreement with theory. This identifies the wave as the gravest $s = 1$ Rossby–Haurwitz mode. Because it is also a Lamb wave it is perhaps better referred to as a Lamb–Rossby (or Lamb–Rossby–Haurwitz) wave.

Free waves are easy to excite and maintain. They may be the selective response to random fluctuations, excited by wave–wave interactions and parametric instabilities, or driven by normal mode instabilities (especially baroclinic instabilities, i.e. instabilities in which pressure depends on both density and temperature; Plumb, 1982).

15.7 Forced waves

Forced waves are waves for which the forcing determines the wave frequency or the structure of the wave. Here, we are concerned with the latter (i.e. tides). Unlike free waves, forced waves may have internal wave structure. Two classes of forced waves are considered. The first are oscillations on a sphere and are treated within the framework of Laplace's tidal theory. The second are quasi-stationary waves in zonal flow. These waves require a different framework to describe propagation that depends on winds.

15.7.1 Laplace's tidal theory for forced waves

Rather than one equivalent depth and an infinite set of associated frequencies, there is an infinite set of equivalent depths associated with a given frequency. The equivalent depths are found from the horizontal structure equation (Eq. 15.24) with frequency given. The vertical structure is found subject to the forcing and depends on the equivalent depth h as a parameter.

This is illustrated in Fig. 15.13 where again we have used the results of Longuet-Higgins shown in Fig. 15.10.

Given the frequency of the forcing, one draws a horizontal line at that frequency, for this example taken to be the diurnal frequency. The intersection of this line with the solution curves gives a set of Lamb's parameters (subset of the infinite set). Given a planet's rotation rate and radius this defines a set of equivalent depths.

In this example, all of the equivalent depths are positive. However, for waves with frequencies lower than the semidiurnal frequency, negative equivalent depths exist. The eigenfunctions for both positive and negative equivalent depths are required to form a complete set (Flattery, 1967; Chapman and Lindzen, 1970). Waves with small positive equivalent depths are mainly Class-1 waves (gravity waves), while waves with large equivalent depths are mainly Class-2 (Rossby–Haurwitz waves). Waves with negative equivalent depths have a different classification (see Longuet-Higgins, 1968).

It should be noted that the term equivalent depth for forced waves (although ingrained) is not as apt as it is for free waves. For free waves, the oscillations of an atmosphere have the same infinite set of frequencies and horizontal structures as those for an ocean of depth h, and h is a property of the atmosphere. However, for forced waves the equivalent depths are independent of the atmosphere and only one frequency (the forcing frequency) is associated with each h.

15.7.1.1 Vertical propagation of forced waves

For forced waves, the refractive index (Eq. 15.21) may be positive (internal waves) or negative (external or evanescent waves), depending on the atmospheric thermal structure and the value of h. For a given value of S, vertical propagation is favored by small values of equivalent depth and large values favor evanescence. Negative equivalent depths always give evanescence. For the Earth's atmosphere, values of h smaller than about 8 km give internal waves. It is far easier for internal waves to propagate energy away from the sources than for external waves. Internal waves propagate vertically with nearly constant energy density, while external waves can be confined close to the source (Chapman and Lindzen, 1970).

15.7.1.2 Tides

In general terms, tides are the periodic response to periodic astronomical forcing. In the atmosphere, by far the dominant forcing agent is thermal excitation by solar radiation, although forcing by latent heat release can also be important (Chapman and Lindzen, 1970; Lindzen, 1978). The dominant atmospheric tides are the diurnal tide and the semidiurnal tide at double the frequency. In the lower and middle atmosphere, tides are excited primarily by the absorption of solar UV radiation by stratospheric ozone and solar near-IR radiation by tropospheric water vapor. The diurnal tide is forced about one-third by water vapor absorption and about two-thirds by ozone absorption. The semidiurnal tide is predominately forced by ozone absorption. Although the diurnal component of the diurnal variation of solar heating is stronger than the semidiurnal component, there is a rough parity between the two because the semidiurnal tide responds more efficiently to ozone forcing than does the diurnal tide. This is because the region of ozone forcing is fairly deep and main semidiurnal modes with their comparatively long vertical wavelengths respond in phase over the forcing regions, while the diurnal tide with its fairly short wavelengths experiences a degree of phase cancellation (Chapman and Lindzen, 1970).

15.7.2 Small-scale gravity waves in an atmosphere

Small-scale gravity waves ($ka \gg 1$) in an atmosphere differ from ones in a shallow wave system (see Section 15.4.1.1) in that they have vertical structure. The dispersion relation reflects the vertical structure of both the wave and the background state. The dispersion relation for waves that are not too fast or too deep is given approximately as

$$m^2 = \frac{N^2 - \omega_I^2}{\omega_I^2 - f^2} - \frac{1}{4H^2}. \tag{15.26}$$

Here m is dimensional, $N^2 = s/H^2$ is the square of the Brunt–Väisälä frequency[†] and f is the Coriolis parameter introduced above. The quantity $\omega_I = k(c - \bar{u})$ is the intrinsic phase velocity (the frequency seen in the frame of the background flow). As $c - \bar{u}$ becomes very large, m^2 becomes negative and the wave evanescent. As c approaches $\bar{u}$ (the wave approaches a critical level), m^2 becomes very large. When a critical level is encountered a strong interaction between the wave and background flow is possible (Andrews and McIntyre, 1976; Andrews *et al.*, 1987).

15.7.3 Stationary waves

Flow over orographical features and land–sea boundaries forces waves with zero frequency in a coordinate system that is fixed relative to the planet. These waves are Rossby waves that can exist because they have non-zero frequencies in the frame of the background wind (the intrinsic frame). These waves require a different formalism than the one that has been used for tides.

15.7.3.1 Quasi-geostrophic dynamics

The framework for theoretical studies of Rossby wave propagation through background flow in middle and high latitudes is the quasi-geostrophic system. In this system the motion is non-divergent except when coupled to planetary vorticity and the stream function is geostrophic (i.e. as when Coriolis and pressure-gradient forces balance). Wave propagation is described by the equation describing conservation of potential vorticity along a parcel's trajectory projected on a horizontal plane or constant pressure surface. The attribute "potential" refers to the fact that vorticity, though not conserved, will return to its initial value whenever the initial background conditions are re-encountered along the horizontal projection of its trajectory. The conservation equation is called the quasi-geostrophic potential vorticity equation. The attribute "quasi" comes from the fact that conservation occurs along the horizontal projection rather than along the total trajectory accounting for vertical motion (Charney and Stern, 1962).

[†] The Brunt–Väisälä frequency, or buoyancy frequency, is the frequency at which a vertically displaced parcel of air oscillates adiabatically and remains in pressure balance with its surroundings.

15.7.3.2 Charney–Drazin theory

With significant simplifications (constant background conditions) Charney and Drazin (1961) derived an equation that describes the vertical propagation of linear Rossby waves through mean zonal flow. These authors obtained a refractive index that governs whether the waves are internal or external. Those that are external may be regarded as being blocked from propagating into the stratosphere and above by the zonal flow. The refractive index is

$$ m^2 = \frac{S}{f_0^2}\left[\frac{\beta}{\bar{u}} - (k^2 + l^2)\right] - \frac{1}{4}, \tag{15.27} $$

where l is the latitudinal wave number.

In eastward flow, vertical propagation is favored by long horizontal wavelengths (small k). The stronger the zonal winds, the smaller k has to be to give propagation. In westward flow all waves are blocked. These indications are consistent with the observations shown in Fig. 15.3. During winter, only the longest waves can propagate into the strong eastward flow. During summer, all waves are blocked and the flow is very nearly zonally symmetric.

15.7.4 Sources of stationary waves

Excited mainly by flow over large continental mountain areas and secondarily by land–sea contrasts, stationary waves are locked to topography. The strongest waves are the very long ones ($s = 1$–3). Planetary waves are stronger in the Northern Hemisphere because of the existence of two extensive mountain systems there (Trenberth, 1980).

15.8 Atmospheric waves on other planets

Wave features are seen in the permanent cloud cover surrounding Venus. Among these are the persistent "Y" pattern with the stem oriented along the equator. The Y pattern is seen in the UV images of the clouds. The pattern appears as the ribbed structures that open to the left at low latitudes in Fig. 15.15. These structures have been interpreted as Rossby waves and have been reproduced in simulations (Smith *et al.*, 1993). Other waves (e.g. Kelvin waves) have been seen in the Venus atmosphere (Covey and Schubert, 1982). Rossby waves also have been seen in other planetary atmospheres, notably Jupiter (Li *et al.*, 2006), and in the solar atmosphere (Kuhn *et al.*, 2000).

Fig. 15.15. The persistent "Y" pattern seen in this negative of a UV image of the permanent cloud cover surrounding Venus. The pattern (on this negative image) appears as the ribbed structures that open to the left at low latitudes. The stem is oriented along the equator. These structures have been interpreted as Rossby waves. (NSSDC photo gallery: http://nssdc.gsfc.nasa.gov/photo_gallery/photogallery-venus.html.)

15.9 Transports and wave forcing

15.9.1 Non-interaction theorem

When certain conditions are met, wave transports will not force a change in the mean zonal state. These conditions are that the waves be steady-state, conservative, linear, and not encounter a critical level (a level where the phase speed in the frame of the mean zonal winds vanishes; Eliassen and Palm, 1961; Andrews and McIntyre, 1976). When one or more of these conditions is not met, wave transports may force changes (sometimes dramatic) in the mean state.

15.9.2 Role in sudden stratospheric warmings

Sudden warmings are dramatic warmings of the winter polar stratosphere, sometimes by 50 K or more (Scherhag, 1960; Quiroz, 1969; Matsuno, 1971). They are forced by a sudden amplification of stationary waves in the high-latitude winter stratosphere, thus violating the steady-state non-interaction condition. This causes a deceleration of the polar westerlies and heat transport northward. The circumpolar vortex becomes highly distorted, breaks down, and the mean zonal winds change from westerlies to easterlies. When this occurs, a critical level

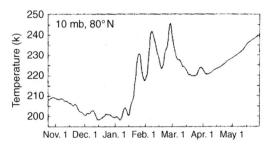

Fig. 15.16. Example of high-latitude warming (at 80° N) associated with strato-spheric sudden warmings. The warming at 80° N rises from its midwinter value of ~200 K to ~245 K in a series of three events between mid January and early March before relaxing to a more typical temperature of ~225 K. (From Andrews *et al.*, 1987, after Gille and Lyjak, 1984.)

forms, violating another non-interaction condition and the process accelerates. The high-latitude stratospheric warming is shown in Fig. 15.16. The warming at 80° N rises from its mid-winter value of ~200 K to ~245 K in a series of three events between mid-January and early-March before relaxing to a more typical temperature of ~225 K.

The manifestation in the upper mesosphere and lower thermosphere is cooling rather than warming (Myrabo, 1984; Walterscheid *et al.*, 2000; Liu and Roble, 2002). This may be due to gravity waves that have greater access to the upper atmosphere in the wind conditions that prevail during sudden warmings or because the stationary waves induce a circulation in this region that is opposite to the one in the stratosphere (Dunkerton, 1978; Walterscheid, 2001).

15.10 Climatic effects of waves

Not only can wave transports force transient phenomena, they can also force the climatic state of atmospheres (Holton, 1983). Wave-transport effects (including those due to convection and turbulence generated when large-amplitude waves break down) may alter the long-term distributions of winds, temperature and minor constituents (Lindzen, 1981; Walterscheid, 1981; Holton, 1983; Andrews *et al.*, 1987; Garcia and Solomon, 1985). Probably the most profound example in the Earth's atmosphere is the effect that waves have in the upper mesosphere. Here, wave forcing is thought to be responsible for maintaining the mean state far out of radiative equilibrium (Holton, 1983).

The summer polar mesopause as the coldest place in the Earth's atmosphere
The summer polar mesopause is coldest during summer, opposite to what would obtain under radiative equilibrium (Wehrbein and Leovy, 1982). This is shown

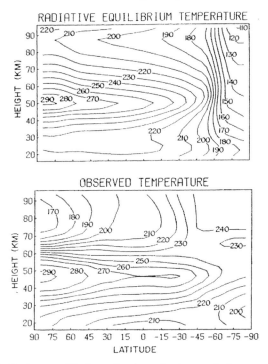

Fig. 15.17. Radiative-equilibrium (top) and observed (bottom) temperatures (K) in the stratosphere, mesosphere and lower thermosphere. The mesopause, the temperature minimum and transition between the mesosphere and thermosphere, is coldest during summer, opposite to what would obtain under radiative equilibrium. The upper panel shows the zonally averaged temperature versus latitude and altitude calculated for radiative equilibrium and the lower panel shows the observed value during solstice. (From Wehrbein and Leovy, 1982.)

in Fig. 15.17. The upper panel shows the zonally averaged temperature versus latitude and altitude calculated for radiative equilibrium during solstice and the lower panel shows the observed values. Figures 15.18 and 15.19 show the zonal winds versus altitude and latitude under the assumption of radiative equilibrium and as observed, respectively (Geller, 1983; Andrews *et al.*, 1987). Under radiative equilibrium conditions the mesospheric jets do not close in the mesosphere, while the observed winds peak near 65 km and decrease upwards. The decreasing wind speed in the upper mesosphere and the latitudinal gradient of temperature with lower temperatures in the summer mesosphere is consistent with thermal wind balance. The strong departure from radiative equilibrium along with the reversal of the wind shear must be maintained by some dynamical process. It is thought that this is accomplished by wave drag exerted by dissipating gravity waves. The direct effect of the drag is to decelerate the winds in the upper mesosphere. The latitudinal

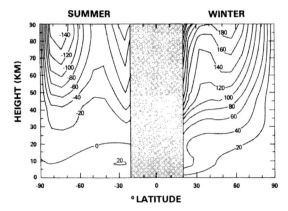

Fig. 15.18. Geostrophic mean zonal winds calculated from radiative equilibrium temperatures. Units are m s^{-1}. Eastward winds are positive. (From Geller, 1983.)

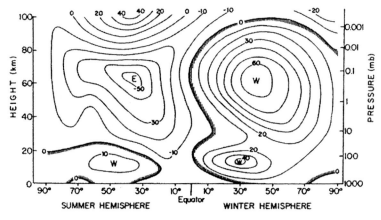

Fig. 15.19. Same as Fig. 15.18 but for observed mean zonal winds. (From Andrews *et al.*, 1987; courtesy R. J. Reed.)

thermal gradient required by thermal wind balance is brought about by a secondary wave-driven circulation in the altitude–latitude plane (Holton, 1983; Lindzen, 1981; Fritts, 1984; Walterscheid, 1995).

The quasi-biennial oscillation (QBO)   A prominent feature of the Earth's stratosphere at low latitudes is the quasi-biennial oscillation (QBO; Reed *et al.*, 1961). The QBO is a long-period oscillation of the circulation at low latitudes with a variable period averaging approximately 28 months. It is seen as a downward-propagating system of easterly and westerly wind regimes. The amplitude is $\sim$30 m s^{-1}. The effects of the QBO are not confined to atmospheric winds. Temperature and chemical constituents, such as ozone, water vapor, and methane are affected by the QBO (Baldwin *et al.*, 2001). The QBO is thought to be forced by an interaction of the mean zonal winds and waves coming up from below (Lindzen and Holton, 1968; Holton and Lindzen, 1972; Plumb, 1977; Mayr *et al.*, 1997). Various waves have been suggested including Kelvin waves, mixed Rossby–gravity waves, and gravity waves (Lindzen and Holton, 1968; Holton and Lindzen, 1972; Andrews *et al.*, 1976, 1987; Mayr *et al.*, 1997). The mix has been debated, but most likely includes gravity waves (Baldwin *et al.*, 2001). The downward descent of the QBO is caused by a feedback between the mean zonal winds and waves owing to the selective absorption of westward-propagating waves in westward winds and eastward-propagating waves in eastward winds (Lindzen and Holton, 1968; Holton and Lindzen, 1972).

The two examples given above are just two of a number of examples of wave–mean-flow interactions. Some others are the Earth's semiannual oscillations (Andrews *et al.*, 1987) and the super-rotation of the Venus atmosphere (Schubert *et al.*, 1980).

15.11 Waves in the ionosphere–thermosphere (IT) system

Waves forced from below grow in amplitude as they propagate upwards and waves of very small amplitude in the lower atmosphere can become large-amplitude waves in the thermosphere. This applies especially to internal tides and gravity waves, though there is evidence of traveling planetary waves in the thermosphere (Garcia *et al.*, 2005). Waves may be excited by regular periodic forcing such as tides, or they may be excited by impulsive events such as troposphere convection or the thermospheric response to the sudden onset of geomagnetic activity.

Waves may interact with the ionosphere and perturb the plasma, giving a disturbance with the same frequency and phase speed and the same horizontal scale. The magnitude of the response depends on the amplitude of the underlying neutral waves and the details of the interaction between neutrals and ions. The interaction depends strongly on the vertical gradient of ion density (Belashova *et al.*, 2007).

15.11.1 Ionosphere–thermosphere response to periodic or stationary forcing

Tides in the lower thermosphere propagate up from the troposphere where they are forced by absorption of solar radiation by water vapor in the troposphere and ozone in the stratosphere. Forcing by latent heat release is also important. The main tides are those that are Sun-synchronous (i.e. follow the Sun; Chapman and Lindzen, 1970). These are called the migrating tides. Tides that are not Sun-synchronous (non-migrating tides) are also significant and can dominate in some places at some times (Angelats and Forbes, 2002; Aso, 2007). Non-migrating tides may be forced by longitudinal asymmetries in forcing and by non-linear interactions (Lindzen, 1978; Williams and Avery, 1996; Angelats and Forbes, 2002). *In-situ* forcing by the absorption of radiation in the Schumann–Runge continuum of molecular oxygen in the lower thermosphere also makes a significant contribution (Chapman and Lindzen, 1970).

Figure 15.20 shows the daily estimates of diurnal tidal wind amplitude for an altitude of 95 km. The tide is modulated with a semiannual period reflecting the twice yearly crossing of the equator by the Sun. The amplitude of the tide is bounded by ~10 m/s for the extreme minima to ~100 m/s for the extreme maxima for the semiannual cycles.

Tides in the upper and lower atmosphere are forced mainly by absorption of solar radiation by atmospheric gases. The tides forced in the troposphere and stratosphere are damped by viscosity, thermal diffusion, and ion drag when they propagate into the thermosphere. The main thermal forcing in the thermosphere is due to the absorption of solar extreme ultraviolet radiation by atomic oxygen (Chapman and Lindzen, 1970).

There is little if any evidence of quasi-stationary waves in the middle thermosphere and above. The main difficulty is getting the waves through the summer westward winds in the stratosphere and mesosphere, and through the strong eastward winds in the winter stratosphere and mesosphere and then through the westward wind above (Garcia *et al.*, 2005).

15.11.2 Ionosphere–thermosphere response to impulsive forcing

Nearly all gravity waves, whether they are excited in the lower or upper atmosphere, are excited by imbalances in the flow. These are often associated with

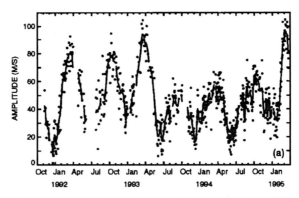

Fig. 15.20. Daily estimates of diurnal tidal wind amplitude from HRDI data for an altitude of 95 km. The data cover the period from October 1991 to March 1995. (From Burrage *et al.*, 1995.)

transient or impulsive events. Free Rossby modes are easily excited by impulsive events, which may be viewed as a ringing phenomenon. Tides, on the other hand, should not respond strongly to impulsive events of limited duration because they depend on a sustained change in the forcing to produce a corresponding change in the periodic response (Walterscheid, 1997).

Tides in the lower thermosphere Gravity waves excited in the lower atmosphere dominate the gravity wave field in the lower thermosphere. In the middle and high thermosphere there is also a large (perhaps dominant) population of waves excited in the lower atmosphere; however, waves excited in the auroral zone can also make a significant contribution.

Gravity waves generated in the lower atmosphere Waves generated by convection, flow over topography, geostrophic adjustment, and shear instabilities may propagate into the thermosphere if they can avoid breakdown, filtering by the background winds, and dissipation by eddy and molecular diffusion and ion drag (Lindzen, 1970, 1981; Pitteway and Hines, 1963). The waves that are able to avoid breakdown are those with sufficiently small initial amplitudes. The background winds may filter out waves by causing them to become evanescent or by causing them to be absorbed due to scale-dependent diffusion. Waves that avoid filtering by background winds are those whose intrinsic frequency does not become either too slow – whence they may be subject to critical level absorption or scale-dependent diffusion – or too fast – whence they may become evanescent (compare Eq. 15.26). The waves that may propagate high in the thermosphere are those that are fast enough to escape severe attenuation by molecular diffusion (Pitteway and Hines, 1963; Walterscheid and Hickey, 2005). Fast waves have long vertical wavelengths and are less subject to scale-dependent diffusion. In addition, high-frequency waves suffer less dissipation over a wave cycle than low-frequency waves subject to the same rate of dissipation.

Seeding of equatorial spread-F Equatorial plasma depletions on the bottom side of the F region appear in ionosondes as a spreading of the returned signal in range or frequency (Farley *et al.*, 1970; Kelley, 1989, and references therein). They typically occur in two bands encircling the Earth at low magnetic latitude on either side of the dip equator known as the Appleton anomaly (Aarons, 1993). Equatorial plasma depletions are widely believed to be caused by a gravitational Rayleigh–Taylor instability (Sultan, 1996). Some theoretical considerations (Scannapieco and Ossakow, 1976; Ossakow, 1981) indicate that growth from thermal plasma fluctuations is too slow to account for the observations, implying the need for seeding by wave-like plasma perturbations. Atmospheric gravity waves are thought to provide this seeding (Huang and Kelley, 1996; Huang *et al.*, 1993). Singh *et al.*, (1997) analyzed Atmospheric Explorer-E data and identified cases where wave-like horizontal winds, indicative of gravity waves, were closely correlated with plasma drift perturbations. Data from successive orbits showed the development of equatorial spread-F, presumably from these seeds. The most likely source of the gravity waves is low-latitude convection (thunderstorms).

Aurorally generated gravity waves (traveling ionospheric disturbances)
Gravity waves are excited when there is a rapid change in the heating of the thermosphere by Joule heating. This can be caused by the dissipation of auroral currents in the ionosphere (Joule heating) and by the kinetic heating caused by precipitating particles when they collide with air molecules (particle heating; Banks and Kokarts, 1973; Walterscheid *et al.*, 1985; Lu *et al.*, 1995). Large-amplitude large-scale waves may be launched during the sudden onset of elevated levels of geomagnetic activity, such as during magnetic storms (Richmond and Matsushita, 1975). These waves may cause perturbations in the ion density and propagate great distances. The ionospheric manifestations are referred to as traveling ionospheric disturbances (TIDs). Figure 15.21 shows an example of a traveling atmospheric disturbance seen in density near 400 km measured by the accelerometer on the CHAMP satellite in connection with a geomagnetic disturbance. The disturbances appear to penetrate into opposite hemispheres from their origins in the Northern

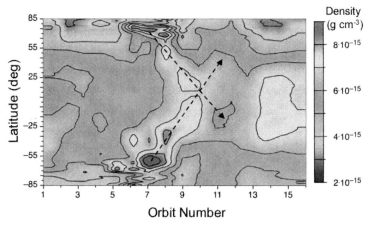

Fig. 15.21. An example of a traveling atmospheric disturbance seen in density near 400 km measured by the accelerometer on the CHAMP satellite in connection with a geomagnetic disturbance. The disturbances appear to penetrate into opposite hemispheres from their origins in the Northern and Southern Hemisphere auroral zones. The simultaneous appearance in both hemispheres is due to conjugate activity. (From Forbes, 2007.) See Color Plate 20.

and Southern Hemisphere auroral zones. The simultaneous appearance in both hemispheres is due to conjugate activity.

15.11.3 Traveling planetary waves in the ionosphere

Traveling planetary waves in the ionosphere can be the upward extension of waves in the lower atmosphere or they can be waves excited within the thermosphere either by impulsive events such as a magnetic storm or by instabilities (Pancheva *et al.*, 2006). Figure 15.22 shows a wavelet analysis of two indices of geomagnetic activity (ap and Dst). The figure shows wavelet amplitude as a function of frequency and day number. Around day 20 one sees the broad-band response to impulsive forcing due to geomagnetic disturbances with significant power near two days. The vertical solid dark lines at days 26 and 70 enclose the occurrence of a 2-day wave in the upper mesosphere and lower thermosphere (MLT). The

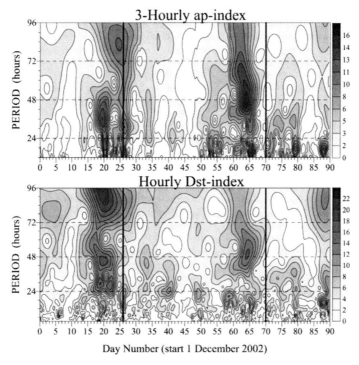

Fig. 15.22. A wavelet analysis of two indices of geomagnetic activity, ap and Dst. The figure shows wavelet amplitude as a function of frequency and day number. Around day 20 one sees the broad-band response to impulsive forcing due to geomagnetic disturbances with significant power near 2 days. The vertical solid dark lines at days 26 and 70 enclose the occurrence of a 2-day wave in the upper mesosphere and lower thermosphere (MLT); the ionospheric manifestation is seen as a series of wavelet power maxima near 48 h. (From Pancheva *et al.*, 2006.)

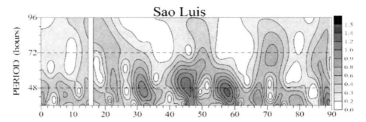

Fig. 15.23. A wavelet analysis similar to Fig. 15.22 but for F_2-region critical frequencies, $f_0 F_2$. (From Pancheva *et al.*, 2006.)

response in terms of F_2-region critical frequencies, $f_0 F_2$, is shown in Fig. 15.23. There is another event close to day 65. This event shows significant power at 2 days. However, there is no 2-day wave response evident in Fig. 15.23, indicating that the response does not extend to low latitudes for this event.

15.12 Consequences of changing the planetary rotation

For the Earth, the Brunt–Väisälä frequency (see below Eq. 15.26) is much larger than the planetary vorticity (or Coriolis parameter, Eq. 15.1), i.e. $N \gg f$: the inertial frequency for Earth is $f \approx 2\Omega \sin(\varphi) \sim 10^{-4}\,\mathrm{s}^{-1}$, while an estimate for N based on isothermal conditions gives $N \sim g/\sqrt{c_p T} \sim 2 \times 10^{-2}\,\mathrm{s}^{-1}$.

From Eq. (15.26) it is easy to see that for gravity waves $m^2 > 0$ only for $N > \omega_1 > f$. Gravity waves cannot propagate vertically outside of this frequency range. Suppose that a planet is rotating so rapidly that $N = f$. Then $m^2 < 0$ and vertically propagating gravity waves do not exist. If this were the case on Earth, the climatic state of the upper atmosphere would be much different than it is. The mesospheric jets would be higher and stronger and the summer polar mesopause would be not nearly so cold, and might be warmer than the winter pole. The quasi-biennial oscillation would be altered and either would exhibit a much different periodicity, or not exist at all. A more rapid rotation yet would allow gravity waves to exist but for $f > \omega_1 > N$. (Note that waves where the Coriolis force is significant are called inertial gravity waves.)

The Earth would have to spin ~ 200 times faster for f to be comparable to N. For Jupiter's high atmospheric layers, $f \sim 2.5 \times 10^{-4}\,\mathrm{s}^{-1}$ (given that it has about 2.5 times the Earth's angular rotation rate), and the corresponding Jovian values of N are larger than Earth's because gravity is also about 2.5 times larger, while the denominator in the above expression for N is about the same. However, in Jupiter's adiabatic interior N is close to zero through a deep layer (with a thickness of $\sim 220\,\mathrm{km}$) below ~ 1 bar, and in this layer it is possible that $N < f$, so that gravity waves are evanescent in that layer.

15.12.1 Planetary Rossby and gravity waves

In the same spirit one can examine what happens to planetary-scale gravity and Rossby waves for planets that do not rotate and planets that rotate very fast. This has already been touched on in Section 15.5.2. For zero rotation only irrotational gravity waves exist. The exact dispersion relation for gravity waves on an ideal ocean is given by Eq. (15.12). Normal modes are all Class-1 waves. Rotational flow is stationary geostrophic flow. On a non-rotating planet the length of a day is the same as the orbital period (a year). Thus tides on a non-rotating planet are all gravity waves with periods of a year and its harmonics. Tides on very slowly rotating planets are very high frequency in the sense of normalized frequency $f_* = \omega/2\Omega$ and are associated with small values of Lamb's parameter ϵ (see Fig. 15.10). Observe from Fig. 15.11 that this implies that the tides are very broad functions of latitude, whereas tides with large values of ϵ are increasingly concentrated at low latitudes. Large values of ϵ are associated with rapid rotation.

For years there was a theoretical effort to explain the observed parity between the diurnal and semidiurnal tides (despite the fact that the thermal forcing for the former was much greater) in terms of resonance (Chapman and Lindzen, 1970). These efforts were fruitless (and in the end unnecessary), but resonances at tidal frequencies are hypothetically possible. A planet is despun toward synchronous rotation by gravitational torques acting on the tidal bulge of the solid planet. Eventually the 5, 8, and 15-day normal modes of the Earth's atmosphere (altered in period by the changing rotation rate) may come into resonance with the solar thermal forcing. The resonantly amplified tides would dominate the motion field over a much greater region of the atmosphere than they presently do. The effects of vertical wave transports, however, would be mitigated by the fact that the tides would be Lamb waves which have evanescent vertical structure and for which the vertical velocity is small.

The ultra-large-scale undulations of the Earth's extra-tropical jet steam are the quasi-stationary Rossby waves forced by zonal flow over topography. These undulations of the jet stream are evident in Fig. 15.3 at 500 hPa during winter (where the contours are most closely packed). The amplitudes of the ultra-long waves are especially pronounced at 10 hPa where the latitudinal amplitude of the waves during winter is tens of degrees. The amplitudes of these waves on Earth exceed those for other bodies in the solar system. The large gaseous bodies lack a comparable topographical source. Rossby waves exist because of the latitudinal gradient in rotation and Venus is a slowly rotating body. The rapid rotation of the Venus atmosphere makes up for this to a large extent, but is still rather slow (an effective rotation rate of something on the order of an Earth week). In addition, though Venus has very prominent orographical features, the zonal winds near the surface are thought to be rather weak (Schubert *et al.*, 1980).

15.12.2 Tidal effects on planetary rotation

Planets with thick atmospheres can feel effects in the opposite direction, that is, atmospheric tides can affect rotation. It is speculated that all planets formed with similar rotation rates and spun in the prograde sense (aligned with the total angular momentum of the solar system). Gravitational torques can despin rotation toward synchronous rotation, but cannot produce retrograde rotation. The torques acting on the solar tidal bulge and coupling with the solid planet can however cause retrograde rotation and this is what may have produced the retrograde rotation of Venus. The present state of Venus is thought to be an equilibrium between gravitational and thermal atmospheric tidal torques (Gold and Soter, 1969; Kundt, 1977). Clearly the resonances supported by planetary atmospheres can affect where equilibrium states might be found and thus the speed of retrograde rotation.

15.12.3 The variation of other planetary parameters (an exercise)

The parameters that control tides and free waves are the non-dimensional parameters ϵ and the normalized frequency f_*. These parameters depend on rotation rate Ω, planetary radius R_p, and h, which may be the undisturbed depth of an ideal ocean or an equivalent depth, which for free waves in an atmosphere depends on the scale height H. We have examined how the properties of waves change for changes in Ω. It is an instructive exercise to examine how waves are changed as a and h are varied.

16

Solar variability, climate, and atmospheric photochemistry

Guy P. Brasseur, Daniel Marsch, and Hauke Schmidt

16.1 Introduction

The possible link between solar variability and climate remains an intriguing and controversial issue. Solar physicists have shown that the total solar irradiance ($S_0 = L_\odot/4\pi d_\odot^2$), whose value is close to $1366\,\mathrm{W\,m^{-2}}$ (Fröhlich, 2004), varies typically by $1.5\,\mathrm{W\,m^{-2}}$ (slightly more than 0.1%) over an 11-year solar cycle (see Chapter 10 and Fig. 10.9). This change is considerably smaller than the radiative forcing produced by enhanced concentrations of greenhouse gases since the beginning of the industrial era. Figure 16.1 (from the Fourth Assessment conducted by the Intergovernmental Panel on Climate Change, or IPCC) highlights a possible longer term trend in the solar irradiance, but, unless some amplification mechanisms occur, this forcing remains small compared, for example, to the effect of carbon dioxide and other radiatively active gases, whose atmospheric concentrations have increased as a result of human activity.

Different mechanisms have been proposed to explain the relations between the state of the atmosphere and the 11-year solar cycle. One of them is the absorption of short-wave solar radiation by ozone in the stratosphere with possible effects on the diabatic heating, temperature, and the general circulation of the atmosphere. Another mechanism refers to the impact of galactic cosmic rays on the formation of cloud condensation nuclei and hence on cloudiness and surface temperature. The cosmic-ray intensity in the atmosphere is anti-correlated with solar activity (Chapters 9 and 11). Kristjansson *et al.* (2008), however, have shown that observed cloudiness does not seem to be affected by rapid decreases in cosmic-ray intensity occurring occasionally as the result of large coronal mass ejections on the Sun. Thus, this mechanism seems unlikely to play an important role, but more work is needed before definitive conclusions can be reached.

In the following sections, we briefly review the atmospheric processes that govern the behavior of the climate system, present the observed variations in the

Heliophysics: Evolving Solar Activity and the Climates of Space and Earth, eds. Carolus J. Schrijver and George L. Siscoe. Published by Cambridge University Press. © Cambridge University Press 2010.

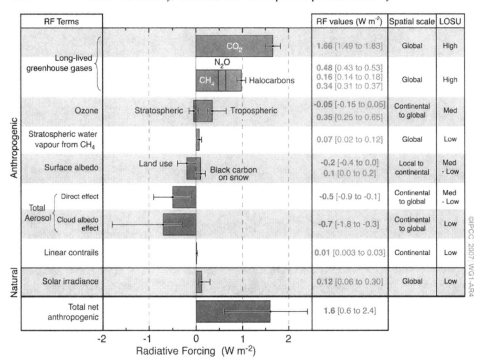

Fig. 16.1. Radiative forcing (RF, in $W\,m^{-2}$) relative to year 1750 associated with natural and anthropogenic changes in the atmosphere, as determined by IPCC (2007). The final column gives the IPCC's assessment of the level of scientific understanding (LOSU).

spectral distribution of the solar flux, summarize our understanding of the processes that govern the response of the upper and middle atmosphere, and discuss a possible dynamical response of the troposphere to solar variability.

16.2 The climate system

Perhaps the best way to represent the exchanges of radiative energy in the atmosphere is to refer to Fig. 16.2. This figure shows that, averaged over the Earth's surface and over a solar day, the Earth intercepts $342\,W\,m^{-2}$ (S_0 multiplied by the ratio of the Earth's disk to its area). The Earth reflects $107\,W\,m^{-2}$ back to space (with a planetary albedo of 0.31), and absorbs $235\,W\,m^{-2}$ ($168\,W\,m^{-2}$ by the surface and $67\,W\,m^{-2}$ by the atmosphere). The Earth's surface, whose average temperature is 288 K (15 °C), emits $390\,W\,m^{-2}$ as infrared radiation, in addition to $24\,W\,m^{-2}$ and $78\,W\,m^{-2}$ released as sensible (thermals) and latent heat (evaporation), respectively. A large fraction of the emitted infrared radiation ($350\,W\,m^{-2}$) is absorbed by greenhouse gases in the atmosphere. These gases, whose temperature

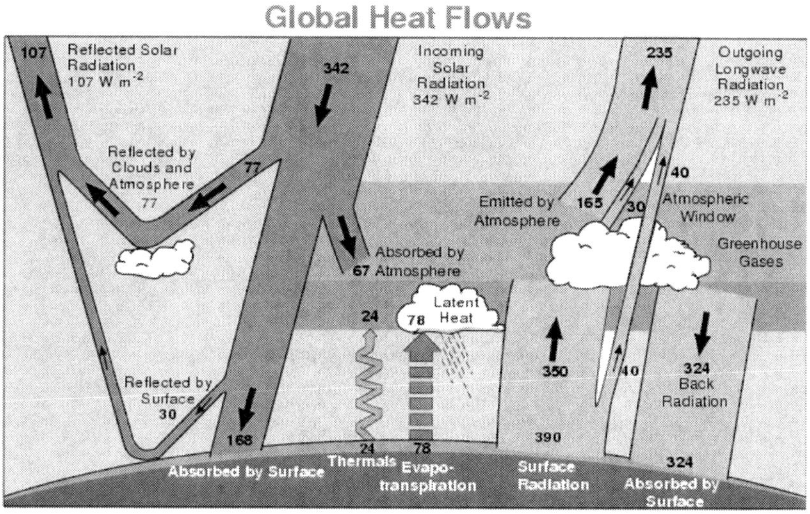

Global Heat Flows

Fig. 16.2. Exchanges of solar (short-wave) and terrestrial (long-wave) energy in the Earth's atmosphere. The flow of energy is expressed in $W\,m^{-2}$, averaged over the entire Earth surface (i.e. averaged over the day–night cycle). (From Kiehl and Trenberth, 1997.)

is lower than the surface temperature, re-emit radiation both towards space and towards the Earth's surface. After integration over all atmospheric layers, the resulting infrared flux amounts to $165\,W\,m^{-2}$ at the top of the atmosphere and to $324\,W\,m^{-2}$ at the Earth's surface. At the top of the atmosphere, the incoming solar energy ($342\,W\,m^{-2}$) is balanced by the reflected short-wave energy ($107\,W\,m^{-2}$) and the outgoing terrestrial energy ($235\,W\,m^{-2}$ from which $165\,W\,m^{-2}$ are emitted by atmospheric gases, $30\,W\,m^{-2}$ by clouds and $40\,W\,m^{-2}$ directly by the surface, including the oceans).

The first question that we address here is to assess the response of the climate system to a change in the total solar irradiance (TSI; compare Section 11.2.3). For this purpose, we develop a simple model of the radiative transfer processes described above; we represent the atmosphere by a single layer of radiatively active gases whose optical transmission is noted T_S and T_L for short-wave (solar) and long-wave (terrestrial) radiation, respectively. The radiative short-wave solar and the long-wave surface fluxes are noted by symbol F_S and F_G, respectively. F_A represents the radiative flux emitted by the atmospheric layer and T_G, an indicator of the Earth's climate, represents the surface (ground) temperature. From Fig. 16.3, we derive the energy balance at the top of the atmosphere,

$$F_S = (1-a)S_0/4 = F_G T_L + F_A, \qquad (16.1)$$

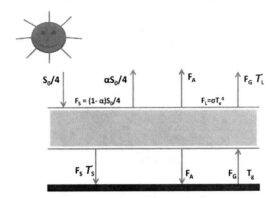

Fig. 16.3. Simplified model of radiative exchange in the Earth's atmosphere. The optical transmission in the atmosphere is represented by $\mathcal{T}_S$ in the short-wave (solar) spectral region and by $\mathcal{T}_L$ in the long-wave (terrestrial) part of the spectrum.

and at the Earth's surface,

$$F_S\mathcal{T}_S = F_G - F_A. \tag{16.2}$$

We deduce that the surface temperature is given by

$$T_g = T_e \left(\frac{1+\mathcal{T}_S}{1+\mathcal{T}_L}\right)^{1/4}, \tag{16.3}$$

where the planetary equilibrium temperature T_e (temperature of the Earth that would occur in the absence of an atmosphere; Eq. 11.33)

$$T_e = [S_0(1-a)/4\sigma]^{1/4} \tag{16.4}$$

is equal to 255 K ($-18\,°\mathrm{C}$) for $S_0/4 = 342\,\mathrm{W\,m^{-2}}$ and for an albedo $a = 0.31$. Assuming that the atmosphere is approximately transparent to solar radiation, so that the short-wave transmission $\mathcal{T}_S$ is close to 1.0, and adopting a long-wave transmission $\mathcal{T}_L$ of 0.2, the surface temperature becomes

$$T_g = T_e \left(\frac{2.0}{1.2}\right)^{1/4} = 289\,\mathrm{K}. \tag{16.5}$$

The value calculated by this simple model, tuned by approximating choices for $\mathcal{T}_S$ and $\mathcal{T}_L$, is in agreement with the observed temperature $T_{e,obs}$ (288 K). More refined models account in greater detail for wavelength-dependent radiative transfer, vertical and horizontal heat transport in the atmosphere, and energy and water exchanges at the Earth's surface. Absorption coefficients for different molecules in different spectral regions are measured in the laboratory. Prominent atmospheric absorbers of solar ultraviolet and visible radiation are ozone (Hartley, Huggins, and Chappuis bands) and molecular oxygen (Schumann–Runge continuum and bands;

Herzberg continuum). Absorption of terrestrial infrared radiation results primarily from the 15 μm band of carbon dioxide, from the 9.6 μm band of ozone and from various spectral features of water vapor. The simple conceptual model presented here can, however, be used to estimate to a first approximation the change in the surface temperature that would result, for example, from a relative change in the solar input S_0 of 0.1%. We derive easily (see Eq. 11.34) that, for constant T_S and T_L,

$$\frac{\Delta T_g}{T_g} = \frac{\Delta S_0}{4S_0}.$$ (16.6)

For $T_g = 288$ K, we obtain a surface temperature change ΔT_g of 0.07 K for a solar-cycle TSI variation of 1.5 W m^{-2}. The amplitude of the solar variation is therefore a factor of 10 smaller than the surface temperature trend observed since the beginning of the industrial era (see Fig. 12.2). However, over a period of a decade or so, the solar signal should be significant compared to human-driven temperature trends, and should therefore be taken into consideration in the analysis of temperature records. Shindell *et al.* (1999) have shown that, even if the global temperature variation associated with solar forcing is small, changes in temperature patterns become significant at the regional scale.

A more accurate treatment requires that the transmission functions and the atmospheric emissivity change with the chemical composition of the atmosphere in response to Sun-induced climatic changes, that dynamical feedbacks be taken into account and that the influence of the ocean be considered. Meehl *et al.* (2009), for example, have shown that, as the Sun reaches maximum activity, it heats cloud-free parts of the Pacific Ocean enough to increase evaporation, intensify tropical rainfall and the trade winds, and cool the eastern tropical Pacific. Coupled three-dimensional ocean–atmosphere models that include a detailed formulation of the coupling between radiative transfer, photochemistry, and dynamics in the whole atmosphere must be used to account for possible mechanisms (feedbacks) that could potentially amplify the response to solar variability. This aspect is discussed in greater detail in Section 16.6. In the following sections, we first examine how solar variability affects the chemical composition, the temperature, and the dynamics of the middle atmosphere. In later sections, we review possible mechanisms by which stratospheric disturbances propagate downwards to the troposphere and produce a signal that is larger than can be expected from direct radiative forcing.

16.3 Atmospheric photochemistry

The photolysis of molecules by solar radiation initiates a number of chemical reactions in the atmosphere, and has a direct impact on the atmospheric chemical composition (Brasseur and Solomon, 2005). An example of photolysis is provided by the photodissociation of molecular oxygen (O_2):

$$O_2 + h\nu \rightarrow O + O, \tag{16.7}$$

which leads to the formation of two oxygen atoms. These atoms react with molecular oxygen to produce ozone molecules (O_3):

$$O + O_2 + M \rightarrow O_3 + M. \tag{16.8}$$

Here, M represents a "third body" (e.g. N_2, O_2, Ar), which removes the thermal energy released by this exothermic reaction. In this example, the rate of ozone production is directly proportional to the frequency at which oxygen molecules are photodissociated:

$$P(O_3) = 2J_{O2}[O_2], \tag{16.9}$$

if J_{O2} represents of the photodissociation coefficient of O_2 and $[O_2]$ the number density of this molecule. The photodissociation frequency depends on the intensity of the incoming solar radiation and on the ability of the molecule to absorb solar photons at particular wavelengths. This last parameter is generally expressed as a wavelength-dependent absorption cross section $\sigma_X(\lambda)$, which can also vary with temperature. In more general terms, the photolysis frequency of a molecule X is expressed as an integral over all wavelengths that contribute to the decomposition of the molecule. The upper bound of this integral corresponds to the minimum energy required to break the molecular bond. The probability that the absorption of a photon leads to the dissociation of molecule X is expressed by the quantum efficiency ϵ_X, which also varies with wavelength and in some cases with temperature. Thus,

$$J_X = \int_\lambda q(\lambda, z, \chi)\, \sigma_X(\lambda, T(z))\, \epsilon_X(\lambda, T(z))\, d\lambda. \tag{16.10}$$

The solar actinic flux q must be calculated by a radiative transfer model that accounts for (1) absorption processes, mainly by oxygen and ozone molecules, (2) multiple scattering by air molecules and atmospheric particles, (3) cloud radiative transfer, and (4) surface reflection. When considering upper and middle atmosphere processes, the most important contribution to photolysis is the direct solar flux, so that the value of the actinic flux can be approximated by considering only absorption processes. In the lower atmosphere, multiple scattering and specifically cloud effects cannot be ignored. Several radiative schemes with different levels of accuracy are available to calculate the radiation field in the different layers of the atmosphere for specified values of the key atmospheric parameters. Numerical modules for the calculation of photodissociation coefficients are also available (Madronich and Flocke, 1999; van Loon *et al.*, 2007; Bian and Prather, 2002).

The depth of penetration of solar radiation varies substantially with wavelength (Fig. 13.3). Most X-ray and EUV radiation is absorbed in the thermosphere above 100 km altitude. The intense solar Lyman-α line (121.6 nm) penetrates down to about 70 km altitude. In the wavelength range 120–180 nm, radiation is effectively absorbed above 60 km through the Schumann–Runge continuum of molecular oxygen. Between 180 and 200 nm, the most effective absorption occurs in the mesosphere and upper stratosphere through the Schumann–Runge bands. At wavelengths between 200 and 300 nm, radiation penetrates into the stratosphere, where it is absorbed primaryily by ozone molecules. Beyond 300 nm, most of the solar radiation reaches the surface; it is, however, affected by scattering processes and by the weak absorption due to the presence of ozone and other gases.

The amplitude of the changes in the solar flux over the 11-year solar cycle or the 27-day mean synodic solar rotation period decreases with increasing wavelengths (see Fig. 10.1) and, as a result, the influence of solar variability is considerably more pronounced in the upper atmosphere than in the lower layers. Strong solar signals associated with the solar cycle are visible in the thermospheric temperature and air density, with impacts, for example, on satellite drag (Qian *et al.*, 2006; see Vol. II, Section 2.5). Substantial changes have also been reported in the concentration of nitric oxide (NO); these changes, however, are also related to the modulation of energetic particle precipitation associated with geomagnetic activity (Marsh *et al.*, 2004). Solar-related changes in the temperature, water vapor, and polar mesospheric clouds have also been reported in the mesosphere (e.g. Beig *et al.*, 2003; Chandra *et al.*, 1997; and DeLand *et al.*, 2003, respectively). In the stratosphere, solar-driven changes in temperature and ozone concentrations have been observed (e.g. Soukharev and Hood, 2006; Randel and Wu, 2007; Randel *et al.*, 2009). The influence of solar variability in the troposphere is less well established although small variations with an 11-year period have been identified in, for example, zonal mean temperature (Crooks and Gray, 2005), surface pressure in the North Pacific (van Loon *et al.*, 2007), and global average surface temperature (Camp and Tung, 2007). A major forcing function for many of these changes is the variation in photolysis rates. Together with the solar-induced changes in atmospheric heating resulting from the absorption of solar radiation by ozone and molecular oxygen, atmospheric models designed to simulate the response of the atmosphere account for the changes in the photolysis coefficients of the different chemical compounds.

16.4 Ozone chemistry in the stratosphere

As indicated in Section 16.2, the photodissociation of molecular oxygen followed by the reaction between oxygen atoms and oxygen molecules leads to the formation of ozone. This photochemical process constitutes the only significant ozone production mechanism above 20 km altitude. Ozone can also be photodissociated:

$$O_3 + h\nu \rightarrow O + O_2, \qquad (16.11)$$

but, in most cases, this reaction does not constitute a net loss for stratospheric ozone because the oxygen atoms that result from this photodecomposition usually recombine with molecular oxygen (reaction 16.8) to reproduce ozone. The net loss of ozone results from the reaction between oxygen atoms and ozone molecules that produces two oxygen molecules:

$$O + O_3 \rightarrow 2O_2. \qquad (16.12)$$

The simple scheme presented here, imagined by Chapman (1930), provides a first-order description of the ozone chemistry in the stratosphere and mesosphere. Photochemical models that account only for the Chapman reactions tend to substantially overestimate the concentration of ozone in the middle atmosphere, as shown by numerous atmospheric observations. The discrepancy can be eliminated by considering several additional reactions that catalyze (i.e. accelerate) the net loss mechanism represented by reaction (16.12). Bates and Nicolet (1950), for example, have shown that the presence of the hydrogen atoms and hydroxyl radicals, produced in the upper atmosphere from the photolysis of water vapor (H_2O), could generate an efficient catalytic cycle such as

$$H + O_3 \rightarrow OH + O_2, \qquad (16.13)$$
$$OH + O \rightarrow H + O_2. \qquad (16.14)$$

More recently, Crutzen (1970) showed that the most effective ozone destruction in the stratosphere results from a catalytic cycle involving nitrogen oxides:

$$NO + O_3 \rightarrow NO_2 + O_2, \qquad (16.15)$$
$$NO_2 + O \rightarrow NO + O_2. \qquad (16.16)$$

NO is produced in the stratosphere by the oxidation of nitrous oxide (N_2O), a long-lived compound released from soils by bacterial activity. It can also be produced in the upper layers of the atmosphere by the dissociation and ionization of molecular nitrogen (N_2) by energetic particles.

Additional destruction mechanisms must be considered, including catalytic processes involving halogen compounds including chlorine (Cl; see Stolarski and Cicerone, 1974) and bromine (Br; see Wofsy *et al.*, 1975). For example

$$Cl + O_3 \rightarrow ClO + O_2, \tag{16.17}$$

$$ClO + O \rightarrow Cl + O_2. \tag{16.18}$$

Before the 1960s, the contribution of this cycle was relatively small. However, its importance has grown in more recent decades as the atmospheric abundance of chlorine has increased steadily due to the production of industrially manufactured chlorofluorocarbons (CFCs; see Molina and Rowland, 1974). The atmospheric lifetime of CFCs varies typically from 50 to 100 years, so that anthropogenic chlorine will remain for several decades in the stratosphere. In the cold polar regions, and specifically in Antarctica, ozone can be efficiently destroyed in a layer between 12 km and 25 km where polar stratospheric clouds are formed. The solid or liquid tiny particles inside these thin and often invisible clouds that are present during winter provide surfaces for heterogeneous chemical reactions to operate. Chemical chlorine reservoirs such as HCl and $ClONO_2$, which are very slow to react in the gas phase, are rapidly converted on the surface of these cloud particles to form less stable molecules such as Cl_2 or HOCl. Large quantities of reactive chlorine atoms (Cl) are liberated via photolysis as soon as the Sun returns in early spring. This chlorine activation leads to rapid ozone destruction with the formation of the springtime Antarctic ozone hole in September and October. These mechanisms are less efficient in the Arctic, where the winter temperature is usually 10–15 °C warmer than at the opposite pole, and the presence of polar stratospheric clouds is therefore less frequent.

A full description of the ozone behavior requires that large-scale transport processes be taken into consideration, specifically in the lower stratosphere, where the photochemical lifetime of this molecule becomes much longer than the time constant associated with transport. Below approximately 25 km altitude, ozone can be regarded as a quasi-inert tracer that is more sensitive to advection and mixing processes than to photochemical transformations. This highlights why the global ozone distribution in the atmosphere is strongly affected by the meridional circulation, and specifically why the ozone column abundance reaches a maximum value at high latitudes at the end of the winter. The poleward meridional circulation transports ozone towards the Arctic where it accumulates from December to April before it is slowly destroyed by photochemical processes after the Sun returns in early spring. This process is clearly visible in the ozone column measurements made by the Total Ozone Mapping Spectrometer (TOMS) satellite instrument and presented in Fig. 16.4. The same dynamical process occurs in the Southern Hemisphere with a lag of 6 months. However, ozone does not easily penetrate poleward of 60 ° S due to the existence of a strong dynamical barrier provided by the intense southern polar vortex. The ozone maximum is therefore located in a latitude band located at about 60° S. Large-scale planetary waves (Chapter 15) that characterize

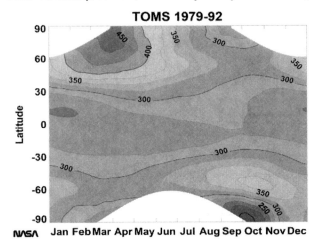

Fig. 16.4. Vertically integrated ozone concentration (expressed in Dobson units or DU; at 1 DU the column depth of ozone only would equal 10 μm at sea-level pressure and average temperature) represented as a function of latitude and month of the year. The distribution is established on the basis of observations made by the spaceborne TOMS instrument between 1972 and 1992. High values at the end of the winter are visible in the Arctic (March and April) and around 60°S in September and October. The presence of the Antarctic ozone hole is visible in the Antarctic in September–November. The value of 300 DU corresponds to an ozone layer of 3 mm under STP conditions. (From NASA.) See Color Plate 21.

the Northern Hemisphere winter dynamics do not allow the Northern Hemisphere polar region to be isolated from lower latitudes as is the case in the less dynamically disturbed Southern Hemisphere stratosphere. The ozone maximum in the Northern Hemisphere is thus located near the pole.

16.5 Response of ozone to solar variability

The response of chemical compounds to a variation in the solar flux can be calculated by first estimating the solar-induced changes in photolysis and in temperature. This is generally accomplished using a numerical model that accounts simultaneously for radiative, photochemical, and dynamical processes occurring in the entire atmosphere. Examples of three-dimensional coupled models that have been designed to quantify the atmospheric response to solar variability are the Whole Atmosphere Community Climate Model (WACCM) developed at the National Center for Atmospheric Research (NCAR) in Boulder, Colorado (Marsh *et al.*, 2007), the Hamburg Model for Neutral and Ionized Atmosphere (HAMMONIA) developed at the Max Planck Institute for Meteorology (MPI-M) in Hamburg, Germany (Schmidt *et al.*, 2006, Schmidt and Brasseur, 2006), the Climate Middle Atmosphere Model developed at the Free University of Berlin (Matthes *et al.*, 2004), the chemistry–climate model initiated at the University of Illinois at Urbana-Champaign (Rozanov *et al.*, 2004), the coupled chemistry–climate model of Austin *et al.* (2007, 2008), and several others. Results from the first two of these are used here to discuss the atmospheric response to solar variability.

16.5.1 Ozone and temperature responses to the 11-year cycle

An important input to chemistry–climate models is the wavelength-dependent solar flux at the top of the atmosphere and its estimated variability over a solar cycle. This information is derived from satellite and rocket observations followed by a careful analysis of the data to reduce unavoidable discrepancies in instrument calibration and to account for existing instrument drifts. Figure 10.1 shows the spectral irradiance and its relative change over a solar cycle as derived by Lean and Woods. It is clear from this figure that solar variability decreases with increasing wavelength; it reaches more than a factor two for wavelengths shorter than 50 nm, is close to 80% in the case of the intense Lyman-α line, is typically of the order of 8–10% at 200 nm, and becomes insignificant at wavelengths beyond 300 nm (and therefore in the visible). As indicated earlier, the effect of solar variability is expected to decrease gradually as solar radiation penetrates deeper in the atmosphere because only the longest wavelengths reach the lowest layers of the atmosphere.

We first show two quite similar estimates of the temperature response to the solar cycle. The global distribution presented in Fig. 16.5 is provided by WACCM, while the distribution shown in Fig. 16.6(left) is obtained by HAMMONIA. The largest changes over the solar cycle are seen in the lower thermosphere (above $\sim$110 km). As can be seen from Fig. 13.3, this is the region of the atmosphere where most of the solar energy absorbed is in the wavelength region of the Schumann–Runge continuum of O_2. Fluxes in this spectral range can vary 10 to 20%, with a subsequent effect on heating rates and temperatures. In the mesosphere, the global mean response in WACCM ranges from 0.4 K to 2 K. Near the stratopause (around 1 hPa or 50 km) there is a localized maximum of around 1 K, which occurs where heating from ozone absorption is greatest. Below the stratopause the response drops off with height, and only a few regions of significant temperature response are predicted in the troposphere.

The initial modeling studies of the response of ozone to solar variability were performed with simple 1D and 2D models (e.g. Garcia *et al.*, 1984; Brasseur, 1993; Huang and Brasseur, 1993; Fleming *et al.*, 1995). The 2D model of Brasseur (1993), for example, reproduces quite well (Fig. 16.8(right)) the tropical response of stratospheric ozone to the 11-year cycle as derived from observations by the Halogen Occultation Experiment (HALOE) onboard the Upper Atmosphere Research Satellite (UARS). More elaborated chemistry–climate models that derive

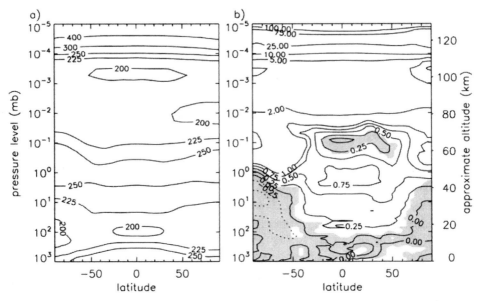

Fig. 16.5. Zonal mean distribution of the atmospheric temperature between the surface and approximately 130 km altitude (left) and response of temperature to the 11-year solar variability derived by the WACCM model (right). Units are in kelvin. The shaded areas show areas where the calculated values are not significant. (From Marsh *et al.*, 2007.)

simultaneously dynamical and chemical patterns in the atmosphere provide a more detailed view and account for the interactions between middle atmosphere dynamics, temperature, and chemistry.

Figure 16.6(right) shows the response of ozone (zonal average) to the 11-year solar cycle derived by such a model between the surface and approximately 110 km altitude. Here, we refer to the HAMMONIA model, and show the percentage change in the zonal and annual mean ozone mixing ratio from solar minimum to solar maximum conditions. We note that the maximum change (approximately 3%) in the stratosphere is located near 50 km altitude (5 hPa) in response to enhanced photodissociation of molecular oxygen and hence of the ozone production rate. In the middle mesosphere (near 60 km or 0.01 hPa; see Fig. 12.1 in Vol. I), the model suggests a negative ozone response attributed to enhanced photolysis of water vapor and therefore enhanced concentration of hydrogen-containing radicals (OH, HO_2) that lead to the destruction of the ozone molecule. In the upper mesosphere, where the effect of water vapor on the ozone loss decreases, the ozone response is again positive (due to an increase in O_2 photolysis). In the thermosphere ($\gtrsim 90$ km), however, although the total amount of odd oxygen (O and O_3) increases with the intensity of the solar irradiance, the response of ozone becomes negative because the enhanced O_3 photolysis tends to reduce the O_3/O concentration ratio.

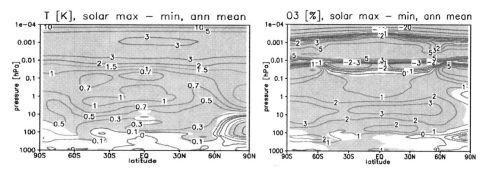

Fig. 16.6. Response of the atmospheric zonal mean, annual mean temperature (left; kelvin) and the relative ozone variation (right; percentage) to the 11-year solar cycle between the surface and approximately 110 km altitude as derived by the HAMMONIA model. Shaded areas correspond to regions where results are significant. (After Schmidt *et al.*, 2010.)

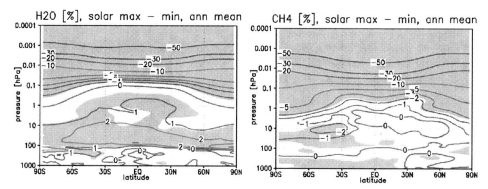

Fig. 16.7. Response of the atmospheric zonal mean, annual mean water vapor (left) and methane (right) concentration (percentage) to the 11-year solar cycle between the surface and approximately 110 km altitude as derived by the HAMMONIA model. (From the model of Schmidt *et al.*, 2010.)

The results provided by WACCM are quite similar. As shown by Fig. 16.8(left), a maximum in the ozone response is also found near 40 km altitude, and also reaches about 3 percent.

Satellite observations (e.g. Soukharev and Hood, 2006) confirm the existence of this maximum although it is simulated at slightly lower altitude than observed. The observed latitudinal structure is often much less homogeneous than indicated in Fig. 16.8(right). However, the analysis of observations is complicated by the simultaneous influence of other processes. As noted by Lee and Smith (2003), the signal provided by the quasi-biennial oscillation (QBO) and by volcanic eruptions affects the observed ozone time series, and may introduce biases in the retrieval of the solar signal. Because of decadal-scale dynamical interactions affecting ozone

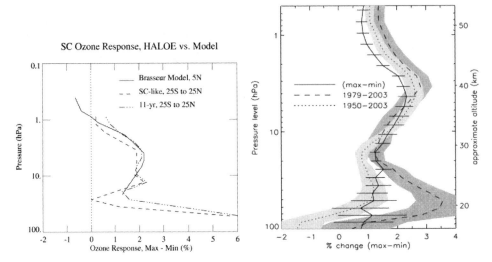

Fig. 16.8. (Left) Comparison of calculated to observed ozone response (percentage) in the stratosphere from minimum to maximum in the solar cycle. The data are retrieved from observations by the HALOE instrument on board UARS. (From Remsberg, 2008.) (Right) Percentage change in tropical ozone density between solar minimum and solar maximum conditions averaged between 24° S and 24° N. Solid line, with 2σ error bars, calculated from constant solar forcing "time slice" simulations. Lines with shaded regions are calculated using a multiple linear regression from the transient simulations of Garcia *et al.* (2007) for the time periods indicated. Shaded regions indicate 2σ errors in trend estimate. (From Marsh *et al.*, 2007.)

in the lower stratosphere, Lee and Smith (2003) cautioned that the ozone response may depend on the specific decade during which the data are reported.

Another maximum in ozone response is simulated in the lower equatorial stratosphere at around 70 hPa. The origin of this feature in HAMMONIA is a reduction in equatorial upwelling caused by solar forcing (see Section 16.6.1). Such a maximum appears in some observational analyses; however, it is under discussion as to what extent this feature is indeed a pure solar signal. Marsh and Garcia (2007) have suggested that some decadal variability in tropical ozone previously attributed to solar variability may be instead related to the occurrence of El Niño Southern Oscillation (ENSO) events. In WACCM simulations, this lower equatorial maximum is significant only in a period when solar forcing and ENSO are correlated (1979–2003, Fig. 16.8(left)). Extending the analysis period to cover the years 1950 to 2003 significantly reduces the modeled ozone response.

Chemical compounds other than ozone are also subject to solar-induced variability. This is the case, for example of water vapor and methane (Fig. 16.7), which are effectively photodissociated by the highly variable Lyman-α line in the upper atmosphere. Significant changes over the 11-year cycle are produced by the model above 60 km altitude (0.1 hPa) with changes larger than 20% at approximately 0.01 hPa and than 50% above 0.001 hPa. The slight increase of water vapor produced in the lower tropical stratosphere is likely due to dynamical changes (see Section 16.6.1).

16.5.2 Response of atmospheric ozone and temperature to the 27-day solar variation

The rotation period of the Sun apparent to an observer on Earth is 27 days. As disturbances in the solar atmosphere are not uniformly distributed in space, and specifically not along the solar longitude, a slight quasi-periodic variation in the solar irradiance with a 27-day period is observed. Figure 16.9 shows the superposition of the 27-day signal on the longer 11-year variation in the wavelength range of 160–170 nm (Rottman, 2000). In the case of the 27-day variability, the amplitude of the solar variation also decreases with wavelength.

For time scales of months, the analysis of time series for atmospheric parameters such as temperature or ozone density is considerably easier to perform than when considering decadal variability because issues associated with instrument drift and instrument changes can be ignored. In addition, the information provided by the time lag in the response of these atmospheric parameters relative to the solar signal provides an important constraint on our description of photochemical processes in the atmosphere. Keating *et al.* (1987) analyzed space observations of ozone and temperature in the tropics and have derived the amplitude and phase of the response of these quantities as a function of altitude. They show that the amplitude

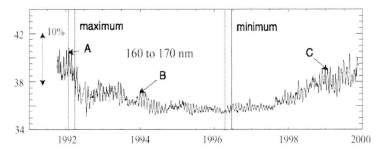

Fig. 16.9. Evolution of the solar irradiance (normalized intensity) at the top of the atmosphere in the spectral range 160–170 nm. These measurements by the SOL-STICE instrument onboard the Upper Atmosphere Research Satellite (UARS) highlight the variation associated with the 11-year solar cycle and with the 27-day synodic rotation of the Sun. Three periods of clear 27-day rotational modulation are marked by symbols A, B, and C. Pairs of vertical lines mark periods of solar maximum in early 1992 and solar minimum in 1996. (Rottman, 2000.)

of the ozone and temperature response decreases as one penetrates deeper in the atmosphere, while the phase lag increases. Brasseur *et al.* (1987) and Zhu *et al.* (2003) have analyzed satellite observations of the 27-day ozone and temperature signals.

A study by Gruzdev *et al.* (2009) using the HAMMONIA model (Fig. 16.10) shows that the ozone and solar signals are best correlated when the time lag of ozone relative to the solar quasi-periodic variation is close to zero at 40 km (1 hPa) (ozone and solar variations are in phase), but reaches 4 days at about 30 km (10 hPa). A negative response is retrieved in the model above 0.1 hPa. Gruzdev *et al.* (2009) note, however, that their analysis provides non-zero correlation fields of ozone with a 27-day sinusoidal function, when the 27-day solar forcing is turned off in the model. Thus, if one is not cautious when conducting the analysis of model results, atmospheric signals resulting from the internal variability generated by the atmospheric model could be erroneously interpreted as solar signals.

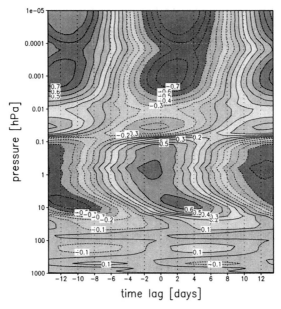

Fig. 16.10. Correlation coefficient between the ozone concentration and the 27-day solar variation as a function of atmospheric pressure (hPa) for different time lags (days) as calculated by the HAMMONIA model. One notes that, at 1 hPa (about 50 km altitude) for example, the highest correlation is found when ozone and solar radiation are in phase. At 10 hPa (about 30 km), the ozone signal with a 4-day phase lag is best correlated with the 27-day solar signal. Above 0.1 hP (∼65 km), the ozone signal appears to be out of phase with the solar periodic variation. (From Gruzdev *et al.*, 2009.) See Color Plate 22.

16.6 Response of atmospheric dynamics to solar variability

In order to establish a potential link between solar variability and climate, it is important to search for potential amplification mechanisms that could be produced primarily by changes in the meridional circulation and in the zonal wind. One of the key questions is the identification of plausible mechanisms by which stratospheric perturbations propagate downward and induce changes in the meteorological patterns below the tropopause. Enhanced solar-related heating of the equatorial lower stratosphere with a polar shift of the subtropical jets (Haigh *et al.*, 2005; Simpson *et al.*, 2009) and a solar-induced modulation of the polar jets with the development of winter patterns resembling the Arctic Oscillation (Matthes *et al.*, 2006; Kuroda *et al.*, 2007) are two mechanisms that have been proposed to explain the existence of dynamical disturbances in the troposphere. Baldwin and Dunkerton (2001) have shown that stratospheric anomalies preceded mean-flow disturbances and affect the Arctic oscillation.

16.6.1 Meridional circulation

As we consider the dynamics of the middle atmosphere, one should recall that the Earth-atmospheric meridional circulation (called the Brewer–Dobson circulation) is produced by the momentum deposited as a result of planetary wave absorption and gravity wave breaking (Holton, 1992; see Chapter 15). Planetary waves propagate upwards in the middle atmosphere during winter time, when the mean zonal wind is directed eastwards (westerlies). Gravity waves propagate during all seasons. In general, their amplitude grows during upward propagation. Gravity waves, however, break when the resulting vertical temperature gradient reaches unstable conditions.

Kodera and Kuroda (2002) have suggested that planetary wave propagation could be affected by changes in solar ultraviolet radiation: an increase in the stratospheric heating rate near the stratopause should lead to a strengthening of the subtropical jet and hence a deflection of the waves towards the winter pole (Fig. 16.11). As a result, momentum deposition should be reduced (positive anomaly in the Eliassen–Palm-flux divergence – see Fig. 16.12), and the strength of the Brewer–Dobson circulation reduced. As a result, equatorial upwelling is reduced, the adiabatic heating rate increases, and the temperature in the tropical lower stratosphere increases slightly when solar activity increases. At the same time, the intrusion of water vapor from the troposphere to the tropical stratosphere could be somewhat facilitated because the "filter" that prevents water vapor penetrating into the stratosphere becomes somewhat less efficient.

These figures indicate that the processes involved in the propagation of the solar signal from the upper to the lower stratosphere as proposed by Kodera and

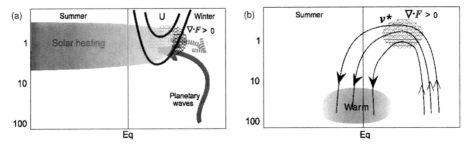

Fig. 16.11. Conceptual representation of a possible mechanism that links lower stratospheric variations in dynamical quantities with solar variability. Increased solar heating in the upper stratosphere, primarily in the summer hemisphere (highlighted on the left panel in dark grey), and consequently in the mean zonal wind (U), in the planetary wave propagation (arrow) and associated Eliassen–Palm flux (F) in the winter hemisphere tend to reduce the strength of the mean meridional (poleward) circulation with impacts on the zonal mean temperature structure. The small winter time "counter" meridional circulation associated with a solar-induced increase in stratospheric heating is represented in the right panel. Such disturbance (reduced upward motions in the tropics) should lead to a slight adiabatic warming of the lower equatorial tropopause. (From Kodera and Kuroda, 2002.)

Kuroda (2002) can be identified eventually in HAMMONIA. Positive anomalies in the zonal wind at high latitudes during late fall (November) associated with enhanced solar heating at lower latitudes are retrieved from observations and from general circulation models (e.g. Matthes *et al.*, 2004). These anomalies propagate poleward and downward during winter (Kodera and Kuroda, 2002). Negative anomalies in the zonal winds appear near the stratopause during late winter and also propagate downward. The wind field is related to the temperature field via the thermal wind relation (geostrophic balance). Accordingly, also high-latitude temperature anomalies propagate slowly downward during winter. Positive temperature signals dominate the high-latitude northern stratosphere in late winter. It should be stressed that these solar-induced signals at mid to high latitudes occur on a highly variable background. They are difficult to disentangle from other forcings as provided in particular by the QBO (see Section 16.6.2), volcanic eruptions (e.g. Lee and Smith, 2003), ENSO (e.g. Calvo *et al.*, 2009), and greenhouse gases (e.g. Kodera *et al.*, 2008). The reality of the solar signature in stratospheric dynamical quantities remains therefore an open question.

Based on the results of a general circulation model, Simpson *et al.* (2009) have highlighted the role of eddy momentum fluxes in driving the tropospheric response to solar-induced stratospheric heating perturbations. Heating of the lower stratosphere, preferentially in the equatorial region, produces changes in the zonal wind distribution and hence in the propagation of eddies and in the dynamical forcing

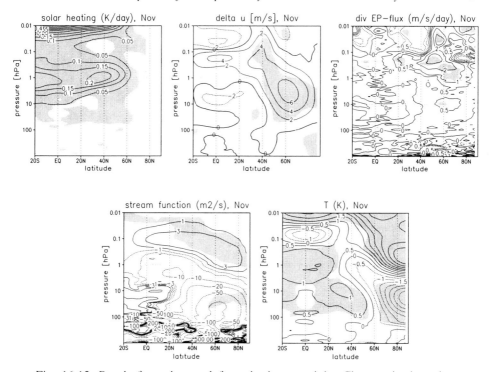

Fig. 16.12. Panels from the top left to the bottom right: Changes in the solar heating, mean zonal wind (m/s), Eliassen–Palm-flux divergence (or momentum deposition from planetary-scale waves) (m/s per day), stream function (m^2/s) and temperature (K) during the month of November derived by the HAMMO-NIA model in response to increased solar irradiance from solar minimum to solar maximum conditions. Shaded areas show the regions in which the statistical significance exceeds 95%. (From Schmidt *et al.*, 2009.)

of the meridional circulation. A positive feedback between eddy momentum fluxes and the zonal wind is identified, resulting in a possible amplification of the solar signal generated in the stratosphere.

16.6.2 Solar cycle-QBO interactions

It has been suggested by Labitzke and van Loon (1988) that the dynamical response of the atmosphere to solar variability is affected by the phase of the quasi-biennial oscillation (QBO), in particular in the winter hemisphere. For example, the number of wintertime sudden warmings observed in the stratosphere appears to be significantly higher during solar maximum conditions than during solar minimum conditions, when the QBO is in its westerly phase. In addition, as highlighted by Salby and Callaghan (2000), the structure of a QBO itself seems to vary with solar variability. For example, the westerly phase of the QBO is slightly shorter during solar maximum than during the solar minimum periods. Very few general circulation models currently treat the QBO fully interactively without imposing an artificial forcing, so that this question requires more detailed investigations. Schmidt *et al.* (2009) have simulated the atmospheric response to solar cycle forcing in a model with internally produced QBO. The model shows some dependence of the solar signals on the QBO, but the observed relation could not be reproduced satisfactorily.

16.6.3 Tropospheric dynamics

Finally, one should consider the possible effects of solar forcing on the dynamics of the lower atmosphere with possible consequences on weather patterns, and specifically on dynamical modes such as El Niño, the Arctic Oscillation, the Madden Julian Oscillation, etc. Shindell *et al.* (1999), using a global climate model with an interactive parameterization of stratospheric chemistry, have reproduced quite successfully the relatively long record of the 11-year oscillation in the geopotential height. They find that solar variability affects the troposphere indirectly through a redistribution of its own energy. They also show that solar variability plays a significant role in regional surface temperatures even though the amplitude of the signal on the globally average temperature (0.07 K) is small in the Northern Hemisphere winter. Coughlin and Tung (2004) found a statistically significant 11-year cycle in temperature and geopotential height reanalyis data throughout the troposphere over the period 1958 to 2003. Van Loon *et al.* (2007) have isolated dynamical signals in the lower troposphere, highlighting a significant Sun–weather relationship. The data analysis of Crooks and Gray (2005), based on the ERA-40 analysis conducted at the European Centre for Medium-Range Weather Forecasts (ECMWF) concluded that a small solar-inducted signal is present in these dynamical quantities, but only in specific areas and often with limited statistical significance. Models, such as WACCM and HAMMONIA reveal similar signals in the longitudinally averaged zonal mean winds and temperature, but also only with limited statistical significance (see Fig. 16.13). As stated above, modeling studies by Haigh *et al.* (2005) have indicate that a slight poleward shift of the upper tropospheric jet streams, as observed by Crooks and Gray (2005), may be caused by a positive temperature anomaly in the lower equatorial stratosphere. Another pathway to transfer the stratospheric signal to the troposphere may be the downward propagation of polar night jet anomalies (also described above) into the upper troposphere. More detailed analyses and model investigations are needed to address this question and to identify the mechanisms involved.

An interesting modeling study by Meehl *et al.* (2009) suggests (Fig. 16.14) that the 11-year signal in surface ozone and temperature can be – at least qualitatively –

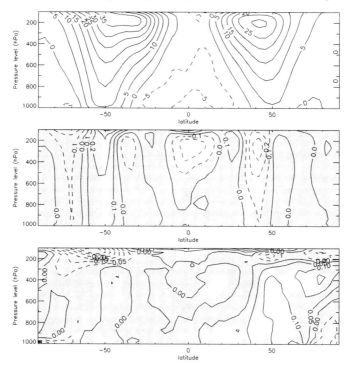

Fig. 16.13. (Top) Tropospheric annually and zonally averaged zonal winds (m/s) for solar minimum conditions. (Lower panels) Solar cycle changes (maximum–minimum) in annually and zonally averaged zonal winds (middle panel, in m/s) and zonal mean temperatures (bottom panel, in K). Unshaded regions are significant at the 95% level. (From Marsh *et al.*, 2007.)

reproduced in a comprehensive climate model if one considers simultaneously (1) the downward propagation of the solar-induced stratospheric disturbance and (2) the upward propagation of perturbations caused by the excess heat storage in the ocean during high solar activity. This work highlights the importance of the ocean in the solar–climate relation. The proposed mechanism assumes that during intense solar forcing, greater amounts of energy stored by the ocean in the relatively cloud-free areas of the subtropics produce enhanced evaporation. The resulting water vapor in the atmosphere that is carried by the trade winds and that converges to the equatorial region produces intensified Hadley and Walker circulations with stronger precipitation in this region and more intense subsidence in the subtropics. The reduced cloudiness in the subtropics allows even more solar absorption by the ocean (positive feedback). Over time scales of typically a decade, these solar-induced perturbations are probably larger than those resulting from increasing concentrations of carbon dioxide over such a limited time period. This highlights the need to include the solar forcing effects in decadal-type climate

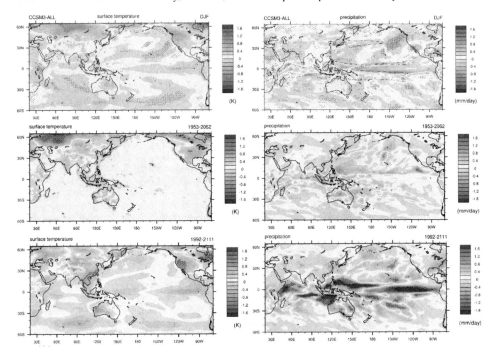

Fig. 16.14. Composites (left temperature; right precipitation) for simulated peaks in the 11-year solar cycle. Bottom-up coupled air–sea mechanism (top panels) and top-down stratospheric ozone mechanism (middle panels) are additive to strengthen convection in the tropical Pacific and produce a stronger La Niña-like response to peaks in solar forcing (bottom panels). (From Meehl *et al.*, 2009.) See Color Plate 23.

predictions. Over longer periods of time, such as centuries, the effect of carbon dioxide increase is expected to be substantially larger than the solar effects.

16.7 Conclusions

Variations in solar radiation over the 11-year cycle as well as over the 27-day solar rotation period have substantial effects in the upper atmosphere where energetic photons penetrate and directly initiate photochemical effects. In the stratosphere and the troposphere, above which shortwave radiation is absorbed, the direct impact of solar variability becomes less pronounced. Solar signals in ozone and temperature, however, can be derived from observations above approximately 25 km altitude. Below this height, the situation becomes more complex because other dynamical signals such as those produced by climatic modes of variability (e.g. El Niño) interfere with possible variations resulting from solar variability.

Several mechanisms have been proposed to explain a plausible relation between solar variability and the observed 11-year dynamical variability in the lower atmosphere. One of them is associated with disturbances produced in the upper atmosphere and resulting from ozone variations generated by changes in short-wave solar radiation. A second mechanism is linked to the ocean-surface response to 11-year changes in the total solar irradiance. Observed weather patterns correlated with solar forcing could result from both downward-propagating disturbances produced in the stratosphere and upward-propagating perturbations generated at the surface of the ocean. To capture the amplifying mechanisms producing a dynamical response of the troposphere to solar variability, atmospheric models must therefore account for photochemical processes in the upper atmosphere and, at the same time, must be coupled to an ocean module. Despite many remaining uncertainties, much progress has been made in recent years to better understand how solar variability could potentially affect the climate system, particularly on decadal time scales.

Appendix I Authors and editors

Jürg Beer
Surface Waters
Eawag
PO Box 611
Ueberlandstrasse 133
Duebendorf 8600
Switzerland

Guy P. Brasseur
NCAR ESSL
PO Box 3000
Boulder, CO 80301

Donald E. Brownlee
Department of Astronomy
University of Washington
Box 351580
331 Physics Astronomy Building
Seattle, WA 98195

Paul Charbonneau
Physics Department
Université de Montreal
Bureau B-418
C.P. 6128, Centre-Ville
Montreal, QC H3C 3J7
Canada

Ulrich R. Christensen
Max Planck Institute for Solar System
Research
Max-Planck-Strasse 2
Katlenburg-Lindau 37191
Germany

Thomas J. Crowley
School of Geosciences
The University of Edinburgh
Room 343, Grant Institute
The King's Buildings, West Mains
Road
Edinburgh EH9 3JW
Scotland

John T. Gosling
Laboratory for Atmospheric and Space
Physics
University of Colorado
1234 Innovation Drive
Boulder, CO 80303

Lee W. Hartmann
Department of Astronomy
University of Michigan
955 Dennison Building
500 Church Street
Ann Arbor, MI 48109-1042

J. R. Jokipii
Lunar and Planetary Laboratory
University of Arizona
Space Sciences 411
Tucson, AZ 85721

Judith L. Lean
Naval Research Laboratory
4555 Overlook Avenue, SW
Washington, DC 20375

Daniel Marsch
National Center for Atmospheric
Research
PO Box 3000
Boulder, CO 80307-3000

Mark S. Miesch
NCAR HAO
PO Box 3000
Boulder, CO 80307

Hauke Schmidt
Max Planck Institute for Meteorology
Hamburg
Germany

Carolus J. Schrijver (editor)
Solar and Astrophysics Laboratory
Lockheed Martin Advanced Technical
Center
3251 Hanover Street, Building 252
Palo Alto, CA 94304-1191

George L. Siscoe (editor)
Boston University
725 Commonwealth Avenue
Boston, MA 02215

Jan J. Sojka
Center for Atmospheric and Space
Sciences
Utah State University
4405 Old Main Hill
Logan, UT 84322-4405

Stanley C. Solomon
NCAR High Altitude Observatory
3450 Mitchell Lane
Boulder, CO 80301

Richard L. Walterscheid
Space Sciences Department
The Aerospace Corporation
PO Box 92957
Los Angeles, CA 90009

Thomas N. Woods
Laboratory for Atmospheric and Space
Physics
University of Colorado
1234 Innovation Drive
Boulder, CO 80303

List of illustrations

List of tables

460

References

Aarons, J.: 1993, Space Sci. Rev. 63, 209, doi:10.1007/BF00750769

Abbett, W. P.: 2007, ApJ 665, 1469

Abbot, C. G.: 1958, Smithsonian Contributions to Astrophysics 3, 13

Adhémar, J.A.: 1842, *Revolucion des Mers, Défluges periodique*, private publication, Paris

Alexakis, A., Mininni, P. D., & Pouquet, A.: 2006, ApJ 640, 335

Alexander, R. D., Clarke, C. J., & Pringle, J. E.: 2006a, MNRAS 369, 216, doi:10.1111/j.1365-2966.2006.10293.x

Alexander, R. D., Clarke, C. J., & Pringle, J. E.: 2006b, MNRAS 369, 229, doi:10.1111/j.1365-2966.2006.10294.x

Amit, H. & Christensen, U.R.: 2008, Geophys. J. Int. 175, 913

Anderson, J. M., Li, Z.-Y., Krasnopolsky, R., & Blandford, R. D.: 2003, ApJL 590, L107, doi:10.1086/376824

Andrews, D. G. & McIntyre, M. E.: 1976, J. Atmos. Sci. 33, 2031, doi:10.1175/1520-0469(1976)033

Andrews, D. G., Holton, J. R., & Leovy, C. B.: 1987, *Middle Atmosphere Dynamics*, Academic Press

Angelats i Coll, M. & Forbes, J. M.: 2002, JGR (Space Physics) 107, 1157, doi:10.1029/2001JA900179

Archontis, V., Moreno-Insertis, F., Galsgaard, K., Hood, A., & O'Shea, E.: 2004, A&A 426, 1047, doi:10.1051/0004-6361:20035934

Arlt, R., Hollerbach, R., & Rüdiger, G.: 2003, A&A 401, 1087, doi:10.1051/0004-6361:20030251

Armitage, P. J., Livio, M., & Pringle, J. E.: 2001, MNRAS 324, 705, doi:10.1046/j.1365-8711.2001.04356.x

Aso, T.: 2007, Earth, Planets, and Space 59

Asplund, M.: 2005, ARA&A 43, 481

Audard, M., Güdel, M., Drake, J. J., & Kashyap, V. L.: 2000, ApJ 541, 396

Aurnou, J., Heimpel, M., & Wicht, J.: 2007, Icarus 190, 110

Austin, J., Hood, L. L., & Soukharev, B. E.: 2007, Atmos. Chem. Phys. 7, 1693

Austin, J., Tourpali, K., Rozanov, E., *et al.*: 2008, JGR 113, D11306, doi:10.1029/2007JD009391

Avrett, E. H.: 1981, in R. M. Bonnet & A. K. Dupree (eds.), *Solar Phenomena in Stars and Stellar Systems*, Proceedings of the BONAS NATO Advanced Study Institute, D. Reidel, p. 173

Babcock, H. W.: 1961, ApJ 133, 572, doi:10.1086/147060

Bacciotti, F., Ray, T. P., Mundt, R., Eislöffel, J., & Solf, J.: 2002, ApJ 576, 222, doi:10.1086/341725

Bacmann, A., André, P., Puget, J.-L., Abergel, A., Bontemps, S., & Ward-Thompson, D.: 2000, A&A 361, 555

Bahcall, J. N.: 2000, J. R. Astron. Soc. Can. 94, 219

Balbus, S. A.: 1995, ApJ 453, 380

Balbus, S. A. & Hawley, J. F.: 1998, Rev. Mod. Phys. 70, 1, doi:10.1103/RevModPhys.70.1

Baldwin, M. P. & Dunkerton, T. J.: 2001, Science 294, 581

Baldwin, M. P., Gray, L. J., Dunkerton, T. J., *et al.*: 2001, Rev. Geophys. 39, 179

Baliunas, S. & Jastrow, R.: 1990, Nature 348, 520, doi:10.1038/348520a0

Baliunas, S. L., Donahue, R. A., Soon, W., & Henry, G. W.: 1998, in R. A. Donahue & J. A. Bookbinder (eds.), *Cool Stars, Stellar Systems, and the Sun*, ASP Conference Series 154, Astronomical Society of the Pacific, p. 153

Ballesteros-Paredes, J., Klessen, R. S., Mac Low, M.-M., & Vazquez-Semadeni, E.: 2007, in B. Reipurth, D. Jewitt, & K. Keil (eds.), *Protostars and Planets V*, University of Arizona Press, p. 63

Balogh, A., Gosling, J. T., Jokipii, J. R., Kallenbach, R., & Kunow, H. (eds.): 1999, Space Sci. Rev. 89, 1

Banks, P. M. & Kokarts, G.: 1973, *Aeronomy, part B*, Academic Press, 203

Bard, E. & Frank, M.: 2006, Earth Planet. Sci. Lett. 248, 1, doi:10.1016/j.epsl.2006.06.016

Bard, E., Raisbeck, G., Yiou, F., & Jouzel, J.: 2000, Tellus B 52, 985

Baross, J. A. & Hoffman, S. E.: 1985, Orig. Life Evol. Biosph. 15, 327, doi:10.1007/BF01808177

Barrow, C. H. & Carr, T. D.: 1992, Adv. Space Res. 12, 155

Basri, G.: 2009, in *Cool Stars, Stellar Systems and the Sun 15*, AIP Conference Series

Bates, D. R. & Massey, H. S. W.: 1947, Proc. Roy. Soc (London) A192, 1

Bates, D. R. & Nicolet, M.: 1950, JGR 55, 301

Baumann, I., Schmitt, D., & Schüssler, M.: 2006, A&A 446, 307

Baumgartner, S., Beer, J., Masarik, J., *et al.*: 1998, Science 279(5355), 1330

Beer, J., Blinov, A., Bonani, G., Hofmann, H. J., & Finkel, R. C.: 1990, Nature 347, 164, doi:10.1038/347164a0

Beer, J., Raisbeck, G. M., & Yiou, F.: 1991, in C. P. Sonett, *et al.* (eds.), *The Sun in Time*, University of Arizona Press, p. 343

Beer, J., *et al.*: 1992, in Proc. Erice NATO Workshop, Springer Verlag

Beer, J., Baumgartner, S. T., Dittrich-Hannen, B., *et al.*: 1994, in J. M. Pap, C. Fröhlich, H. S. Hudson, & S. K. Solanki (eds.), *The Sun as a Variable Star: Solar and Stellar Irradiance Variations*, Cambridge University Press, p. 291

Beer, J., Tobias, S., & Weiss, N.: 1998, Solar Phys. 181, 237

Beig, G., Keckhut, P., Lowe, R. P., *et al.*: 2003, Rev. Geophys. 41, 1015, doi:10.1029/2002RG000121

Belashova, E. S., Belashov, V. Y., & Vladimirov, S. V.: 2007, JGR (Space Physics) 112(A11), 7302, doi:10.1029/2006JA012220

Belcher, J. W. & Davis, Jr., L.: 1971, JGR 76, 3534, doi:10.1029/JA076i016p03534

Berger, A.: 1978, J. Atmos. Sci. 35, 2362

Berger, E., Basri, G., Gizis, J. E., *et al.*: 2008, ApJ 676, 1307, doi:10.1086/529131

Berggren, A.-M., Beer, J., Possnert, G., *et al.*: 2009, GRLe 36, 11801, doi:10.1029/2009GL038004

Bian, H. & Prather, M. J.: 2002, J. Atmos. Chem. 41, 281

Bigazzi, A. & Ruzmaikin, A.: 2004, ApJ 604, 944, doi:10.1086/381932

Bilitza, D. (ed.): 1990, *International Reference Ionosphere 1990*, NASA publication NSSDC 90-22

Bilitza, D.: 2001, Radio Sci. 36(2), 261, doi:10.1029/2000RS002432

Blandford, R. D. & Payne, D. G.: 1982, MNRAS 199, 883

Bloxham, J.: 1989, Phil. Trans. R. Soc. Lond. 87, 669

Boldyrev, S. & Cattaneo, F.: 2004, Phys. Rev. Lett. 92, 144501

Boley, A. C., Mejía, A. C., Durisen, R. H., *et al.*: 2006, ApJ 651, 517, doi:10.1086/507478

Bond, G., Kromer, B., Beer, J., *et al.*: 2001, Science 294, 2130, doi:10.1126/science.1065680

Bonnell, I. A., Larson, R. B., & Zinnecker, H.: 2007, in B. Reipurth, D. Jewitt, & K. Keil (eds.), *Protostars and Planets V*, University of Arizona Press, p. 149

Bopp, B. W. & Stencel, R. E.: 1981, ApJL 247, L131, doi:10.1086/183606

Borrini, G., Wilcox, J. M., Gosling, J. T., Bame, S. J., & Feldman, W. C.: 1981, JGR 86, 4565, doi:10.1029/JA086iA06p04565

Boss, A. P.: 2003, ApJ 599, 577, doi:10.1086/379163

Bouvier, J., Forestini, M., & Allain, S.: 1997, A&A 326, 1023

Brain, D. A., Halekas, J. S., & Eastwood, J. P.: 2008, AGU Fall Meeting Abstracts B1320

Braithwaite, J.: 2006, A&A 449, 451, doi:10.1051/0004-6361:20054241

Braithwaite, J.: 2009, MNRAS 397, 763

Brandenburg, A.: 2001, ApJ 550, 824

Brandenburg, A.: 2005, ApJ 625, 539, doi:10.1086/429584

Brandenburg, A.: 2009, Space Sci. Rev. 144, 87, doi:10.1007/s11214-009-9490-0

Brandenburg, A. & Subramanian, K.: 2005, Phys. Rep. 417, 1

Brasseur, G.: 1993, JGR 98, 23079, doi:10.1029/93JD02406

Brasseur, G., de Rudder, A., Keating, G. M., & Pitts, M. C.: 1987, JGR 92, 903, doi:10.1029/JD092iD01p00903

Brasseur, G. P. & Solomon, S.: 2005, *Aeronomy of the Middle Atmosphere*, 3rd edn., Springer

Bremer, J.: 1992, J. Atmos. Terr. Phys. 54, 1505

Bremer, J.: 1998, Ann. Geophys. 16, 989

Breuer, D., Hauck II, S. A., Buske, M., Pauer, M., & Spohn, T.: 2007, Space Sci. Rev. 229-260, 132

Brooke, J., Moss, D., & Phillips, A.: 2002, A&A 395, 1013, doi:10.1051/0004-6361:20021320

Brown, B. P., Browning, M. K., Brun, A. S., Miesch, M. S., & Toomre, J.: 2008, ApJ 689, 1354

Brown, B. P., Browning, M. K., Miesch, M. S., Brun, A. S., & Toomre, J.: 2009, arXiv: 0906.2407

Browning, M. K., Miesch, M. S., Brun, A. S., & Toomre, J.: 2006, ApJ 648, L157

Brueckner, G. E., Edlow, K. L., Floyd, IV, L. E., Lean, J. L., & Vanhoosier, M. E.: 1993, JGR 98, 10695, doi:10.1029/93JD00410

Brummell, N. H., Hurlburt, N. E., & Toomre, J.: 1996, ApJ 473, 494

Brun, A. S., Miesch, M. S., & Toomre, J.: 2004, ApJ 614, 1073

Brun, A. S., Browning, M. K., & Toomre, J.: 2005, ApJ 629, 461, doi:10.1086/430430

Bruno, R., Villante, U., Bavassano, B., Schwenn, R., & Mariani, F.: 1986, SPh 104, 431, doi:10.1007/BF00159093

Buick, R.: 2007, in W. T. Sullivan III & J. A. Baross (eds.), *Planets and Life: The Emerging Science of Astrobiology*, Cambridge University Press, p. 237

Burlaga, L. F.: 1974, JGR 79, 3717, doi:10.1029/JA079i025p03717

Burlaga, L. F.: 1983, JGR 88, 6085, doi:10.1029/JA088iA08p06085

Burlaga, L. F.: 1984, Space Sci. Rev. 39, 255, doi:10.1007/BF00173902

Burlaga, L. F., Lepping, R. P., Behannon, K. W., Klein, L. W., & Neubauer, F. M.: 1982, JGR 87, 4345, doi:10.1029/JA087iA06p04345

Burlaga, L. F., Lepping, R. P., & Jones, J. A.: 1990, in C. T. Russell, E. R. Priest, & L. C. Lee (eds.), *Physics of Magnetic Flux Ropes*, Geophys. Monogr. 58, AGU, p. 373

Burrage, M. D., Hagan, M. E., Skinner, W. R., Wu, D. L., & Hays, P. B.: 1995, GRLe 22, 2641, doi:10.1029/95GL02635

Burrows, C. J., Stapelfeldt, K. R., Watson, A. M., *et al.*: 1996, ApJ 473, 437, doi:10.1086/178156

Bushby, P. J.: 2006, MNRAS 371, 772, doi:10.1111/j.1365-2966.2006.10706.x

Bushby, P. J. & Tobias, S. M.: 2007, ApJ 661, 1289, doi:10.1086/516628

Busse, F. H.: 1970, J. Fluid Mech. 44, 441

Busse, F. H.: 1975, Geophys. J. R. Astron. Soc. 42, 437

Busse, F. H.: 2002, Phys. Fluids 14, 1301

Butler, J., Johnson, B. C., Rice, J. P., Shirley, E. L., & Barnes, R. A.: 2008, J. Res. Natl. Inst. Stand. Technol. 113

Caballero-Lopez, R. A., Moraal, H., Mewaldt, R. A., McDonald, F. B., & Wiedenbeck, M. E.: 2007, ApJ 663, 1335, doi:10.1086/518394

Calvet, N.: 1998, in S. S. Holt & T. R. Kalman (eds.), *Disk Accretion in Pre-Main Sequence Stars*, American Institute of Physics Conference Series, Vol. 431, p. 495

Calvet, N. & Gullbring, E.: 1998, ApJ 509, 802, doi:10.1086/306527

Calvo, N., Giorgetta, M. A., Garcia-Herrera, R., & Manzini, E.: 2009, JGR, 114, D13109, doi: 10.1029/2008JD011445

Cameron, R. & Schüssler, M.: 2007, ApJ 659, 801, doi:10.1086/512049

Cameron, R. & Schüssler, M.: 2008, ApJ 685, 1291, doi:10.1086/591079

Camp, C. D. & Tung, K. K.: 2007, GRLe 34, 14703, doi:10.1029/2007GL030207

Camuffo, D.: 2001, Earth Moon and Planets 85, 99

Carr, J. S.: 2007, in J. Bouvier & I. Appenzeller (eds.), *IAU Symposium*, Vol. 243, p. 135

Carrington, R.: 1860, MNRAS 20, 13

Cattaneo, F. & Hughes, D. W.: 2001, Astron. Geophys. 42, 3.18

Cattaneo, F. & Hughes, D. W.: 2006, J. Fluid Mech. 553, 401

Cattaneo, F., Lenz, D., & Weiss, N.: 2001, ApJ 563, L91

Cattaneo, F., Brummell, N. H., & Cline, K. S.: 2006, MNRAS 365, 727

Chamberlain, J. W. & Hunten, D. M.: 1987, *Theory of Planetary Atmospheres*, Academic Press

Chandra, S., Jackman, C. H., Fleming, E. L., & Russell, III, J. M.: 1997, GRLe 24, 639, doi:10.1029/97GL00546

Chandrasekhar, S.: 1961, *Hydrodynamic and Hydromagnetic Stability*, Oxford University Press

Chapelle, F. H., O'Neill, K., Bradley, P. M., *et al.*: 2002, Nature 415, 312

Chapman, M. R. & Shackleton, N. J.: 2001, The Holocene 10, 287

Chapman, S.: 1930, Mem. R. Meteorol. Soc. 3

Chapman, S.: 1931, Proc. Phys. Soc. 43, 26

Chapman, S. & Lindzen, R.: 1970, *Atmospheric Tides: Thermal and Gravitational*, Reidel

Charbonneau, P.: 2001, Solar Phys. 199, 385

Charbonneau, P.: 2005, Solar Phys. 229, 345

Charbonneau, P. & MacGregor, K. B.: 1996, ApJ 473, L59

Charbonneau, P. & MacGregor, K. B.: 2001, ApJ 559, 1094, doi:10.1086/322417

Charbonneau, P., Schrijver, C. J., & MacGregor, K. B.: 1997, in J. R. Jokipii, C. P. Sonnett, & M. S. Giampapa (eds.), *Cosmic Winds and the Heliosphere*, University of Arizona Press, p. 677

Charbonneau, P., Christensen-Dalsgaard, J., Henning, R., *et al.*: 1999, ApJ 527, 445, doi:10.1086/308050

Charbonneau, P., Blais-Laurier, G., & St-Jean, C.: 2004, ApJL 616, L183, doi:10.1086/426897

Charbonneau, P., St-Jean, C., & Zacharias, P.: 2005, ApJ 619, 613, doi:10.1086/426385

Charbonneau, P., Beaubien, G., & St-Jean, C.: 2007, ApJ 658, 657, doi:10.1086/511177

Charney, J. G.: 1973, in P. Morel (ed.), *Dynamic Meteorology*, Reidel, p. 97

Charney, J. G. & Drazin, P. G.: 1961, JGR 66, 83, doi:10.1029/JZ066i001p00083

Charney, J. G. & Stern, M. E.: 1962, J. Atmos. Sci. 19, 159

Charvátová, I.: 2000, Ann. Geophys. 18, 399, doi:10.1007/s005850050897

Chen, R. H., Cravens, T. E., & Nagy, A. F.: 1978, JGR 83, 3871

Cheung, M. C. M., Schüssler, M., Tarbell, T. D., & Title, A. M.: 2008, ApJ 687, 1371

Childress, S. & Gilbert, A. D.: 1995, *Stretch, Twist, Fold: The Fast Dynamo*, Springer

Choudhuri, A. R.: 1992, A&A 253, 277

Choudhuri, A. R., Schussler, M., & Dikpati, M.: 1995, A&A 303, L29

Choudhuri, A. R., Chatterjee, P., & Jiang, J.: 2007, Phys. Rev. Lett. 98(13), 131103, doi:10.1103/PhysRevLett.98.131103

Christensen, U. R.: 2006, Nature 444, 1056

Christensen, U. R.: 2009, Space Sci. Rev. 71, doi:doi:10.1007/511214-009-9553-2

Christensen, U. R. & Aubert, J.: 2006, Geophys. J. Int. 166, 97

Christensen, U. R. & Olson, P.: 2003, Phys. Earth Planet. Inter. 138, 39

Christensen, U. R. & Wicht, J.: 2007, in P. Olson (ed.), *Treatise of Geophysics, Volume 8: Core Dynamics*, Elsevier, p. 245

Christensen, U. R. & Wicht, J.: 2008, Icarus 196, 16

Christensen, U. R., Olson, P., & Glatzmaier, G. A.: 1998, GRLe 25, 1565

Christensen, U. R., Holzwarth, V., & Reiners, A.: 2009, Nature 457, 167

Christensen-Dalsgaard, J.: 2002, Rev. Mod. Phys. 74, 1073

Christensen-Dalsgaard, J., Gough, D. O., & Thompson, M. J.: 1991, ApJ 378, 413

Christensen-Dalsgaard, J., Däppen, W., Ajukov, S. V., *et al.*: 1996, Science 272, 1286

Claire, M. W., Catling, D. C., & Zahnle K. J.: 2006, Geobiology 4, 239

Clapperton, C. M. & Sugden, D. E.: 1988, Quat. Sci. Rev. 7, 185

Clarke, C. J., Gendrin, A., & Sotomayor, M.: 2001, MNRAS 328, 485, doi:10.1046/j.1365-8711.2001.04891.x

Cliver, E. W. & Svalgaard, L.: 2004, Solar Phys. 224, 407, doi:10.1007/s11207-005-4980-z

Coffey, D., Bacciotti, F., Ray, T. P., Eislöffel, J., & Woitas, J.: 2007, ApJ 663, 350, doi:10.1086/518100

Coles, W. A. & Maagoe, S.: 1972, JGR 77, 5622, doi:10.1029/JA077i028p05622

Coles, W. A., Rickett, B. J., Rumsey, V. H., *et al.*: 1980, Nature 286, 239, doi:10.1038/286239a0

Connerney, J. E. P.: 2007, in T. Spohn (ed.), *Treatise of Geophysics, Volume 10: Planets and Moons*, Elsevier, p. 243

Cook, E. R., Meko, D. M., & Stockton, C. W.: 1997, J. Climate 10, 1343, doi:10.1175/1520-0442(1997)010

Coughlin, K. & Tung, K. K.: 2004, JGR 109, D21105, doi:10.1029/2004JD004873

Covey, C. & Schubert, G.: 1982, J. Atmos. Sci. 39, 2397, doi:10.1175/1520-0469 (1982)039

Covey, K. R., Greene, T. P., Doppmann, G. W., & Lada, C. J.: 2005, AJ 129, 2765, doi:10.1086/429736

Cox, A. N., Livingston, W. C., & Matthews, M. S.: 1991, *Solar Interior and Atmosphere*, University of Arizona Press

Cranmer, S. R.: 2008, ApJ 689, 316, doi:10.1086/592566

Croll, J.: 1875, *Climate and Time in their Geological Relations: A Theory of Secular Changes of the Earth's Climate*, Daldy, Ibister & Co.

Crommelynck, D., Fichot, A., Lee, III, R. B., & Romero, J.: 1995, Adv. Space Res. 16, 17, doi:10.1016/0273-1177(95)00261-C

Crooker, N. U., Siscoe, G. L., Shodhan, S., *et al.*: 1993, JGR 98, 9371, doi:10.1029/93JA00636

Crooker, N., Joselyn, J. A., & Feynman, J.: 1997, *Coronal Mass Ejections*, Geophys. Monogr. 99, AGU, p. 299

Crooker, N. U., Gosling, J. T., Bothmer, V., *et al.*: 1999, Space Sci. Rev. 89, 179, doi:10.1023/A:1005253526438

Crooker, N. U., Gosling, J. T., & Kahler, S. W.: 2002, JGR (Space Physics) 107, 1028, doi:10.1029/2001JA000236

Crooks, S. A. & Gray, L. J.: 2005, J. Climate 18, 996

Crowley, T. J.: 2000, Science 289, 270

Crowley, T. J. & Kim, K.-Y.: 1993, Quat. Sci. Rev. 12, 375

Crowley, T. J. & Lowery, T. S.: 2000, Ambio 29, 51

Crowley, T. J., Baum, S. K., Kim, K.-Y., Hegerl, G. C., & Hyde, W. T.: 2003, GRLe 30(18), 180000, doi:10.1029/2003GL017801

Crowley, T. J., Zielinski, G., Vinther, B., *et al.*: 2008, PAGES News 16, 22

Crowley, T. J., Payette, S., & Popa, I.: 2010, in preparation

Crutzen, P. J.: 1970, Quart. J. R. Met. Soc. 96, 320

Currie, R. G. & Fairbridge, R. W.: 1985, Quat. Sci. Rev. 4, 109

Cuzzi, J. N., Hogan, R. C., Paque, J. M., & Dobrovolskis, A. R.: 2001, ApJ 546, 496, doi:10.1086/318233

D'Alessio, P., Calvet, N., & Hartmann, L.: 2001, ApJ 553, 321, doi:10.1086/320655

Danilov, A. D. & Mikhailov, A. V.: 1999, Ann. Geophys. 17, 1239

Dansgaard, W., Johnsen, S. J., Clausen, H. B., *et al.*: 1993, Nature 364, 218, doi:10.1038/364218a0

Davidson, P. A.: 2001, *An Introduction to Magnetohydrodynamics*, Cambridge University Press

Davidson, P. A.: 2004, *Turbulence: An Introduction for Scientists and Engineers*, Oxford University Press

De Jager, C. & Versteegh, G. J. M.: 2005, Solar Phys. 229(1), 175

Debret, M., Bout-Roumazeilles, V., Grousset, F., *et al.*: 2007, Clim. Past 3, 569

DeLand, M. T. & Cebula, R. P.: 1998, Solar Phys. 177, 105

DeLand, M. T., Shettle, E. P., Thomas, G. E., & Olivero, J. J.: 2003, JGR 108(D8), 8445, doi:10.1029/2002JD002398

Denissenkov, P. A., Pinsonneault, M., Terndrup, D. M., & Newsham, G.: 2010, ApJ 716, 1269

Denton, G. H. & Karlen, W.: 1973, Quat. Res. 3, 155

DeRosa, M. L. & Toomre, J.: 2004, ApJ 616, 1242

Desch, S. J.: 2007, ApJ 671, 878, doi:10.1086/522825

Dewitte, S., Crommelynck, D., & Joukoff, A.: 2004a, JGR (Space Physics) 109, A02102, doi:10.1029/2002JA009694

Dewitte, S., Crommelynck, D., Mekaoui, S., & Joukoff, A.: 2004b, Solar Phys. 224, 209, doi:10.1007/s11207-005-5698-7

Dickey, J. O., Bender, P. L., Faller, J. E., *et al.*: 1994, Science 265, 482

Dikpati, M.: 2005, Adv. Space Res. 35, 322, doi:10.1016/j.asr.2005.04.061

Dikpati, M. & Charbonneau, P.: 1999, ApJ 518, 508, doi:10.1086/307269

Dikpati, M. & Choudhuri, A. R.: 1995, Solar Phys. 161, 9, doi:10.1007/BF00732081

Dikpati, M. & Gilman, P. A.: 1999, ApJ 512, 417

Dikpati, M. & Gilman, P. A.: 2001, ApJ 559, 428, doi:10.1086/322410

Dikpati, M., Gilman, P. A., & Rempel, M.: 2003, ApJ 596, 680, doi:10.1086/377708

Dikpati, M., de Toma, G., Gilman, P. A., Arge, C. N., & White, O. R.: 2004, ApJ 601, 1136, doi:10.1086/380508

Dikpati, M., de Toma, G., & Gilman, P. A.: 2006, GRLe 33, 5102, doi:10.1029/2005GL025221

Dobler, W., Stix, M., & Brandenburg, A.: 2006, Solar Phys. 638, 336, doi:10.1086/498634

Dominik, C., Blum, J., Cuzzi, J. N., & Wurm, G.: 2007, in B. Reipurth, D. Jewitt, & K. Keil (eds.), *Protostars and Planets V*, University of Arizona Press, p. 783

Donati, J. & Landstreet, J.: 2009, ARA&A 47, 333

Donati, J. F., Jardine, M. M., Gregory, S. G., *et al.*: 2008a, MNRAS 380, 1297

Donati, J.-F., Morin, J., Petit, P., *et al.*: 2008b, MNRAS 390, 545, doi:10.1111/j.1365-2966.2008.13799.x

Donnelly, R. F., Grubb, R. N., & Crowley, F. C.: 1977, *Solar X-ray Measurements from SMS-1, SMS-2, and GOES-1: Information for Data Users*, NOAA Technical Memorandum, ERL SEL-48, NOAA

Dormy, E., Soward, A. M., Jones, C. A., Jault, D., & Cardin, P.: 2004, J. Fluid Mech. 501, 43

Dullemond, C. P. & Dominik, C.: 2005, A&A 434, 971, doi:10.1051/0004-6361: 20042080

Dunkerton, T.: 1978, J. Atmos. Sci. 35, 2325, doi:10.1175/1520-0469(1978)035

Durney, B. R.: 1995, Solar Phys. 160, 213, doi:10.1007/BF00732805

Durney, B. R.: 2000, Solar Phys. 196, 421, doi:10.1023/A:1005285315323

Durney, B. R., De Young, D. S., & Roxburgh, I. W.: 1993, Solar Phys. 145, 207, doi:10.1007/BF00690652

Eather, R. H.: 1980, *Majestic Lights: The Aurora in Science, History, and the Arts*, American Geophysical Union

Eddy, J. A.: 1976, Science 192, 1189

Egbert, G. D. & Ray, R. D.: 2000, Nature 405, 775

Eisner, J. A., Hillenbrand, L. A., White, R. J., *et al.*: 2007, ApJ 669, 1072, doi:10.1086/521874

Eisner, J. A., Graham, J. R., Akeson, R. L., & Najita, J.: 2009, ApJ 692, 309, doi:10.1088/0004-637X/692/1/309

Eliasen, E. & Machenhauer, B.: 1965, Tellus 17, 220

Eliassen, A. & Palm, E.: 1961, Geophys. Publ. 22(3), 1

Emmert, J. T., Picone, J. M., & Lean, J. L.: 2004, JGR 109, A02301, doi:10.1029/2003JA010176

Espaillat, C., Calvet, N., D'Alessio, P., *et al.*: 2007, ApJL 670, L135, doi:10.1086/524360

Etheridge, D. M., Steele, L. P., Langenfelds, R. L., *et al.*: 1996, JGR 101, 4115, doi:10.1029/95JD03410

Fan, Y.: 2004, Living Rev. Solar Phys. 1, 1

Fan, Y.: 2008, ApJ 676, 680

Fan, Y.: 2009, ApJ 697, 1529

Fan, Y., Zweibel, E. G., & Lantz, S. R.: 1998, ApJ 493, 480

Farley, D. T., Balsley, B. B., Woodman, R. F., & McClure, J. P.: 1970, JGR 75, 7199

Ferriz-Mas, A., Schmitt, D., & Schuessler, M.: 1994, A&A 289, 949

Fisk, L. A.: 1996, JGR 101, 15547, doi:10.1029/96JA01005

Fisk, L. A. & Schwadron, N. A.: 2001, ApJ 560, 425, doi:10.1086/322503

Fisk, L. A., Jokipii, J. R., Simnett, G. M., von Steiger, R., & Wenzel, K.-P.: 1998a, Space Sci. Rev. 83

Fisk, L. A., Schwadron, N. A., & Zurbuchen, T. H.: 1998b, Space Sci. Rev. 86, 51, doi:10.1023/A:1005015527146

Flattery, T. W.: 1967, Ph.D. thesis, University of Chicago

Fleming, E. L., Chandra, S., Jackman, C. H., Considine, D. B., & Douglass, A. R.: 1995, J. Atmos. Terr. Phys. 57, 333, doi:10.1016/0021-9169(94)E0013-D

Fleming, T. A., Giampapa, M. S., & Garza, D.: 2003, ApJ 594, 982, doi:10.1086/376968

Florinski, V. & Jokipii, J. R.: 1999, ApJL 523, L185, doi:10.1086/312273

Fludra, A. & Ireland, J.: 2008, A&A 483, 609, doi:10.1051/0004-6361:20078183

Font, A. S., McCarthy, I. G., Johnstone, D., & Ballantyne, D. R.: 2004, ApJ 607, 890, doi:10.1086/383518

Fontenla, J. M., Avrett, E., Thuillier, G., & Harder, J.: 2006, ApJ 639, 441, doi:10.1086/499345

Fontenla, J. M., Balasubramaniam, K. S., & Harder, J.: 2007, ApJ 667, 1243, doi:10.1086/520319

Forbes, J. M.: 2007, J. Meteor. Soc. 85B, 193

Forsyth, R. J., Balogh, A., Smith, E. J., Erdös, G., & McComas, D. J.: 1996, JGR 101, 395, doi:10.1029/95JA02977

Foster, S. S.: 2004, Ph.D. thesis, University of Southampton

Foukal, P.: 1990, *Solar Astrophysics*, Wiley-Interscience

Foukal, P. V.: 2004, *Solar Astrophysics*, 2nd, revised edition, Wiley-VCH

Foukal, P. & Eddy, J.: 2007, Solar Phys. 245, 247, doi:10.1007/s11207-007-9057-8

Foukal, P. & Vernazza, J.: 1979, ApJ 234, 707, doi:10.1086/157547

Fox, J. L. & Bougher, S. W.: 1991, Space Sci. Rev. 55, 357

Franck, S., von Bloh, W., Bounama, C., et al.: 2000, JGR 105, 1651, doi:10.1029/1999JE001169

Frick, P., Soon, W., Popova, E., & Baliunas, S.: 2004, New Astron. 9, 599

Fritts, D. C.: 1984, Rev. Geophys. Space Phys. 22, 275

Fritts, D. C., Vadas, S. L., & Andreassen, O.: 1998, A&A 333, 343

Fröhlich, C.: 2004, in *Solar Variability and its Effects on Climate*, Geophysical Monograph 141, American Geophysical Union, p. 97

Fröhlich, C.: 2006, Space Sci. Rev. 125, 53

Fröhlich, C.: 2009, A&A 501(3), L27-U508

Fröhlich, C. & Lean, J. L.: 1998, Geophys. Res. Lett. 25, 4377, doi:10.1029/1998GL900157

Fröhlich, C. & Lean, J. L.: 2004, Astron. Astrophys. Rev. 12, 273, doi:10.1007/s00159-004-0024-1

Furlan, E., Hartmann, L., Calvet, N., et al.: 2006, ApJSS 165, 568, doi:10.1086/505468

Fusco, A. & Salby, M. L.: 1999, J. Climate 12, 1619

Gaidos, E. J., Güdel, M., & Blake, G. A.: 2000, GRLe 27, 501

Gammie, C. F.: 1996, ApJ 457, 355, doi:10.1086/176735

Gammie, C. F.: 2001, ApJ 553, 174, doi:10.1086/320631

Garcia, R. & Solomon, S.: 1985, JGR 90, 3850

Garcia, R. R., Solomon, S., Roble, R. G., & Rusch, D. W.: 1984, Planet. Space Sci. 32, 411

Garcia, R. R., Lieberman, R., Russell J. M., & Mlynczak M. G.: 2005, J. Atmos. Sci. 62, 4384

Garcia, R. R., Marsh, D. R., Kinnison, D. E., Boville, B. A., & Sassi, F.: 2007, JGR 112, D09301, doi:10.1029/2006JD007485

Gary, G. A.: 2001, Solar Phys. 203, 71

Geller, M. A.: 1970, J. Atmos. Sci. 27(2), 202

Geller, M. A.: 1983, Space Sci. Rev. 34, 359, doi:10.1007/BF00168828

Giacalone, J. & Jokipii, J. R.: 1999, ApJ 520, 204, doi:10.1086/307452

Gille, J. C. & Lyjak, L. V.: 1984, in *Dynamics of the Middle Atmosphere*, D. Reidel, p. 289

Gilman, P. A. & Fox, P. A.: 1997, ApJ 484, 439

Gilman, P. A., Morrow, C. A., & DeLuca, E. E.: 1989, ApJ 338, 528

Gizon, L.: 2004, Solar Phys. 224, 217, doi:10.1007/s11207-005-4983-9

Gizon, L. & Birch, A. C.: 2005, Living Rev. Solar Phys. 2, http://www.livingreviews. org/lrsp-2005-6

Glassmeier, K.-H., Auster, H. U., & Motschmann, U.: 2007, GRLe 34, L22201

Glatzmaier, G. A.: 1985, ApJ 291, 300

Glatzmaier, G. A. & Coe, R. S.: 2007, in P. Olson (ed.), *Treatise of Geophysics, Volume 8: Core Dynamics*, Elsevier, p. 283

Gleeson, L. J. & Axford, W. I.: 1967, ApJ 149, L115

Gold, T. & Soter, S.: 1969, Icarus 11, 356, doi:10.1016/0019-1035(69)90068-2

Gomes, R., Levison, H. F., Tsiganis, K., & Morbidelli, A.: 2005, Nature 435, 466, doi:10.1038/nature03676

Gómez-Pérez, N. & Heimpel, M.: 2007, Geophys. Astrophys. Fluid Dyn. 101, 371

Goodson, A. P. & Winglee, R. M.: 1999, ApJ 524, 159, doi:10.1086/307780

Goody, R. M. & Walker, J. C. G.: 1972, *Atmospheres*, Prentice-Hall

Gosling, J. T. & Skoug, R. M.: 2002, JGR (Space Physics) 107, 1327, doi:10.1029/2002JA009434

Gosling, J. T., Hundhausen, A. J., Pizzo, V., & Asbridge, J. R.: 1972, JGR 77, 5442, doi:10.1029/JA077i028p05442

Gosling, J. T., Hundhausen, A. J., & Bame, S. J.: 1976, JGR 81, 2111, doi:10.1029/ JA081i013p02111

Gosling, J. T., Asbridge, J. R., Bame, S. J., & Feldman, W. C.: 1978, JGR 83, 1401, doi:10.1029/JA083iA04p01401

Gosling, J. T., Asbridge, J. R., Bame, S. J., *et al.*: 1981, JGR 86, 5438, doi:10.1029/ JA086iA07p05438

Gosling, J. T., Baker, D. N., Bame, S. J., *et al.*: 1987, JGR 92, 8519, doi:10.1029/ JA092iA08p08519

Gosling, J. T., Bame, S. J., McComas, D. J., *et al.*: 1993, GRLe 20, 2789, doi:10.1029/ 93GL03116

Gosling, J. T., Bame, S. J., McComas, D. J., *et al.*: 1994a, GRLe 21, 237, doi:10.1029/ 94GL00001

Gosling, J. T., McComas, D. J., Phillips, J. L., *et al.*: 1994b, GRLe 21, 2271, doi:10.1029/ 94GL02245

Gosling, J. T., Bame, S. J., Feldman, W. C., *et al.*: 1995a, GRLe 22, 3329, doi:10.1029/ 95GL02163

Gosling, J. T., Birn, J., & Hesse, M.: 1995b, GRLe 22, 869, doi:10.1029/95GL00270

Gosling, J. T., Feldman, W. C., McComas, D. J., *et al.*: 1995c, GRLe 22, 3333, doi:10.1029/95GL03312

Gosling, J. T., McComas, D. J., Phillips, J. L., *et al.*: 1995d, GRLe 22, 1753, doi:10.1029/95GL01776

Gosling, J. T., Skoug, R. M., McComas, D. J., & Smith, C. W.: 2005, JGR (Space Physics) 110(A9), 1107, doi:10.1029/2004JA010809

Gough, D. O.: 1981, Solar Phys. 71, 21

Gough, D. O., Kosovichev, A. G., Toomre, J., et al.: 1996, Science 272, 1296

Gough, D. O. & McIntyre, M. E.: 1998, Nature 394, 755

Gregory, S. G., Matt, S. P., Donati, J.-F., & Jardine, M.: 2008, MNRAS 389, 1839, doi:10.1111/j.1365-2966.2008.13687.x

Griemeier, J.-M., Stadelmann, A., Penz, T., et al.: 2004, A&A 425, 753, doi:10.1051/0004-6361:20035684

Grinsted, A.: 2002–4, *Continuous Wavelet Transform*, Matlab download

Grove, J. M.: 1988, *The Little Ice Age*, Methuen, New York

Gruzdev, A. N., Schmidt, H., & Brasseur, G. P.: 2009, Atmos. Chem. Phys. 9, 595

Gubbins, D. & Bloxham, J.: 1987, Nature 325, 509

Güdel, M.: 2007, Living Rev. Solar Phys. 4, 3

Güdel, M., Guinan, E. F., & Skinner, S. L.: 1997, ApJ 483, 947

Güdel, M. & Telleschi, A.: 2007, A&A 474, L25, doi:10.1051/0004-6361:20078143

Guillot, T. & Gautier, D.: 2007, in T. Spohn (ed.), *Treatise of Geophysics, Volume 10: Planets and Moons*, Elsevier, p. 439

Guinan, E. F. & Engle, S. G.: 2009, ArXiv:0901.1860

Guinan, E. F., Ribas, I., & Harper, G. M.: 2003, ApJ 594, 561

Günther, H. M. & Schmitt, J. H. M. M.: 2008, A&A 481, 735, doi:10.1051/ 0004-6361:20078674

Hagenaar, H. J., Schrijver, C. J., & Title, A. M.: 2003, ApJ 584, 1107

Haigh, J., Blackburn, M., & Day, R.: 2005, J. Climate 18, 3672

Hale, G. E.: 1908, ApJ 28, 315, doi:10.1086/141602

Hall, J. C. & Lockwood, G. W.: 2004, ApJ 614, 942

Hall, J. C., Henry, G. W., Lockwood, G. W., Skiff, B. A., & Saar, S. H.: 2009, AJ 138, 312, doi:10.1088/0004-6256/138/1/312

Hansen, C. J. & Kawaler, S. D.: 1994, *Stellar Interiors: Physical Principles, Structure, and Evolution*, Springer-Verlag

Harder, J., Lawrence, G., Fontenla, J., Rottman, G., & Woods, T.: 2005, Solar Phys. 230, 141, doi:10.1007/s11207-005-5007-5

Harder, J. W., Fontenla, J. M., Pilewskie, P., Richard, E. C., & Woods, T. N.: 2009, GRLe 36, 7801, doi:10.1029/2008GL036797

Hartmann, L.: 2009a, *Accretion Processes in Star Formation*, Cambridge University Press

Hartmann, L.: 2009b, in K. Tsinganos, T. P. Ray, & M. Stute (eds.), *Protostellar Jets in Context*, Springer, p. 23

Hartmann, L. & Kenyon, S. J.: 1996, ARA&A 34, 207, doi:10.1146/annurev.astro.34.1.207

Hartmann, L. & Stauffer, J. R.: 1989, AJ 97, 873, doi:10.1086/115033

Hartmann, L., Hewett, R., & Calvet, N.: 1994, ApJ 426, 669, doi:10.1086/174104

Hartmann, L., Cassen, P., & Kenyon, S. J.: 1997, ApJ 475, 770, doi:10.1086/303547

Hartmann, L., Calvet, N., Gullbring, E., & D'Alessio, P.: 1998, ApJ 495, 385, doi:10.1086/305277

Hartmann, L., D'Alessio, P., Calvet, N., & Muzerolle, J.: 2006, ApJ 648, 484, doi:10.1086/505788

Harvey, K. L. & Zwaan, C.: 1993, Solar Phys. 148, 85

Hathaway, D. H.: 1996, ApJ 460, 1027, doi:10.1086/177029

Hathaway, D. H.: 2009, Space Sci. Rev. 144, 401, doi:10.1007/s11214-008-9430-4

Hathaway, D. H., Beck, J. G., Bogart, R. S., et al.: 2000, Solar Phys. 193, 299

Hathaway, D. H., Wilson, R. M., & Reichmann, E. J.: 1999, JGR 104, 22375, doi:10.1029/1999JA900313

Hedin, A. E.: 1984, JGR 89, 9828, doi:10.1029/JA089iA11p09828

Hedin, A. E.: 1987, JGR 92, 4649

Hedin, A. E.: 1991, JGR 96(A2), 1159

Hedin, A. E., Reber, C. A., Newton, G. P., *et al.*: 1977, JGR 82, 2148, doi:10.1029/JA082i016p02148

Hedin, A. E., Niemann, H. B., Kasprzak, W. T. & Seif, A.: 1983, JGR 88, 73

Hegerl, G. C., Crowley, T. J., Baum, S. K., Kim, K.-Y., & Hyde, W. T.: 2003, GRLe 30(5), 050000, doi:10.1029/2002GL016635

Hegerl, G. C., Crowley, T. J., Hyde, W. T., & Frame, D. J.: 2006, Nature 440, 1029

Hegerl, G. C., Crowley, T. J., Allen, M., *et al.*: 2007, J. Climate 20, 650

Heikkila, U., Beer, J., & Feichter, J.: 2008, Atmos. Chem. Phys. 8(10), 2797

Heimpel, M. & Aurnou, J.: 2007, Icarus 187, 540

Herbst, W., Eislöffel, J., Mundt, R., & Scholz, A.: 2007, in B. Reipurth, D. Jewitt, & K. Keil (eds.), *Protostars and Planets V*, University of Arizona Press, p. 297

Hernández, J., Calvet, N., Briceño, C., *et al.*: 2007, ApJ 671, 1784, doi:10.1086/522882

Hickey, J. R., Alton, B. M., Kyle, H. L., & Hoyt, D.: 1988, Space Sci. Rev. 48, 321, doi:10.1007/BF00226011

Hillenbrand, L., Mamajek, E., Stauffer, J., *et al.*: 2009, in E. Stempels (ed.), *American Institute of Physics Conference Series*, Vol. 1094, p. 800

Hindman, B. W., Haber, D. A., & Toomre, J.: 2009, ApJ 698, 1749, doi:10.1088/ 0004-637X/698/2/1749

Hinteregger, H. E., Fukui, K., & Gilson, B. R.: 1981, GRLe 8, 1147, doi:10.1029/ GL008i011p01147

Hirooka, T.: 2000, J. Atmos. Sci. 57, 1277

Hirzberger, J., Gizon, L., Solanki, S. K., & Duvall, Jr., T. L.: 2008, Solar Phys. 251, 417

Hollenbach, D., Johnstone, D., Lizano, S., & Shu, F.: 1994, ApJ 428, 654, doi:10.1086/174276

Holme, R.: 2007, in P. Olson (ed.), *Treatise of Geophysics, Volume 8: Core Dynamics*, Elsevier, p. 107

Holt, J. M. & Zhang, S. R.: 2008, GRLe 35, L05813, doi:10.1029/2007GL031148

Holt, S. S. & Kallman, T. R. (eds.): 1998, *Disk Accretion in Pre-Main Sequence Stars*, American Institute of Physics Conference Series, Vol. 431

Holton, J. R.: 1983, J. Atmos. Sci. 40, 2497

Holton, J. R.: 1992, *An Introduction to Dynamic Meteorology*, 3rd edn., Academic Press

Holton, J. R. & Lindzen, R. S.: 1972, J. Atmos. Sci. 29, 1076, doi:10.1175/ 1520-0469(1972)029

Holton, J. R. & Wehrbein, W. M.: 1980, J. Atmos. Sci. 37, 1968

Holton, J. R., Haynes, P. H., McIntyre, M. E., *et al.*: 1995, Rev. Geophys. 33, 403

Holzhauser, H., Magny, M., & Zumbuhl, H. J.: 2005, Holocene 15(6), 789

Holzwarth, V., Mackay, D. H., & Jardine, M.: 2007, Astron. Nachr. 328, 1108, doi:10.1002/asna.200710854

Hormes, A., Beer, J., & Schluchter, C.: 2006, Geogr. Ann. A, Phys. Geogr. 88(4), 281

Hough, S. S.: 1898, Phil. Trans. R. Soc. Lond. A191, 139

Howard, R. A., Sheeley, Jr., N. R., Michels, D. J., & Koomen, M. J.: 1985, JGR 90, 8173, doi:10.1029/JA090iA09p08173

Hoyng, P.: 1993, A&A 272, 321

Hoyt, D. V. & Schatten, K. H.: 1993, JGR 98, 18895, doi:10.1029/93JA01944

Hoyt, D. V. & Schatten, K. H.: 1996, Solar Phys. 165, 181, doi:10.1007/BF00149097

Huang, C.-S. & Kelley, M. C.: 1996, JGR 101, 283, doi:10.1029/95JA02211

Huang, C.-S., Kelley, M. C., & Hysell, D. L.: 1993, JGR 98, 15631, doi:10.1029/93JA00762

Huang, T. Y. W. & Brasseur, G. P.: 1993, JGR 98, 20,413, doi:10.1029/93JD02187

Hubickyj, O., Bodenheimer, P., & Lissauer, J. J.: 2005, Icarus 179, 415, doi:10.1016/j.icarus.2005.06.021

Hudson, H. S., Silva, S., Woodard, M., & Willson, R. C.: 1982, Solar Phys. 76, 211, doi:10.1007/BF00170984

Hughes, D. W., Rosner, R., & Weiss, N. O. (eds.): 2007, *The Solar Tachocline*, Cambridge University Press

Hulot, G., Eymin, C., Langlais, B., Mandea, M., & Olson, N.: 2002, Nature 416, 620

Hundhausen, A. J.: 1973, JGR 78, 1528, doi:10.1029/JA078i010p01528

Hundhausen, A. J.: 1977, in J. Zirker (ed.), *Coronal Holes and High Speed Wind Streams*, Colorado Associated University Press, p. 225

Hundhausen, A. J.: 1985, in R. G. Stone & B. T. Tsurutani (eds.), *Collisionless Shocks in the Heliosphere: A Tutorial Review*, Geophysical Monograph 34, AGU, p. 37

Hundhausen, A. J.: 1988, in V. J. Pizzo, T. E. Holzer, & D. G. Sime (eds.), *Proceedings of the Sixth International Solar Wind Conference*, Tech Note 306+, p. 181

Hundhausen, A. J. & Gosling, J. T.: 1976, JGR 81, 1436, doi:10.1029/JA081i007p01436

Hundhausen, A. J., Burkepile, J. T., & St. Cyr, O. C.: 1994, JGR 99, 6543, doi:10.1029/93JA03586

Hurlburt, N. & DeRosa, M.: 2008, ApJL 684, 123, doi:10.1086/591736

Hussain, G. A. J., Cameron, A. C., Jardine, M., & Donati, J.-F.: 2009, in *IAU Symposium*, Vol. 259, p. 447

Indermuhle, A., Stocker, T. F., Joos, F., *et al.*: 1999, Nature 398, 121

Intergovernmental Panel on Climate Change (IPCC): 2007, *Climate Change 2007, The Physical Science Basis*, S. Solomon *et al.* (eds.), Cambridge University Press

Iskakov, A. B., Schekochihin, A. A., Cowley, S. C., McWilliams, J. C., & Proctor, M. R. E.: 2007, Phys. Rev. Lett. 98, 208501

Jacchia, L. G.: 1970, SAO Special Report 313

Jackson, A. & Finlay, C. C.: 2007, in M. Kono (ed.), *Treatise of Geophysics, Volume 5: Geomagnetism*, Elsevier, p. 147

Jackson, A., Jonkers, A. R. T., & Walker, M. R.: 2000, Phil. Trans. R. Soc. Lond. A358, 957

Jardine, M. & Unruh, Y. C.: 1999, A&A 346, 883

Järvinen, S. P., Berdyugina, S. V., Korhonen, H., Ilyin, I., & Tuominen, I.: 2007, A&A 472, 887, doi:10.1051/0004-6361:20077551

Jarvis, M. J., Jenkins, B., & Rogers, G. A.: 1998, JGR 103, 20,775

Joerin, U. E., Stocker, T. F., & Schluchter, C.: 2006, The Holocene 16(5), 697

Johns-Krull, C. M.: 2009, in *IAU Symposium*, Vol. 259, p. 345

Johnson, B. M. & Gammie, C. F.: 2003, ApJ 597, 131, doi:10.1086/378392

Jokipii, J. R.: 1971, Rev. Geophys. Space Phys. 27, 9

Jokipii, J. R.: 1986, in C. P. Sonett (ed.), *The Sun in Time*, University of Arizona Press, p. 205

Jokipii, J. R.: 1989, Adv. Space Res. 9, 105

Jokipii, J. R. & Kota, J.: 1989, GRLe 16, 1, doi:10.1029/GL016i001p00001

Jokipii, J. R. & Marti, K.: 1986, in R. Smoluchowski (ed.), *The Galaxy and the Solar System*, University of Arizona Press, p. 116

Jokipii, J. R. & Owens, A. J.: 1973, ApJ 181, L147

Jokipii, J. R. & Parker, E. N.: 1968, Phys. Rev. Lett. 21, 44, doi:10.1103/PhysRevLett.21.44

Jokipii, J. R. & Thomas, B.: 1981, ApJ 243, 1115

Jokipii, J. R., Levy, E. H., & Hubbard, W. B.: 1977, ApJ 213, 861

Jokipii, J. R., Kóta, J., & Merenyi, E.: 1993, JGR 405, 782

Jones, C. A.: 2007, in P. Olson (ed.), *Treatise of Geophysics, Volume 8: Core Dynamics*, Elsevier, p. 131

Jones, P. D. & Mann, M. E.: 2003, Rev. Geophys. 42, RG2002, doi:doi:10.1029/ 2003RG000143

Jones, P. D., Briffa, K. R., Barnett, T. P., & Tett, S. F. B.: 1998, The Holocene 8, 455

Jose, P. D.: 1965, AJ 70, 193, doi:10.1086/109714

Judge, D. L., McMullin, D. R., Ogawa, H. S., *et al.*: 1998, Solar Phys. 177, 161

Judge, P. G. & Saar, S. H.: 2007, ApJ 663, 643, doi:10.1086/513004

Judge, P. G., Solomon, S. C., & Ayres, T. R.: 2003, ApJ 593, 534, doi:10.1086/376405

Kageyama, A., Miyagoshi, T., & Sato, T.: 2008, Nature 454, 1106

Kahler, S. & Lin, R. P.: 1994, GRLe 21, 1575, doi:10.1029/94GL01362

Käpylä, P. J., Korpi, M. J., Ossendrijver, M., & Stix, M.: 2006, A&A 455, 401, doi:10.1051/0004-6361:20064972

Käpylä, P. J., Korpi, M. J., & Brandenburg, A.: 2008, A&A 491, 353

Kasting, J. F.: 1988, Icarus 74, 472, doi:10.1016/0019-1035(88)90116-9

Kasting, J. F., Whitmire, D. P., & Reynolds, R. T.: 1993, Icarus 101, 108, doi:10.1006/icar.1993.1010

Kawamura, K., Parrenin, F., Lisiecki, L., *et al.*: 2007, Nature 448(7156), 912

Keating, G. M., Pitts, M. C., Brasseur, G. P., & De Rudder, A.: 1987, JGR 92, 889

Keating, G. M., Tolson, R. H., & Bradford, M. S.: 2000, GRLe 27, 1523

Keeling, C. D. & Whorf, T. P.: 2000, Proc. Nat. Acad. Sci. USA 97(8), 3814

Kelley, M. C.: 1989, *The Earth's Ionosphere: Plasma Physics and Electrodynamics*, Academic Press

Kernthaler, S. C., Toumi, R., & Haigh, J. D.: 1999, GRLe 26, 863, doi:10.1029/ 1999GL900121

Kiehl, J. T. & Trenberth, K. E.: 1997, Bull. Amer. Meteor. Soc. 78, 197

Kimball, D. S.: 1960, *A Study of the Aurora of 1859*, Rep. 6, Geophysical Institute, University of Alaska, Fairbanks

King, J. H.: 1979, JGR 84, 5938, doi:10.1029/JA084iA10p05938

Klahr, H. & Bodenheimer, P.: 2006, ApJ 639, 432, doi:10.1086/498928

Knobloch, E., Tobias, S. M., & Weiss, N. O.: 1998, MNRAS 297, 1123, doi:10.1046/ j.1365-8711.1998.01572.x

Knudsen, M., Riisager, P., Donadini, F., *et al.*: 2008, Earth Planet. Sci. Lett. 272(1-2), 319, doi:10.1016/j.epsl.2008.04.048

Kodera, K. & Kuroda, Y.: 2002, JGR 107, doi:10.1029/2002JD002224

Kodera, K., Hori, M. E., Yukimoto, S., & Sigmond, M.: 2008, GRLe 35, 3704, doi:10.1029/2007GL031958

Koenigl, A.: 1991, ApJL 370, L39, doi:10.1086/185972

Konigl, A.: 1989, ApJ 342, 208, doi:10.1086/167585

Kopp, G. & Lawrence, G.: 2005, Solar Phys. 230, 91, doi:10.1007/s11207-005-7446-4

Kopp, G., Heuerman, K., & Lawrence, G.: 2005a, Solar Phys. 230, 111, doi:10.1007/ s11207-005-7447-3

Kopp, G., Lawrence, G., & Rottman, G.: 2005b, Solar Phys. 230, 129, doi:10.1007/ s11207-005-7433-9

Kosovichev, A. G. & Stenflo, J. O.: 2008, ApJL 688, L115, doi:10.1086/595619

Kóta, J. & Jokipii, J. R.: 1995, Science 268, 1024, doi:10.1126/science.268.5213.1024

Krause, F. & Rädler, K.-H.: 1980, *Mean-Field Magnetohydrodynamics and Dynamo Theory*, Pergamon Press

Krieger, A. S., Timothy, A. F., & Roelof, E. C.: 1973, Solar Phys. 29, 505, doi:10.1007/BF00150828

Kristjansson, J. E., Stjern, C. W., Stordal, F., *et al.*: 2008, Atmos. Chem. Phys. 8, 73

Krivova, N. A., Solanki, S. K., Fligge, A., & Unruh, Y. C.: 2003, A&A 399(1), L1

Krivova, N. A., Solanki, S. K., & Floyd, L.: 2006, A&A 452, 631, doi:10.1051/0004-6361:20064809

Krivova, N. A., Balmaceda, L., & Solanki, S. K.: 2007, A&A 467, 335, doi:10.1051/0004-6361:20066725

Kuhn, J. R.: 1988, ApJ 331, L131, doi:10.1086/185251

Kuhn, J. R. & Libbrecht, K. G.: 1991, ApJ 381(1), L35

Kuhn, J. R., Armstrong, J. D., Bush, R. I., & Scherrer, P.: 2000, Nature 405, 544

Küker, M., Arlt, R., & Rüdiger, G.: 1999, A&A 343, 977

Küker, M., Rüdiger, G., & Schultz, M.: 2001, A&A 374, 301, doi:10.1051/0004-6361:20010686

Kulikov, Y. N., Lammer, H., Lichtenegger, H., *et al.*: 2007, Space Sci. Rev. 129, 207

Kundt, W.: 1977, A&A 60, 85

Kunow, H., Crooker, N. U., Linker, J. A., Schwenn, R., & von Steiger, R. (eds.): 2006, *Coronal Mass Ejections*, Springer

Kuroda, Y., Deushi, M., & Shibata, K.: 2007, GRLe 34, L21704, doi:10.1029/2007GL030983

Labitzke, K. & van Loon, H.: 1988, J. Atmos. Sol. Terr. Phys. 64, 203

Laj, C., Kissel, C., Mazaud, A., Channell, J. E. T., & Beer, J.,: 2000, Phil. Trans. R. Soc. Lond. A358(1768), 1009

Lammer, H., Bredehöft, J. H., Coustenis, A., *et al.*: 2009, Astron. Astrophys. Rev. 17, 181, doi:10.1007/s00159-009-0019-z

Laskar, J., Robutel, P., Joutel, F., *et al.*: 2004, A&A 428(1), 261

Lay, T., Hernlund, J., & Buffett, B. A.: 2008, Nature Geosci. 1, 25

Lean, J. L.: 2000, GRLe 27, 2425, doi:10.1029/2000GL000043

Lean, J. L.: 2005, Physics Today, June, 32

Lean, J. L. & Rind, D. H.: 2009, GRLe 36, 15708, doi:10.1029/2009GL038932

Lean, J. L., Skumanich, A., & White, O.: 1992, GRLe 19, 1591, doi:10.1029/92GL01578

Lean, J. L., Beer, J., & Bradley, R.: 1995, GRLe 22, 3195, doi:10.1029/95GL03093

Lean, J. L., Rottman, G. J., Kyle, H. L., *et al.*: 1997, JGR 102, 29939, doi:10.1029/97JD02092

Lean, J. L., Cook, J., Marquette, W., & Johannesson, A.: 1998, ApJ 492, 390, doi:10.1086/305015

Lean, J. L., White, O. R., Livingston, W. C., & Picone, J. M.: 2001, JGR 106(A6), 10,645

Lean, J. L., Wang, Y.-M., & Sheeley, N. R.: 2002, GRLe 29(24), 240000, doi:10.1029/2002GL015880

Lean, J. L., Warren, H. P., Mariska, J. T., & Bishop, J.: 2003, JGR (Space Physics) 108, 1059, doi:10.1029/2001JA009238

Lean, J. L., Rottman, G., Harder, J., & Kopp, G.: 2005, Solar Phys. 230, 27, doi:10.1007/s11207-005-1527-2

Lean, J. L., Picone, J. M., & Emmert, J. T.: 2009, JGR (Space Physics) 114(A13), 7301, doi:10.1029/2009JA014285

Lee, H. & Smith, A. K.: 2003, JGR 108(D2), 4049, doi:10.1029/2001JD001503

Leighton, R. B.: 1964, ApJ 140, 1547, doi:10.1086/148058

Leighton, R. B.: 1969, ApJ 156, 1, doi:10.1086/149943

Leighton, R. B., Noyes, R. W., & Simon, G. W.: 1962, ApJ 135, 474

Letfus, V.: 2000, Solar Phys. 197, 203

Levison, H. F., Morbidelli, A., Gomes, R., & Backman, D.: 2007, in B. Reipurth, D. Jewitt, & K. Keil (eds.), *Protostars and Planets V*, University of Arizona Press, p. 669

Levy, E. H. & Rose, W. K.: 1974, ApJ 193, 419, doi:10.1086/153177

Li, L., Ingersoll, A. P., Vasavada, A. R., *et al.*: 2006, Icarus 185, 416

Liebert, J., Kirkpatrick, J. D., Reid, I. N., & Fisher, M. D.: 1999, ApJ 519, 345, doi:10.1086/307349

Lighthill, J.: 1978, *Waves in Fluids*, Cambridge University Press

Lindzen, R.: 1970, Geophys. Astrophys. Fluid Dyn. 1, 303, doi:10.1080/03091927009365777

Lindzen, R. S.: 1978, Mon. Wea. Rev. 106, 526

Lindzen, R. S.: 1981, JGR 86, 9707

Lindzen, R. S. & Holton, J. R.: 1968, J. Atmos. Sci. 25, 1095, doi:10.1175/ 1520-0469(1968)025

Lindzen, R. S., Straus, D. M., & Katz, B.: 1984, J. Atmos. Sci. 41, 1320

Lineweaver, C. H.: 2001, Icarus 151, 307, doi:10.1006/icar.2001.6607

Lineweaver, C. H., Fenner, Y., & Gibson, B. K.: 2004, Science 303, 59, doi:10.1126/science.1092322

Linsky, J. L.: 1985, Solar Phys. 100, 333, doi:10.1007/BF00158435

Lisiecki, L. E. & Raymo, M. E.: 2005, Paleoceanography 20(2), PA1003

Lissauer, J. J. & Stevenson, D. J.: 2007, in B. Reipurth, D. Jewitt, & K. Keil (eds.), *Protostars and Planets V*, University of Arizona Press, p. 591

Lites, B. W.: 2008, Space Sci. Rev. 156, doi:10.1007/s11214-008-9437-x

Lites, B. W., Kubo, M., Socas-Navarro, H., *et al.*: 2008, ApJ 672, 1237

Liu, H.-L. & Roble, R. G.: 2002, JGR (Atmospheres) 107, 4695, doi:10.1029/2001JD001533

Lockwood, M.: 2006, Space Sci. Rev. 125, 95, doi:10.1007/s11214-006-9049-2

Lockwood, M. & Stamper, R.: 1999, GRLe 26, 2461, doi:10.1029/1999GL900485

Long, M., Romanova, M. M., & Lovelace, R. V. E.: 2005, ApJ 634, 1214, doi:10.1086/497000

Long, M., Romanova, M. M., & Lovelace, R. V. E.: 2007, MNRAS 374, 436, doi:10.1111/j.1365-2966.2006.11192.x

Longcope, D. W., McKenzie, D. E., Cirtain, J., & Scott, J.: 2005, ApJ 630, 596

Longuet-Higgins, M. S.: 1968, Phil. Trans. R. Soc. Lond. A262, 511

Lorenz, R. D., Lunine, J. I., & McKay, C. P.: 1997, GRLe 24, 2905, doi:10.1029/97GL52843

Low, B. C.: 2001, JGR 106, 25,141

Lu, G., Richmond, A. D., Emery, B. A., & Roble, R. G.: 1995, JGR 100, 19643, doi:10.1029/95JA00766

Lundin, R., Lammer, H., & Ribas, I.: 2007, Space Sci. Rev. 129, 245, doi:10.1007/s11214-007-9176-4

Lunine, J. I., O'Brien, D. P., Raymond, S. N., *et al.*: 2009, ArXiv:0906.4369v1

Lynden-Bell, D. & Boily, C.: 1994, MNRAS 267, 146

Lynden-Bell, D. & Pringle, J. E.: 1974, MNRAS 168, 603

Lyra, W., Johansen, A., Klahr, H., & Piskunov, N.: 2008, A&A 491, L41, doi:10.1051/0004-6361:200810626

MacGregor, K. B. & Cassinelli, J. P.: 2003, ApJ 586, 480, doi:10.1086/346257

MacGregor, K. B. & Charbonneau, P.: 1997, ApJ 486, 484, doi:10.1086/304484

MacGregor, K. B. & Charbonneau, P.: 1999, ApJ 519, 911

Mackay, D. H. & van Ballegooijen, A. A.: 2006, ApJ 641, 577, doi:10.1086/500425

Madden, R. A.: 1978, J. Atmos. Sci. 35, 1605

Madronich, S. & Flocke, S.: 1999, in P. Boule (ed.), *Handbook of Environmental Chemistry*, Springer, p. 1

Maeder, A. & Meynet, G.: 1988, A&AS 76, 411

Maeder, A. & Meynet, G.: 2003, A&A 411, 543, doi:10.1051/0004-6361:20031491

Maeder, A. & Meynet, G.: 2005, A&A 440, 1041, doi:10.1051/0004-6361:20053261

Marin, D., Mikhailov, A. V., de la Morera, B. A., & Herraiz, M.: 2001, Ann. Geophys. 19, 761

Marsch, E.: 1991, in R. Schwenn & E. Marsch (eds.), *Physics of the Inner Heliosphere 2: Particles, Waves and Turbulence*, Springer-Verlag, p. 159

Marsh, D. R. & Garcia. R. R : 2007, GRLe 34, L21807, doi:10.1029/2007GL030935

Marsh, N. & Svensmark, H.: 2003, JGR 108, 4195, doi:10.1029/2001JD001264

Marsh, D. R., Solomon, S. C., & Reynolds, A. E.: 2004, JGR 109, A07301, doi:10.1029/2003JA010199

Marsh, D. R., Garcia, R. R., Kinnison, D. E., *et al.*: 2007, JGR 112, D23306, doi:10.1029/2006JD008306

Masarik, J. & Beer, J.: 1999, JGR 104, 12099, doi:10.1029/1998JD200091

Masarik, J. & Beer, J.: 2009, JGR 114(D11103)

Mason, J., Hughes, D. W., & Tobias, S. M.: 2008, MNRAS 391, 467, doi:10.1111/j.1365-2966.2008.13918.x

Matsuno, T.: 1971, J. Atmos. Sci. 28, 1479

Matt, S. & Pudritz, R. E.: 2004, ApJL 607, L43, doi:10.1086/421351

Matt, S. & Pudritz, R. E.: 2005, ApJL 632, L135, doi:10.1086/498066

Matt, S. & Pudritz, R. E.: 2008a, ApJ 678, 1109, doi:10.1086/533428

Matt, S. & Pudritz, R. E.: 2008b, ApJ 681, 391, doi:10.1086/587453

Matt, S., Goodson, A. P., Winglee, R. M., & Böhm, K.-H.: 2002, ApJ 574, 232, doi:10.1086/340896

Matthes, K., Langematz, U., Gray, L. J., Kodera, K., & Labitzke, K.: 2004, JGR 109, doi:10.1029/2003JD004012.

Matthes, K., Kuroda, Y., Kodera, K., & Langematz, U.: 2006, JGR 111, D06108, doi:10.1029/2005JD006283

Maus, S., Rother, M., Stolle, C., *et al.*: 2006, Geochem. Geophys. Geosys. 7, Q07008

Mayr, H. G., Mengel, J. G., Hines, C. O., *et al.*: 1997, JGR 102, 26,093

McClintock, W. E., Rottman, G. J., & Woods, T. N.: 2005, Solar Phys. 230, 225, doi:10.1007/s11207-005-7432-x

McComas, D. J., Gosling, J. T., & Phillips, J. L.: 1992, JGR 97, 171, doi:10.1029/91JA02370

McComas, D. J., Goldstein, R., Gosling, J. T., & Skoug, R. M.: 2001, Space Sci. Rev. 97, 99, doi:10.1023/A:1011826111330

McComas, D. J., Elliott, H. A., Gosling, J. T., *et al.*: 2002, GRLe 29(9), 090000, doi:10.1029/2001GL014164

McComas, D. J., Ebert, R. W., Elliott, H. A., *et al.*: 2008, GRLe 35, 18103, doi:10.1029/2008GL034896

McCracken, K. G. & Beer, J.: 2007, JGR (Space Physics) 112(A11), A10101, doi:10.1029/2006JA012117

McCracken, K. G., Dreschhoff, G. A. M., Zeller, E. J., Smart, D. F., & Shea, M. A.: 2001, JGR 106, 21585, doi:10.1029/2000JA000237

McCracken, K. G., McDonald, F. B., Beer, J., Raisbeck, G., & Yiou, F.: 2004, JGR (Space Physics) 109(A18), 12103, doi:10.1029/2004JA010685

McIntyre, M. E.: 1998, Prog. Theor. Phys. Suppl. 130, 137, Corrigendum, Prog. Theor. Phys., 101, 189 (1999).

McIntyre, M. E.: 2007, in D. W. Hughes, R. Rosner, & N. O. Weiss (eds.), *The Solar Tachocline*, Cambridge University Press, p. 183

Meehl, G. A., Arblaster, J. M., Matthes, K., Sassi, F., & van Loon, H.: 2009, Science 325, 1114

Mekaoui, S. & Dewitte, S.: 2008, Solar Phys. 247, 203, doi:10.1007/s11207-007-9070-y

Merrill, R. T., McElhinny, M. W., & McFadden, P. L.: 1998, *The Magnetic Field of the Earth*, Academic Press

Miesch, M. S.: 2005, Living Rev. Solar Phys. 2, http://www.livingreviews.org/lrsp-2005-1

Miesch, M. S.: 2007, Astron. Nachr. 328, 998

Miesch, M. S. & Toomre, J.: 2009, Ann. Rev. Fluid Mech. 41, 317

Miesch, M. S., Brun, A. S., & Toomre, J.: 2006, ApJ 641, 618

Miesch, M. S., Gilman, P. A., & Dikpati, M.: 2007, ApJSS 168, 337

Miesch, M. S., Brun, A. S., DeRosa, M. L., & Toomre, J.: 2008, ApJ 673, 557

Miesch, M. S., Browning, M. K., Brun, A. S., Toomre, J., & Brown, B. P.: 2009, in *Proc. GONG 2008/SOHO XXI Meeting on Solar-Stellar Dynamos as Revealed by Helio- and Asteroseismology*, ASP Conference Series

Mikhailov, A. V. & Marin, D.: 2001, Ann. Geophys. 19, 733

Milankovich, M.: 1930, in W. Köppen & R. Geiger (eds.), *Handbuch der Klimatologie*, Vol. 1, Gebrüder Borntraeger, p. 1

Mininni, P. D. & Gómez, D. O.: 2002, ApJ 573, 454, doi:10.1086/340495

Mischna, M. A., Kasting, J. F., Pavlov, A., & Freedman, R.: 2000, Icarus 145, 546, doi:10.1006/icar.2000.6380

Miyahara, H., Masuda, K., Muraki, Y., *et al.*: 2004, Solar Phys. 224, 317, doi:10.1007/s11207-005-6501-5

Miyahara, H., Yokoyama, Y., & Masuda, K.: 2008, Earth Planet. Sci. Lett. 272, 290

Moffatt, H. K.: 1978, *Magnetic Field Generation in Electrically Conducting Fluids*, Cambridge University Press

Mohanty, S., Basri, G., Shu, F., Allard, F., & Chabrier, G.: 2002, ApJ 571, 469

Molina, M. J. & Rowland, F. S.: 1974, Nature 249, 810

Monchaux, R., Berhanu, M., Bourgoin, M., *et al.*: 2007, Phys. Rev. Lett. 98, 044502

Moore, T. E. & Horwitz, J. L.: 2007, Rev. Geophys. 45, 3002, doi:10.1029/2005RG000194

Morin, J., Donati, J.-F., Petit, P., *et al.*: 2008, MNRAS 390, 567

Moss, D. & Brooke, J.: 2000, MNRAS 315, 521, doi:10.1046/j.1365-8711.2000.03452.x

Moss, D., Brandenburg, A., Tavakol, R., & Tuominen, I.: 1992, A&A 265, 843

Müller, P.: 1995, Rev. Geophys. 33, 67

Murray, M. J., Hood, A. W., Moreno-Insertis, F., Galsgaard, K., & Archontis, V.: 2006, A&A 460, 909, doi:10.1051/0004-6361:20065950

Muscheler, R., Beer, J., Kubik, P. W., & Synal, H.-A.: 2005, Quat. Sci. Rev. 24, 1849

Muzerolle, J., Calvet, N., & Hartmann, L.: 2001, ApJ 550, 944, doi:10.1086/319779

Muzerolle, J., Calvet, N., Hartmann, L., & D'Alessio, P.: 2003, ApJL 597, L149, doi:10.1086/379921

Myrabo, H. K.: 1984, Planet. Space Sci. 32, 249, doi:10.1016/0032-0633(84)90159-4

Myrabo, H. K., Deehr, C. S., & Lybekk, B.: 1984, Planet. Space Sci. 32, 853

Nagasawa, M., Thommes, E. W., Kenyon, S. J., Bromley, B. C., & Lin, D. N. C.: 2007, in B. Reipurth, D. Jewitt, & K. Keil (eds.), *Protostars and Planets V*, University of Arizona Press, p. 639

Nagy, A. F., Balogh, A., Cravens, T. E., Mendillo, M., & Muller-Wodarg, I.: 2008, Space Sci. Rev. 139, 1, doi:10.1007/s11214-008-9353-0

Nandy, D. & Choudhuri, A. R.: 2001, ApJ 551, 576, doi:10.1086/320057

Narain, U. & Ulmschneider, P.: 1996, Space Sci. Rev. 75, 453

Nastrom, G. D. & Belmont, A. D.: 1976, J. Atmos. Sci. 33, 315

Nellis, W. J., Weir, S. T., & Mitchell, A. C.: 1999, Phys. Rev. B 59, 3434

Nelson, R. P.: 2005, A&A 443, 1067, doi:10.1051/0004-6361:20042605

Neugebauer, M., Gloeckler, G., Gosling, J. T., *et al.*: 2007, ApJ 667, 1262, doi:10.1086/521019

Newkirk, G.: 1980, Geochim. Cosmochim. Acta Suppl. 13

Newkirk, G.: 1983, ARA&A 21, 429, doi:10.1146/annurev.aa.21.090183.002241

Nier, A. O. & McElroy, M. B.: 1977, JGR 82, 4341

Nimmo, F.: 2007, in P. Olson (ed.), *Treatise of Geophysics, Volume 8: Core Dynamics*, Elsevier, p. 31

Nisbet, E. G. & Sleep, N. H.: 2001, Nature 409, 1083

Nordlund, A.: 1985, Solar Phys. 100, 209

North, G. R. & Stevens, M. J.: 1998, J. Climate 11, 563

North, G. R., Mengel, J. G., & Short, D. A.: 1984, in J. E. Hansen & T. Takahashi (eds.), *Climate Processes and Climate Sensitivity*, Geophysical Monograph 29, AGU, p. 164

Noyes, R. W., Hartmann, L., Baliunas, S. L., Duncan, D. K., & Vaughan, A. H.: 1984a, ApJ 279, 763

Noyes, R. W., Weiss, N. O., & Vaughan, A. H.: 1984b, ApJ 287, 769, doi:10.1086/162735

Obrochta, S. P., Lotti-Bond, R., & Crowley, T. J.: 2010, in preparation

O'dell, C. R.: 1998, AJ 115, 263, doi:10.1086/300178

O'dell, M. A., Panagi, P., Hendry, M. A., & Collier Cameron, A.: 1995, A&A 294, 715

Odstrčil, D. & Pizzo, V. J.: 1999, JGR 104, 493, doi:10.1029/1998JA900038

Oeschger, H., Siegenthaler, U., Schotterer, U., & Gugelmann, A.: 1975, Tellus 27(2), 168

Ogilvie, K. W.: 1972, in P. J. Coleman, C. P. Sonnet, & J. M. Wilcox (eds.), *Solar Wind*, NASA SP 308, p. 30

Olsen, N., Hulot, G., & Sabaja, T. J.: 2007, in M. Kono (ed.), *Treatise of Geophysics, Volume 5: Geomagnetism*, Elsevier, p. 33

Olson, P. & Aurnou, J.: 1999, Nature 402, 170

Olson, P. & Christensen, U. R.: 2006, Earth Planet. Sci. Lett. 250, 561

Olson, P., Christensen, U. R., & Glatzmaier, G. A.: 1999, JGR 104, 10,383

Ossakow, S. L.: 1981, J. Atmos. Terr. Phys. 43, 437

Ossendrijver, A. J. H. & Hoyng, P.: 1996, A&A 313, 959

Ossendrijver, A. J. H., Hoyng, P., & Schmitt, D.: 1996, A&A 313, 938

Ossendrijver, M.: 2003, Astron. Astrophys. Rev. 11, 287, doi:10.1007/s00159-003-0019-3

Ossendrijver, M., Stix, M., & Brandenburg, A.: 2001, A&A 376, 713, doi:10.1051/ 0004-6361:20011041

Ossendrijver, M. A. J. H.: 2000a, A&A 359, 364

Ossendrijver, M. A. J. H.: 2000b, A&A 359, 1205

O'Sullivan, D. & Salby, M. L.: 1990, J. Atmos. Sci. 47, 650

Otmianowska-Mazur, K., Rudiger, G., Elstner, D., & Arlt, R.: 1997, Geophys. Astrophys. Fluid Dyn. 86, 229, doi:10.1080/03091929708245463

Ott, E.: 1998, Phys. Plasmas 5, 1636

Owen, J. E., Ercolano, B., Clarke, C. J., & Alexander, R. D.: 2010, MNRAS 401, 1415

Owens, M. J. & Crooker, N. U.: 2006, JGR (Space Physics) 111(A10), 10104, doi:10.1029/2006JA011641

Owens, M. J., Crooker, N. U., Schwadron, N. A., *et al.*: 2008, GRLe 35, 20108, doi:10.1029/2008GL035813

Paardekooper, S.-J. & Papaloizou, J. C. B.: 2009, MNRAS 394, 2283, doi:10.1111/j.1365-2966.2009.14511.x

Pancheva, D. V., Mukhtarov, P. J., Shepherd, M. G., *et al.*: 2006, JGR (Space Physics) 111(A10), 7313, doi:10.1029/2005JA011562

Papaloizou, J. C. B., Nelson, R. P., Kley, W., Masset, F. S., & Artymowicz, P.: 2007, in B. Reipurth, D. Jewitt, & K. Keil (eds.), *Protostars and Planets V*, University of Arizona Press, p. 655

Parfrey, K. P. & Menou, K.: 2007, ApJ 667, L207

Parker, E. N.: 1955a, ApJ 122, 293, doi:10.1086/146087

Parker, E. N.: 1955b, ApJ 121, 491, doi:10.1086/146010

Parker, E. N.: 1958, ApJ 128, 664, doi:10.1086/146579

Parker, E. N.: 1963, *Interplanetary Dynamical Processes*, Interscience Publishers

Parker, E. N.: 1965, Planet. Space Sci. 13, 9

Parker, E. N.: 1993, ApJ 408, 707

Parker, E. N.: 2008, Space Sci. Rev. 144, 15

Patten, B. M. & Simon, Th.: 1996, ApJSS 106, 489

Pedlosky, J.: 1987, *Geophysical Fluid Dynamics*, Springer-Verlag

Pesnell, W. D.: 2008, Solar Phys. 252, 209, doi:10.1007/s11207-008-9252-2

Petit, P., Dintrans, B., Solanki, S. K., *et al.*: 2008, MNRAS 388, 80

Petrovay, K. & Kerekes, A.: 2004, MNRAS 351, L59, doi:10.1111/j.1365-2966.2004.07971.x

Pevtsov, A. A. & Balasubramaniam, K. S.: 2003, Adv. Space Res. 32, 1867

Pevtsov, A. A., Fisher, G. H., Acton, L. W., *et al.*: 2003, ApJ 598, 1387

Pietrinferni, A., Cassisi, S., Salaris, M., & Castelli, F.: 2004, ApJ 612, 168, doi:10.1086/422498

Pitteway, J. L. V. & Hines, C. O.: 1963, Can. J. Phys. 41, 1935

Pizzo, V. J.: 1980, JGR 85, 727, doi:10.1029/JA085iA02p00727

Pizzo, V. J.: 1982, JGR 87, 4374, doi:10.1029/JA087iA06p04374

Pizzo, V. J.: 1991, JGR 96, 5405, doi:10.1029/91JA00155

Pizzo, V. J.: 1994, JGR 99, 4185.

Pizzo, V. J. & Gosling, J. T.: 1994, GRLe 21, 2063, doi:10.1029/94GL01581

Plumb, R. A.: 1977, J. Atmos. Sci. 34, 1847

Plumb, R. A.: 1982, J. Atmos. Sci. 40, 262

Pneuman, G. W. & Kopp, R. A.: 1971, Solar Phys. 18, 258, doi:10.1007/BF00145940

Pollack, H. N. & Smerdon, J. E.: 2004, JGR 109, D11106, doi:10.1029/2003JD004163

Porter, D. H. & Woodward, P. R.: 2000, ApJSS 321, 323

Pouquet, A., Frisch, U., & Léorat, J.: 1976, J. Fluid Mech. 77, 321

Preibisch, T., Kim, Y.-C., Favata, F., *et al.*: 2005, ApJSS 160, 401, doi:10.1086/432891

Pudritz, R. E. & Norman, C. A.: 1983, ApJ 274, 677, doi:10.1086/161481

Qian, L., Roble, R. G., Solomon, S. C., & Kane, T. J.: 2006, JGR 33, L23705, doi:10.1029/2006GL027185

Quartly, G. D., Cipollini, P., Cromwell, D., & Challenor, P. G.: 2003, Phil. Trans. R. Soc. Lond. A 361, 57

Quiroz, R. S.: 1969, Mon. Wea. Rev. 97, 541

Radick, R. R., Lockwood, G. W., Skiff, B. A., & Baliunas, S. L.: 1998, ApJSS 118, 239

Rafikov, R. R.: 2005, ApJL 621, L69, doi:10.1086/428899

Raisbeck, G. M., Yiou, F., Bourles, D., *et al.*: 1987, Nature, 326, 273

Randel, W. J. & Wu, F.: 2007, JGR 112(D6), doi:10.1029/2006JD007339

Randel, W. J., Shine, K. P., Austin, J., *et al.*: 2009, JGR 114, 107, doi:10.1029/2008JD010421

Rast, M. P.: 1995, ApJ 443, 863

Rast, M. P.: 2003, ApJ 597, 1200

Rast, M. P. & Toomre, J.: 1993, ApJ 419, 240

Rast, M. P., Ortiz, A., & Meisner, R. W.: 2008, ApJ 673, 1209

Reed, R. J., Campbell, W. J., Rasmussen, L. A., & Rogers, D. G.: 1961, JGR 66, 813, doi:10.1029/JZ066i003p00813

Reid, I. N., Kirkpatrick, J. D., Gizis, J. E., & Liebert, J.: 1999, ApJL 527, 105

Reimer, P. J., Baillie, M. G. L., Bard, E., *et al.*: 2004, Radiocarbon 46(3), 1029

Reiners, A. & Basri, G.: 2009, A&A 496, 787

Reiners, A., Basri, G., & Browning, M.: 2009, ApJ 692, 538

Reipurth, B., Jewitt, D., & Keil, K. (eds.): 2007, *Protostars and Planets V*, University of Arizona Press

Reisenfeld, D. B., Gosling, J. T., Forsyth, R. J., Riley, P., & St. Cyr, O. C.: 2003, GRLe 30(19), 190000, doi:10.1029/2003GL017155

Rempel, M.: 2005, ApJ 622, 1320

Rempel, M.: 2006, ApJ 647, 662, doi:10.1086/505170

Rempel, M.: 2008, J. Phys. Conf. Ser. 118:012032, Proc. HELAS II Int. Conf. on Helioseismology, Asteroseismology and MHD Connections

Rempel, M. & Schüssler, M.: 2001, ApJ 552, L171

Remsberg, E. E.: 2008, JGR 113, D22304, doi:10.1029/2008JD010189

Rhines, P. B.: 1975, J. Fluid Mech. 69, 417

Ribas, I., Guinan, E. F., Güdel, M., & Audard, M.: 2005, ApJ 622, 680

Ribes, J. C. & Nesme-Ribes, E.: 1993, A&A 276, 549

Rice, W. K. M., Armitage, P. J., Bonnell, I. A., *et al.*: 2003, MNRAS 346, L36, doi:10.1111/j.1365-2966.2003.07317.x

Richards, P. G., Fennelly, J. A., & Torr, D. G.: 1994, JGR 99, 8981, doi:10.1029/94JA00518

Richards, P. G., Woods, T. N., & Peterson, W. K.: 2006, Adv. Space Res. 37, 315, doi:10.1016/j.asr.2005.06.031

Richmond, A. D. & Matsushita, S.: 1975, JGR 80, 2839, doi:10.1029/JA080i019p02839

Riley, P. & Gosling, J. T.: 2007, JGR (Space Physics) 112(A11), 6115, doi:10.1029/2006JA012210

Riley, P., Linker, J. A., & Mikić, Z.: 2001, JGR 106, 15889, doi:10.1029/2000JA000121

Riley, P., Linker, J. A., Mikić, Z., *et al.*: 2006, ApJ 653, 1510, doi:10.1086/508565

Rincon, F., Lignières, F., & Rieutord, M.: 2005, A&A 430, L57

Rishbeth, H.: 1990, Planet. Space Sci. 38, 945

Roberts, D. A., Goldstein, M. L., Klein, L. W., & Matthaeus, W. H.: 1987, JGR 92, 12023, doi:10.1029/JA092iA11p12023

Roberts, P. H.: 1987, in J. A. Jacobs (ed.), *Geomagnetism, Vol. 2*, Academic Press p. 251

Roberts, P H.: 2007, in P. Olson (ed.), *Treatise of Geophysics, Volume 8: Core Dynamics*, Elsevier, Amsterdam, p. 57

Roberts, P. H. & Stix, M.: 1972, A&A 18, 453

Roble, R. G.: 1995, in R. M. Johnson & T. L. Killeen (eds.), *The Upper Mesosphere and Lower Thermosphere*, Geophysical Monograph Series 87, AGU, p. 1

Roble, R. G. & Dickinson, R. E.: 1989, Geophys. Res. Lett. 16, 1441

Roble, R. G., Ridley, E. C., & Dickinson, R. E.: 1987, JGR 92(A8), 8745

Rogachevskii, I. & Kleeorin, N.: 2003, Phys. Rev. E 68, 036301

Rogers, T. M., MacGregor, K. B., & Glatzmaier, G. A.: 2008, MNRAS 387, 616

Rose, W. K.: 1998, *Advanced Stellar Astrophysics*, Cambridge University Press

Rosenbauer, H., Schwenn, R., Marsch, E., *et al.*: 1977, J. Geophys. Z. Geophys. 42, 561

Rossby, C.-G.: 1939, J. Marine Res. 38

Rottman, G.: 1999, J. Atmos. Terr. Phys. 61, 37

Rottman, G.: 2000, Space Sci. Rev. 94, 83

Rottman, G. J., Woods, T. N., & Sparn, T. P.: 1993, JGR 98, 10667, doi:10.1029/93JD00462

Rozanov, E. V., Schlesinger, M. E., Egorova, T. A., *et al.*: 2004, JGR 109, D01110, doi:10.1029/2003JD003796

Rüdiger, G., Elstner, D., & Ossendrijver, M.: 2003, A&A 406, 15, doi:10.1051/0004-6361:20030738

Rutten, R. G. M.: 1987, A&A 177, 131

Ryan, R. D., Neukirch, T., & Jardine, M.: 2005, A&A 433, 323, doi:10.1051/0004-6361:20047076

Ryu, D. & Goodman, J.: 1992, ApJ 388, 438, doi:10.1086/171165

Saar, S. H. & Brandenburg, A.: 1999, ApJ 524, 295

Sackmann, I.-J. & Boothroyd, A. I.: 2003, ApJ 583, 1024

Sackmann, I.-J., Boothroyd, A. I., & Kraemer, K. E.: 1993, ApJ 418, 457

Salby, M. & Callaghan, P.: 2000, J. Climate 13, 328

Sanderson, T. R., Marsden, R. G., Wenzel, K.-P., *et al.*: 1995, Space Sci. Rev. 72, 291, doi:10.1007/BF00768793

Sano, T., Miyama, S. M., Umebayashi, T., & Nakano, T.: 2000, ApJ 543, 486, doi:10.1086/317075

Sato, M., Hansen, J. E., McCormick, M. P., & Pollack, J. B.: 1993, JGR 98, 22987, doi:10.1029/93JD02553

Scannapieco, A. J. & Ossakow, S. L.: 1976, GRLe 3, 451

Schaefer, B. E., King, J. R., & Deliyannis, C. P.: 2000, ApJ 529, 1026

Schaefer, J. M., Denton, G. H., Kaplan, M., *et al.*: 2009, Science 324, 622

Schatten, K.: 2005, GRLe 32, 21106, doi:10.1029/2005GL024363

Schatten, K. H.: 2009, Solar Phys. 255, 3, doi:10.1007/s11207-008-9308-3

Schatten, K. H. & Orosz, J. A.: 1990, Solar Phys. 125, 179, doi:10.1007/BF00154787

Schatten, K. H., Scherrer, P. H., Svalgaard, L., & Wilcox, J. M.: 1978, GRLe 5, 411, doi:10.1029/GL005i005p00411

Schatten, K. H., Mayr, H. G., Omidvar, K., & Maier, E.: 1986, ApJ 311, 460, doi:10.1086/164786

Schekochihin, A. A., Cowley, S. C., Taylor, S. F., Maron, J. L., & McWilliams, J. C.: 2004, ApJ 612, 276

Scherhag, R.: 1960, J. Meteorol. 17, 572

Schmidt, H. & Brasseur, G. P.: 2006, Space Sci. Rev. 125(1–4), 345, doi:10.1007/s11214-006-9068-z

Schmidt, H., Brasseur, G. P., & Giorgetta, M.: 2010, JGR, 115, doi:1029/2009JD012542

Schmidt, J., Brasseur, G. P., Charron, M., *et al.*: 2006, J. Climate 19(16), 3903

Schmitt, D., Schuessler, M., & Ferriz-Mas, A.: 1996, A&A 311, L1

Schrijver, C. J.: 1992, A&A 258, 507

Schrijver, C. J.: 1993, A&A 269, 446

Schrijver, C. J.: 1995, Astron. Astrophys. Rev. 206, 181

Schrijver, C. J.: 2001, ApJ 547, 475

Schrijver, C. J.: 2002, Astron. Nachr. 323, 157

Schrijver, C. J.: 2005, in *Solar Wind 11/SOHO 16, Connecting Sun and Heliosphere*, ESA Special Publication, Vol. 592, p. 213

Schrijver, C. J.: 2007, ApJL 655, 117, doi:10.1086/511857

Schrijver, C. J.: 2009, ApJL 699, L148, doi:10.1088/0004-637X/699/2/L148

Schrijver, C. J. & DeRosa, M. L.: 2003, Solar Phys. 212, 165

Schrijver, C. J. & Harvey, K. L.: 1994, Solar Phys. 150, 1

Schrijver, C. J. & Title, A. M.: 2001, ApJ 551, 1099

Schrijver, C. J. & Title, A. M.: 2005, ApJ 619, 1077

Schrijver, C. J. & Van Ballegooijen, A. A.: 2005, ApJ 630, 552

Schrijver, C. J. & Zwaan, C.: 2000, *Solar and Stellar Magnetic Activity*, Cambridge University Press

Schrijver, C. J., Hagenaar, H. J., & Title, A. M.: 1997, ApJ 475, 328

Schrijver, C. J., DeRosa, M. L., & Title, A. M.: 2002, ApJ 577, 1006

Schrijver, C. J., Derosa, M. L., Metcalf, T. R., *et al.*: 2006, Solar Phys. 235, 161

Schröder, K.-P. & Smith, R. C.: 2008, MNRAS 386, 155, doi:10.1111/ j.1365-2966.2008.13022.x

Schubert, G., Covey, C., del Genio, A., *et al.*: 1980, JGR 85, 8007, doi:10.1029/ JA085iA13p08007

Schuessler, M. & Paehler, A.: 1978, A&A 68, 57

Schuessler, M., Caligari, P., Ferriz-Mas, A., Solanki, S. K., & Stix, M.: 1996, A&A 314, 503

Schunk, R. W. & Nagy, A. F.: 1980, Rev. Geophys. Space Phys. 18, 813

Schüssler, M. & Vögler, A.: 2008, A&A 481, L5

Schwabe, M.: 1844, Astron. Nachr. 21, 233

Schwadron, N. A.: 2002, GRLe 29(14), 140000, doi:10.1029/2002GL015028

Schwadron, N. A. & McComas, D. J.: 2005, GRLe 32, 3112, doi:10.1029/2004GL021579

Schwenn, R.: 1990, in R. Schwenn & E. Marsch (eds.), *Physics of the Inner Heliosphere I*, Springer, 99

Seehafer, N.: 1996, Phys. Rev. E 53, 1283

Selsis, F., Kasting, J. F., Levrard, B., *et al.*: 2007, A&A 476, 1373, doi:10.1051/ 0004-6361:20078091

Shea, M. A. & Smart, D. F.: 2006, Adv. Space Res. 38(2), doi:10.1016/j.asr.2006.07.005

Sheeley, N. R.: 1999, in S. R. Habbal, R. Esser, J. V. Hollweg, & P. A. Isenberg (eds.), *Solar Wind Nine*, American Institute of Physics Conference Series, Vol. 471, p. 41

Sheeley, N. R., Boris, J. P., Young, T. R., DeVore, C. R., & Harvey, K. L.: 1983, in J. O. Stenflo (ed.), *Solar and Stellar Magnetic Fields: Origins and Coronal Effects*, IAU Symp. 102, D. Reidel, p. 273

Shibata, K. & Yokoyama, T.: 2002, ApJ 577, 422

Shindell, D., Rind, D., Balachandran, N., Lean, J., & Lonergan, P.: 1999, Science 284, 305

Shu, F., Najita, J., Ostriker, E., *et al.*: 1994, ApJ 429, 781, doi:10.1086/174363

Siggia, E. D.: 1994, Annu. Rev. Fluid Mech. 26, 137

Sills, A., Pinsonneault, M. H., & Terndrup, D. M.: 2000, ApJ 534, 335, doi:10.1086/ 308739

Simnett, G. M., Sayle, K. A., Tappin, S. J., & Roelof, E. C.: 1995, Space Sci. Rev. 72, 327, doi:10.1007/BF00768799

Simpson, I. R., Blackburn, M., & Haigh, J. D.: 2009, J. Atmos. Sci. 66, 1347

Singh, S., Johnson, F. S., & Power, R. A.: 1997, JGR 102, 7399, doi:10.1029/96JA03998

Siscoe, G. L.: 1980, Rev. Geophys. Space Phys. 18, 647

Skumanich, A.: 1972, ApJ 171, 565

Smith, C. W. & Phillips, J. L.: 1997, JGR 102, 249, doi:10.1029/96JA02678

Smith, D. S., Scalo, J., & Wheeler, J. C.: 2004, Orig. Life Evol. Biosphere 34, 513, doi:10.1023/B:ORIG.0000043120.28077.c9

Smith, E. J.: 1979, Rev. Geophys. Space Phys. 17, 610

Smith, E. J. & Wolfe, J. H.: 1976, GRLe 3, 137, doi:10.1029/GL003i003p00137

Smith, E. J., Neugebauer, M., Balogh, A., *et al.*: 1995, Space Sci. Rev. 72, 165, doi:10.1007/BF00768773

Smith, M. D., Gierasch, P. J., & Schinder, P. J.: 1993, J. Atmos. Sci. 50, 4090

Smithtro, C. G.: 2004, Ph.D. thesis, Utah State University, Logan

Smithtro, C. G. & Sojka, J. J.: 2005a, JGR 110, A08306, doi:10.1029/2004JA010782

Smithtro, C. G. & Sojka, J. J.: 2005b, JGR 110, A08305, doi:10.1029/2004JA010781

Snyder, C. W., Neugebauer, M., & Rao, U. R.: 1963, JGR 68, 6361

Sofia, S. & Li, L. H.: 2006, In *Conference on Solar Variability and Earths Climate* 2005, Rome, 76(4), 768

Sohl, F. & Schubert, G.: 2007, in T. Spohn (ed.), *Treatise of Geophysics, Volume 10: Planets and Moons*, Elsevier p. 27

Sokoloff, D. & Nesme-Ribes, E.: 1994, A&A 288, 293

Solanki, S. K. & Fligge, M.: 2002, J. Atmos. Sol. Terr. Phys. 64(5–6), 677

Solanki, S. K., Usoskin, I. G., Kromer, B., Schüssler, M., & Beer, J.: 2004, Nature 431, 1084, doi:10.1038/nature02995

Sonett, C. P. & Finney, S. A.: 1990, Phil. Trans. R. Soc. Lond. A330, 413

Sonett, C. P., Giampapa, M. S., & Matthews, M. S. (eds.): 1991, *The Sun in Time*, University of Arizona Press

Soukharev, B. E. & Hood, L. L.: 2006, JGR 111, doi:10.1029/2006JD007107

Spiegel, E. A. & Zahn, J.-P.: 1992, A&A 265, 106

Spitzer, L.: 1962, *Physics of Fully Ionized Gases*, Interscience

Spruit, H. C.: 1976, Solar Phys. 50, 269

Spruit, H. C.: 1999, A&A 349, 189

Spruit, H. C.: 2000, Space Sci. Rev. 94, 113

Spruit, H. C.: 2002, A&A 381, 923, doi:10.1051/0004-6361:20011465

Spruit, H. C., Nordlund, Å, & Title, A. M.: 1990, ARA&A 28, 263

Stahler, S. W.: 1988, ApJ 332, 804, doi:10.1086/166694

Stahler, S. W., Shu, F. H., & Taam, R. E.: 1980a, ApJ 241, 637, doi:10.1086/158377

Stahler, S. W., Shu, F. H., & Taam, R. E.: 1980b, ApJ 242, 226, doi:10.1086/158459

Stanley, S. & Bloxham, J.: 2004, Nature 428, 151

Stanley, S. & Bloxham, J.: 2006, Icarus 184, 556

Stanley, S., Bloxham, J., & Hutchison, W. E.: 2005, Earth Planet. Sci. Lett. 234, 341

Starchenko, S. V. & Jones, C. A.: 2002, Icarus 157, 426

Stein, R. F. & Nordlund, Å.: 1989, ApJ 342, L95

Stein, R. F. & Nordlund, Å.: 1998, ApJ 499, 914

Stein, R. F. & Nordlund, Å.: 2006, ApJ 642, 1246

Steiner, O. & Ferriz-Mas, A.: 2005, Astron. Nachr. 326, 190, doi:10.1002/ asna.200410375

Steinhilber, F., Abreu, J. A., & Beer, J.: 2008, Astrophys. Space Sci. Trans. 4, 1

Steinhilber, F., Beer, J., & Fröhlich, C.: 2009, GRLe, 36, 12

Steinhilber, F., Abreu, J. A., Beer, J., & McCracken, K. G.: 2010, JGR (Space Physics) 115, 1104, doi:10.1029/2009JA014193

Stephenson, F. R., Willis, D. M., & Hallinan, T. J.: 2004, Astron. Geophys. 45:6, 6.15

Stevens, T. O. & McKinley, J. P.: 1995, Science 270, 450, doi:10.1126/ science.270.5235.450

Stevenson, D. J.: 1980, Science 208, 746

Stevenson, D. J.: 1982, Geophys. Astrophys. Fluid Dyn. 21, 113

Stevenson, D. J.: 1983, Rep. Prog. Phys. 46, 555

Stevenson, D. J.: 1987, Earth Planet. Sci. Lett. 82, 114

Stevenson, D. J.: 2003, Earth Planet. Sci. Lett. 208, 1

Stix, M.: 1976, A&A 47, 243

Stokes, D. E.: 1997, *Pasteur's Quadrant: Basic Science and Technological Innovation*, Brookings Institution Press

Stolarski, R. S. & Cicerone, R. J.: 1974, Can. J. Chem. 52, 1610

Stone, J. M. & Balbus, S. A.: 1996, ApJ 464, 364, doi:10.1086/177328

Stott, P. A., Jones, G. A., & Mitchell, J. F. B.: 2003, J. Climate 16, 4079

Strassmeier, K. G.: 2009, in *IAU Symposium*, Vol. 259, p. 363

Stuiver, M. & Quay, P. D.: 1980, Science 207, 11, doi:10.1126/science.207.4426.11

Stuiver, M., Reimer, P. J., Bard, E., *et al.*: 1998, Radiocarbon 40, 1041

Sultan, P. J.: 1996, JGR 101, 26875, doi:10.1029/96JA00682

Svalgaard, L. & Cliver, E. W.: 2007, ApJL 661, L203, doi:10.1086/518786

Svalgaard, L., Cliver, E. W., & Kamide, Y.: 2005, GRLe 32, 1104, doi:10.1029/2004GL021664

Svensmark, H.: 1998, Phys. Rev. Lett. 81, 5027, doi:10.1103/PhysRevLett.81.5027

Takahashi, F. & Matsushima, M.: 2006, GRLe 33, L10202

Talon, S., Kumar, P., & Zahn, J.-P.: 2002, ApJ 574, L175

Tanaka, H., Himeno, Y., & Ida, S.: 2005, ApJ 625, 414, doi:10.1086/429658

Tao, L., Weiss, N. O., Brownjohn, D. P., & Proctor, M. R. E.: 1998, ApJ 496, L39

Tapping, K. F., Boteler, D., Charbonneau, P., *et al.*: 2007, Solar Phys. 246, 309, doi:10.1007/s11207-007-9047-x

Tarduno, J. A., Cottrell, R. D., Watkeys, M. K., & Bauch, D.: 2007, Nature 466, 657

Tassoul, J. L.: 1978, *Theory of Rotating Stars*, Princeton University Press

Taylor, B. N. & Kuyatt, C. E.: 1994, *Guidelines for Evaluating and Expressing the Uncertainty of NIST Measurement Results*, NIST Technical Note 1297

Taylor, H. A., Jr., Brinton, H. C., Bauer, S. J., *et al.*: 1980, JGR 85, 7765

Telleschi, A., Güdel, M., Briggs, K., *et al.*: 2005, ApJ 622, 653

Thomas, B. T. & Smith, E. J.: 1980, JGR 85, 6861, doi:10.1029/JA085iA12p06861

Thompson, L. G., Yao, T., Davis, M. E., *et al.*: 1997, Science, 276(5320), 1821

Thompson, M. J., Christensen-Dalsgaard, J., Miesch, M. S., & Toomre, J.: 2003, ARA&A 41, 599

Thomson, D. J., Lanzerotti, L. J., Vernon, F. L., Lessard, M. R., & Smith, L. T. P.: 2007, Proc. IEEE 95, 1085

Thomson, W.: 1862, Macmillan's Magazine 5, 288

Thuillier, G., Sofia, S., & Haberreiter, M.: 2005, Adv. Space Res. 35(3), 329

Thuillier, G., Dewitte, S., & Schmutz, W.: 2006, in M. Shay, M. Wiltberger, M. Lester, W. Liu (eds.) *Magnetospheric Dynamics and the International Living with a Star Program* 38(8), 1792, 35th COSPAR Scientific Assembly July 18-25, 2004 Paris

Tian, F., Kasting, J. F., Liu, H.-L., & Roble, R. G.: 2008a, JGR 113, E05008, doi:10.1029/2007JE002946

Tian, F., Solomon, S. C., Qian, L., Lei, J., & Roble, R. G.: 2008b, JGR 113, E07005, doi:10.1029/2007JE003043

Timmreck, C., Lorenz, S. J., Crowley, T. J., *et al.*: 2009, GRLe 36, L21708, doi:10.1029/2009GL040083

Tinsley, B. A.: 1996, JGR 101(D23), 29701

Tinsley, B. A.: 2000, Space Sci. Rev. 94, 231

Title, A. M. & Schrijver, C. J.: 1998, in R. A. Donahue & J. A. Bookbinder (eds.), *Cool Stars, Stellar Systems, and the Sun*, ASP Conference Series, Vol. 154, p. 345

Tobias, S. M.: 1997, A&A 322, 1007

Tobias, S. M.: 1998, MNRAS 296, 653, doi:10.1046/j.1365-8711.1998.01412.x

Tobias, S. M. & Cattaneo, F.: 2008, J. Fluid Mech. 601, 101

Tobias, S. M., Brummell, N. H., Clune, T. L., & Toomre, J.: 1998, ApJ 502, L177

Tobias, S. M., Diamond, P. H., & Hughes, D. W.: 2007, ApJ 667, L113

Tobias, S. M., Cattaneo, F., & Brummell, N. H.: 2008, ApJ 685, 596

Tobiska, W. K., Woods, T., Eparvier, F., *et al.*: 2000, J. Atmos. Sol. Terr. Phys. 62, 1233

Torr, D. G. & Torr, M. R.: 1978, Rev. Geophys. Space Phys. 16, 327

Trenberth, K. E.: 1980, Mon. Wea. Rev. 108, 1378

Tworkowski, A., Tavakol, R., Brandenburg, A., *et al.*: 1998, MNRAS 296, 287, doi:10.1046/j.1365-8711.1998.01342.x

Uchida, Y. & Shibata, K.: 1984, PASJ 36, 105

Ulich, T. & Turunen, E.: 1997, GRLe 24, 1103

Umebayashi, T. & Nakano, T.: 1990, MNRAS 243, 103

Unruh, Y. C., Solanki, S. K., & Fligge, M.: 1999, A&A 345, 635

Unruh, Y. C., Krivova, N. A., Solanki, S. K., Harder, J. W., & Kopp, G.: 2008, A&A 486, 311, doi:10.1051/0004-6361:20078421

Upadhyay, H. O. & Mahajan, K. K.: 1998, GRLe 25, 3375

Usoskin, I. G.: 2008, Living Rev. Sol. Phys. 5, 3

Usoskin, I. G., Marsh, N., Kovaltsov, G. A., Mursula, K., & Gladysheva, O. G.: 2004, GRLe 31, L16109, doi:10.1029/2004GL019507

Vaiana, G. S., Krieger, A. S., Timothy, A. F., & Zombeck, F.: 1976, Astrophys. Space Sci. 39, 75

Valet, J.-P., L, Meynadier, & Guyodo, Y.: 2005, Nature 435(7043), 802, doi:10.1038/nature03674

van Ballegooijen, A. A.: 1994, Space Sci. Rev. 68, 299, doi:10.1007/BF00749156

van Ballegooijen, A. A. & Choudhuri, A. R.: 1988, ApJ 333, 965, doi:10.1086/166805

van Ballegooijen, A. & Martens, P. C. H.: 1990, ApJ 361, 283

van Loon, H., Meehl, G. A., & Shea, D. J.: 2007, JGR 112 (D2), doi:10.1029/2006JD007378

Vaquero, J. M.: 2007, Adv. Space Res. 40, 929, doi:10.1016/j.asr.2007.01.087

Vasavada, A. R. & Showman, A. P.: 2005, Rep. Prog. Phys. 68, 1935

Viereck, R. A., Floyd, L. E., Crane, P. C., *et al.*: 2004, Space Weather 2, 5, doi:10.1029/2004SW000084

Vilhu, O.: 1987, in J. L. Linsky & R. E. Stencel (eds.), *Cool Stars, Stellar Systems, and the Sun*, Springer-Verlag, p. 110

Vilhu, O. & Walter, F. M.: 1987, ApJ 321, 958

Vögler, A. & Schüssler, M.: 2007, A&A 465, L43

Voigt, G.-H.: 1981, Planet. Space Sci. 29, 1

von Bloh, W., Franck, S., Bounama, C., & Schellnhuber, H.-J.: 2002, in B. H. Foing & B. Battrick (eds.), *Earth-like Planets and Moons*, ESA Special Publication, Vol. 514, p. 289

von Bloh, W., Bounama, C., Cuntz, M., & Franck, S.: 2007, A&A 476, 1365, doi:10.1051/0004-6361:20077939

von Bloh, W., Bounama, C., Cuntz, M., & Franck, S.: 2008, in Y.-S. Sun, S. Ferraz-Mello, & J.-L. Zhou (eds.), *IAU Symposium*, Vol. 249, p. 503

von Bloh, W., Cuntz, M., Schröder, K.-P., Bounama, C., & Franck, S.: 2009, Astrobiology 9, 593, doi:10.1089/ast.2008.0285

von Hardenberg, J., Parodi, A., Passoni, G., Provenzale, A., & Spiegel, E. A.: 2007, Phys. Lett. A 372, 2223

von Hoerner, S., Burbidge, G. R., Kahn, F. D., Ebert, R., & Temesvary, S.: 1959, *Die Entstehung von Sternen durch Kondensation diffuser Materie*, Springer-Verlag

Vonmoos, M., Beer, J., & Muscheler, R.: 2006, JGR (Space Physics) 111(A10), A10105, doi:10.1029/2005JA011500

Wadhwa, M., Amelin, Y., Davis, A. M., *et al.*: 2007, in B. Reipurth, D. Jewitt, & K. Keil (eds.), *Protostars and Planets V*, University of Arizona Press, p. 835

Wagner, G., *et al.*: 2000a, Earth Planet. Sci. Lett. 181(1–2), 1

Wagner, G., Masarik, J., Beer, J., *et al.*: 2000b, Nucl. Instrum. Methods Phys. Res. B 172, 597

Wagner, G., Livingstone, D. M., Masarik, J., Muscheler, R., & Beer, J.: 2001, J. Geophys. Res. Atmos. 106(D4), 3381

Walterscheid, R. L.: 1980, Pure Appl. Geophys. 118, 239

Walterscheid, R. L.: 1981, GRLe 8, 1235, doi:10.1029/GL008i012p01235

Walterscheid, R. L.: 1995, in R. M. Johnson & T. L. Killeen (eds.), *The Upper Mesosphere and Lower Thermosphere: A Review of Experiment and Theory*, Geophysical Monograph 87, American Geophysical Union

Walterscheid, R. L.: 1997, JGR 102, 25,807

Walterscheid, R. L.: 2001, Adv. Space Res. 27, 1713, doi:10.1016/ S0273-1177(01)00298-8

Walterscheid, R. L. & Hickey, M. P.: 2005, JGR (Space Physics) 110(A9), 10307, doi:10.1029/2005JA011166

Walterscheid, R. L. & Schubert, G.: 1989, GRLe 16, 719, doi:10.1029/GL016i007p00719

Walterscheid, R. L. & Vincent, R. A.: 1996, JGR 101, 26,567

Walterscheid, R. L., Lyons, L. R., & Taylor, K. E.: 1985, JGR 90, 12235, doi:10.1029/ JA090iA12p12235

Walterscheid, R. L., Sivjee, G. G., & Roble, R. G.: 2000, GRLe 27, 2897, doi:10.1029/2000GL003768

Wang, Y.-M. & Sheeley, Jr., N. R.: 1991, ApJ 375, 761, doi:10.1086/170240

Wang, Y.-M. & Sheeley, N. R.: 1993, ApJ 414, 916

Wang, Y.-M., Nash, A. G., & Sheeley, Jr., N. R.: 1989, Science 245, 712, doi:10.1126/ science.245.4919.712

Wang, Y.-M., Sheeley, Jr., N. R., & Nash, A. G.: 1991, ApJ 383, 431, doi:10.1086/170800

Wang, Y.-M., Sheeley, N. R., & Lean, J.: 2002, ApJ 580, 1188

Wang, Y.-M., Lean, J. L., & Sheeley, Jr., N. R.: 2005, ApJ 625, 522, doi:10.1086/429689

Wanner, H., Beer, J., Btikofer, J., *et al.*: 2008, Quat. Sci. Res. 27(19–20), 1791

Ward, W. R.: 1997, ApJL 482, L211, doi:10.1086/310701

Ward, P. D.: 2007, in W. Sullivan & J. A. Baross (eds.), *Planets and Life: The Emerging Science of Astrobiology*, Cambridge University Press, p. 335

Ward, P. & Brownlee, D.: 2000, *Rare Earth: Why Complex Life is Uncommon in the Universe*, Copernicus

Warren, H. P.: 2005, ApJSS 157, 147, doi:10.1086/427171

Warren, H. P., Mariska, J. T., & Lean, J. L.: 2001, JGR 106, 15745, doi:10.1029/ 2000JA000282

Webb, D. F. & Howard, R. A.: 1994, JGR 99, 4201, doi:10.1029/93JA02742

Webb, M. & Webb, M.: 1952, *The Northern Lights of Old Aberdeen*, song lyrics, D. Davis and Company

Webber, W. R. & Higbie, P. R.: 2003, JGR (Space Physics) 108, 1355, doi:10.1029/ 2003JA009863

Weber, S. L., Crowley, T. J., & van der Schrier, G.: 2004, Clim. Dynam. 22, 539

Wehrbein, W. M. & Leovy, C. B.: 1982, J. Atmos. Sci. 39, 1532

Weiss, N. O. & Tobias, S. M.: 2000, Space Sci. Rev. 94, 99

Wenzler, T., Solanki, S. K., Krivova, N. A., & Fröhlich, C.: 2006, A&A 460, 583, doi:10.1051/0004-6361:20065752

White, O. R. (ed.): 1977, *The Solar Output and its Variation*, Colorado Associated University Press

White, O. R. & Livingston, W. C.: 1981, ApJ 249, 798, doi:10.1086/159338

Wigley, T. M. L. & Kelly, P. M.: 1990, Phil. Trans. R. Soc. Lond. A330, 547

Wild, O., Xin, Z., & Prather, M. J.: 2000, J. Atmos. Chem. 37, 245

Williams, C. R. & Avery, S. K.: 1996, JGR 101, 4079

Willson, R. C.: 1979, Appl. Opt. 18, 179

Willson, R. C. & Mordvinov, A. V.: 2003, GRLe 30(5), 1199, doi:10.1029/2002GL016038

Willson, R. C., Gulkis, S., Janssen, M., Hudson, H. S., & Chapman, G. A.: 1981, Science 211, 700, doi:10.1126/science.211.4483.700

Wilson, O. C.: 1978, ApJ 226, 379, doi:10.1086/156618

Wimmer-Schweingruber, R. F., von Steiger, R., & Paerli, R.: 1997, JGR 102, 17407, doi:10.1029/97JA00951

Wofsy, S. C., McElroy, M. B., & Yung, Y. L.: 1975, GRLe 2, 215

Wood, B. E., Muller, H.-R., Zank, G., & Linsky, J. L.: 2002, ApJ 574, 412

Woods, T. N. & Rottman, G. J.: 2002, in M. Mendillo, A. Nagy, & J. H. Waite (eds.), *Atmospheres in the Solar System: Comparative Aeronomy*, AGU, Geophysical Monograph Series, Vol. 30, p. 221

Woods, T. N., Prinz, D. K., Rottman, G. J., et al.: 1996, JGR 101, 9541, doi:10.1029/96JD00225

Woods, T. N., Tobiska, W. K., Rottman, G. J., & Worden, J. R.: 2000, JGR 105, 27195, doi:10.1029/2000JA000051

Woods, T. N., Eparvier, F. G., Fontenla, J., et al.: 2004, GRLe 31, 10802, doi:10.1029/2004GL019571

Woods, T. N., Eparvier, F. G., Bailey, S. M., et al.: 2005, JGR (Space Physics) 110(A9), 1312, doi:10.1029/2004JA010765

Wright, J. T.: 2004, AJ 128, 1273

Wright, J. T., Upadhyay, S., Marcy, G. W., et al.: 2009, ApJ 693, 1084, doi:10.1088/0004-637X/693/2/1084

Wunsch, C.: 2000, Nature 405(6788), 743

Xu, Z.-W., Wu, J., Igarashi, K., Kato, H., & Wu, Z.-S: 2004, JGR 109, A09307, doi:10.1029/2004JA010572

Yeates, A. R., Nandy, D., & Mackay, D. H.: 2008, ApJ 673, 544, doi:10.1086/524352

Yndestad, H.: 1999, ICES J. Mar. Sci. 63, 401

Yoshimura, H.: 1975, ApJ 201, 740, doi:10.1086/153940

Yue, X., Wan, W., Liu, L., Ning, B., & Zhao, B.: 2006, JGR 111, A10303, doi:10.1029/2005JA011577

Zahnle , K., Claire, M. W., & Catling, D. C.: 2006, Geobiology 4, 271

Zahnle, K., Arndt, N., Cockell, C., et al.: 2007, Space Sci. Rev. 129, 35, doi:10.1007/s11214-007-9225-z

Zatman, S. & Bloxham, J.: 1997, Nature 388, 760

Zhang, M. & Low, B. C.: 2005, ARA&A 43, 103

Zhao, X.-P. & Hundhausen, A. J.: 1981, JGR 86, 5423, doi:10.1029/JA086iA07p05423

Zhu, X., Yee, J.-H. & Talaat, E. R.: 2003, J. Atmos. Sci. 60, 491

Zhu, Z., Hartmann, L., Calvet, N., et al.: 2007, ApJ 669, 483, doi:10.1086/521345

Zhu, Z., Hartmann, L., & Gammie, C.: 2009a, ApJ 694, 1045, doi:10.1088/ 0004-637X/694/2/1045

Zhu, Z., Hartmann, L., Gammie, C., & McKinney, J. C.: 2009b, ApJ 701, 620

Zielinski, G. A.: 1995, JGR 100, 20,937

Zurbuchen, T. H. & Richardson, I. G.: 2006, Space Sci. Rev. 123, 31, doi:10.1007/ s11214-006-9010-4

Zwaan, C. & Cram, L. E.: 1989, in L. E. Cram & L. V. Kuhi (eds.), *FGK Stars and T Tauri Stars*, NASA SP-502, p. 215

Index

CPSIA information can be obtained at www.ICGtesting.com
Printed in the USA
BVOW06s0437250713

326944BV00005B/66/P

9 780521 130202